Integrated Food and Bioprocess Engineering

This book covers different aspects of food bioprocess engineering for value-added food production and waste management. The various topics presented include food biotechnology and bioprocess engineering, fermenter and bioreactor design, studies related to unit operations, and food safety and regulatory issues for safe and nutritious food product development using novel biotechnology and bioprocessing technologies. The book also addresses food industry waste utilization for first-, second-, and third-generation biofuel production, as well as biodegradable packaging materials. This book:

- Includes the basic principles of both biotechnology and bioprocess engineering related to food preservation and product development.
- Covers unit operations such as membrane filtration, chromatography, co-crystallization and nano-encapsulation, supercritical fluid extraction, DNA recombination, and PCR technology.
- Discusses pertinent regulatory constraints.
- Reviews recent developments, challenges, and opportunities in food biotechnology and bioprocess engineering, including fermenter and bioreactor design.
- Addresses food safety and regulatory issues for safe and nutritious food product development.
- Is aimed at graduate students and researchers in food and chemical engineering, bioprocessing, and food biotechnology.

Integrated Food and Bioprocess Engineering

Edited by Anwesa Sarkar
and Kshirod Kumar Dash

CRC Press
Taylor & Francis Group
Boca Raton London New York

CRC Press is an imprint of the
Taylor & Francis Group, an **informa** business

Designed cover image: Getty Images

First edition published 2026
by CRC Press
2385 NW Executive Center Drive, Suite 320, Boca Raton FL 33431

and by CRC Press
4 Park Square, Milton Park, Abingdon, Oxon, OX14 4RN

CRC Press is an imprint of Taylor & Francis Group, LLC

The electronic version of this book was funded to publish Open Access through Taylor & Francis' Pledge to Open, a collaborative funding open access books initiative. The full list of pledging institutions can be found on the Taylor & Francis Pledge to Open webpage.

For Product Safety Concerns and Information please contact our EU representative GPSR@taylorandfrancis.com. Taylor & Francis Verlag GmbH, Kaufingerstraße 24, 80331 München, Germany.

Library of Congress Cataloging-in-Publication Data
Names: Sarkar, Anwesa editor | Dash, Kshirod Kumar editor
Title: Integrated food and bioprocess engineering / edited by Anwesa Sarkar and Kshirod Kumar Dash.
Description: First edition. | Boca Raton FL : CRC Press, 2026. | Includes bibliographical references.
Identifiers: LCCN 2025048243 (print) | LCCN 2025048244 (ebook) | ISBN 9781032599274 hardback | ISBN 9781032650609 paperback | ISBN 9781032650616 ebook
Subjects: LCSH: Food–Biotechnology | Food industry and trade
Classification: LCC TP248.65.F66 I6577 2026 (print) | LCC TP248.65.F66 (ebook)
LC record available at https://lccn.loc.gov/2025048243
LC ebook record available at https://lccn.loc.gov/2025048244

ISBN: 9781032599274 (hbk)
ISBN: 9781032650609 (pbk)
ISBN: 9781032650616 (ebk)

DOI: 10.1201/9781032650616

Typeset in Times
by Deanta Global Publishing Services, Chennai, India

Contents

About the Editors

Anwesa Sarkar is currently working as an Assistant Professor in the Department of Food Processing Technology, Ghani Khan Choudhury Institute of Engineering & Technology (GKCIET), Malda, West Bengal, India. She earned her B.Tech. degree in Agricultural Engineering from Uttar Banga Krishi Vishwavidyalaya, and did her M.Tech. in Agricultural Engineering with a major in Food Biotechnology Engineering at Gobind Ballabh Pant University of Agriculture and Technology, Pantnagar, where she was also awarded a PhD in Agricultural Engineering with a major in Process and Food Engineering. She is also a life member of professional bodies such as ISAE and AFSTI. Her areas of interest include food biotechnology, food bioprocessing, fermentation, biofuels, food industry waste management, and by-product development, etc. She has published several research papers in peer-reviewed journals.

Kshirod Kumar Dash, PhD, is currently working as a Professor in the Department of Food Processing Technology, Ghani Khan Choudhury Institute of Engineering & Technology (GKCIET), Malda, West Bengal, India. Previously he worked as an Assistant Professor in the Department of Food Engineering and Technology, Tezpur University, Tezpur, Assam. Dr Dash completed his B.Tech in Agricultural Engineering at the the College of Agricultural Engineering & Technology, OUAT, Bhubaneswar; and his M.Tech. in Dairy and Food Engineering and his PhD in Food Process Engineering at IIT Kharagpur. Dr Dash obtained his Post-Doc from Ohio State University, Columbus, Ohio, USA. He has guided several postgraduate and doctoral students and published material in over 50 publications, including peer-reviewed international journals and book chapters in edited volumes.. He is a life member of various professional bodies in India and a member of IFT, USA. Dr Dash's research interests include food process engineering, unit operations in food processing, fluid mechanics, and heat and mass transfer in food processing.

List of Contributors

Ahmad Sadaf
Centre for Food Science and Technology
Aligarh Muslim University
Aligarh, India

Ahmed Waseem
Symbiosis Law School NOIDA
Symbiosis International (Deemed University)
Pune, India

Ali Mudasir
College of Agricultural Engineering and Technology
Sher-e-Kashmir University of Agricultural Sciences and Technology
Shalimar, Srinagar, India

Anand Anupama
Department of Food Science and Technology
Dr. YS Parmar University of Horticulture and Forestry
Nauni, Solan, India

Dash Kshirod Kumar
Department of Food Engineering and Technology
Ghani Khan Chowdhury Institute of Engineering and Technology
Narayanpur, Malda, India

Desai Shivani
Department of Food Engineering
National Institute of Food Technology Entrepreneurship and Management
Kundli, Sonipat, India

Dobhal Neetu
Department of Foods and Nutrition
College of Community Science
G. B. Pant University of Agriculture and Technology
Pantnagar, India

Gaibimei Palmei
Indian Agricultural Research Institute
Gogamukh, Dhemaji, India

Gintu Sara George
Karunya Institute of Technology and Sciences
Coimbatore

Gupta Rachna
Department of Food Engineering
National Institute of Food Technology Entrepreneurship and Management
Kundli, Sonipat, India

Hussain Ashraque
Department of Food Technology
Jain (Deemed-to-Be-University)
Ramanagar, India

Jayarajan Megharaj
Department of Food Technology
Jain (Deemed-to-Be-University)
Ramanagar, India

Joshi Deepa
College of Community Science
G. B. Pant University of Agriculture and Technology
Pantnagar, India

Joshi Sweta
Department of Food Technology
School of Interdisciplinary Sciences and Technology
Jamia Hamdard University
New Delhi, India

Kalpana Devi
Department of Agricultural and Food Engineering, Indian Institute of Technology
Kharagpur, WB, India

Khan Irfan
Department of Bioengineering
Integral University
Lucknow, India

Khan Mohammad Umar
Delvi Natural
Delhi, India

Khan Zober Alam
National Institute of Food Technology Entrepreneurship and Management
Centre for Food Research & Analysis
India

Kohli Deepika
Department of Food Technology
Vignan's Foundation for Science, Technology & Research
(Deemed-to-Be-University)
Vadlamudi, Guntur, India

Kumar Sumit
Department of Food Technology
Jain (Deemed-to-Be-University)
Ramanagar, India

Kumari Purnima
Department of Food Engineering
National Institute of Food Technology Entrepreneurship and Management
Kundli, Sonipat, India

Mahajan Monika
Punjab Agricultural University (PAU)
Regional Research Station
Bathinda, India

Mane Sheetal
Department of Food Engineering
National Institute of Food Technology Entrepreneurship and Management
Kundli, Sonipat, India

Mobeen Ahmed
Informatics and Big Data
CSIR – Institute of Genomics and Integrative Biology
Sukhdev Vihar, New Delhi, India

Moirangthem
Department of Agricultural and Food Engineering, Indian Institute of Technology Kharagpur, WB, India

Naqash Farah
Department of Food Technology
Islamic University of Science and Technology
Awantipora, Pulwama, India

Nazir Sadaf
Department of Food Technology
School of Interdisciplinary Sciences and Technology
Jamia Hamdard University
New Delhi, India

Ningthoujam Manda Devi
Department of Agricultural Engineering and Technology
School of Engineering and Technology
Nagaland University
Dimapur, India

Parida Shweta
School of CIST
KISS-Deemed to be a University
Bhubaneswar, India

Parvez Sadaf
Department of Food Science & Technology
University of Kashmir
Srinagar, India

Prabhakar Baraskar Ruchi
Department of Food Technology
Faculty of Engineering and Technology
Jain (Deemed-to-Be-University)
Ramanagar, India

Reena
Department of Food Engineering
National Institute of Food Technology Entrepreneurship and Management
Sonipat, India

Richa Rishi
College of Agricultural Engineering and Technology
Sher-e-Kashmir University of Agricultural Sciences and Technology
Shalimar, Srinagar, India

Saini Harpreet Kaur
Department of Food Science and Technology
Dr YS Parmar University of Horticulture and Forestry
Nauni, Solan, India

Sarkar Anwesa
Department of Food Engineering and Technology
Ghani Khan Chowdhury Institute of Engineering and Technology
Narayanpur, Malda, India

Showkat Shabnum
Division of Food Science and Technology
Sher-e-Kashmir University of Agricultural Science and Technology
Kashmir, India

Siddiqui Tohfa
Galgotias College of Pharmacy
Greater Noida, India

Singh Anupama
Department of Food Engineering
National Institute of Food Technology Entrepreneurship and Management
Kundli, Sonipat, India

Singla Prabhjot
Department of Biochemistry
Punjab Agriculture University PAU
Ludhiana, India

Soni Akanksha
Faculty of Agricultural Sciences
Swami Vivekananda University
Sagar, India

Srivalli Radha
College of Agricultural Engineering
Kandi, PJTSAU, India

Tabassum Nazia
Department of Post-Harvest Engineering and Technology
Aligarh Muslim University
Aligarh, India

Vilas Valekar Swapnil
Department of Food Technology
Jain (Deemed-to-Be-University)
Ramanagar, India

Vivek Kumar
Department of Food Engineering and Technology
Ghani Khan Chowdhury Institute of Engineering and Technology
Narayanpur, Malda, India

Wasiya Farzana
Karunya Institute of Technology and Sciences, Coimbatore

Yadav Shivam
Teagasc Food Research Centre
Dublin, Ireland

Z. R. A. A. Azad
Department of Post-Harvest Engineering and Technology
Aligarh Muslim University
Aligarh, India

Preface

The book *Integrated Food and Bioprocess Engineering* emphasizes the importance of maintaining food quality and safety throughout the processing chain. Bioprocess engineering involves optimizing various parameters to improve the efficiency conditions for microbial fermentation, enzymatic reactions, and other biological processes used in food processing. Comprehending food bioprocess engineering becomes crucial in light of global food concerns such as population increase, food security, and shifting dietary preferences. By applying the knowledge readers learn from this book, readers can assist in addressing these issues by creating sustainable and effective food processing solutions.

In particular, this book explores, explains, and defines bioprocessing in relation to food production, processing, and human nutrition. This resource, which offers comprehensive descriptions of the wide variety of issues that constitute the area of Food Bioprocess Engineering, is written for undergraduate, postgraduate, and researchers. With an emphasis on applications to the food industry and strategies for large-scale use of the technology, readers will have a thorough grasp of the principles and practices of food bioprocess engineering. It also looks at hazards, other ethical issues, and how these technologies can impact food security.

The book covers different aspects related to Food Bioprocess Engineering. It consists of four major areas covering the basic aspects as well as the emerging trends of new research for the development of various value-added food and waste management. The first four chapters cover recent developments, challenges, and opportunities in Food Bioprocess Engineering. Chapters 5–8include different studies related to new emerging technologies like membrane filtration, chromatography, co-crystallization and nano-encapsulation, DNA recombination and PCR technology, biosensors, nano-biotechnology, and cell-free technology. Chapters 9–11 are about food safety and regulatory issues for safe and nutritious food product development using novel technologies of bioprocessing. The last five chapters include a study of food industry waste utilization for the production of first-, second-, and third-generation biofuel production as well as biodegradable packaging material by different researchers. The book develops an understanding of the potential contributions of food bioprocess engineering to the food industry and towards improved food safety and public health.

1 Recent Developments in Food Biotechnology

Sadaf Ahmad, Sweta Joshi, Ahmed Mobeen, Nazia Tabassum, Z. R. A. A. Azad, and Zober Alam Khan

1.1 INTRODUCTION

Biotechnology, a dynamic and rapidly advancing field, lies at the intersection of biology, technology, and innovation. It encompasses a wide range of scientific disciplines and utilizes living organisms or their components to develop products and processes that benefit society. By harnessing the power of biology, biotechnology has the potential to revolutionize various sectors, including healthcare, agriculture, energy, and environmental conservation. At its core, biotechnology involves manipulating biological systems and organisms to create novel solutions to complex challenges. This can be achieved through genetic engineering, cellular and tissue culture, molecular biology, and other cutting-edge techniques. Biotechnologists employ their expertise to study and understand the intricate mechanisms within living organisms, unlocking their secrets and translating them into practical applications.

In agriculture, biotechnology has revolutionized crop improvement and food production. Genetically modified organisms (GMOs) have been engineered to possess desirable traits, such as increased yield, enhanced nutritional content, and improved resistance to pests and diseases. These genetically modified crops can be used to address issues with food security, lower the use of pesticides, and enhance farmer livelihoods. Moreover, biotechnology contributes to the development of sustainable and eco-friendly solutions. It enables the production of biofuels, which offer alternatives to fossil fuels and help mitigate climate change. Bioremediation techniques utilize microorganisms to clean up polluted environments, while industrial biotechnology utilizes enzymes and microorganisms to create environmentally friendly processes and reduce waste generation.

However, as biotechnology continues to advance, it also raises ethical, social, and regulatory considerations. Controversial topics such as genetic modification, cloning, and gene editing prompt discussions about the boundaries and responsible use of biotechnological advancements. Striking a balance between scientific progress, ethical considerations, and societal well-being is crucial for the sustainable development of biotechnology.

DOI: 10.1201/9781032650616-1

With the potential to solve some of the most important problems facing humanity, biotechnology is a fascinating area of scientific inquiry and invention. By harnessing the power of living organisms, biotechnologists are pushing the boundaries of what is possible in fields ranging from healthcare to agriculture and environmental conservation. However, it is essential to foster dialogue, transparency, and responsible practices to ensure the ethical and beneficial application of biotechnology for the betterment of our world.

Food biotechnology is a field within the larger realm of biotechnology that focuses specifically on the application of biological techniques and processes to improve the production, quality, safety, and sustainability of food. By utilizing genetic engineering, molecular biology, and other biotechnological tools, food scientists and technologists are able to develop innovative solutions to address the challenges faced by the global food system.

The world's population continues to grow, placing increasing demands on agricultural resources and food production. At the same time, factors such as climate change, limited arable land, and the need for sustainable farming practices pose significant challenges to ensuring food security. Food biotechnology offers a range of approaches to overcome these obstacles and create a more resilient and efficient food supply.

One of the primary applications of food biotechnology is the genetic modification of crops. Through the use of genetic engineering, scientists are able to put certain genes into plants, giving them the ability to produce more, resist diseases and pests better, have better nutritional value, and be more resilient to environmental shocks. These genetically altered plants have the potential to boost agricultural output, decrease post-harvest losses, and enhance food nutrition.

Additionally, food biotechnology plays a crucial role in food safety. Rapid and accurate methods for detecting foodborne pathogens and contaminants have been developed using biotechnological techniques. These methods enable early identification of harmful substances and help prevent outbreaks of foodborne illnesses. Biotechnology also contributes to the development of food-processing techniques that preserve the nutritional value of food while ensuring its safety and extending its shelf life.

Furthermore, food biotechnology has the potential to address nutritional deficiencies and improve public health. Through biofortification, essential nutrients can be enhanced in crops to combat micronutrient deficiencies, particularly in regions where access to a diverse diet is limited. This approach has the potential to alleviate malnutrition and improve overall health outcomes.

However, the use of food biotechnology also raises questions and concerns. Public perception, ethical considerations, and regulatory frameworks surrounding genetically modified organisms (GMOs) continue to be topics of debate. It is essential to balance scientific advancements with rigorous safety assessments, transparent labeling, and comprehensive risk assessments to ensure consumer confidence and address potential environmental and health implications.

Food biotechnology holds tremendous potential for transforming the way we produce, process, and consume food. By applying biotechnological tools and techniques, we can address the challenges of a growing population, climate change, and limited resources. The responsible and ethical application of food biotechnology, coupled with robust regulatory frameworks, can contribute to a sustainable and resilient food system, ensuring food security, safety, and improved nutrition for all.

1.2 CULTURED MEAT

1.2.1 Production Technology

Cultured meat, also known as lab-grown meat, cell-based meat, or clean meat, refers to meat products that are produced by culturing animal cells in a lab rather than by traditional animal agriculture methods. It is an emerging field of research and technology that aims to provide an alternative to conventional meat production, which is associated with environmental concerns, animal welfare issues, and resource inefficiencies.

There are various ways to produce fibrous products through different processing routes, which can be categorized into two strategies: bottom-up and top-down. The bottom-up approach involves the creation of individual unit structures from molecules or units like proteins. These components are then combined together to form larger structures and products. Techniques like in vitro cell culturing, mycoproteins, wet-spinning, and electrospinning are examples of bottom-up methods. On the other hand, the top-down strategy involves blending of proteins and/or polysaccharides to generate anisotropic, fibrous materials by application of shear forces. The resulting structure becomes visible when the material is deformed or stretched. Techniques such as extrusion, protein and hydrocolloid mixing, freeze structuring, and the shear cell approach are part of the top-down approach.

From a structural standpoint, the bottom-up approach has the potential to closely mimic meat. However, from the perspective of process design, the top-down approach is more sturdy, scalable, and economical. While significant progress has been made in characterizing structured foods in recent decades, further advancement in top-down processing strategies will require new methods of analysis that can track the structuring process in real time. The evolution of both analytical methods and structuring processes will enable the creation of meat analogues with desired sensory properties and excellent efficiency in terms of resources (Dekkers et al., 2018).

The process of producing cultured meat typically involves taking a small sample of animal cells, such as muscle cells, and keeping them in a culture medium that provides the necessary nutrients for the cells to grow and multiply. These cells are then stimulated to differentiate into muscle fibers, which eventually form muscle tissue that can be harvested and processed into meat products.

This is an exciting field of innovation that shows promise in addressing concerns related to the environment, ethics, and sustainability associated with traditional meat production. Cultured meat, as originally proposed, is a tissue-engineered combination of muscle fibers and fat tissue obtained from tissue cultures and mammalian cells. At least two types of cells are involved: skeletal muscle cells and adipocytes. In the case of cultured beef, specific cells called bovine satellite cells and fibro-adipogenic progenitor cells are obtained from muscle samples that contain fat within the muscles. These primary cells can be cultured directly after extraction from living animal tissue, or they can be genetically modified to have a specified lifetime duration and role (Hubalek et al., 2022).

Myoblasts are the cells that depend on attachment, and they require a substrate or scaffold for cell proliferation and differentiation. An ideal scaffold should provide a large surface area for growth and attachment. They should also allow contraction of myoblasts, facilitate medium diffusion, and easily detach from the meat culture. It is preferable for scaffolds to be degradable, edible, safe, and easily available for the production of cultured meat at a large scale. While myoblasts naturally undergo sudden contraction, a pliable substrate is necessary to prevent the separation of developing myotubes. However, developing a scaffold that can stimulate differentiation by mechanically stretching the attached cells remains a significant challenge. Currently, Cytodex-3 microcarrier beads are used as scaffolds in rotary bioreactors, but these beads lack the potential for stretching of cells. Alternatively, mechanical stretching of myoblasts can be done using edible and stimulus-sensitive porous microspheres made from materials like cellulose, alginates, pullulan, starch, chitosan, and hyaluronic acid. It is important to avoid alterations in materials obtained from biological sources such as collagen since collagen cannot be replicated exactly and requires considerable livestock production, which would increase production costs (Kashim et al., 2023).

Lastly, to achieve tissues that closely resemble conventional meat in terms of nutrition and sensory qualities, it is crucial to differentiate precursor cells into mature muscle fibers and fat tissues. Traditionally, fetal bovine serum derived from animals, horse serum, or human platelet lysate are favored protein sources for culturing mammalian cells. Serum provides growth factors, amino acids, lipids, and hormones, serving various types of cells. However, it is a costly byproduct of conventional meat production and subject to batch variation among batches, which can negatively influence muscle phenotype. As cultured meat production scales up successfully, alternative and more affordable sources of proteins and growth factors need to be explored and their appropriate concentrations determined.

Medium composition also plays a vital role in cell performance. Nutrients, such as amino acids and glucose, and their concentrations should be tailored to meet the specific demands of the cell types. The same applies to proteins and growth factors, where cost-effective substitutes need to be identified and studied at proper concentrations. Failure to adapt the medium composition properly can result in inefficient nutrient utilization, leading to inefficient carbon and nitrogen metabolism and potentially an excessive production of waste products like lactate and ammonium. Also, the composition of the medium varies depending on the stage of cell culture, as the requirements differ significantly during proliferation and differentiation phases.

Additionally, it is important to replace expensive media compounds with more affordable alternatives, such as those derived from plants or produced sustainably through fermentation, and that are of food or feed grade. This is particularly crucial for plentiful proteins like albumin, insulin, and transferrin and less crucial for proteins present in lesser amounts in nature. Hydrolyzed plants and microalgae can serve as a rich source of nutrients and bioactive compounds, playing a significant role in reducing costs and enhancing the sustainability of the production processes of cultured meat. Plants and algae have relatively low media compound requirements, and hydrolyzed biomass has the ability to replace costly components like serum, albumin, and growth factors in culture media for mammalian cells.

I. The recycling and enhancement of the lifetime duration of culture media can be achieved in several ways.
II. Prevention or reduction of the metabolic wastes produced (e.g., ammonia and lactate) that can harm cell survival and growth.
III. Dilution or removal of metabolic waste and subsequent utilization.
IV. Introduction of valuable components, such as growth factors, from spent media back into the culture media.
V. Successive and symbiotic growth of different cell types using each other's conditioned media (Hubalek et al., 2022).

The stirred tank bioreactor is the most commonly incorporated bioreactor for the proliferation of cells due to its availability and established protocols. Microcarriers are required in this type of bioreactor to provide attachment surfaces for the cells. The choice of microcarrier determines the optimal rotating speed of the impeller, which depends on the geometry of the impeller and vessel. Sufficient stirring is necessary for proper mixing, but excessive stress can induce cell injury or differentiation.

The wave or rocking bioreactor offers a regulated, low-shear substitute to the stirred tank bioreactor. It consists of a rocking platform on which a container or bag is mounted. This set-up helps in providing a mild and subtle environment for cell growth in suspensions with microcarriers. In this bioreactor, rocking angle and rocking speed can be regulated for controlled mixing. However, it is primarily used for small- to medium-scale production as the physical footprint increases significantly when operating bags at half of its capacity to ensure proper mixing. Careful monitoring of the interface is necessary in this bioreactor due to potential high stress caused by wave motion.

Hollow fiber bioreactors mimic the functionality of blood vessels by supplying nutrients and removing metabolites. They can be used for the proliferation and differentiation of mammalian cells. They contain semi-permeable membranes that facilitate the flow of water and nutrients while providing an adherent surface for cell growth. Cells can be seeded on the surface of the fibers or in a gel in the extra-capillary space between the fibers. This offers flexibility as each seeding area experiences different shear stresses.

Fluidized bed bioreactors utilize a fluid to suspend solid particles, creating a fluid-like behavior. They provide output with higher cell densities as compared to cell densities in static cell cultures. This might be due to low shear stress and continuous media perfusion. Cells may be grown on microcarriers, or they may be encapsulated in gel beads. They are suspended by the continuous flow of media while a secondary flow helps in removing the metabolites, ensuring a stable and well-mixed environment.

Airlift bioreactors employ sparged gas from the bottom side to generate an upward flow in the riser part. Then the fluid recirculates downward in the downcomer section parallel to the riser. This recirculation creates a more homogeneous distribution of shear stress compared to the stirred tank bioreactor, along with equivalent rates of mixing and mass transfer. Airlift bioreactors (ALB) have been commonly used in wastewater treatment, bacterial fermentation, and plant cell culture. Unlike fluidized bed bioreactors, they can work with lower concentrations of solids and they generally have a recirculating loop.

The Rotary Wall Vessel and High-Aspect Ratio Vessel are two low-shear alternatives. This bioreactor consists of two concentric cylinders, and cells are grown in the annular space between them. When both cylinders rotate at equal speeds, it results in high oxygen mass transfer rates and significantly reduced shear stresses compared to the stirred tank bioreactor. High-Aspect Ratio Vessels are cylindrical containers with a high aspect ratio of diameter to length. This helps in the spreading of media in a thin layer. On the other hand, the Rotary Wall Vessel is a closed cylinder in which one of the walls rotates (Rivera et al., 2022).

Cultured meat is capable of offering several advantages over traditional meat production. Firstly, it can help in reducing the environmental impact of meat production. It requires fewer resources, such as land, feed, and water, and produces fewer greenhouse gas emissions compared to raising animals for meat. Secondly, cultured meat has the potential to eliminate the need for traditional animal slaughter, which could lead to improvements in animal welfare. Lastly, cultured meat could potentially reduce the risks associated with foodborne illnesses and antibiotic resistance, as it can be produced in a controlled environment without the need for antibiotics.

1.2.2 Challenges

However, there are still significant challenges to overcome before cultured meat can be commercially viable and widely available. One of the main challenges is scaling up production and reducing costs. Currently, the production of cultured meat is expensive, and the technology is not yet capable of producing meat at a competitive price point compared to conventionally produced meat. Researchers and companies are actively working on improving the scalability and cost-effectiveness of cultured meat production.

Upscaling cultured meat-compatible cells from planar systems to dynamic systems, such as spinner flask cultures and tabletop bioreactor cultures at liter scale, has advanced significantly. Several types of bioreactors mentioned earlier can be taken into account for scaling up. It is essential for successful scaling up to strike the correct balance between sufficient mixing to avoid nutrients, temperature, dissolved oxygen, and pH gradients, as well as to regulate hydrodynamic forces. The particular cell type and production volume determine the type of bioreactor to use. The advantages of stirred tank bioreactors and rocking bioreactors are scalability and the capacity to provide adequate mixing at greater scales. It is crucial to remember that, particularly when employed with microcarriers, these bioreactors might also introduce hydrodynamic forces that could impede cell growth.

As per the Hayflick limit, unaltered progenitor cells frequently have a limited ability for proliferating—less than 50 doublings, on average. By removing variability among donors, which might obstruct commercial manufacturing, genetically modified cells, such as induced pluripotent stem cells (iPSCs) or induced myogenic progenitors, offer advantages over primary cells. However, using induced progenitor cells necessitates the use of pricey substances for induction and calls for closer regulation of the processes of proliferation and differentiation. It may also spark regulatory questions and could affect consumer acceptance.

Regulatory frameworks and public acceptance are also important factors to consider. Cultured meat products need to undergo rigorous safety assessments and gain regulatory approval before they can be sold to consumers. Additionally, public perception and acceptance of cultured meat will play a crucial role in its adoption and market success.

While customizing media, ethical issues should be taken into account as well, such as what prospective customers would want. Utilizing components without animal testing and supplied sustainably may be necessary. Specific requirements must be fulfilled so that the Muslim community accepts cultured meat as halal. Animals must be slaughtered in accordance with Shariah law, and stem cells to be used in the manufacturing of cultured meat should be from halal animals permitted by Shariah. Furthermore, Halalan toyyiban substances (Halal foods that are pure, approved by fitrah, free from questionable ingredients, sanitary, safe, and nourishing) should be employed as the medium and biomaterials for scaffolds in the synthesis of cultured meat (Kashim et al., 2023).

1.3 NUTRITIONAL GENOMICS

Nutritional genomics is a branch of science that explores the interaction between nutrition, genetics, and gene expression. It aims to understand how nutrients and bioactive compounds in food can affect gene expression and, consequently, individual responses to diet. By studying the relationship between nutrition and genetics, nutrigenomics seeks to provide personalized dietary recommendations for optimal health and disease prevention.

I. Precision Nutrition: Nutrigenomics aims to enable personalized dietary recommendations based on an individual's genetic profile. With advancements in genetic sequencing technology and bioinformatics, researchers are exploring how specific genetic variations impact nutrient metabolism and dietary requirements. This knowledge could lead to tailored dietary guidelines and interventions.

II. Disease Prevention and Management: Obesity, diabetes, cardiovascular disease, and some forms of cancer are examples of chronic diseases that nutritional genomics may help control and prevent. By identifying genetic factors that influence an individual's susceptibility to certain diseases, researchers can develop targeted dietary strategies to mitigate risk and improve health outcomes.

III. Nutrigenetics and Nutrigenomics Testing: Advances in genetic testing have made it more accessible and affordable for individuals to obtain information about their genetic predispositions. Nutrigenetics testing can provide insights into how an individual's genetic variations may impact their response to specific nutrients. Combined with nutrigenomics, this information can guide personalized dietary recommendations.

IV. Gut Microbiota and Gene-Nutrient Interactions: In our digestive system, there is a population of microbes known as the gut microbiota, and it is crucial for the metabolism of nutrients. The gut flora may have an impact on how people react to varied diets, according to recent studies. In order to comprehend how these variables interact and impact health outcomes, nutrigenomics is looking into the intricate relationship between the gut bacteria, genetics, and nutrition.

V. Epigenetics and Nutrigenomics: The term "epigenetics" describes changes in gene expression that don't include DNA sequence changes. Nutrients and bioactive compounds in food can influence epigenetic modifications, potentially impacting gene expression and long-term health. Nutrigenomics research aims to uncover how specific dietary components can modulate epigenetic processes and influence health and disease.

VI. Nutrigenetics, which focuses on the interaction between genetics and nutrition, can indeed have implications in the field of food engineering. Here are a few ways in which nutrigenetics can be relevant:

   a. Personalized Food Formulation: Nutrigenetics can contribute to personalized food formulation by considering an individual's genetic variations. By understanding how different genetic profiles respond to specific nutrients, food engineers can develop customized food products that optimize nutritional content and meet individual needs. This could involve modifying macronutrient ratios, fortifying with specific bioactive compounds, or reducing allergenic components based on genetic predispositions.

   b. Targeted Functional Foods: Functional foods are designed to provide health benefits beyond basic nutrition. Nutrigenetics can help identify specific genetic variations that affect the metabolism of certain nutrients or bioactive compounds. This knowledge can be used to develop functional foods that are targeted toward individuals with particular genetic profiles, maximizing their potential benefits.

VI Nutritional Labeling and Product Information: With nutrigenetics, food engineers can develop more informative and personalized nutritional labels. By considering genetic variations related to nutrient metabolism, they can provide tailored information about how individuals with specific genotypes may respond to the food product. This can empower consumers to make more informed decisions about their diet and help them choose foods that are better suited to their genetic makeup.

VII Nutrigenetics and Food Allergies: Nutrigenetics research can provide insights into the genetic factors influencing food allergies and intolerances. Food engineers can utilize this information to develop allergen-free or hypoallergenic food products that are specifically designed to cater to individuals with particular genetic predispositions.

VIII Genetically Informed Food Safety: Nutrigenetics can play a role in assessing the safety of certain food components for individuals with specific genetic variations. By understanding how genetic factors influence the metabolism of certain substances, food engineers can determine appropriate safety thresholds for different populations and develop guidelines to minimize potential adverse effects.

It's important to note that the field of nutrigenetics is still relatively new, and the practical implementation of these concepts in food engineering may still be in the early stages. However, as our understanding of the interaction between genetics and nutrition advances, it is likely that nutrigenetics will continue to influence food engineering practices and lead to the development of more personalized and targeted food products.

1.3.1 Direct-to-Consumer Genetic Testing in Nutrigenomics

There are two complementary approaches in the field of nutritional genomics: nutrigenetics, which examines how genetic variation affects the body's response to nutrients, and nutrigenomics, which investigates how nutrients affect gene function. Direct-to-consumer genetic testing is now commercially available and offers information on ancestry, genetic risk for diseases such as type 2 diabetes, cancer, and cardiovascular disease. Companies offering this service of genetic testing also provide information on monogenic diseases, macronutrient and energy metabolism, weight control, obesity, and vitamin and mineral requirements. The clinical application of nutritional genomics has been particularly successful in treating monogenic diseases like galactosemia, phenylketonuria, and maple syrup urine disease, which can be managed through dietary interventions. However, polygenic diseases, which involve multiple genetic loci, are more complex and challenging to address using direct-to-consumer genetic testing. Polygenic traits and diseases are influenced by numerous genes and can interact with various physiological processes, making it difficult to provide tailored genetic recommendations based on single variants.

Generally, it is assumed that providing individuals with genetic information can motivate them to change their unhealthy behaviors. However, sometimes this is not true, and earlier studies have yielded mixed results in this regard. To enhance dietary assessment, technology-driven methods such as mobile applications have been developed to track food consumption accurately. Advancements in genotyping coverage, whole-genome sequencing, and integration with other omics disciplines like epigenomics, metagenomics, and metabolomics hold promise for further advancements in the field.

However, the proclamation and circulation of genomic information to consumers lacks substantial scientific evidence, and a greater comprehension of the relevance of genetics in promoting human health is crucial. Genetic information is complicated and random. It is influenced by factors such as family health history, ethnic background, and health status. Genetic counseling aims to help individuals understand and adapt to the connections of genetic contributions to disease.

Nutritional genomics has the potential to provide customized nutritional suggestions based on genetic vulnerability to polygenic diseases, but more precise techniques for determining dietary consumption and genetic vulnerability are needed. The integration of additional omics information, such as metabolomics, epigenetic modifications, transcriptomics, and gut microbiota composition, can provide a more extensive understanding of the interaction between dietary intake and genetic susceptibility.

Rudimentary steps toward precision nutrition have been taken, but more studies are required to develop a strong scientific model for its practical application to benefit individuals and societies (Guash-Ferré et al., 2018).

The Single Nucleotide Polymorphisms (SNP) Consortium is mapping polymorphic regions of the genome responsible for individual phenotypic differences among the human population. Specific genetic polymorphisms can alter metabolic responses to diet and influence the possibility of the occurrence of disease. SNPs can change gene recipes, resulting in variations in protein production or structure. These genetic variations affect the body's reaction to the components of the diet by influencing absorption and metabolism. Epigenetic events, such as DNA methylation pattern changes, can also impact gene expression and be changed in response to the components of the diet. Various dietary elements can influence post-translation events, and many of them contribute to the variation in response to dietary elements (Gaboon, 2011).

1.3.2 Relationship between Diet and Diseases

The connection between diet and diseases is well-established, with obesity being the most common nutrition-related disorder. Obesity occurs when excessive body fat accumulates and negatively impacts health, leading to a decrease in life expectancy and an increased risk of health problems. It is often measured using various indices such as body mass index, waist-to-hip ratio, etc. Obesity is a primary component of metabolic syndrome, a cluster of metabolic abnormalities that includes insulin resistance, diabetes mellitus (insulin-independent), hyperinsulinemia, high blood pressure, and impaired glucose tolerance. Additionally, obesity and these associated metabolic abnormalities significantly increase the risk of chronic diseases such as cardiovascular disease and cancer. However, the susceptibility to obesity is strongly influenced by genetic factors for controlling energy balance.

Treatment of obesity typically involves dietary interventions, physical activity, pharmacotherapy, diuretics, and surgical methods. The fat-mass and obesity-associated gene plays a crucial role in obesity development and contributes to various metabolic diseases. Targeting FTO's m6A demethylase activity or finding plant extracts that inhibit FTO expression could provide potential remedies for obesity and related diseases. Natural products can help in one or more ways by inhibiting the activity of the lipase enzyme, suppressing appetite, stimulating energy expenditure, inhibiting the differentiation of adipocytes, controlling lipid metabolism, and providing laxatives and general body cleansing effects can also be beneficial in obesity treatment. Their actions may include dissolving fat, enhancing secretions from glands, reducing cravings, and removal of water and other wastes from the body. However, their use should be combined with regular exercise, dietary modifications, and behavioral changes (Mohamed et al., 2014; Gaboon, 2011; Huang et al., 2022).

Cardiovascular disorder is the primary persistent disease associated with diet in recent times. Inflammation is a common fundamental factor for various chronic diseases, including cardiovascular disorders. Cardiovascular disorder encompasses a range of conditions linked to obesity, hypertension, atherosclerosis, and thrombosis. These pathological conditions have both genetic and environmental influences. Diet composition has a well-established relationship with CVD risk. Obesity itself is a significant cardiovascular risk factor, and genes involved in regulating energy balance can contribute to the development of cardiovascular disorders. Atherosclerosis, characterized by lipid transport and metabolism disorders along with persistent inflammation, plays a central role in cardiovascular disorder pathogenesis. Genetic variations in genes encoding apolipoproteins as well as some hormones and enzymes can influence individual susceptibility to cardiovascular diseases. Dietary factors also play a role in arterial hypertension, which is a crucial element in cardiovascular disorders. Sodium chloride is a well-defined dietary risk factor for hypertension, although blood pressure responses to salt intake can vary among individuals. Nutrigenomics aims to understand why some individuals can regulate high blood pressure with diet alone, while others need medication.

Thrombosis, particularly in arteries with atherosclerosis, is a key mechanism leading to acute coronary and cerebrovascular syndromes. Imbalances in blood coagulation factors can increase the likelihood of hypercoagulation and thrombosis. Genetic factors, along with dietary factors such as high-fat intake, can stimulate a hypercoagulable state. Gene polymorphisms affecting hemostasis-related genes have been suggested in thrombosis.

Genetic variants can also influence the response to dietary factors and modulate the risk of disease development. The epigenome, which is affected by environmental variations and lifestyle, can also play a role. Sufficient vitamin D intake and supplementation, along with dietary interventions, can impact blood pressure control and cardiovascular disease prevention. The evaluation of interactions between genes and nutrients is essential for public health policies to identify prevention techniques and promote healthy eating (Nunes et al., 2020).

The Rho-associated protein kinase is involved in both pathological and physiological cardiac hypertrophy. Many factors such as knowledge of fundamental mechanisms of its activation or inhibition, isoforms, modulation, kinase activity, expression, etc. can help in developing new methods for treating cardiovascular disorders, improving cardiac function, and increasing lifespan (Anaruma et al., 2020).

Cancer is a complex process influenced by gene expression, protein function, and metabolite function. Inherited gene mutations can increase susceptibility to cancer, and gene-diet interactions can further modify the risk. Dietary factors can act as either carcinogens or protective agents. Carcinogen metabolism-related gene polymorphisms can modify the probability of contact between carcinogens and target cells, affecting cancer initiation. Polymorphisms in genes encoding factors involved in the modulation of hormones are particularly relevant in hormone-dependent tumors. Obesity and physical inactivity contribute to the development of several common cancers, including breast, colon, endometrial, kidney, and esophageal cancers. Certain dietary factors, such as food additives, have been found to be carcinogenic, leading to gastric cancer. Nutrient intake can affect various pathways involved in carcinogenesis, including carcinogen metabolism, cell proliferation, differentiation, inflammation, DNA repair, oxidant/antioxidant balance, and angiogenesis. Dietary fibers and long-chain polyunsaturated fatty acids have protective effects against certain types of cancer, while bioactive compounds in fruits and vegetables can prevent cancer through multiple pathways (Elsamanoudy et al., 2016; Gaboon, 2011).

Gestational diabetes mellitus biomarkers require more detailed studies to understand their connection with the development of gestational diabetes fully. Weight loss through lifestyle changes and a healthier diet can mitigate the adverse effects of these biomarkers and improve insulin sensitivity. For obese and overweight women, especially with other risk factors such as genetic predisposition, weight loss before pregnancy can significantly reduce the risk of gestational diabetes mellitus (Heath et al., 2023).

Specialized pro-resolving mediators (SPM) have demonstrated anti-inflammatory and pro-resolving actions that can contribute to maintaining tissue homeostasis and controlling excessive inflammation. ω-3 fatty acids play a role in the resolution of inflammation, and their interaction with nutritional genomics suggests a link between acute inflammation, resolution, and nutrition. SPM may serve as agonists for new therapeutic approaches (Serhan, 2010).

Nutrition has a significant impact on immune function, with nutrients influencing both innate and acquired immune functions. Animal models have provided insights into the interrelationships between immunity and nutrition, and plant products offer potential resources for pharmaceutical development. Nutritional genomics, proteomics, and metabolomics are valuable tools for determining the nutrient requirements of animals, preventing and treating diseases, and testing functional ingredients and supplements in the pet food industry (Cooper and Ma, 2017).

In recent years, the human genome has been extensively mapped, providing detailed information about the genetic order of polymorphic markers and single-nucleotide polymorphisms (SNPs). This high-resolution recombination map has significantly expanded our knowledge of the human genome. The aim is to utilize the SNP map to gain valuable insights into the impact of nutrition on human health and disease, and to establish optimal diets. Consequently, the field of nutrigenomics considers food as a crucial environmental factor in the gene-environment interaction, and its main objective is to personalize food and nutrition to tailor it to individual genotypes (Gaboon, 2011).

The human-animal relationship, known as the human-animal interface, plays a multifaceted role. Improvements in the well-being of animals and the quality of animal products can lead to the betterment of human life. This includes providing equitable access to high-quality protein globally and reducing cardiovascular problems in society by offering low-cholesterol meat. In the future, it may be possible to modify the diets of livestock and poultry based on their genotypes, allowing nutrients to influence gene expression. Enhancing these relationships and changing them accordingly to improve health and competitiveness across all age groups will be a difficult endeavor. Numerous studies have been done in the past exploring the effects of nutrigenomics in mice and humans. However, research on nutrigenomics in food-producing organisms is still in its early stages.

By employing omics approaches like metabolomics, proteomics, transcriptomics, and molecular genetics, technology can be applied to heritable characteristics such as growth rate, feed intake, body weight, and carcass quality in animals. Nutrigenomics focuses on an animal's physiological and nutritional condition and adjusts metabolism according to nutrient levels in body fluids like blood, feces, urine, saliva, etc., with the aim of promoting animal growth. The application of other omics approaches and the identification of nutrient imbalances in livestock feeding research provide valuable insights and knowledge for assessing and regulating nutrient balance in farm animals. The findings from human investigations can also be applied to these hypotheses. Nutrigenomics assists scientists in identifying genes and DNA in animal cells or tissues, enabling them to select appropriate nutrients. Consequently, nutrigenomics plays a crucial role in developing feed and food that matches an animal's genotype, selecting nutrients tailored to individual genes, and comprehending how management in nutrition affects animal performance (Haq et al., 2022).

All in all, the connection between diet and diseases is multifaceted, with genetic factors, environmental influences, and gene-diet interactions playing significant roles. Understanding these interactions can lead to personalized approaches to nutrition and health, with the aim of preventing and managing various diseases (Swanson et al., 2003).

1.3.3 Functional Genomics

It combines information from genome sequence, gene and protein expression, and metabolite profiles using computational tools. This approach aims to understand the genotype and biological processes that contribute to the phenotype. By integrating data from genomic, proteomic, and metabolic domains, functional genomics provides insights into complex biological systems. Poultry breeding programs have successfully modified chicken genotypes to improve traits without knowing the underlying genes. Transcriptional profiling and genetic genomics help identify candidate genes that influence production characteristics. The integration of transcriptomic, metabolic, and proteomic data into gene networks and pathways enhances our understanding of genetic control in chickens. Emerging technologies like RNA silencing and transgenics offer possibilities for modifying chicken phenotypes. However, rigorous testing is necessary to ensure that desired improvements do not compromise performance and health (Cogburn et al., 2007).

When translating results across species, it's important to consider the differences between developmental models and human systems, including variations in gene function due to evolutionary trends. Standardization of diets used in non-mammalian models is needed to reduce variability in nutrition research. Genetic scoring of disease risks or treatment outcomes relies on specific gene polymorphisms. Developmental models can help identify genes related to adiposity and contribute to a comprehensive index of risk of obesity. Understanding gene function in combination with dietary interventions may lead to targeted interventions for preventing childhood obesity (Chowanadisai et al., 2020).

1.4 BIOTECHNOLOGY FOR IMPROVING CROP PRODUCTION

In essence, plant breeders are practical scientists involved in the development of agricultural products. Due to the lengthy duration required for creating new plant varieties, it is essential for breeders to strive for accelerated breeding processes. This enables them to generate products more rapidly and adapt to unforeseen shifts in crop production environments, climate conditions, and market demands. One key challenge that breeders must address is the emergence of new races and biotypes of pathogens and pests that can overcome the resistance found in current varieties, typically known as major gene or "vertical" resistance. Furthermore, there is a growing need for breeders to adopt consumer-oriented approaches that can cater to evolving preferences and quality standards in both domestic and international markets. Consequently, breeders should explore methods to reduce the time required for variety development and breeding cycles, while ensuring that the performance of new varieties remains uncompromised by the use of technologies aimed at expediting the breeding process (Lenaerts et al., 2019).

Single Seed Descent (SSD) is a plant breeding technique used to improve crop varieties. It is a method of selecting and propagating desirable traits in plants by continuously self-pollinating or selfing a single superior individual, known as the "seed parent" or "donor plant." A superior individual with desirable traits, such as high yield, disease resistance, or specific quality characteristics, is identified as the seed parent. The seed parent is self-pollinated, meaning it is fertilized using its own pollen. This ensures that the offspring inherit the traits of the seed parent. The seeds produced from self-pollination are harvested. Each seed represents an individual offspring with a unique combination of genes from the seed parent. From the harvested seeds, a subset of superior individuals is selected based on the desired traits. This selection process may involve evaluating various phenotypic or genotypic characteristics. The selected individuals are grown to maturity, self-pollinated again, and their seeds are harvested. This process of selfing and selection is repeated for multiple generations. After several generations of selfing and selection, the population stabilizes, and the offspring become genetically uniform. This means that most individuals in the population possess the desired traits. Single Seed Descent is an efficient breeding method that accelerates the development of improved crop varieties by focusing on a single superior individual. It reduces the time and space required compared to traditional breeding methods that involve crossing multiple plants. However, it is essential to maintain genetic diversity by periodically introducing new genetic material into the breeding population to avoid inbreeding depression and improve long-term adaptability.

In essence, this approach allows for expedited line fixation by manipulating the growth conditions of plants. This manipulation accelerates the process of flowering and seed production, surpassing the time it would typically take in a regular field environment during a typical crop growing season. On the other hand, the bulk-population method, while effective and resource-efficient, does not offer time savings. The key benefits of this Rapid Generation Advancement (RGA) method, when compared to other breeding approaches, are its speed, technical simplicity, minimal resource requirements, and low costs (Lenaerts et al., 2019).

1.5 BIOTECHNOLOGY FOR REDUCTION IN FOOD WASTE

The term "food waste" encompasses both agricultural waste that is generated before cooking and biodegradable waste from kitchens, including leftovers. This waste is produced by various industries, especially the food-processing sector and households. Food waste consists of different chemical compositions depending on its source, but it can serve as an excellent sustainable raw material for various bioprocesses that produce bioenergy and other commercially valuable bioproducts. Composting and turning food waste into animal feed with the help of living organisms are two ways to recycle the nutrients found in food waste for use in agriculture. Fruit and vegetable waste, for example, contains starch, cellulose, soluble sugars, and organic acids. Yeasts, particularly those found in citrus fruit waste, can utilize the sugars and acids, resulting in the production of high-protein biomass (Ng et al., 2020; Stabnikova et al., 2005).

Food waste includes peels, kitchen leftovers, slaughterhouse waste, fats and oils, bones, scales, and food wrappers and packaging. Some components, such as bones and packaging, may not be effectively digested through anaerobic processes. Legislation and regulations have been implemented to divert food waste from landfills and incineration due to the presence of high amounts of biodegradable materials and moisture (>70%). Organic matter in food waste can lead to environmental issues like leaching, fugitive emissions, and eutrophication. High moisture content can also reduce the efficiency of incineration. Consequently, suitable treatment methods are required (Narisetty et al., 2022).

Finding novel and environmentally friendly bioprocesses to transform food waste into useful bioproducts has been the focus of countless investigations in recent years. This is driven by the economic potential of transforming food-processing and agricultural waste into bio-based products, as compared to the decomposition issues associated with landfills. Biotechnology plays a significant role in reducing food waste across the entire food supply chain. It offers solutions such as the use of beneficial microorganisms (biocontrol agents) to prevent spoilage-causing pathogens and reduce post-harvest losses. Enzymes derived from biotechnological processes can enhance efficiency, quality, and waste reduction in food processing. Fermentation processes utilizing microorganisms can also minimize waste and improve raw material utilization (Ng et al., 2020).

Various bioprocesses like fermentation, solventogenesis, acidogenesis, photosynthesis, methanogenesis, bioelectrogenesis, and oleaginous processes can be employed to convert food waste into usable products such as chemicals, biofuels, bioelectricity, biofertilizer, animal feed, and biomass. Additionally, a multi-feedstock biorefinery design could be developed to obtain the same product from different types of food waste and biomass. These bioprocesses offer profitable alternatives to transforming food waste into hydrogen, ethanol, enzymes, chemicals, and methane, surpassing the conversion to animal feed or transportation fuel in terms of profitability (Nguyen et al., 2020).

Furthermore, biotechnology contributes to the development of biodegradable and compostable bioplastics made from renewable sources, which can replace conventional plastic packaging, thereby reducing environmental impact and waste. Biotechnology also enables the valorization of food waste through processes like anaerobic digestion or composting, where microorganisms convert organic waste into biogas (methane) or nutrient-rich compost. These waste-to-energy or waste-to-fertilizer approaches help extract value from food waste while reducing its volume. Additionally, biotechnology aids in the development of sensors and monitoring devices that detect food spoilage, ensuring food safety. Real-time monitoring systems can track the freshness and quality of food throughout the supply chain, enabling timely interventions to prevent waste.

Biotechnological tools, including genetic engineering and molecular breeding techniques, contribute to the development of crops resistant to pests, diseases, and environmental stresses. By improving crop yield and quality, biotechnology minimizes losses due to crop damage and increases overall food production. Genetically modified organisms with extended shelf life can resist spoilage, thereby reducing food waste resulting from premature deterioration. Biotechnological tools also offer solutions for preserving harvested crops through techniques like controlled atmosphere storage, which regulates temperature, humidity, and gas composition to extend the shelf life of fruits and vegetables.

In summary, biotechnology plays a crucial role in reducing food waste by enhancing crop productivity, preservation techniques, waste valorization, and providing sustainable packaging solutions. These advancements contribute to a more efficient and sustainable food system, reducing the environmental impact associated with food waste and enhancing global food security.

Researchers face challenges when using food waste as raw materials; for example, the unknown chemical composition percentages of different food waste types make it difficult to predict the bioproducts synthesized during bioconversion processes. Furthermore, converting food waste into food ingredients may not always lead to reduced environmental impacts or greenhouse gas emissions, as other waste management methods may be more environmentally beneficial. Utilizing food waste to feed animals offers environmental advantages by reducing the use of conventional feeds and greenhouse gas emissions. Furthermore, anaerobic digestion of food waste is responsible for significant greenhouse gas emissions, and alternative energy sources also exhibit high greenhouse gas emissions when based on fossil fuels (Ng et al., 2020; Scherhaufer et al., 2020).

Consequently, ongoing research focuses on investigating and optimizing innovative biotechnological techniques for the production of desirable bioproducts and renewable bioenergy, aiming for optimal reuse and recycling of food and agricultural wastes (Ng et al., 2020).

1.5.1 Fermentation

Microbial fermentation is recognized as a promising method for converting food waste into valuable end products. The choice of fermentation strategy depends on the type of food waste used as the raw material. Solid-state fermentation is commonly employed for solid substrates, aiming for maximum utilization of nutrients, while submerged fermentation is more suitable for liquid substrates. It is widely used in large-scale industrial practices due to its low processing cost and high productivity. Moreover, submerged fermentation is favored for various enzyme production processes because it is simple to control the process and sterilization.

Food waste is considered an excellent and sustainable raw material for microbial fermentation to produce a range of bioactive products. Food waste is rich in fatty acids, phenolic compounds, minerals, and proteins, making it ideal for biotransformation into beneficial bioproducts with enhanced functionality. This approach reduces the cost of production and pollution of the environment by utilizing food waste and eliminating the need for separate waste treatment. Agricultural waste, in particular, can be used as an inexpensive source of carbon and nitrogen for microbial growth and the fermentation of bioactive compounds. Proteins, enzymes, antioxidants, and pigments are among the numerous bioproducts that can be produced through microbial fermentation using food waste as the raw material. Lactic acid and ethanol are common products obtained from the fermentation of food wastes.

Bioethanol, a renewable biofuel, offers an alternative to petroleum-based fuels and helps mitigate the environmental issues associated with fossil fuel consumption. Carbohydrate-rich food waste serves as a suitable substrate for bioethanol production through fermentation. This typically involves a two-step process: enzymatic digestion of carbohydrates followed by the fermentation of the digested simple sugars using the yeast *Saccharomyces cerevisiae*. Pretreatment of food waste by heating or hydrolysis is necessary to break down the complex food constituents that are hard to digest and utilize in the fermentation process. Food waste sources rich in energy content, such as sugarcane bagasse, apple pomace, corn residue, rice, potato, and sugarcane, are commonly used for ethanol production.

Electro-fermentation is an innovative technique that increases biogas production by incorporating electrodes into microbial fermentation processes. Electro-fermentation improves the stability of anaerobic digestion and methane production. By placing electrodes and applying voltage, the metabolism rate of bioproducts and electron kinetics is enhanced, leading to increased biogas production. The electric field reduces activation energy and heat of activation, thereby improving the heat of adsorption and overall fermentation process (Ng et al., 2020).

In recent years, various new types of bioreactors have been developed and widely used in microbial cultivation processes. These include batch process bioreactors, continuous single- and double-phase bioreactors (such as acid phase and methane phase bioreactors), continuous stirred tank reactors, hybrid anaerobic solid-liquid bioreactors, tubular bioreactors, up-flow anaerobic sludge blanket bioreactors, anaerobic baffled bioreactors, piston flow anaerobic bioreactors, sequential batch anaerobic composting bioreactor systems, and solid-state stratified bed bioreactors. These bioreactors utilize different microorganisms to mitigate fruit and vegetable waste and generate desired value-added products (Sharma et al., 2021).

1.5.2 Anaerobic Digestion

Due to the ideal circumstances for anaerobic microbial development in the organic matter of food waste, anaerobic digestion is a well-recognized and environmentally acceptable approach for the effective treatment of food waste that also produces biogas. Food waste is a great substrate for anaerobic digestion since it is readily biodegradable and contains a lot of water (Ng et al., 2020).

In the absence of oxygen, anaerobic digestion is a multistage, complicated microbial process that decomposes organic materials. Biogas, which mostly consists of methane and carbon dioxide, is the main output of anaerobic digestion. The exact feedstock that is utilized for digestion determines the methane to carbon dioxide ratio in the biogas. The four steps of anaerobic digestion are hydrolysis, acidogenesis, acetogenesis, and methanogenesis. In the first stage, anaerobic hydrolytic bacteria hydrolyze complicated polymers to release their soluble monomers. Then, acidogenic bacteria subsequently use these soluble monomers during the acidogenesis stage to create volatile fatty acids, including acetic acid, butyric acid, and propionic acid. These higher volatile fatty acids are converted during the acetogenesis stage into acetic acid, which is the main precursor for the formation of biogas. In the acidogenic and acetogenic stages, hydrogen and carbon dioxide are created as by-products. Methanogenesis, the last phase of anaerobic digestion, involves the use of acetic acid by methanogenic archaea to generate biogas by acetoclastic methanogenesis. It is also possible for hydrogenotrophic methanogens to create methane using hydrogen and carbon dioxide produced at various phases. Furthermore, based on the microbial makeup and feedstock, other products such as lactates and alcohols may also be produced (Narisetty et al., 2022).

Methane and carbon dioxide make up the majority of biogas, which makes it a suitable replacement for fossil fuels in a variety of applications because of its excellent thermal and energy efficiency. For the production of biogas from food waste, a variety of bioprocess technologies, such as anaerobic digestion and electro-fermentation, can be used. Methane production is increased when food waste has a high concentration of water-soluble organic matter and a low concentration of carbohydrates. The hydrolysis and overall performance of anaerobic digestion can be further enhanced by pretreatment techniques such as mechanical grinding, pressurization-depressurization, ultrasound, microwave, biological, thermal, and chemical treatments (Ng et al., 2020).

Continuous stirred tank reactors and fluidized bed reactors have been utilized to produce lactic acid from food waste and waste-activated sludge in the conversion by repeated batch fermentation of industrial waste into biogas. Repeated batch fermentation is an efficient and quick process using lactic acid bacteria such as

Bifidobacterium, *Enterococcus*, *Alkaliphilus*, and *Dysgonomonas* for the fermentation of food waste.

Biohythane, a gaseous biofuel made of a mixture of 10–30% by volume hydrogen and 70–90% by volume methane, has also been made from food waste. This is accomplished through a procedure that involves anaerobic digestion followed by dark fermentation. Methane can be replaced with biohythane in the automotive industry (Sharma et al., 2021).

1.6 BIOTECHNOLOGY FOR PRODUCTION OF USEFUL SUBSTANCES

1.6.1 Antioxidants

The extraction of antioxidants from fruit and vegetable waste is a common practice due to the high content of these beneficial compounds in seeds, peels, and kernels. Additionally, functional lipids, proteins, and starch are other valuable substances that can be obtained from various food wastes, including dairy, meat, and marine industries. Solvent extraction is the traditional method used to recover soluble natural compounds from food wastes. Organic solvents and environmentally friendly solvents like water (known as green solvents) are utilized in this process. In this process, solvents penetrate the sample, solubilize the biomolecules which then diffuse out of the sample leading to their separation. Techniques such as high pressure, microwave energy, and ultrasound can accelerate the extraction process of bioactive compounds from food waste.

Supercritical fluid extraction is an environmentally friendly substitute for organic solvents, as carbon dioxide in its supercritical state is non-toxic, inexpensive, and non-explosive. CO_2 is commonly chosen as the supercritical fluid due to its critical point (304 K, 7.4 MPa), which allows extraction to occur at low temperatures, preventing thermal degradation of biomolecules. The low polarity of carbon dioxide facilitates the extraction of non-polar or slightly polar biomolecules, such as lipids and volatile oils, under supercritical conditions. Enzyme-assisted extraction is a technique that relies on the catalytic properties of enzymes. Various enzymes like cellulase, pectinase, xylanase, and glucosidase can be used to break down the structure of the cell wall. This helps in releasing the intracellular substances from the cells. This enzymatic disruption eases the transfer of desired compounds from food waste.

Electric treatment is an effective method for extracting lipids from various sources, offering advantages such as low energy consumption and moderate operating conditions when a low voltage electric field is applied. In recent studies, the use of aqueous biphasic systems has gained popularity for the recovery of bioactive compounds from agricultural waste. These types of systems are characterized by high water content and low interfacial tension, providing a biocompatible environment for extraction. This approach allows for single-step operation, leading to high recovery yields and purification of target biomolecules. Consequently, ABS is considered a cost-effective and time-saving separation method for industrial-scale production of various biomolecules (Ng et al., 2020).

1.6.2 Biosurfactants

Biosurfactants are surfactant compounds produced by microbial cell factories such as bacteria, fungi, and yeast. They offer a suitable alternative to chemical surfactants and have gained significant popularity due to their non-toxicity, biodegradability, and effectiveness over a wide range of pH and temperature conditions. Biosurfactants are generally amphipathic, capable of reducing surface tension at oil-water interfaces and enhancing the miscibility of water-insoluble substances. These surfactants also function as emulsifiers for both water-in-oil and oil-in-water mixtures, and they can produce stable gels and foams.

The biological generation of surfactants has become a viable alternative option in recent years, delivering improved structural and functional characteristics while being safer for the environment. In general, biosurfactants can be divided into two groups on the basis of molecular weight. Biosurfactants are molecules with a lower molecular weight, often between 0.5 and 1.5 kDa. Examples of these molecules include the glycolipids (such as sophorolipids, rhamnolipids, trehalose lipids, and mannosylerythritol lipids) and lipopeptides (such as surfactin, fengycin). Bioemulsans are high-molecular-weight substances that have a molecular weight greater than 500 kDa and are used to stabilize oil-in-water emulsions. Lipopolysaccharides, lipoproteins, and polymers of heteropolysaccharides are a few examples of bioemulsans. Microorganisms like yeast (*Candida* and *Rhodotorula* species), bacteria (*Acinetobacter*, *Arthrobacter*, *Bacillus*, *Enterobacter*, *Halomonas*, *Pseudomonas*, and *Rhodococcus*), and molds (*Aspergillus* and *Penicillium*) are commonly used in biosurfactant production (Narisetty et al., 2022).

1.6.3 Microbial-Produced Biopolymers

Sugars obtained from food waste or other organic wastes can be used as carbon sources to synthesize biopolymers generated by microorganisms, notably polyhydroxyalkanoates. Polyhydroxyalkanoates are biodegradable plastics that can replace synthetic plastics in some situations. They have numerous uses in the biomedical and utility industries as well as in the packaging of food and other products. PHAs have recently found widespread application in therapeutic settings, including the creation of vascular tissue, heart valves, bones, and drug delivery carrier matrices. Biopolymers, unlike synthetic polymers, have a distinct three-dimensional structure and are made up of long chains of monomers that are joined together by covalent bonds. A variety of value-added goods, including biopolymers that are both biodegradable and biocompatible, can be produced using food waste. Food waste can be treated by the addition of acid or enzyme, drying, etc., to obtain fermentable sugars. The biopolymer-producing bacteria subsequently metabolize the available carbon to create PHAs. Biopolymers such as polyhydroxyalkanoates, polylactides, polysaccharides, and aliphatic polyesters have been successfully tested as bioplastics because of how closely their chemical and physical characteristics match those of conventional synthetic polymers. The search for less expensive substrates, enhanced microbe growth techniques, and streamlined downstream processing techniques are the main prospects for PHA production because they are all necessary to lower production costs. PHAs and PHB can be made from a variety of feedstocks, such as dairy, potato processing, sugarcane industry, fish, poultry, waste containing fat (like wastewater from olive mills, waste cooking oil, and waste from palm oil mills), spent coffee grounds, rice mill waste, grist mill waste, malt house waste, etc., all of which are easily available at low cost.

Polyhydroxyalkanoates are the most well-known biopolymer among biodegradable plastics, and this is owing to their biodegradability and their thermoplastic capabilities, which are on par with those of polymers derived from petroleum. Polyhydroxyalkanoates can be accumulated by a wide variety of bacteria, including *Cupriavidus necator*, *Haloferax* species, *Rhodobacter* species, *Sphaerotilus* species, *Azotobacter* species, *Azospirillum* species, *Burkholderia* species, *Pseudomonas* species, recombinant E. coli, *Bacillus* species, *Haloferax* species, *Acinetobacter* species, and *Aeromonas* species. By utilizing food wastes as the only carbon source, biopolymers and other precursors for bioplastics, such as lactic acid, can be created. The type and content of food waste, the requirement for physical, chemical, or enzymatic pretreatment procedures for saccharification, and the choice of strains resistant to raw carbon sources are some restrictions, nevertheless. A variety of native and non-native strains have been explored to produce these important chemicals because of breakthroughs in molecular and microbial techniques. The isolation and purification of these biopolymers for their targeted uses still face difficulties in terms of yields, scalability, and separation (Narisetty et al., 2022).

1.6.4 Immobilized Enzymes

As an alternative to current chemical catalyst systems, immobilized enzyme systems have been presented in the bioconversion of food waste. This is because immobilized enzymes are much more stable and recyclable than chemical catalysts, which have low selectivity, large energy requirements, and are not ecologically friendly. Advantages of enzyme catalysts include their high selectivity, low energy needs, and catalytic power. They do, however, have drawbacks, such as poor stability, recovery, and temperature sensitivity. Enzyme immobilization strategies have been used to improve the characteristics of biocatalysts in order to get beyond these restrictions. By trapping biological catalysts or enzymes onto or inside of a matrix, enzyme immobilization increases the stability, adaptability, reusability, and resilience to environmental changes of the enzymes. It is important to select a proper supporting matrix, which can include leftovers from the vegetable and food-processing sectors, waste streams from dairies, used cooking oil, etc. The complicated chemical makeup of food waste streams makes them unsuitable for enzyme reactions and can lower the overall productivity of bioprocesses. It has been recognized that enzyme immobilization techniques are a well-engineered tool for assisting biocatalysts in catalyzing bioconversion activities even under less-than-ideal circumstances. Enzyme immobilization is not only a more economical and environmentally friendly method than using traditional chemical catalysts because biocatalysts can be recycled and reused, but it also gives the biocatalyst efficacy, thermal stability, and physicochemical stability, resulting in the bioconversion process's peak productivity. Mono-, di-, and polysaccharides, chitin, pectin, proteins, carotenoids, and phenolic chemicals are abundant in food waste streams. Food waste has been turned into a variety of useful products, including biodiesel, biofuels, sweeteners, food colorants, food supplements, nutraceuticals, fat substitutes, antioxidants, stabilizers, and thickeners, through the use of enzyme conversion processes like hydrolysis, oxidation, acylation, phosphorylation, glycosylation, deamination, and esterification (Ng et al., 2020; Sharma et al., 2021).

Fermentation technology has also been utilized to produce a number of enzymes from food waste. Food waste thus encourages the idea of a circular economy and no-waste generation. *Caldicellulosiruptor saccharolyticus* has been used in a two-stage method for the synthesis of biohydrogen and biomethane from food waste that entails hyperthermophilic dark fermentation and mesophilic anaerobic digestion. They discovered that the conversion of food waste produces 24.4 MJ/kg of energy in addition to leftover organic waste and gases that need to be processed further. A three-stage procedure (Hydrolysis/Acidogenesis-Methanogenesis-Composting) has been utilized to improve resource recovery effectiveness. Co-digestion was carried out in a Leach Bed Reactor as part of the three-stage processing procedure before hydrogenation and acidogenesis. The methanogenesis process has been found to benefit from airlift bioreactors, which increase methane recovery by using solid waste and off-gases from Leach Bed Reactors. Zero waste generation is the goal of the three-stage procedure (Sharma et al., 2021).

1.7 CONCLUSION

This chapter has covered several significant biotechnological techniques, along with current developments and application-related difficulties. Acellular compound production through biotechnology is already having an impact on a wide range of industries, as it makes it easier to replace outdated methods with more cost-effective and sustainable ones. Emerging cellular products have the potential to revolutionize the production of materials, food, and cosmetics. However, it is crucial to guarantee customer acceptance because these products are more noticeable to end users.

Cultured meat is a novel meat product that can lessen its adverse effects on the environment. It is a long-term, secure substitute for meat production. We can produce meat analogues with the appropriate sensory qualities and high resource efficiency thanks to the joint development of analytical techniques and structuring procedures. Cultured meat appears to have a big effect on the food and livestock industries.

Food waste reduction biotechnology research can focus on two main areas: converting food waste into higher-value products and preventing food waste by extending shelf life. By increasing TRL and improving models that integrate technical, economic, and environmental parameters, food waste transformation technologies have reached a point where environmental impact analysis is sufficiently advanced to understand the net impact of biotechnologies. Nonetheless, to fully comprehend the environmental effects of industrial-scale food waste reduction processes, boundary conditions must be carefully considered.

For certain people, there may be significant health benefits associated with nutrigenomics. Medical geneticists are in great demand and are hard to come by, and primary care physicians have little training in nutrition and genetics. Dietitians are specialists in nutrition science, and as a professional group, they are becoming more and more interested in nutrigenomics. To incorporate nutrigenomics into their practice, dietetics practitioners would need to undergo extensive training, just like doctors do. Therefore, by treating food as a significant environmental factor in the gene-environment interaction, nutrigenomics aims to ultimately personalize strategies for maintaining health by adjusting nutrition and food to everyone's genotype.

Various techniques for transforming food waste have been developed to create sustainable biofuels and other substances. Up until now, a number of novel strategies have been researched in an effort to lower the cost of the bioprocess and achieve a high production yield. Food waste is currently being valued using a variety of methods, such as dark fermentation for the production of biohydrogen, anaerobic digestion for the production of biomethane, co-fermentation for the simultaneous production of biohydrogen and biomethane, transesterification for the production of biodiesel, acidogenic fermentation for the production of bioplastics, and composting for the creation of biofertilizers. Modern biotechnological concerns are focused on finding the most productive and economical ways to valorize food waste while maintaining a high level of efficiency.

Ensuring the safety of these biotechnological products requires adherence to the corresponding food and cosmetic regulatory frameworks. Once all of these requirements are satisfied, sustainability is established, and financial arguments are made, the technology may actually experience a breakthrough.

REFERENCES

Anaruma, Chadi Pellegrini; Pereira, Rodrigo Martins; Rodrigues, Kellen Cristina da Cruz; Silva, Adelino Sanchez Ramos da; Cintra, Dennys Esper; Ropelle, Eduardo Rochete; Pauli, José Rodrigo; Moura, Leandro Pereira de. Rock protein as cardiac hypertrophy modulator in obesity and physical exercise, *Life Sciences*, 2020, 254, 116955.

Chowanadisai, Winyoo; Hart, Matthew D.; Strong, Morgan D.; Graham, David M.; Rucker, Robert B.; Smith, Brenda J.; Keen, Carl L.; Messerli, Mark A. Genetic and genomic advances in developmental models: Applications for nutrition research, *Advances in Nutrition*, 2020, 11, 971–978.

Cogburn, L.A.; Porter, T.E.; Duclos, M.J.; Simon, J.; Burgess, S.C.; Zhu, J.J.; Cheng, H.H.; Dodgson, J.B.; Burnside, J. Functional genomics of the chicken-a model organism, *Poultry Science*, 2007, 86, 2059–2094.

Cooper, Edwin L.; Ma, Melissa J. Understanding nutrition and immunity in disease management, *Journal of Traditional and Complementary Medicine*, 2017, 7, 386–391.

Dekkers, Birgit L.; Boom, Remko M.; Van der Goot, Atze Jan. Structuring processes for meat analogues, *Trends in Food Science and Technology*, 2018, 81, 25–36.

Elsamanoudy, Ayman Zaky; Neamat-Allah, Moustafa Ahmed Mohamed; Mohammad, Fatma Azzahra' Hisham; Hassanien; Mohammed, Nada, Hoda Ahmed. The role of nutrition related genes and nutrigenetics in understanding the pathogenesis of cancer, *Journal of Microscopy and Ultrastructure*, 2016, 4, 115–122.

Gaboon, Nagwa E.A. Nutritional genomics and personalized diet, *The Egyptian Journal of Medical Human Genetics*, 2011, 12, 1–7.

Guasch-Ferré , Marta; Dashti, Hassan S.; Merino, Jordi. Nutritional genomics and direct-to-consumer genetic testing: An overview, *Advances in Nutrition*, 2018, 9, 128–135, https://doi.org/10.1093/advances/nmy001

Haq, Zulfqar ul; Saleem, Afnan; Khan, Azmat Alam; Dar, Mashooq Ahmad; Ganaie, Abdul Majeed; Beigh, Yasir Afzal; Hamadani, Heena; Ahmad, Syed Mudasir. Nutrigenomics in livestock sector and its human-animal interface-a review, *Veterinary and Animal Science*, 2022, 17, 100262.

Heath, Hannah; Degreef, Kelsey; Rosario, Rodrigo; Smith, MaryKate; Mitchell, Isabel; Pilolla, Kari; Phelan, Suzanne; Brito, Alex,; Frano, Michael R. La. Identification of potential biomarkers and metabolic insights for gestational diabetes prevention: A review of evidence contrasting gestational diabetes versus weight loss studies that may direct future nutritional metabolomics studies, *Nutrition*, 2023, 107, 111898.

Huang, Chaoqun; Chen, Wei; Wang, Xinxia Studies on the fat mass and obesity-associated (FTO) gene and its impact on obesity-associated diseases, *Genes & Diseases*, 2022, https://doi.org/10.1016/j.gendis.2022.04.014

Hubalek, Sophie; Post, Mark J.; Moutsatsou, Panagiota. Towards resource-efficient and cost-efficient cultured meat, *Current Opinion in Food Science*, 2022, 47, 100885.

Kashim, Mohd Izhar Ariff Mohd; Abdul Haris, Alia Aryssa; Abd. Mutalib, Sahilah; Anuar, Nurina; Shahimi, Safiyyah. Scientific and Islamic perspectives in relation to the Halal status of cultured meat, *Saudi Journal of Biological Sciences*, 2023, 30, 103501.

Lenaerts, Bert; Collard, Bertrand C.Y.; Demont, Matty. Review: Improving global food security through accelerated plant breeding, *Plant Science*, 2019, 287, 110207.

Mohamed, Gamal A.; Ibrahim, Sabrin R.M.; Elkhayat, Ehab S.; Salah El Dine, Riham. Natural anti-obesity agents, *Bulletin of Faculty of Pharmacy*, Cairo University, 2014, 52, 269–284.

Narisetty, Vivek; Adlakha, Nidhi; Singh, Navodit Kumar; Dalei, Sudipt Kumar; Prabhu, Ashish A.; Nagarajan, Sanjay; Kumar, A. Naresh; Nagoth, Joseph Amruthraj; Kumar, Gopalakrishnan; Singh, Vijai; Kumar, Vinod. Integrated biorefineries for repurposing of food wastes into value-added products, *Bioresource Technology*, 2022, 363, 127856.

Ng, Hui Suan; Kee, Phei Er; Yim, Hip Seng; Chen, Po-Ting; Wei, Yu-Hong; Lan, John Chi-Wei. Recent advances on the sustainable approaches for conversion and reutilization of food wastes to valuable bioproducts, *Bioresource Technology*, 2020, 302, 122889.

Nguyen, X. Cuong; Tran, Thi Phuong Quynh; Nguyen, T. Thanh Huyen; La, D. Duc; Nguyen, V. Khanh; Nguyen, T. Phuong; Nguyen, X.H.; Chang, S.W.; Balasubramani, R.; Chung, W. Jin; Nguyen, D. Duc. Call for planning policy and biotechnology solutions for food waste management and valorization in Vietnam, *Biotechnology Reports*, 2020, 28, e00529.

Nunes, Ivone F.O.C.; Cavalcante, Ana A.C.M.; Alencar, Marcus V.O.B.; Carvalho, Marcos D.F.; Sarmento, José L.R.; Teixeira, Nayra S.C.C.A., Paiva, Adriana A.; Carvalho, Lídia R.; Nascimento, Leopoldo F.M.; Cruz, Maria S.P.; Rogero, Marcelo M.; Lima, Andréia C.B.; Carvalho, Cecilia M.R.G. Meta-analysis of the association between the rs228570 vitamin D receptor gene polymorphism and arterial hypertension risk, *Advances in Nutrition*, 2020, 11, 1211–1220, https://doi.org/10.1093/advances/nmaa076

Rivera, Fernando José Cantarero; Chen, Jiajia. Computational fluid dynamics modeling of cell cultures in bioreactors and its potential for cultivated meat production—A mini-review, *Future Foods*, 2022, 6, 100195.

Scherhaufer, S.; Davis, J.; Metcalfe, P.; Gollnow, S.; Colin, F.; Menna, F. De; Vittuari, M.; Östergren, K. Environmental assessment of the valorisation and recycling of selected food production side flows, *Resources, Conservation & Recycling*, 2020, 161, 104921.

Serhan, Charles N. Novel lipid mediators and resolution mechanisms in acute inflammation: To resolve or not?, *The American Journal of Pathology*, 2010, 177(4), 1576–1591.

Sharma, Poonam; Gaur, Vivek K.; Sirohi, Ranjna; Varjani, Sunita; Kim, Sang Hyoun; Wong, Jonathan W.C. Sustainable processing of food waste for production of bio-based products for circular bioeconomy, *Bioresource Technology*, 2021, 325, 124684.
Stabnikova, Olena; Wang, Jing-Yuan; Ding, Hong Bo; Tay, Joo-Hwa. Biotransformation of vegetable and fruit processing wastes into yeast biomass enriched with selenium, *Bioresource Technology*, 2005, 96, 747–751.
Swanson, Kelly S.; Schook, Lawrence B.; Fahey Jr., George C. Nutritional genomics: Implications for companion animals, *J. Nutr.*, 2003, 133, 3033–3040.

2 Biotechnology
Improving Food Nutrition

Tohfa Siddiqui, Mohammad Umar Khan, Sweta Joshi, Sadaf Nazir, Ahmed Mobeen, Sadaf Ahmad, Nazia Tabassum, and Z. R. A. A. Azad

2.1 INTRODUCTION

Biotechnology plays a critical role in increasing food nutrition in the context of integrated food and bioprocess engineering. This multidisciplinary strategy incorporates ideas from biology, chemistry, engineering, and other scientific domains to provide novel solutions for improving food quality, safety, and nutritional value (Ranjha et al., 2022). Products derived from natural sources have been utilised for ages. Processing natural materials to get considerable advantages has been a goal throughout history (McCullum et al., 2003). Biotechnology is a sophisticated, yet still developing, technique that creates or alters a product for a specific use by using live organisms and/or compounds derived from them. It may be extended to all organism genera, ranging from simpler genera such as viruses and bacteria to more complex genera such as plants and animals (Amin, 2018). As a result, biotechnology has become an important component of contemporary industry, agriculture, and medicine (Fernie and Sonnewald, 2021). Scientists employ modern biotechnology to recognise and regulate the genetic structure of species for application in agricultural product production or processing (Benkeblia and Hefferon, 2020). The dynamic and multidisciplinary discipline of biotechnology uses biological systems, organisms, and cellular processes to create novel products and services that advance several industries and facets of human existence (Dahiya, 1938; Goulart et al., 2014). It includes a wide range of scientific methods for manipulating and using living things for useful purposes, such as genetic engineering, molecular biology, microbiology, and bioinformatics (Bhatia et al., 2019; Godbey, 2022). The creation of biopharmaceuticals, gene therapies, and diagnostic tools in this sector has revolutionised healthcare by enabling more precise diagnosis and personalised medication (Bhatia et al., 2019). Genetically engineered crops with increased yield, nutritional value, and insect resistance have been produced in agriculture thanks to biotechnology, helping to secure the world's food supply (Delaney et al., 2018; National Academies of Sciences, Engineering, and Medicine, 2017; Romeis et al., 2019). Additionally, biotechnology is essential for environmental protection since it provides methods for

DOI: 10.1201/9781032650616-2

waste management, pollution reduction, and the creation of biofuels from renewable resources.

Biotechnology has profoundly changed a number of sectors and has the potential to solve some of the most urgent problems facing the world today, including the elimination of illness, food security, and sustainable energy generation (Cornelissen et al., 2021; DaSilva et al., 2002; Holei et al., 2022). However, when dealing with genetically modified organisms and other potentially harmful manipulations in the natural world, its uses also create ethical and safety concerns (Tamis et al., 2009). The field's continued development will probably have a significant impact on how medicine, agriculture, business, and the environment develop in the future.

Here's a more detailed description of how biotechnology helps to improve food nutrition in this integrated framework.

2.2 ENHANCED NUTRITIONAL CONTENT

Enhancing nutritional content in food through biotechnology and bioprocess engineering entails tactics targeted at raising the amounts of vitamins, minerals, antioxidants, and other beneficial substances in diverse food items (Nicholson et al., 2021). This strategy is motivated by the desire to treat nutritional inadequacies, enhance public health, and give customers more healthy alternatives (Glover and Poole, 2019). Enhanced nutrient content in food is achieved through a combination of genetic engineering, selective breeding, optimised growing conditions, and innovative processing techniques (Gaikwad et al., 2020; Giri et al., 2023). Scientists can use genetic engineering to put certain genes into plants and crops to increase their nutritional content (Giri et al., 2023; Hameed et al., 2018). This entails finding and transferring the genes responsible for the production of desired nutrients into target crops. As an example, Golden rice: rice that has been genetically modified to generate beta-carotene, a precursor of vitamin A, aids in the fight against vitamin A deficiency, which is a serious public health concern in many poor nations (Brooks, 2013; Glover et al., 2020; Zimmermann, 2004). Crops that have been biofortified, such as maize, wheat, and beans, have greater quantities of vital minerals such as iron, zinc, and vitamins (Chaudhary et al., 2022; Mogendi et al., 2016). Plants are encouraged to accumulate larger quantities of vitamins, minerals, and antioxidants via the introduction of specialised genes, conventional breeding, and nutrient-rich soil management (Clarke and Zhang, 2013). Controlled growth settings, sustainable practices, and biostimulants all play a part. Minimal processing, nanoencapsulation, and fermentation are used to maintain or improve nutritional content. Finally, these multidisciplinary techniques result in more nutritious food items that treat nutritional inadequacies and promote public health.

2.3 NUTRIENT BIOAVAILABILITY

The term "nutrient bioavailability" describes how well a food's nutrients are absorbed and used by the body after eating (Melse-Boonstra, 2020). A food may have significant amounts of nutrients, but if those nutrients are chemically bonded, interfered with by anti-nutritional elements, or impacted by processing techniques, they may not be effectively absorbed (Alowo et al., 2018). For instance, various heating or fermentation methods might increase the availability of nutrients by dissolving anti-nutritional substances or facilitating the release of nutrients from the food matrix (Samtiya et al., 2021). Enhancing nutrient bioavailability is essential to ensuring that the body can efficiently use the nutrients included in the diet for a variety of physiological processes (Arshad et al., 2021; Bouis and Saltzman, 2017). The bioavailability of nutrients is increased using a variety of methods. The use of enzymes to break down complex nutrients, the matching of nutrients that facilitate absorption, the fortification of meals, the promotion of gut health through the use of prebiotics and probiotics, and the use of novel delivery systems are a few examples of how to reduce anti-nutritional factors (McClements, 2015; Mcclements et al., 2015). Optimising nutrient absorption also involves the use of genetic engineering and personalised dietary strategies. Overall, these techniques from bioprocess engineering make sure that the body can absorb and use the nutrients in food more effectively for greater health (Kong et al., 2017; Van der Klis and Jansman, 2002).

2.4 FUNCTIONAL FOODS

Functional foods go beyond basic nutrition to provide particular health advantages. When included in a healthy diet, these items can improve physiological processes, boost well-being, and lower the chance of developing chronic illnesses (Clare M Hasler, 2002). Functional foods contain bioactive substances, which are either naturally occurring or artificially added ingredients that have a favourable effect on health (Clare M. Hasler, 2002). These substances may consist of phytochemicals, polyphenols, flavonoids, carotenoids, and antioxidants (Aguiar et al., 2019). They can promote numerous body processes and fight oxidative stress, inflammation, and other negative effects. This type of food also contains vitamins, minerals, and other helpful components that are frequently added to functional meals as fortifiers (Cencic and Chingwaru, 2010). Bioactive substances, vitamins, and minerals are added to give unique health advantages above and beyond those of basic nutrition. Functional foods are created to treat certain health issues, providing targeted health benefits (Kaur and Das, 2011). For instance, diets containing probiotics or prebiotics may boost immunity and intestinal health (Arevalo-Villena et al., 2017; Singh et al., 2022), while foods containing omega-3 fatty acids may target cardiovascular health and other advantageous substances that promote general health and aid in the management or prevention of chronic illnesses (NIH, 2019). Functional foods are created by interdisciplinary teams and contribute to consumer nutrition support while helping to promote health.

2.5 NUTRACEUTICALS AND BIOACTIVE COMPOUNDS

Nutraceuticals and bioactive compounds are bioactive ingredients found in food that provide a variety of health advantages. They have the ability to bridge the nutritional-medicine divide by providing a natural alternative to enhance health and wellness

(Sharma et al., 2020). Continuous research, creative formulations, and good communication are critical to realising their promise for improving health outcomes.

2.6 HEALTH ADVANTAGES

Nutraceuticals and bioactive substances have been associated with several health benefits. Many bioactive compounds, such as vitamins C and E and polyphenols, exhibit antioxidant properties that help protect cells from oxidative stress and cellular damage. (Das et al., 2012; Durazzo et al., 2020; Puri et al., 2022). Heart Health: omega-3 fatty acids, which are included in foods like fatty fish and flaxseeds, are associated with cardiovascular health by lowering inflammation, enhancing lipid profiles, and promoting normal blood pressure levels (Chaddha and Eagle, 2015; Christensen, 2011; Kang, 2012). Cognitive performance: some bioactive substances, such as the flavonoids found in dark chocolate and berries, may enhance cognitive performance and lower the risk of age-related cognitive decline. Bone Health: vitamins D and K and calcium are nutrients that support healthy bones and help ward off diseases like osteoporosis (Ashton et al., 2022; Luo et al., 2023; Szot et al., 2022).

In personalised nutrition, where individual genetic, dietary, and health aspects are taken into account to prescribe certain bioactive compounds to meet particular health needs, nutraceuticals can play a role (Chhabra et al., 2021; Kajmi et al., 2023). Research on nutraceuticals is underway in an effort to better understand their modes of action and possible advantages. Researchers are examining how these substances interact with the body, determining the best doses, and assessing how well they work to prevent and treat various illnesses (Mannucci et al., 2021). Regarding the labelling and marketing of nutraceutical goods, several national laws apply. One of the most important aspects of product development and marketing is making sure that consumers are accurately informed about the health advantages. The composition and distribution of nutraceutical goods must be taken into account while developing them. To maintain their stability and bioavailability, nutraceuticals can be added to meals, drinks, supplements, or even functional foods (Aronson, 2017; Borodina et al., 2020; Chopra et al., 2022; Souyoul et al., 2018).

2.7 BIOLOGICAL FORTIFICATION

The goal of biological fortification, often referred to as biofortification, is to increase the nutritional content of crops during their growth and development (Ofori et al., 2022). It is a revolutionary method in agriculture and food science. Biological fortification concentrates on naturally raising the nutritional levels in the crops themselves, in contrast to conventional fortification, which adds nutrients to processed meals (Walton, 2019; Listman et al., 2019). This technique has a tremendous deal of potential for addressing micronutrient shortages and enhancing the nutritional value of common food. In order to boost vital vitamins and minerals and solve widespread nutritional inadequacies, it incorporates selective breeding and genetic engineering. Focusing on basic foods ensures long-term effects in this sustainable strategy. Collaboration between scientists, agronomists, and politicians is necessary to develop biofortified crops, which offer a remedy for starvation and enhance public health internationally (Bouis et al., 2019; Galani et al., 2022; Roriz et al., 2020). In order to treat widespread nutritional inadequacies, biofortification frequently focuses on vital micronutrients such as iron, zinc, vitamin A, and folate. Consider this: To address anaemia and zinc deficiency, beans and grains are biofortified with iron and zinc. Sweet potatoes and maize with added vitamin A are intended to lower immune system susceptibility and impairment caused by vitamin A insufficiency (Yang et al., 2007). Plant breeders, agronomists, food scientists, nutritionists, and politicians must work together to fortify biologically. Partnerships that work well make sure that fortified crops are not only nutritionally improved but also successfully incorporated into regional agricultural systems.

2.8 QUALITY CONTROL AND SAFETY OF FOOD

In the context of bioprocess engineering, quality control and safety are crucial for preserving the nutritive content of foods while preserving consumer health (Varghese et al., 2022). A variety of technologies and procedures are included in bioprocess engineering, which is centred on enhancing processing procedures to maintain vital nutrients and avoid contamination (Yamei et al., 2020). A few techniques, such as evaluating the raw material, evaluate the safety and quality of the raw ingredients used in the manufacture of food. These substances, which include agricultural products, additives, and ingredients, must adhere to regulations and be checked for pollutants or adulterants (Junejo et al., 2022; Li et al., 2021). Another way of enhancing the nutrient value of food is a systematic method for locating and managing possible risks in the food manufacturing process, known as HACCP (Hazard Analysis and Critical Control Points). To reduce hazards to consumer health, it entails identifying crucial control points, keeping an eye on them, and taking remedial action when necessary (Najah and Putri, 2019; Umi Raihanah and Norazmir, 2021). Throughout the course of manufacturing, regular analyses and tests are carried out. This entails sensory assessment to preserve uniformity in flavour, texture, and appearance, chemical analysis to assure ingredient compliance, and microbiological testing to identify dangerous infections. These evaluation methods ensure that food products are safe, consistent, and of the highest quality, meeting regulatory requirements and consumer expectations.

2.9 HARMONISING THE PAST AND FUTURE: BIOTECHNOLOGY'S EVOLUTION IN FOOD PRODUCTION

Traditional and contemporary biotechnology are two different but related areas of technological advancement that have revolutionised food production (Hugenholtz, 2013). Utilising the natural processes of bacteria, plants, and animals is what traditional biotechnology entails in order to create novel food kinds, increase agricultural yields, and better food preservation. Since ancient times, methods like selective breeding and fermentation have been used to modify the features of crops to produce a variety of meals, including bread, cheese, and yoghurt (Abigaba et al., 2022).

In contrast, modern biotechnology introduces advanced tools such as genetic engineering and molecular biology to manipulate and modify organisms at the genetic level (McCullum et al., 2003; Yılmaz, 2019). Genetic modification allows for the precise introduction of specific traits, such as disease resistance or enhanced nutritional content, into crops (van Esse et al., 2020). This has led to the development of products like genetically modified (GM) crops and biofortified foods, exemplified by the renowned Golden Rice engineered to contain increased levels of vitamin A (Romig, 2019). The synergistic interaction between conventional and contemporary biotechnology and food production is a distinguishing feature. Traditional procedures are expanded upon by modern biotechnology, which gives them cutting-edge accuracy and capability. Innovative goods like probiotic yoghurt enhanced with microorganisms that are good for your health or enzyme-assisted bread with better texture and nutrition are the outcome of this combination (Jorge and Dellagostin, 2017).

Additionally, by combining these biotechnological strategies, important global issues, including food security, malnutrition, and sustainable agriculture, are addressed. Their integration does not, however, come without difficulties. Modern biotechnology is surrounded by regulatory frameworks, consumer perceptions, and ethical issues that call for thorough analysis and open discussion (Shabbir et al., 2021). A continuum of human inventiveness is essentially demonstrated by the merging of conventional and contemporary biotechnology in food production. Together, these methods improve our capacity to create meals that are not just delectable and nutrient-dense but also address important environmental and health issues in a world that is becoming more complicated.

2.10 SYNERGISING FOOD AND BIOPROCESS ENGINEERING: INTEGRATED TECHNIQUES FOR ENHANCED PRODUCTION

In order to improve the production, processing, and quality of food items using biotechnological methods, integrated food and bioprocess engineering techniques combine a variety of scientific ideas, technologies, and methodologies (Hu and Liao, 2013). This multidisciplinary approach seeks to produce food more sustainably and efficiently while preserving or improving the end products' nutritional and sensory qualities (Shuler and Kargi, 1992; Velayudhan, 2014). The following are some essential elements and illustrations of integrated food and bioprocess engineering methods:

- *System Optimisation*: To improve food processing, integrated approaches use engineering, biological, and chemical concepts. For instance, in the dairy industry, optimising the fermentation conditions for yoghurt manufacturing entails managing variables like temperature, pH, and starter cultures to guarantee the development of advantageous bacteria and the proper texture and flavour of the finished product (Duong-Trung et al., 2023; Su et al., 2019).

- *Design and Use of Bioreactors*: Enzymes, probiotics, and other bioactive substances used in food processing are produced by cultivating microorganisms in bioreactors for a variety of reasons. By integrating bioprocess engineering approaches, cultivation is made more effective and regulated, producing larger yields and higher-quality products. As an illustration, consider how microbial fermentation is used to produce citric acid (Amer and Ramsey, 2018; Selden and Fuller, 2018).
- *Enhancing Nutrients*: Bioprocess engineering may be used to make food items more nutritious. This includes strategies like fortification, in which vitamins or minerals are added to food during processing to improve its nutritional worth. For instance, items made from cereal can be vitamin-fortified to treat micronutrient deficits (Sivaramakrishnan et al., 2023; Verni et al., 2019).
- *Enzyme-Assisted Processing*: To enhance the nutritional content and product quality, enzymes are frequently utilised in the food processing industry. Integrated strategies choose the best enzymes, increase their activity, and include them into the processing stages. The manufacturing of bread with the aid of enzymes, as was previously indicated, is a prime example of this strategy (Costa et al., 2021).
- *Utilisation of Waste*: Integrated strategies seek to reduce waste and increase resource efficiency. Biotechnological techniques can convert food processing waste into goods with additional value. Agricultural waste can be transformed into biofuels, animal feed, or bioactive substances (Bala et al., 2023).
- *Biopreservation*: By using methods like fermentation, managed microbial growth, and the utilisation of natural preservatives, bioprocess engineering can help to preserve food goods. Microbial processes are used to preserve fermented foods like kimchi and sauerkraut while also improving flavour and nutritional content (Verni et al., 2019).
- *Durability and Environmental Effect*: Integrated techniques also take the effects of food production on the environment into account. Food processing sewage may be treated using methods like bioremediation, which employs microorganisms to break down contaminants (Sharma, 2021).
- *Real-time monitoring* and management of food product quality is possible with the help of bioprocess engineering approaches. Consistent product quality and safety are made possible through sensors, automation, and data analysis (da Costa et al., 2023).
- *Bioprocess engineering and nanotechnology* can be used to build nutrition delivery systems that are nanoscale in size, enhancing the nutrients' absorption and bioavailability (Arshad et al., 2021).

The overall goal of integrated food and bioprocess engineering approaches is to build more effective, long-lasting, and nutrient-rich food production systems by combining the concepts of engineering, biology, and chemistry. This integration makes it possible to successfully address the issues of dietary quality, environmental sustainability, and food security in a fast-shifting global food system.

2.11 BIOTECHNOLOGY: METHODS AND TECHNIQUES FOR ADVANCEMENT

Biotechnological techniques have played an important role in increasing the nutritional content of food items by allowing for the modification and enhancement of their intrinsic properties. The way we improve the nutritional value of food items has been transformed by biotechnological methods, which provide novel technologies that address certain nutritional deficits and health issues. Genes encoding advantageous features, such as increased vitamin or mineral levels, can be directly inserted into crops by genetic modification (GM) or genetic engineering (Lee and Jungbauer, 2014). This strategy has produced products like Golden Rice, which is fortified with provitamin A to fight vitamin A deficiency. Through conventional breeding or genetic alteration, biofortification aims to enrich crops with vital minerals like iron, zinc, and vitamins to make them more nutrient-dense without only depending on dietary supplements. An age-old method called fermentation uses microbes to break down complex substances, increasing nutritional bioavailability and producing probiotics that support gut health (Jungbauer, 2013). By reducing anti-nutritional elements, enzyme technology helps to increase nutrient absorption and raise the total nutritional value of food ("Biotech Methods and Advances," 2010). While nanotechnology helps in successfully encapsulating and distributing nutrients, microbial biotechnology allows us to create nutrients like vitamins throughout the production process (Figure 2.1). Some of the key biotechnological techniques used for this purpose include the following.

2.11.1 Biofortification

Biofortification is the process of enhancing crop nutritional content by traditional breeding or genetic modification. The idea is to create crops that have high quantities of certain nutrients. Crops, for example, can be altered to have increased iron or zinc content, so targeting areas where these elements are deficient (Dhaliwal et al., 2022; Jha and Warkentin, 2020).

2.11.2 Genetic Engineering

Genetic engineering entails changing an organism's genetic makeup to add desired qualities. By adding genes that encode for desirable features, this approach is used to increase the nutritional value of food. For example, genes can be added to crops to boost the number of vitamins, minerals, or other necessary nutrients (Pyne et al., 2019; Sreenikethanam et al., 2022).

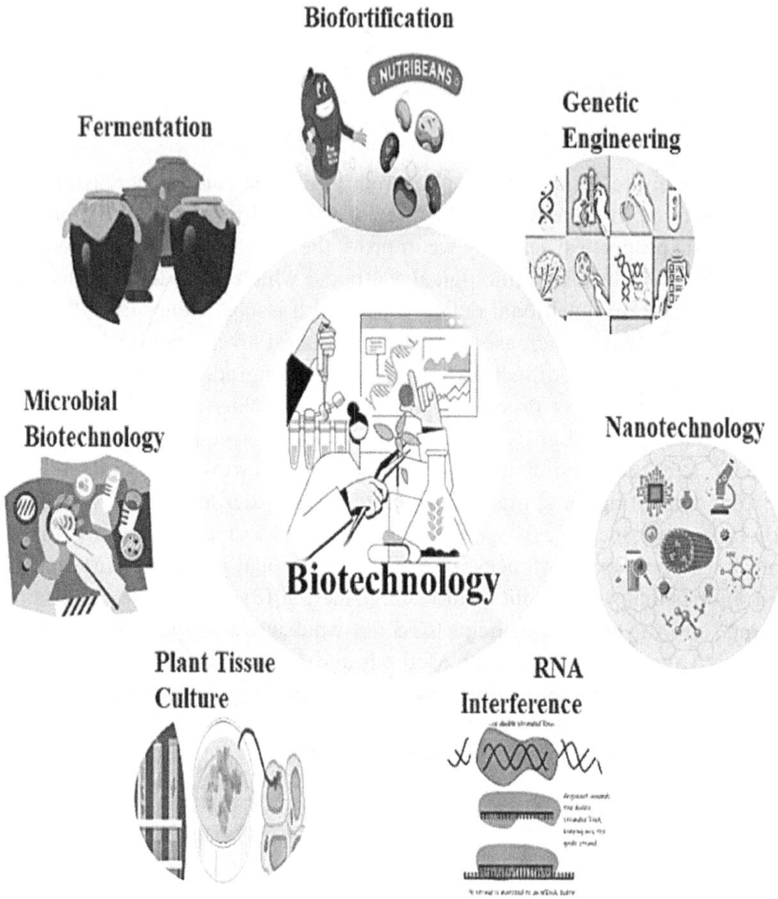

FIGURE 2.1 Enhancing nutritional content through key biotechnological techniques

2.11.3 Nanotechnology

Nanotechnology entails the nanoscale manipulation of materials. It can be used in food to enclose and safeguard nutrients, improving their stability and absorption. Nutritional availability may be increased by integrating nutritional delivery devices that are nanoscale into food products (Saritha et al., 2022).

2.11.4 Plant Tissue Culture

Plant cells are grown under controlled conditions in plant tissue culture. Using this method, desirable chemicals or nutrients may be produced in greater amounts.

Plants with improved nutritional profiles can be restored by modifying cell cultures (Espinosa-Leal et al., 2018; Paula et al., 2020).

2.11.5 Microbial Biotechnology

During the manufacturing of food, microorganisms can be used to synthesise a variety of nutrients. During fermentation processes, certain bacteria may generate vitamins like B12. The nutritional value of the finished goods can be increased by including these bacteria in food manufacturing (Pikaar et al., 2023; Sanchez-Andrea and Jetten, 2018).

2.11.6 Fermentation

Fermentation is a centuries-old biotechnological process that uses microorganisms such as bacteria, yeast, and moulds to metabolise food components. This method has the potential to improve nutritional availability and bioavailability. Fermented foods, such as yoghurt and sauerkraut, are loaded with helpful bacteria and have improved nutritional profiles as a result of the breakdown of complex compounds (Augustin et al., 2023; Pérez-Alva et al., 2022).

These biotechnological techniques provide prospective opportunities for increasing food product nutritional content, correcting nutritional deficiencies, and boosting population health and well-being. However, when applying these approaches to food production, it is critical to address ethical, safety, and regulatory concerns.

2.12 REVOLUTIONISING NUTRITION: BIOTECHNOLOGICAL ADVANCES FOR HEALTHIER FOOD PRODUCTS

Biotechnological methods to better food products include a variety of tactics that take advantage of the positive effects of probiotics, prebiotics, symbiotics, and enzymes to boost gut health and nutrient availability. These techniques are based on the concept that a healthy, diversified gut microbiota is essential for general health and well-being (Kvakova et al., 2021; Markowiak and Ślizewska, 2017; Markowiak and Śliżewska, 2017). Each of these techniques is explained below.

2.12.1 Probiotics

Probiotics are living microorganisms that, when ingested in suitable amounts, give health benefits (Savarino et al., 2022). These beneficial bacteria help to maintain a healthy gut microbiota, which aids digestion and immunological function. Biotechnological procedures involve the identification and selection of certain probiotic strains that may be put into meals such as yoghurt and fermented goods (Staniszewski and Kordowska-Wiater, 2021). These living cultures improve gut health by affecting the digestive system's microbial community (Wieërs et al., 2020).

2.12.2 Prebiotics

Prebiotics are nondigestible chemicals that provide sustenance for healthy gut microorganisms. They foster a favourable environment for probiotic development and action. Biotechnological techniques are centred on integrating prebiotic substances into food items, thus encouraging the growth of healthy bacteria in the colon. This fermentation process produces short-chain fatty acids, which benefit intestinal health and general well-being (Davani-Davari et al., 2019; Whisner and Castillo, 2018).

2.12.3 Symbiotics

Synergistic advantages of probiotics and prebiotics are combined in symbiotics. Biotechnological solutions entail developing products that supply particular strains of probiotics as well as the prebiotic substrates required for growth. Symbiotics boost the survival and colonisation of probiotics in the gut, boosting their health-promoting effects and establishing a varied gut microbiota (Yadav et al., 2022).

2.12.4 Enzymes

Enzymes are essential for converting complicated nutrients into more absorbable forms. Enzymes are used in food preparation to improve nutritional availability in biotechnological applications (Sg et al., 2018). Adding enzymes that target carbs, proteins, or lipids, for example, can increase digestion and nutritional absorption (Torbica et al., 2022). Enzymes improve nutrition utilisation and general health by optimising nutrient degradation. By increasing gut health and nutritional availability, these biotechnology technologies jointly contribute to the development of better food items. These tactics, which harness the power of probiotics, prebiotics, symbiotics, and enzymes, provide novel approaches to improve the nutritional content of meals while also supporting healthy digestion and well-being (ClinicalTrials.gov, 2012).

2.13 ENHANCING FOOD QUALITY AND SAFETY: BIOPROCESSING INNOVATIONS IN FERMENTATION, ENZYME ENGINEERING, MICROENCAPSULATION, AND ALLERGEN MANAGEMENT

Bioprocessing is a multidisciplinary technique that employs biological entities such as bacteria and enzymes to convert raw materials into marketable products. This approach combines biological, chemical, and engineering concepts to optimise diverse processes in businesses such as food, medicines, and biofuels (Xiang et al., 2019). Bioprocessing encompasses techniques such as fermentation, in which microorganisms metabolise substrates to create molecules with specified qualities such as flavour or nutritional improvements (Voidarou et al., 2021). Enzyme engineering is the process of tailoring enzymes to expedite biological processes, improve product quality, or increase nutritional availability. To preserve delicate molecules and manage their release, microencapsulation wraps them within protective materials. Furthermore, allergen elimination alters ingredients or procedures to reduce allergenic components in meals, assuring the safety of allergy sufferers (Molina et al., 2019). :

Distinct approaches within bioprocessing include the following:

Fermentation is a bioprocessing approach in which microorganisms such as bacteria, yeast, or fungi metabolise substrates such as carbohydrates to produce a variety of chemicals. Fermentation improves food preservation, flavour development, and nutritional value; for example, yoghurt fermentation includes microorganisms turning lactose into lactic acid, boosting shelf life and nutritional quality (Sharma et al., 2020).

Enzyme engineering improves the performance of enzymes by changing their structure to improve efficiency, stability, and selectivity in activities such as increasing texture, flavour, or nutritional availability; designed enzymes help carbohydrate breakdown in goods such as bread (Ali et al., 2020). For example, probiotics are protected in dietary supplements so they can survive the digestive system and increase effectiveness (Amara and Shibl, 2015).

Microencapsulation encases small particles in a protective material to shield sensitive compounds, preserving bioactive ingredients like vitamins, flavours, and probiotics in the food industry and ensuring stability during processing and storage (Calderón-Oliver and Ponce-Alquicira, 2022).

Through ingredient, protein, or process modification, allergen removal reduces the likelihood that allergenic components will cause allergies to develop. Techniques include genetic modification, enzymatic hydrolysis, and specialised processing, resulting in allergen-safe foods for people with allergies and promoting food safety (Gallo and Sayre, 2009; Singh and Bhalla, 2008).

These bioprocessing approaches demonstrate biotechnology's adaptability in food production, allowing for the development of unique and better food items with enhanced characteristics, decreased allergenicity, and higher stability. Overall, bioprocessing combines science and technology to generate novel solutions that serve a wide range of industries.

2.14 NAVIGATING BIOTECHNOLOGY: SAFETY, REGULATIONS, AND ETHICAL DIMENSIONS

The field of biotechnology includes a broad framework for safety and regulatory issues designed to ensure the ethical design, manufacture, and use of biotechnological goods and procedures (Trump et al., 2023). This framework takes into account a number of factors to ensure the health of consumers and the environment. Risk assessment entails locating possible risks connected to biotechnological practices or products and assessing their potential consequences (Vasanthi and Bhat, 2018). The process of evaluating a product's safety requires thorough testing, with an emphasis on factors including toxicity, allergenicity, and unforeseen consequences (Bawa and Anilakumar, 2013). In order to inform customers on the nature, composition, and possible hazards of biotechnological goods, labelling regulations demand clear and accurate information on product packaging. As a result, customers are more equipped to decide depending on their preferences and potential sensitivities (Gizaw, 2019). Promoting consumer education is essential because knowledgeable people are better able to comprehend the advantages, hazards, and moral ramifications of biotechnology. This involves informing the public on innovative substances, developing biotechnological techniques, and genetically modified organisms (GMOs). In order to ensure that technological breakthroughs are consistent with social values and do not violate ethical norms, ethical concerns are crucial in biotechnology (Trump et al., 2023). This entails ensuring transparency in the decision-making processes and balancing any potential risks against any potential advantages.

In general, safety and regulatory issues in biotechnology seek to achieve an equilibrium between invention and public safety, supporting sustainable development and use of biotechnological goods while respecting ethical ideals and rights of consumers (Trump et al., 2023).

2.15 CULTIVATING HEALTH: BIOTECHNOLOGICAL TRIUMPHS IN GOLDEN RICE, PROBIOTIC YOGHURT, AND ENZYME-ASSISTED BREAD

The use of biotechnology has produced amazing advancements in the constantly changing world of food production, transforming everyday items into sources of improved nutrition and well-being. Golden rice, probiotic yoghurt, and enzyme-assisted bread are three outstanding examples that highlight the effectiveness of biotechnological applications (Tang et al., 2009). These examples not only show how biotechnology may solve important nutritional issues, but they also show how it can produce meals that are not only delicious but also have positive health effects. These goods serve as examples of how science and technology may work together to redefine the possibilities of our everyday nutrition through the careful regulation of genetic features, microbial cultures, and enzyme interactions (Hefferon, 2015).

Golden Rice: A classic example of how genetic engineering has been applied to improve the nutritional value of a staple crop is Golden Rice. It was created to deal with vitamin A deficiency, a significant health problem in many poor nations. Blindness and visual problems can result from a vitamin A shortage (Mary E. Gearing, 2015). The rice's distinctive golden colour is caused by beta-carotene, which is produced as a result of genetic modification in Golden Rice. Traditional rice lacks beta-carotene, but scientists were able to create Golden Rice by inserting genes from other organisms that synthesise beta-carotene (Tang et al., 2009). This biotechnological advancement sought to offer a long-term remedy for vitamin A insufficiency, particularly when rice is a major food. Golden Rice demonstrates how biotechnology may be utilised to address certain nutritional deficits through crop modification, despite regulatory approval and acceptance facing difficulties (Dawe et al., 2002).

Probiotic Yoghurt: A great illustration of how biotechnology may improve the nutritional and health benefits of a typical food item is probiotic yoghurt. Live beneficial bacteria known as probiotics help to maintain a healthy gut and overall well-being (Hadjimbei et al., 2022). Milk fermentation microorganisms found in yoghurt naturally give it its distinctive flavour and texture. By purposefully adding particular probiotic bacteria strains, such as *Lactobacillus* and *Bifidobacterium*, to yoghurt during the fermentation process, biotechnologists have gone one step further (Savaiano and Hutkins, 2021). These microorganisms support diverse gut microbiomes while also improving digestion. This example shows how biotechnology may alter microbes to create a product that is not only more delicious but also has immediate health advantages, satisfying the growing consumer demand for functional meals (Azad et al., 2018; Rowland et al., 2018).

Enzyme-Assisted Bread: A prime illustration of how enzyme technology may alter a common product to enhance its nutritional content and sensory qualities is enzyme-assisted bread. Complex carbohydrates and proteins may be broken down by enzymes like amylases and proteases, which increases the accessibility of nutrients for digestion (Maghraby et al., 2023). In the case of bread, enzymes are employed to change the structure of the dough, producing bread that is softer, fluffier, and more delectable (Wang et al., 2018). Additionally, by improving nutrient absorption, specific enzymes can lower anti-nutritional factors and increase the nutritional value of the bread. This invention solves problems regarding vitamin bioavailability while also enhancing the dining experience. Bread made with enzyme assistance is an example of how biotechnology may be used to improve the nutritional value and quality of common meals (Gibson et al., 2006; Samtiya et al., 2020).

In each of these case studies, biotechnological methods effectively addressed certain nutritional shortfalls, health issues, or customer preferences by enhancing the nutritional profiles and functional properties of food items. These instances highlight how biotechnology has the ability to develop unique and useful answers for the world's problems with food and nutrition (A. A. E. Sundarraj et al., 2018; Ramón Vidal, 2018).

2.16 TRENDS, CHALLENGES, AND COLLABORATIVE PATHWAYS FOR THE FUTURE

Future perspectives and challenges associated with biotechnology span a fluid environment that represents both exciting possibilities and challenging roadblocks. The development of biotechnological applications such as personalised nutrition, synthetic biology, and sophisticated gene editing methods is highlighted by emerging trends. These developments have the potential to molecularly alter the way we think about food production, nutrition, and health (Ranabhat et al., 2020). Regulatory obstacles do, however, exist because assuring the safety and moral implications of emerging biotech products calls for careful examination and unambiguous standards to achieve a balance between innovation and risk reduction. As biotechnology offers products with changed properties or novel components, consumer approval is crucial. To build trust and understanding among customers and ensure they make educated decisions about what they consume, open communication and education are crucial. Given the rising demand for food in the face of finite resources, sustainability continues to be a major concern. Crop yields may be increased, waste can be reduced, and resource efficiency can be increased thanks to biotechnology, but its long-term effects on the environment must be carefully considered (Jesus et al., 2023; Liu et al., 2021). For current issues to be resolved and future possibilities to be taken advantage of, interdisciplinary cooperation is essential. The success of biotechnology depends on the fusion of disciplines, including biology, engineering, ethics, policy, and communication. Collaboration that works will promote innovation, address issues, and establish a foundation for responsible and advantageous biotechnological breakthroughs.

2.17 CONCLUSION

While the future of biotechnology is extremely promising, overcoming these obstacles calls for a comprehensive team effort to guarantee that its potential is realised for the benefit of all. The intricate challenges posed by the growing global population, diminishing agricultural resources, and nutritional deficiencies demand innovative solutions to ensure food security and human well-being. The fusion of biotechnology and food engineering emerges as a pivotal approach in addressing these multifaceted issues. By harnessing the power of biotechnological advancements, such as genetic modifications, biofortification, and the development of functional foods, we can enhance the nutritional content of food products and contribute to a healthier society.

The success stories of biotechnological applications like Golden Rice, probiotic yoghurt, and enzyme-assisted bread underscore the potential of this field to drive transformative change in the food industry. Through techniques like fermentation, enzyme engineering, and microencapsulation, we can further refine food processing methods to optimise nutrient availability and promote gut health.

However, it is essential to tread this path with cautious consideration of safety, ethical implications, and regulatory frameworks. Robust risk assessments, safety evaluations, clear labelling, and increased consumer awareness should be integral to the implementation of biotechnological innovations in the food sector. Moreover, fostering interdisciplinary collaboration and ensuring sustainable practices are crucial in realising the full potential of biotechnology in integrated food and bioprocess engineering. As we look ahead, the promise of biotechnology to revolutionise the way we produce, enhance, and consume food is undeniable. By embracing this potential, we can not only address the challenges of today but also shape a healthier and more resilient future for global food systems. The journey towards achieving these goals underscores the importance of responsible innovation, ethical considerations, and a steadfast commitment to improving both nutrition and overall human health.

REFERENCES

Abigaba, R., Sianangama, P.C., Nyanga, P.H., Mwenya, W.N.M.M., Mwaanga, E.S., 2022. Attitudes and preferences of traditional farmers toward reproductive biotechnology application for improved indigenous pig production in Zambia. *Vet. World* 15, 403–413. https://doi.org/10.14202/vetworld.2022.403-413

Aguiar, L.M., Geraldi, M.V., Betim Cazarin, C.B., Maróstica Junior, M.R., 2019. Functional food consumption and its physiological effects, in: *Bioactive Compounds: Health Benefits and Potential Applications*, pp. 205–225. https://doi.org/10.1016/B978-0-12-814774-0.00011-6

Ali, M., Ishqi, H.M., Husain, Q., 2020. Enzyme engineering: Reshaping the biocatalytic functions. *Biotechnol. Bioeng.* https://doi.org/10.1002/bit.27329

Alowo, D., Muggaga, C., Ongeng, D., 2018. The effect of traditional malting technology practiced by an ethnic community in northern Uganda on in-vitro nutrient bioavailability and consumer sensory preference for locally formulated complementary food formulae. *Food Sci. Nutr.* 6, 2491–2498. https://doi.org/10.1002/fsn3.856

Amara, A.A., Shibl, A., 2015. Role of Probiotics in health improvement, infection control and disease treatment and management. *Saudi Pharm. J.* https://doi.org/10.1016/j.jsps.2013.07.001

Amer, M., Ramsey, J.D., 2018. Multi-chamber single-use bioreactor– A proof of concept prototype. *Biochem. Eng. J.* 130, 113–120. https://doi.org/10.1016/j.bej.2017.12.002

Amin, T., 2018. Applications of biotechnology in food industry, agriculture and pharmaceutical-a review. *Digitalxplore.Org* 1, 72–83.

Arevalo-Villena, M., Briones-Perez, A., Corbo, M.R., Sinigaglia, M., Bevilacqua, A., 2017. Biotechnological application of yeasts in food science: Starter cultures, probiotics and enzyme production. *J. Appl. Microbiol.* https://doi.org/10.1111/jam.13548

Aronson, J.K., 2017. Defining 'nutraceuticals': neither nutritious nor pharmaceutical. *Br. J. Clin. Pharmacol.* https://doi.org/10.1111/bcp.12935

Arshad, R., Gulshad, L., Haq, I.U., Farooq, M.A., Al-Farga, A., Siddique, R., Manzoor, M.F., Karrar, E., 2021. Nanotechnology: A novel tool to enhance the bioavailability of micronutrients. *Food Sci. Nutr.* https://doi.org/10.1002/fsn3.2311

Ashton, B.J., Thornton, A., Cauchoix, M., Ridley, A.R., 2022. Long-term repeatability of cognitive performance. *R. Soc. Open Sci.* 9. https://doi.org/10.1098/rsos.220069

Augustin, M.A., Hartley, C.J., Maloney, G., Tyndall, S., 2023. Innovation in precision fermentation for food ingredients. *Crit. Rev. Food Sci. Nutr.* https://doi.org/10.1080/10408398.2023.2166014

Azad, M.A.K., Sarker, M., Li, T., Yin, J., 2018. Probiotic species in the modulation of gut microbiota: An overview. *Biomed Res. Int.* https://doi.org/10.1155/2018/9478630

Bala, S., Garg, D., Sridhar, K., Inbaraj, B.S., Singh, R., Kamma, S., Tripathi, M., Sharma, M., 2023. Transformation of agro-waste into value-added bioproducts and bioactive compounds: Micro/nano formulations and application in the agri-food-pharma sector. *Bioengineering.* https://doi.org/10.3390/bioengineering10020152

Bawa, A.S., Anilakumar, K.R., 2013. Genetically modified foods: Safety, risks and public concerns - A review. *J. Food Sci. Technol.* https://doi.org/10.1007/s13197-012-0899-1

Benkeblia, N., Hefferon, K.L., 2020. Modern biotechnologies and mineral biofortification of edible crops, in: *Vitamins and Minerals Biofortification of Edible Plants*, pp. 45–69. https://doi.org/10.1002/9781119511144.ch3

Bhatia, S., Naved, T., Sardana, S., 2019. *Introduction to Pharmaceutical Biotechnology*, Volume 3. Institute of Physics Publishing. https://doi.org/10.1088/2053-2563/aafac0

Borodina, I., Kenny, L.C., McCarthy, C.M., Paramasivan, K., Pretorius, E., Roberts, T.J., Van Der Hoek, S.A., Kell, D.B., 2020. The biology of ergothioneine, an antioxidant nutraceutical. *Nutr. Res. Rev.* 33, 190–217. https://doi.org/10.1017/S0954422419000301

Bouis, H.E., Saltzman, A., 2017. Improving nutrition through biofortification: A review of evidence from HarvestPlus, 2003 through 2016. *Glob. Food Sec.* https://doi.org/10.1016/j.gfs.2017.01.009

Bouis, H.E., Saltzman, A., Birol, E., 2019. Improving nutrition through biofortification introduction: Exploring the potential of biofortification, in: Fan, S., Yosef, S., Pandya-Lorch, R. (Eds.) *Agriculture for Improved Nutrition: Seizing the Momentum*. Routledge, pp. 47–57.

Brooks, S., 2013. Biofortification: Lessons from the golden rice project. *Food Chain* 3, 77–88. https://doi.org/10.3362/2046-1887.2013.007

Calderón-Oliver, M., Ponce-Alquicira, E., 2022. The role of microencapsulation in food application. *Molecules*. https://doi.org/10.3390/molecules27051499

Cencic, A., Chingwaru, W., 2010. The role of functional foods, nutraceuticals, and food supplements in intestinal health. *Nutrients*. https://doi.org/10.3390/nu2060611

Chaddha, A., Eagle, K.A., 2015. Omega-3 fatty acids and heart health. *Circulation* 132, e350–e352. https://doi.org/10.1161/CIRCULATIONAHA.114.015176

Chaudhary, M., Katel, S., Parajuli, M., Pandey, S., Mandal, H.R., 2022. Biofortification of major cereal crops: A review. *Plant Physiol. Soil Chem.* 2, 01–07. https://doi.org/10.26480/ppsc.01.2022.01.07

Chhabra, S., Bakshi, A., Kaur, R., 2021. Role of nutraceuticals in health promotion and disease prevention: A review in the Indian context. *Curr. Top. Nutraceutical Res.* https://doi.org/10.37290/ctnr2641-452X.19:139-145

Chopra, A.S., Lordan, R., Horbańczuk, O.K., Atanasov, A.G., Chopra, I., Horbańczuk, J.O., Jóźwik, A., Huang, L., Pirgozliev, V., Banach, M., Battino, M., Arkells, N., 2022. The current use and evolving landscape of nutraceuticals. *Pharmacol. Res.* https://doi.org/10.1016/j.phrs.2021.106001

Christensen, J.H., 2011. Omega-3 polyunsaturated fatty acids and heart rate variability. *Front. Physiol.* 2 November. https://doi.org/10.3389/fphys.2011.00084

Clarke, J.L., Zhang, P., 2013. Plant biotechnology for food security and bioeconomy. *Plant Mol. Biol.* https://doi.org/10.1007/s11103-013-0097-1

ClinicalTrials.gov, 2012. *Effect of the Consumption of a Cheese Enriched with Probiotic Organisms in Improving Symptoms of Constipation* (NCT01680432). https://clinicaltrials.gov/show/NCT01680432.

Cornelissen, M., Małyska, A., Nanda, A.K., Lankhorst, R.K., Parry, M.A.J., Saltenis, V.R., Pribil, M., Nacry, P., Inzé, D., Baekelandt, A., 2021. Biotechnology for tomorrow's world: Scenarios to guide directions for future innovation. *Trends Biotechnol.* https://doi.org/10.1016/j.tibtech.2020.09.006

Costa, R. dos S., de Almeida, S.S., Cavalcanti, E. d. A.C., Freire, D.M.G., Moura-Nunes, N., Monteiro, M., Perrone, D., 2021. Enzymes produced by solid state fermentation of agro-industrial by-products release ferulic acid in bioprocessed whole-wheat breads. *Food Res. Int.* 140. https://doi.org/10.1016/j.foodres.2020.109843

da Costa, T.P., Gillespie, J., Cama-Moncunill, X., Ward, S., Condell, J., Ramanathan, R., Murphy, F., 2023. A systematic review of real-time monitoring technologies and its potential application to reduce food loss and Waste: Key elements of food supply chains and IoT technologies. *Sustain.* https://doi.org/10.3390/su15010614

Dahiya, B.P., 1938. Bioinformatics impacts on medicine, microbial genome and agriculture. *J. Pharmacogn. Phytochem. JPP* 6, 1938–1942.

Das, L., Bhaumik, E., Raychaudhuri, U., Chakraborty, R., 2012. Role of nutraceuticals in human health. *J. Food Sci. Technol.* 49, 173–183. https://doi.org/10.1007/s13197-011-0269-4

DaSilva, E.J., Baydoun, E., Badran, A., 2002. Biotechnology and the developing world. *Electron. J. Biotechnol.* https://doi.org/10.2225/vol5-issue1-fulltext-1

Davani-Davari, D., Negahdaripour, M., Karimzadeh, I., Seifan, M., Mohkam, M., Masoumi, S.J., Berenjian, A., Ghasemi, Y., 2019. Prebiotics: Definition, types, sources, mechanisms, and clinical applications. *Foods.* https://doi.org/10.3390/foods8030092

Dawe, D., Robertson, R., Unnevehr, L., 2002. Golden rice: What role could it play in alleviation of vitamin A deficiency? *Food Policy* 27, 541–560. https://doi.org/10.1016/S0306-9192(02)00065-9

Delaney, B., Goodman, R.E., Ladics, G.S., 2018. Food and feed safety of genetically engineered food crops. *Toxicol. Sci.* 162, 361–371. https://doi.org/10.1093/toxsci/kfx249

Dhaliwal, S.S., Sharma, V., Shukla, A.K., Verma, V., Kaur, M., Shivay, Y.S., Nisar, S., Gaber, A., Brestic, M., Barek, V., Skalicky, M., Ondrisik, P., Hossain, A., 2022. Biofortification—A frontier novel approach to enrich micronutrients in field crops to encounter the nutritional security. *Molecules.* https://doi.org/10.3390/molecules27041340

Duong-Trung, N., Born, S., Kim, J.W., Schermeyer, M.T., Paulick, K., Borisyak, M., Cruz-Bournazou, M.N., Werner, T., Scholz, R., Schmidt-Thieme, L., Neubauer, P., Martinez, E., 2023. When bioprocess engineering meets machine learning: A survey from the perspective of automated bioprocess development. *Biochem. Eng. J.* 190. https://doi.org/10.1016/j.bej.2022.108764

Durazzo, A., Lucarini, M., Santini, A., 2020. Nutraceuticals in human health. *Foods.* https://doi.org/10.3390/foods9030370

Espinosa-Leal, C.A., Puente-Garza, C.A., García-Lara, S., 2018. In vitro plant tissue culture: Means for production of biological active compounds. *Planta.* https://doi.org/10.1007/s00425-018-2910-1

Fernie, A.R., Sonnewald, U., 2021. Plant biotechnology for sustainable agriculture and food safety. *J. Plant Physiol.* https://doi.org/10.1016/j.jplph.2021.153416

Gaikwad, K.B., Rani, S., Kumar, M., Gupta, V., Babu, P.H., Bainsla, N.K., Yadav, R., 2020. Enhancing the nutritional quality of major food crops through conventional and genomics-assisted breeding. *Front. Nutr.* https://doi.org/10.3389/fnut.2020.533453

Galani, Y.J.H., Orfila, C., Gong, Y.Y., 2022. A review of micronutrient deficiencies and analysis of maize contribution to nutrient requirements of women and children in Eastern and Southern Africa. *Crit. Rev. Food Sci. Nutr.* https://doi.org/10.1080/10408398.2020.1844636

Gallo, M., Sayre, R., 2009. Removing allergens and reducing toxins from food crops. *Curr. Opin. Biotechnol.* https://doi.org/10.1016/j.copbio.2009.03.005

Gibson, R.S., Perlas, L., Hotz, C., 2006. Improving the bioavailability of nutrients in plant foods at the household level. *Proc. Nutr. Soc.* 65, 160–168. https://doi.org/10.1079/pns2006489

Giri, V.P., Shukla, P., Tripathi, A., Verma, P., Kumar, N., Pandey, S., Dimkpa, C.O., Mishra, A., 2023. A review of sustainable use of biogenic nanoscale agro-materials to enhance stress tolerance and nutritional value of plants. *Plants.* https://doi.org/10.3390/plants12040815

Gizaw, Z., 2019. Public health risks related to food safety issues in the food market: A systematic literature review. *Environ. Health Prev. Med.* https://doi.org/10.1186/s12199-019-0825-5

Glover, D., Kim, S.K., Stone, G.D., 2020. Golden Rice and technology adoption theory: A study of seed choice dynamics among rice growers in the Philippines. *Technol. Soc.* 60. https://doi.org/10.1016/j.techsoc.2019.101227

Glover, D., Poole, N., 2019. Principles of innovation to build nutrition-sensitive food systems in South Asia. *Food Policy* 82, 63–73. https://doi.org/10.1016/j.foodpol.2018.10.010

Godbey, W.T., 2022. An introduction to biotechnology, in: *Biotechnology and Its Applications*, pp. 3–9. https://doi.org/10.1016/b978-0-12-817726-6.00001-0

Goulart, L., Dantas, N., Silva, A., Madurro, J.M., Brito-Madurro, A.G., Ueira-Vieira, C., Fujimura, P., Maia, Y., Santos, P., Freschi, A.P., Almeida, J., Goulart, I., 2014. Frontiers of biology in human diseases: Strategies for biomolecule's discovery, nanobiotechnologies and biophotonics. *BMC Proc* 8(Suppl 4), Article O9.

Hadjimbei, E., Botsaris, G., Chrysostomou, S., 2022. Beneficial effects of yoghurts and probiotic fermented milks and their functional food potential. *Foods*. https://doi.org/10.3390/foods11172691

Hameed, A., Zaidi, S.S. e. A., Shakir, S., Mansoor, S., 2018. Applications of new breeding technologies for potato improvement. *Front. Plant Sci.* https://doi.org/10.3389/fpls.2018.00925

Hasler, Clare M., 2002. Functional foods: Benefits, concerns and challenges - A position paper from the American Council on Science and Health. *J. Nutr.* 132, 3772–3781. https://doi.org/10.1093/jn/132.12.3772

Hefferon, K.L., 2015. Nutritionally enhanced food crops; progress and perspectives. *Int. J. Mol. Sci.* https://doi.org/10.3390/ijms16023895

Holei, Y., Stasiuk, Y., Krupskyi, O., 2022. Research on world trends in biotechnology development. *Innov. Econ.* 12–22. https://doi.org/10.37332/2309-1533.2022.1.2

Hu, W.S., Liao, J.C., 2013. Biotechnology and bioprocess engineering: Editorial overview. *Curr. Opin. Chem. Eng.* https://doi.org/10.1016/j.coche.2013.10.004

Hugenholtz, J., 2013. Traditional biotechnology for new foods and beverages. *Curr. Opin. Biotechnol.* https://doi.org/10.1016/j.copbio.2013.01.001

Jesus, A.R., Paiva, A., Duarte, A.R.C., 2023. Current developments and future perspectives on biotechnology applications of natural deep eutectic systems. *Curr. Opin. Green Sustain. Chem.* https://doi.org/10.1016/j.cogsc.2022.100731

Jha, A.B., Warkentin, T.D., 2020. Biofortification of pulse crops: Status and future perspectives. *Plants*. https://doi.org/10.3390/plants9010073

Jorge, S., Dellagostin, O.A., 2017. The development of veterinary vaccines: A review of traditional methods and modern biotechnology approaches. *Biotechnol. Res. Innov.* 1, 6–13. https://doi.org/10.1016/j.biori.2017.10.001

Junejo, I., Siddiqui, M.B., Thebo, J.A., Bhutto, N., 2022. Factors affecting intention to control quality safety: Evidence food supply chain firms. *J. Bus. Soc. Rev. Emerg. Econ.* 8, 547–556. https://doi.org/10.26710/jbsee.v8i2.2365

Jungbauer, A., 2013. Editorial: Biotech methods and advances. *Biotechnol. J.* https://doi.org/10.1002/biot.201200405

Kajmi, J.A., Singh, R., Shukla, D.T.P., 2023. Role of nutraceuticals in health and disease prevention. *Int. J. Res. Appl. Sci. Eng. Technol.* 11, 897–901. https://doi.org/10.22214/ijraset.2023.48692

Kang, J.X., 2012. Reduction of heart rate by omega-3 fatty acids and the potential underlying mechanisms. *Front. Physiol.* https://doi.org/10.3389/fphys.2012.00416

Kaur, S., Das, M., 2011. Functional foods: An overview. *Food Sci. Biotechnol.* https://doi.org/10.1007/s10068-011-0121-7

Kong, D., Wang, J., Zeng, H., Liu, M., Miao, Y., Wu, H., Kardol, P., 2017. The nutrient absorption–transportation hypothesis: Optimizing structural traits in absorptive roots. *New Phytol.* https://doi.org/10.1111/nph.14344

Kvakova, M., Bertkova, I., Stofilova, J., Savidge, T.C., 2021. Co-encapsulated synbiotics and immobilized probiotics in human health and gut microbiota modulation. *Foods* 10. https://doi.org/10.3390/foods10061297

Lee, S.Y., Jungbauer, A., 2014. Editorial: Latest methods and advances in biotechnology. *Biotechnol. J.* https://doi.org/10.1002/biot.201300522

Li, S., Tian, Y., Jiang, P., Lin, Y., Liu, X., Yang, H., 2021. Recent advances in the application of metabolomics for food safety control and food quality analyses. *Crit. Rev. Food Sci. Nutr.* https://doi.org/10.1080/10408398.2020.1761287

Listman, G. M., Guzmán, C., Palacios-Rojas, N., Pfeiffer, W. H., San Vicente, F., Govindan, V., 2019. Improving nutrition through biofortification: Preharvest and postharvest technologies. *Cereal Foods World* 64(3), 1–7. https://doi.org/10.1094/cfw-64-3-0025

Liu, Y., Tang, Y., Gao, H., Zhang, W., Jiang, Y., Xin, F., Jiang, M., 2021. Challenges and future perspectives of promising biotechnologies for lignocellulosic biorefinery. *Molecules.* https://doi.org/10.3390/molecules26175411

Luo, W., Kramer, R., de Kort, Y., Rense, P., Adam, J., van Marken Lichtenbelt, W., 2023. Personal comfort systems and cognitive performance: Effects on subjective measures, cognitive performance, and heart rate measures. *Energy Build.* 278. https://doi.org/10.1016/j.enbuild.2022.112617

Maghraby, Y.R., El-Shabasy, R.M., Ibrahim, A.H., Azzazy, H.M.E.S., 2023. Enzyme immobilization technologies and industrial applications. *ACS Omega.* https://doi.org/10.1021/acsomega.2c07560

Mannucci, C., Casciaro, M., Sorbara, E.E., Calapai, F., Di Salvo, E., Pioggia, G., Navarra, M., Calapai, G., Gangemi, S., 2021. Nutraceuticals against oxidative stress in autoimmune disorders. *Antioxidants.* https://doi.org/10.3390/antiox10020261

Markowiak, P., Ślizewska, K., 2017. Effects of probiotics, prebiotics, and synbiotics on human health. *Nutrients* 9, 193–198. https://doi.org/10.3390/nu9091021

Markowiak, P., Śliżewska, K., 2017. Effects of probiotics, prebiotics, and synbiotics on human health. *Nutrients* 9, 193–198.

Mary E. Gearing, 2015. Good as gold: Can golden rice and other biofortified crops prevent malnutrition? *Science in the News.* Harvard University Science in the News.

McClements, D.J., 2015. Enhancing nutraceutical bioavailability through food matrix design. *Curr. Opin. Food Sci.* https://doi.org/10.1016/j.cofs.2014.12.008

Mcclements, D.J., Zou, L., Zhang, R., Salvia-Trujillo, L., Kumosani, T., Xiao, H., 2015. Enhancing nutraceutical performance using excipient foods: Designing food structures and compositions to increase bioavailability. *Compr. Rev. Food Sci. Food Saf.* 14, 824–847. https://doi.org/10.1111/1541-4337.12170

McCullum, C., Benbrook, C., Knowles, L., Roberts, S., Schryver, T., 2003. Application of modern biotechnology to food and agriculture: Food systems perspective. *J. Nutr. Educ. Behav.* 35, 319–332. https://doi.org/10.1016/S1499-4046(06)60347-3

Melse-Boonstra, A., 2020. Bioavailability of micronutrients from nutrient-dense whole foods: Zooming in on dairy, vegetables, and fruits. *Front. Nutr.* https://doi.org/10.3389/fnut.2020.00101

Mogendi, J.B., De Steur, H., Gellynck, X., Makokha, A., 2016. A novel framework for analysing stakeholder interest in healthy foods: A case study on iodine biofortification. *Ecol. Food Nutr.* 55, 182–208. https://doi.org/10.1080/03670244.2015.1112283

Molina, G., Gupta, V.K., Singh, B.N., Gathergood, N., 2019. *Bioprocessing for Biomolecules Production.* Wiley. https://doi.org/10.1002/9781119434436

Najah, Z., Putri, N.A., 2019. A review of hazard analysis and critical control points implementation in amplang processing. *Food Sci. J.* 1, 54. https://doi.org/10.33512/fsj.v1i1.6187

National Academies of Sciences, Engineering, and Medicine, 2017. *Genetically Engineered Crops: Experiences and Prospects.* The National Academies Press.

Nicholson, C.F., Monterrosa, E., Garrett, J.L., 2021. Food value chain interventions and nutritional outcomes: A review of evidence and recommendations for future assessments. *Curr. Opin. Biotechnol.* https://doi.org/10.1016/j.copbio.2020.11.015

NIH, 2019. *Omega-3 Fatty Acids - Health Professional Fact Sheet* [WWW Document]. Office of Dietary Supplements.

Nkhata, S. G., Ayua, E., Kamau, E. H., Shingiro, J.-B., 2018. Fermentation and germination improve nutritional value of cereals and legumes through activation of endogenous enzymes. *Food Sci. Nutr.* 6(8), 2446–2458.

Ofori, K.F., Antoniello, S., English, M.M., Aryee, A.N.A., 2022. Improving nutrition through biofortification–A systematic review. *Front. Nutr.* https://doi.org/10.3389/fnut.2022.1043655

Paula, B.O., Hitter Buru, T., Cătană, C., Cantor, M., 2020. In vitro plant tissue culture: Means for production of passiflora species. *Int. J. Innov. Approaches Agric. Res.* 4, 505–523. https://doi.org/10.29329/ijiaar.2020.320.12

Pérez-Alva, A., MacIntosh, A.J., Baigts-Allende, D.K., García-Torres, R., Ramírez-Rodrigues, M.M., 2022. Fermentation of algae to enhance their bioactive activity: A review. *Algal Res.* https://doi.org/10.1016/j.algal.2022.102684

Pikaar, I., Hülsen, T., Angenent, L.T., 2023. Editorial overview: The future of food biotechnology is microbial. *Curr. Opin. Biotechnol.* https://doi.org/10.1016/j.copbio.2022.102889

Puri, V., Nagpal, M., Singh, I., Singh, M., Dhingra, G.A., Huanbutta, K., Dheer, D., Sharma, A., Sangnim, T., 2022. A comprehensive review on nutraceuticals: Therapy support and formulation challenges. *Nutrients.* https://doi.org/10.3390/nu14214637

Pyne, M., Sukhija, K., Chou, C.P., 2019. Genetic engineering, in: *Comprehensive Biotechnology*, pp. 22–31. https://doi.org/10.1016/B978-0-444-64046-8.00071-9

Ramón Vidal, D., 2018. Food biotechnology: From genetically modified foods to personalized nutrition. *Nutr. Hosp.* 35, 28–32. https://doi.org/10.20960/NH.2121

Ranabhat, G., Dhakal, Ashmita, Ranabhat, S., Dhakal, Ananta, Aryal, R., 2020. Future perspective of plant biotechnology: A review. *Rev. Food Agric.* 1, 36–41. https://doi.org/10.26480/rfna.01.2020.36.41

Ranjha, M.M.A.N., Shafique, B., Khalid, W., Nadeem, H.R., Mueen-ud-Din, G., Khalid, M.Z., 2022. Applications of biotechnology in food and agriculture: A mini-review. *Proc. Natl. Acad. Sci. India Sect. B - Biol. Sci.* https://doi.org/10.1007/s40011-021-01320-4

Romeis, J., Naranjo, S.E., Meissle, M., Shelton, A.M., 2019. Genetically engineered crops help support conservation biological control. *Biol. Control.* https://doi.org/10.1016/j.biocontrol.2018.10.001

Romig, W.R., 2019. Biotechnologies and food: Assuring the safety of foods produced by genetic modification. *HortScience* 25, 1176a. https://doi.org/10.21273/hortsci.25.9.1176a

Roriz, M., Carvalho, S.M.P., Castro, P.M.L., Vasconcelos, M.W., 2020. Legume biofortification and the role of plant growth-promoting bacteria in a sustainable agricultural era. *Agronomy.* https://doi.org/10.3390/agronomy10030435

Rowland, I., Gibson, G., Heinken, A., Scott, K., Swann, J., Thiele, I., Tuohy, K., 2018. Gut microbiota functions: Metabolism of nutrients and other food components. *Eur. J. Nutr.* https://doi.org/10.1007/s00394-017-1445-8

Samtiya, M., Aluko, R.E., Dhewa, T., 2020. Plant food anti-nutritional factors and their reduction strategies: An overview. *Food Prod. Process. Nutr.* https://doi.org/10.1186/s43014-020-0020-5

Samtiya, M., Aluko, R.E., Puniya, A.K., Dhewa, T., 2021. Enhancing micronutrients bioavailability through fermentation of plant-based foods: A concise review. *Fermentation.* https://doi.org/10.3390/fermentation7020063

Sanchez-Andrea, I., Jetten, M., 2018. Editorial overview: Microbial environmental biotechnology. *Curr. Opin. Biotechnol.* https://doi.org/10.1016/j.copbio.2018.03.004
Saritha, G.N.G., Anju, T., Kumar, A., 2022. Nanotechnology - Big impact: How nanotechnology is changing the future of agriculture? *J. Agric. Food Res.* 10. https://doi.org/10.1016/j.jafr.2022.100457
Savaiano, D.A., Hutkins, R.W., 2021. Yogurt, cultured fermented milk, and health: A systematic review. *Nutr. Rev.* 79, 599–614. https://doi.org/10.1093/nutrit/nuaa013
Savarino, E., Ojetti, V., Saviano, A., 2022. Probiotics, in: *Colonic Diverticular Disease*, pp. 245–259. https://doi.org/10.1007/978-3-030-93761-4_20
Selden, C., Fuller, B., 2018. Role of bioreactor technology in tissue engineering for clinical use and therapeutic target design. *Bioengineering.* https://doi.org/10.3390/bioengineering5020032
Shabbir, R., Javed, T., Afzal, I., El Sabagh, A., Ali, A., Vicente, O., Chen, P., 2021. Modern biotechnologies: Innovative and sustainable approaches for the improvement of sugarcane tolerance to environmental stresses. *Agronomy.* https://doi.org/10.3390/agronomy11061042
Sharma, I., 2021. Bioremediation techniques for polluted environment: Concept, advantages, limitations, and prospects, in: *Trace Metals in the Environment - New Approaches and Recent Advances.* https://doi.org/10.5772/intechopen.90453
Sharma, K., Mishra, K., Senapati, K.K., Danciu, C., 2020. *Bioactive Compounds in Nutraceutical and Functional Food for Good Human Health.* IntechOpen. https://doi.org/10.5772/intechopen.78846
Sharma, R., Garg, P., Kumar, P., Bhatia, S.K., Kulshrestha, S., 2020. Microbial fermentation and its role in quality improvement of fermented foods. *Fermentation.* https://doi.org/10.3390/fermentation6040106
Shuler, M., Kargi, F., 1992. Bioprocess engineering: Basic concepts. *J. Control. Release* 22, 293. https://doi.org/10.1016/0168-3659(92)90106-2
Singh, M.B., Bhalla, P.L., 2008. Genetic engineering for removing food allergens from plants. *Trends Plant Sci.* https://doi.org/10.1016/j.tplants.2008.04.004
Singh, R.P., Shadan, A., Ma, Y., 2022. Biotechnological applications of probiotics: A multifarious weapon to disease and metabolic abnormality. *Probiotics Antimicrob. Proteins* 14, 1184–1210. https://doi.org/10.1007/s12602-022-09992-8
Sivaramakrishnan, R., Ramadoss, G., Pancha, I.Y., Incharoensakdi, A., 2023. Editorial: Methods in industrial biotechnology and bioprocess engineering—Microalgae as a source of valuable compounds. *Front. Bioeng. Biotechnol.* https://doi.org/10.3389/fbioe.2023.1141280
Souyoul, S.A., Saussy, K.P., Lupo, M.P., 2018. Nutraceuticals: A review. *Dermatol. Ther. (Heidelb).* https://doi.org/10.1007/s13555-018-0221-x
Sreenikethanam, A., Raj, S., J, R.B., Gugulothu, P., Bajhaiya, A.K., 2022. Genetic engineering of microalgae for secondary metabolite production: Recent developments, challenges, and future prospects. *Front. Bioeng. Biotechnol.* https://doi.org/10.3389/fbioe.2022.836056
Staniszewski, A., Kordowska-Wiater, M., 2021. Probiotic and potentially probiotic yeasts—characteristics and food application. *Foods* 10. https://doi.org/10.3390/foods10061306
Su, R., Sujarani, M., Shalini, P., Prabhu, N., 2019. A review on bioreactor technology assisted plant suspension culture. *Asian J. Biotechnol. Bioresour. Technol.* 1–13. https://doi.org/10.9734/ajb2t/2019/v5i330062
Sundarraj, A. A. E., Ezhil, R.J.H., Vishaal, S.C., Karan, S.H., Mathew, N., Tony, S., Angeline Rajathi, A., 2018. Biotechnology applications in food processing and safety. *J. Adv. Pharm. Educ. Res.* 10, 598–603.

Szot, M., Karpęcka-Gałka, E., Dróżdż, R., Frączek, B., 2022. Can nutrients and dietary supplements potentially improve cognitive performance also in esports? *Healthc.* https://doi.org/10.3390/healthcare10020186

Tamis, W.L.M., van Dommelen, A., de Snoo, G.R., 2009. Lack of transparency on environmental risks of genetically modified micro-organisms in industrial biotechnology. *J. Clean. Prod.* 17, 581–592. https://doi.org/10.1016/j.jclepro.2008.12.011

Tang, G., Jian, Q., Dolnikowski, G.G., Russell, R.M., Grusak, M.A., 2009. Golden rice is an effective source of vitamin A. *Am. J. Clin. Nutr.* 89, 1776–1783. https://doi.org/10.3945/ajcn.2008.27119

Torbica, A., Radosavljević, M., Belović, M., Tamilselvan, T., Prabhasankar, P., 2022. Biotechnological tools for cereal and pseudocereal dietary fibre modification in the bakery products creation – Advantages, disadvantages and challenges. *Trends Food Sci. Technol.* https://doi.org/10.1016/j.tifs.2022.09.018

Trump, B., Cummings, C., Klasa, K., Galaitsi, S., Linkov, I., 2023. Governing biotechnology to provide safety and security and address ethical, legal, and social implications. *Front. Genet.* 13. https://doi.org/10.3389/fgene.2022.1052371

Umi Raihanah, C.M.N., Norazmir, M.N., 2021. A systematic review on hazard analysis and critical control points (HACCP) in Southeast Asia countries. *Pakistan J. Med. Heal. Sci.* https://doi.org/10.37506/ijphrd.v12i2.14169

Van der Klis, J.D., Jansman, A.J.M., 2002. Optimising nutrient digestion, absorption and gut barrier function in monogastrics: Reality of illusion? in: M.C. Blok, H.A. Vahl, L. de Lange, A.E. van de Braak, G. Hemke, M. Hessing (eds.), Nutrition and Health of the Gastrointestinal Tract. Wageningen Academic Publishers, pp. 15–36.

van Esse, H.P., Reuber, T.L., van der Does, D., 2020. Genetic modification to improve disease resistance in crops. *New Phytol.* https://doi.org/10.1111/nph.15967

Varghese, S.M., Parisi, S., Singla, R.K., Begum, A.S.A., 2022. Food safety and quality control in food industry, in: *Trends in Food Chemistry, Nutrition and Technology in Indian Sub-Continent.* SpringerBriefs in Molecular Science. Springer, pp. 31–44. https://doi.org/10.1007/978-3-031-06304-6_5

Vasanthi, S., Bhat, R. V., 2018. Management of food safety risks in India. *Proc. Indian Natl. Sci. Acad.* https://doi.org/10.16943/ptinsa/2018/49439

Velayudhan, A., 2014. Overview of integrated models for bioprocess engineering. *Curr. Opin. Chem. Eng.* https://doi.org/10.1016/j.coche.2014.09.007

Verni, M., Rizzello, C.G., Coda, R., 2019. Fermentation biotechnology applied to cereal industry by-products: Nutritional and functional insights. *Front. Nutr.* https://doi.org/10.3389/fnut.2019.00042

Voidarou, C., Antoniadou, M., Rozos, G., Tzora, A., Skoufos, I., Varzakas, T., Lagiou, A., Bezirtzoglou, E., 2021. Fermentative foods: Microbiology, biochemistry, potential human health benefits and public health issues. *Foods.* https://doi.org/10.3390/foods10010069

Walton, J. 2019. Improving nutrition through biofortification: From strategy to implementation. *Cereal Foods World* 64(3). https://doi.org/10.1094/cfw-64-3-0026

Wang, X., Pei, D., Teng, Y., Liang, J., 2018. Effects of enzymes to improve sensory quality of frozen dough bread and analysis on its mechanism. *J. Food Sci. Technol.* 55, 389–398. https://doi.org/10.1007/s13197-017-2950-8

Whisner, C.M., Castillo, L.F., 2018. Prebiotics, bone and mineral metabolism. *Calcif. Tissue Int.* https://doi.org/10.1007/s00223-017-0339-3

Wieërs, G., Belkhir, L., Enaud, R., Leclercq, S., Philippart de Foy, J.M., Dequenne, I., de Timary, P., Cani, P.D., 2020. How probiotics affect the microbiota. *Front. Cell. Infect. Microbiol.* https://doi.org/10.3389/fcimb.2019.00454

Xiang, H., Sun-Waterhouse, D., Waterhouse, G.I.N., Cui, C., Ruan, Z., 2019. Fermentation-enabled wellness foods: A fresh perspective. *Food Sci. Hum. Wellness*. https://doi.org/10.1016/j.fshw.2019.08.003

Yadav, M.K., Kumari, I., Singh, B., Sharma, K.K., Tiwari, S.K., 2022. Probiotics, prebiotics and synbiotics: Safe options for next-generation therapeutics. *Appl. Microbiol. Biotechnol.* https://doi.org/10.1007/s00253-021-11646-8

Yamei, D., Xiujuan, W., Minli, Y., Muyi, H., Feng, Z., 2020. Research advances in imaging technology for food safety and quality control. *Chinese J. Chromatogr. (Se Pu)* 38, 741–749. https://doi.org/10.3724/SP.J.1123.2020.03015

Yang, X.E., Chen, W.R., Feng, Y., 2007. Improving human micronutrient nutrition through biofortification in the soil-plant system: China as a case study. *Environ. Geochem. Health* 29, 413–428. https://doi.org/10.1007/s10653-007-9086-0

Yılmaz, R., 2019. Modern biotechnology breakthroughs to food and agricultural research in developing countries. *GM Crop. Food.* https://doi.org/10.1080/21645698.2019.1600969

Zimmermann, R., 2004. Potential health benefits of golden rice: A Philippine case study. *Food Policy* 29, 147–168. https://doi.org/10.1016/s0306-9192(04)00009-0

3 Applications of Microbial Enzymes in Food Industry

Neetu Dobhal and Deepa Joshi

3.1 INTRODUCTION

Enzymes are the biological catalysts that play a crucial role in metabolic and biochemical processes. It is possible to manufacture enzymes from a variety of sources, including plants, animals, food, etc. Microbial enzymes are enzymes specifically produced from microorganisms. Microorganisms, specifically bacteria, multiply rapidly when grown in appropriate culture and subjected to genetic manipulation and are thus able to generate a greater number of enzymes (Anbu et al., 2017). Usage of microbial enzymes for the production of food items is an important technique applied from ancient times. Fermentation leads to the quicker generation of microbial enzymes with less expense and waste, and with higher consistency. Additionally, process improvement and modification are simple to carry out. These enzymes are more stable than animal or plant-based enzymes (Gurung et al., 2013) and are not only in the food industries but also in other sectors like textiles, pulp, leather, paper, pharmaceuticals, beverages, chemicals, biofuels and detergent industries (Pandey et al., 2000). These enzymes are used in the food industry to make products like corn syrup, glucose syrup and maltose syrup and are used as additives in the textile industry to get rid of starch stains and size warps. In the paper industry, microbial enzymes are used for paper coating (Desouza and Megalhaes, 2010).

The use of microbial enzymes is replacing demand for chemicals in industrial processes as the enzymes are eco-friendly in nature and do not produce greenhouse gases. To meet this increasing demand for enzymes in industries, there is a need to produce microbial enzymes at a larger scale. However, the use of microbial enzymes is not new, but with technological advancement, a wide range of enzymes have been specified and their new areas of application are being explored. Strains of yeast, moulds and bacteria are used very often as sources of enzymes to enhance the taste and texture of the product. Apart from this, there are other additional benefits of using microbial enzymes like:

DOI: 10.1201/9781032650616-3

- These enzymes are highly specific in their action, hence avoiding side reactions.
- These enzymes work in mild conditions, requiring no extremes of temperature and pH.

Due to these advantages, microbial enzymes are replacing the physical and chemical alternatives used in various industrial sectors at a rapid speed. This chapter discusses the different types of microbial enzymes used in food industries, along with their sources and the recent advances in the field.

3.2 MICROBIAL ENZYMES AND THEIR CLASSIFICATION

Enzymes are the biological catalysts that speed up many chemical reactions and convert substrates into products. These highly specific enzymes can be catalytic RNA (ribozymes) or proteinaceous macromolecules. These are large macromolecules with amino acids as their monomeric units. Enzymes' molecular weight ranges from 10 to 2000 kDa (Worthington et al., 2019). However, enzymes possess various sites, the most important of which are their active sites that are located deep within a hydrophobic pocket. This enzyme's specificity for its substrate depends on these active sites, which makes each enzyme different from the other (Robinson, 2015). This also served as the basis for the classification, numbering and naming of more than 3,000 enzymes. Based on reactions catalyzed, microbial enzymes are categorized into seven classes (NC-IUBMB, 2021):

I. **Transferases**: Transferases, using a nucleophilic substitution process, facilitate the transfer of functional groups such as methyl, hydroxymethyl, formal, glycosyl, acyl, alkyl, phosphate and sulfate groups. Examples of transferases are transaminase, kinase, peptidyltransferase, formyltransferase, glycosyltransferase, acetyltransferase, methyltransferase, transaldolase, transketolase, sulfotransferase etc.
II. **Isomerases**: Isomerases are the microbial enzymes that make it easier for molecules with the same atomic structure but different chemical group configurations, known as isomers, to be converted into one another (Cuesta et al., 2016). These enzymes are involved in the formation and breakage of intramolecular bonds. Examples include epimerases, racemases, cis-trans isomerases, etc.
III. **Oxidoreductases**: These enzymes catalyze the oxidation or reduction reactions. Examples include oxygenases, oxidases, reductases, peroxidases, dehydrogenases, etc.
IV. **Ligases**: These enzymes help in creating new chemical bonds between two or more molecules to unite them into a single new molecule by using the energy released by the hydrolysis of ATP or another high-energy phosphate bond. Examples include carboxylases, synthetases, DNA ligase, chelatases, etc.

V. **Hydrolases**: These enzymes utilize water to catalyze the bond cleavage in large molecules and are therefore responsible for breaking down the large molecules into smaller ones. Examples of hydrolases are amylase, sacylase, protease or peptidases, glycosidase, phospholipases, lipase, lactase, etc.

VI. **Lyases**: These enzymes do not require oxidation or hydrolysis for addition or elimination reactions. Examples of lyases include adenylyl cyclase, decarboxylase, aldehyde lyase, pectolyase, dehydratases, hydratases, etc.

VII. **Translocases**: This class of enzymes, identified in 2018, is the newest in its origin. These enzymes facilitate the passage of ions or molecules through membranes or their division within the membranes. Examples of translocases include carnitine-acylcarnitine translocase, translocase of the inner and outer membrane, ADP/ATP translocase, etc.

3.3 IMPORTANCE OF MICROBIAL ENZYMES IN INDUSTRIES

Till date, more than 3,000 enzymes have been identified, but only a few are used in food industries. Utilizing these enzymes has decreased the need for energy in various sectors, and the waste produced with the usage of enzymes is non-toxic and biodegradable, therefore preventing environmental pollution to a great extent (Gurung et al., 2013). The microbes are easily available and can be cultured in cheap growth media. They also have high growth rates and their genome can also be manipulated through recombinant DNA technology to produce thermoactive, thermostable, acid and alkaline enzymes that can withstand adverse industrial production conditions (Patel et al., 2017). Apart from so many advantages of microbial enzymes, there are some limitations too. Some industrial conditions can be harsh for proper enzyme performance, viz, pressure, pH, potential inhibitors and temperature. As most of the enzymes perform well at approximately 37°C, in the absence of inhibitors and at neutral pH, changes in these conditions in industries may affect the enzymatic activity (Singh and Bajaj, 2016). The increase in the demand for microbial enzymes in industries is due to the increasing demand for enzymes that can replace both synthetic and traditional chemicals in various industrial processes, as microbial enzymes are eco-friendly in nature and their use can prevent the release of around 700 million kg of carbon dioxide per year (Robinson, 2015). This increased demand for enzymes has led to the growth of many enzyme-producing companies.

Microbial enzymes are being used in various industries. Among these, the food industry utilizes the highest amount of about 55 per cent (Guerrand, 2017). The food industries include dairy, baking, food packaging, beverages, fruit and vegetable juice and oil refinery brewing industries. According to estimates, the global market for food enzymes would be valued at USD 2.3 billion in 2020, or around 40% of the total market for industrial enzymes. By 2026, the global market for food enzymes is expected to reach more than USD 3.3 billion (Research and Markets, 2021). The various types of microbial enzymes used in food industries are summarized in Table 3.1, and the function of different microbial enzymes in food industries has been presented in Table 3.2.

TABLE 3.1
Types of microbial enzymes used in different food industries

Sl. No.	Food Industries	Microbial Enzymes Used
1.	Dairy	Lipase, transglutaminase, protease, catalases, phospholipase, laccase, lactase
2.	Beverage	Lipase, protease, glucoamylase, cellulase, xylanase, pectinase, alpha-amylase, glucose oxidase, catalase, laccase, esterase
3.	Oil refinery	Lipase, phospholipase, pectinase
4.	Confectionery	Alpha-amylase, glucoamylase
5.	Baking	Lipase, protease, transglutaminase, xylanase, alpha-amylase, glucose oxidase, laccase, phospholipase
6.	Fruit and vegetable juice	Cellulose, xylanase, naringinase, alpha-amylase, laccase, pectinase
7.	Animal feed	Protease, pectinase, alpha-amylase, xylanase, phospholipase, cellulase, phytase
8.	Food processing and packaging	Transglutaminase, protease, catalase, cellulase

Let us discuss some of the important microbial enzymes used in various food industries.

3.3.1 Lipases

Lipases are present naturally in the pancreas and stomach of humans and some animal species to digest lipids and fats. These enzymes can also be produced from seeds, leaves, latex and stems of various plants (Melani et al., 2020). Long-chain triglycerides undergo hydrolysis with the help of these enzymes. In addition to bacteria and yeast, fungi also produce microbial lipases. Among bacteria, *Bacillus* species are the best potential source for lipase production. Commercially significant fungi that produce lipase are found in the genera *Penicillium*, *Rhizopus*, *Aspergillus* and *Mucor*. They contribute to about 90 per cent of the global lipase market. Lipases are commonly used in detergents, food, animal feed, biofuel and various other industries. Apart from these, these enzymes are also used in the textile, leather and paper industries (Gerits et al., 2014).

In the food industry, lipases are used in baking, dairy, wine and beer industries, mainly for the development of flavour in dairy products and other fat-containing food processing industries. They hydrolyze milk fat to free fatty acids and improve the flavour of cheese. Lipases from different sources can be used to produce different types of cheese, like cheddar cheese from *Aspergillus oryzae* and *Aspergillus niger* and Romano cheese from lamb pre-gastric lipase, etc. Lipases are used to improve

TABLE 3.2
Role of various microbial enzymes in food industry

Sl. No.	Microbial Enzyme	Role in Food Industry
1.	Lipase	Production of cheddar cheese and development of flavour in cheese
2.	Protease	Improvement of bread quality, meat tenderization, brewing, milk coagulation
3.	Glucoamylase	High fructose and glucose syrups, improvement of bread quality, beer production
4.	Alpha-Amylase	Fruit juice clarification, baking, liquefaction of starch, improvement of bread quality, brewing
5.	Lactase	Main ingredient of prebiotic foods
6.	Esterase	Dietary fibre de-estrification, fragrance and flavour enhancement in fruit juices, short-chain flavour esters production
7.	Phospholipase	Lipolyzed milk fat production, development of cheese flavour
8.	Cellulase	Fruit juice clarification, clarification of fruit juice, beer quality improvement
9.	Glucose oxidase	Improvement of food flavour, increase in shelf life of foods
10.	Pectinase	Fruit juice clarification
11.	Laccase	Baking, removal of polyphenol from wine
12.	Catalase	Hydrogen peroxide removal from milk prior to cheese production, preservation of food
13.	Peroxidase	Increase in nutritional quality, colour and flavour of food
14.	Alpha-Acetolactate dehydrogenase	Beer maturation and shortening
15.	Asparaginase	Reduction in acrylamide formation during baking
16.	Naringinase	Enhancement of wine aroma, removal of bitter taste from fruit juice
17.	Xylanases	Improvement of beer quality, fruit juice clarification

the softness and texture of cheese and the development of flavour in margarine and butter (Jooyendeh et al., 2009). Lipases are used in the baking industry to hydrolyze the lipids in egg white, leading to improvement in product quality (Aravindan et al., 2007).

Lipases are used as degumming agents in oil refining industries in order to remove phosphatides from sunflower, crude soybean and rapeseed oil during the refining process. This degumming process improves the quality, yield and organoleptic properties of the vegetable refined oil (Cong et al., 2019). Lipases can modify aroma in wine and other alcoholic beverages. They are also used to improve the quality of cocoa butter (Sharma et al., 2001). Microbial lipases play many important roles in different industries that cannot be performed with chemicals. Lipases can be used for the production of single-cell protein (SCP), amino acid derivatives, ice cream and carbohydrate esters (Aravindan et al., 2007). Additionally, waste that is frequently produced by the food industry can be processed using lipase (Ghosh et al., 1996).

3.3.2 Proteases

Proteins and polypeptides contain peptide bonds that need to be hydrolyzed by the proteases. These enzymes can be categorized according to their origin, catalytic activity and the type of reactive group that is present at the catalytic site. Based on the site of action, proteases are divided into endopeptidases and exopeptidases. Endopeptidases act in the inner regions and exopeptidases act on the ends of polypeptide chains. Various biological creatures can be used for the production of these enzymes, like archaea, plants, animals, bacteria and fungi. They are used in the pharmaceutical, detergent and food industries. Proteases are mainly used in baking, brewing, food processing, dairy and animal feed industries. Microbial proteases are of three types: acidic, neutral and alkaline. Alkaline proteases are ideal because they can function in conditions of high pH, high temperature and in the presence of chemicals that inhibit them (Gupta et al., 2002). Proteases derived from microorganisms have a higher economic significance than those from plants or animals. The main bacteria producing proteases are *Bacillus*, *Lactobacillus*, *Microbacterium* and *Pseudomonas*. Among fungal strains, *Aspergillus* species, *Entomophthora* and *Cladosporium* are of commercial importance.

In dairy industries, proteases are added to milk to hydrolyze the peptide bond to kappa-casein. This hydrolysis destabilizes casein micelles during curd production. Additionally, proteases enhance the cheese's rheological and organoleptic properties. These enzymes may also reduce the allergenic properties present in certain fermented milk products (Zou et al., 2023). In baking industry, proteases are used

- to regulate the strength of gluten,
- to decrease dough uniformity and consistency,
- to decrease the mixing time and
- to improve the flavour and texture (Miguel, 2013).

During biscuit or cracker production, proteases weaken the structure of gluten protein to make the dough easy to handle. The advantage of employing microbial protease for the preparation of soft dough is that it has no effect on the other nutritional components, in contrast to when a chemical like sodium bisulphite is used for the same purpose (Deng et al., 2016).

Proteases from plant origin like papain, bromelain and ficin are used for meat tenderization, brewing, as a digestive aid and milk coagulation in the food industry. These enzymes are also used to improve the nutritional value, digestibility, flavour and solubility of food proteins along with improvement in functional properties like emulsification and coagulation (Aruna et al., 2014).

3.3.3 Glucoamylases

One of the oldest and most used biocatalysts in the food business is glucoamylase. These enzymes hydrolyze polysaccharides from the non-reducing chain ends by cleaving α-1,4 and α-1,6 glycosidic bonds, consecutively, to release β-glucose. Glucoamylase's main application involves the saccharification of starch or dextrin into glucose, a vital substrate for numerous fermentation processes as well as a range of industries related to food and beverages. Since they secrete substantial amounts of the enzyme extracellularly, a wide range of microorganisms are reported to manufacture glucoamylase; however, filamentous fungi have traditionally been employed to produce the enzyme for commercial purposes. The cost of the procedure is increased by the limitations of the commercially available fungal glucoamylases, including their modest thermostability, need for an acidic pH and slow catalytic activity (James and Lee, 2007).

Microbial glucoamylases have almost replaced the chemical hydrolysis of starch in starch-producing industries due to their numerous benefits, including better yield, specificity, stability of the chemicals produced and lack of neutralizing steps (Singh et al., 2009). Glucoamylases are produced mainly by bacterial strains because of their easy cultivation and fast growth. Fungal producers of these enzymes include *Aspergillus* and *Penicillium* species. Most glucoamylases are stable at low temperatures, but they lose their activity at higher temperatures due to conformational changes. Glucoamylases are widely used in the food industry for high-glucose, high-fructose syrup and alcohol production. These enzymes are also used in baking in order to improve the colour of bread crust, flour quality, baked product quality and reduce dough staling (James et al., 1996). Soy sauce, light beer and shakes are all made with the help of glucoamylases. They break down dextrins into fermentable sugars that have a lower alcohol level and calorific value in the beer. Additionally, glucose is produced using these enzymes, and when it is fermented with *Saccharomyces cerevisiae*, it yields ethanol (Blanco et al., 2014).

3.3.4 A-Amylase

α-Amylases are the starch-degrading enzymes that are involved in short-chain dextrin production through hydrolyzation of α-1,4 glycosidic bonds in polysaccharides. All living things contain a substantial amount of these enzymes. Since most alpha-amylases are metalloenzymes, calcium ions are necessary for the stability, activity and integrity of these enzymes (Sindhu et al., 2016). Alpha-amylases are used for the liquefaction of starch, in brewing and baking in the food industry (Rodriguez and Angeles, 2006). These enzymes are used as antistaling agents and flavour enhancers for the improvement of bread quality in the baking industry. Alpha-amylases convert starch to shorter chain dextrins during baking, which are further fermented by yeast. This improves the toasting qualities, colour of crust and taste of bread (Van der et al., 2002). During the production of powdery food and rice cake, alpha-amylase is also used as a glazing agent (Aiyer, 2005). Alpha-amylases are also used for the conversion of starch into fructose and glucose syrup through the liquefaction of starch in the starch industry (Raveendran et al., 2018). This conversion involves liquefaction, gelatinization and saccharification. Saccharification is responsible for the production of maltose and glucose. Alpha-amylases produced with *Bacillus licheniformis*, *Bacillus stearothermophilus* and *Bacillus amyloliquefacienc* are necessary for starch saccharification. Alpha-amylase is also used in ethanol production and fruit juice clarification. The process requires the presence of pectinases and cellulases along with alpha-amylase to make it more cost-effective and to improve the yield (Garg et al., 2016).

3.3.5 Lactase

Lactose hydrolysis is a crucial biotechnology process in the food companies. Lactase enzyme, also known as β-galactosidase, is involved in the hydrolytic cleavage of lactose to galactose and glucose. The enzyme can be obtained from microbes like archaea, fungi, bacteria and from animals and plants (Panesar et al., 2010). Among all lactase enzymes, the one produced by microbes is most preferred due to its low cost and high productivity. The lactase generated from the dairy yeast *Saccharomyces lactis* is currently the most popular commercial lactase preparation. The ideal processing settings for this lactase are in the range of milk's normal temperature and pH (35–40°C, 6.6–6.7). Even at a low temperature of 40°C, *S. lactis* shows significant activity. The sluggish growth of spoilage microorganisms at this temperature makes this feature crucial for practical application. A fungal lactase generated from *Aspergillus niger* is the other most used commercial lactase product. The ideal processing temperatures are below 50°C and pH 4.5. Therefore, only acid (or acidified) whey or lactose can be used with this lactase (Nijpels, 1981).

Lactase is used in the dairy industry to improve the digestibility of milk and milk-based products. These products prevent tissue dehydration and diarrhoea among people suffering from lactose intolerance. Lactase hydrolyzes lactose in ice cream to improve the creamy texture and digestibility. It also prevents crystallization during the production of certain milk products, which makes the product more soluble and stable (Martarello et al., 2019).

3.3.6 Esterases

Esterases are hydrolase-class enzymes that catalyze the formation of new ester bonds as well as the breakdown of existing ester bonds into alcohol and acid. Interesterification, intraesterification and transesterification processes involve esterases (Kulkarni et al., 2013). These enzymes are different from lipases in a way that the former also catalyzes the hydrolysis of short-chain triglycerides.

Esterases play an important role in the alcoholic beverage industries where they are used to produce flavours and fragrance and for the modification of fat and oil into fruit juices (Panda and Gowrishankar, 2005). The beverage industry uses an esterase called feruloyl esterase to create ferulic acid, which serves as a precursor for vanillin (a phenolic aldehyde aroma molecule) responsible for improving the flavour of the beverages (Gallage et al., 2014). Esterases are also used as flavour improvers in fruit juices.

Ferulyl esterase is also important for waste management as it hydrolyzes lignocellulosic biomass (Faulds, 2010). Industrial-scale esterase production mostly utilizes bacteria, i.e., *Feruloyl esterase* is produced using *Lactobacillus acidophilus, Lactobacillus amylovorus, Lactobacillus fermentum, Lactobacillus farciminis* and *Bacillus licheniformis* (Xu et al., 2017).

3.3.7 Phospholipases

The phospholipase enzyme initiates the conversion of phospholipids into fatty acids and other lipophilic compounds. These enzymes are further divided into two groups, i.e., acyl hydrolases and phosphodiesterases. Phospholipases A_1, A_2 and B constitute acyl hydrolases, which are responsible for replacing and displacing the acyl group in phospholipids through either esterification, hydrolysis or transesterification. On the other hand, phospholipases C and D constitute phosphodiesterases, which modify the polar head group (Borrelli and Trono, 2015). At the industrial level, phospholipases are produced by fungi and other microbes. Some of the microorganisms that produce phospholipase include *Aspergillus oryzae, Bacillus cereus, Streptomyces chromofuscus, Fusarium oxysporum* and *Bacillus cereus.*

According to Virgen-Ortz et al. (2019), phospholipases are employed in the dairy, bakery and oil production industries for a variety of functions. In order to maintain the stability of the fat, these enzymes are used in the preparation of some dairy products. Using phospholipases during cheese production can improve the cheese's flavour. In addition, phospholipase is added to milk during cheese production to stop the substantial loss of milk fat in whey, enhancing cheese yield and texture (Lilbaek et al., 2006). Phospholipases are used in vegetable oil-producing facilities as degumming agents to increase the quantity and quality of refined vegetable oil. The emulsifying properties of egg yolk, which are utilized to make sauces and mayonnaise, are improved by the use of these enzymes in the food processing sector, thus extending the shelf life of the food products (Su et al., 2017).

In the bakery industry, phospholipases are also employed in order to emulsify the dough, improve flavour and boost the nutritional content of animal feed, which includes lipids, such as soybean lecithin. Phospholipases provide desirable advantages when used in the production of specific foods, such as enhanced yield, reduced energy consumption, greater productivity, improved attributes or higher product quality (Casado et al., 2012).

3.3.8 Cellulases

Cellulase enzymes are involved in cellulolysis, i.e., the breakdown of cellulose. These enzymes break down the β-1,4 glycosidic bonds present in cellulose to produce glucose in lichenin, polymeric cellulose, glucans and hemicelluloses (Pandey et al., 2021). The presence of cellulose and hemicellulose in fruits and vegetables leads to cloudiness in juices, making the juices less appealing. Cellulases are used as fruit-softening enzymes that hydrolyze hemicellulose and cellulose in raw vegetable and fruit juice, thus enhancing juice yield, extraction, stabilization and clarification (Bhat, 2000). Cellulases are also used to reduce damage due to heat and viscosity in vegetable and fruit puree production and to improve the yield. These microbial enzymes are used in animal feed processing as these enzymes improve the digestibility of animal feed that is rich in pectin, cellulose, hemicellulose and lignin.

Best sources of cellulase production are fungi and bacteria; however, some metazoans like termites, snails and earthworms can also produce cellulases. In the food industry, commonly used sources of cellulases are fungi like *Trichoderma reesei* and *Aspergillus niger* and bacteria like *Streptomyces* species, *Paenibacillus* species, *Bacillus licheniformis* and *Bacillus subtilis* (Manzum and Mamun, 2018; Yadav et al., 2020).

3.3.9 Glucose Oxidase

This enzyme is a member of the vast family of enzymes known as "oxidoreductases". Muller first identified the flavoprotein glucose oxidase in 1928. Glucose oxidase's function is to change glucose into gluconic acid. This homodimeric enzyme has two 80 kDa polypeptide chain subunits that are identical to one another. Citrus fruits, red algae, bacteria and fungi are just a few examples of the many sources from which glucose oxidase can be obtained. For microbial use, glucose oxidase enzyme is produced from fungal strains, including *Aspergillus species* and *Penicillium species*. The synthesis of gluconic acid, which is employed as a causticity controller, a colour stabilizer, an antioxidant and a chelating agent in bread and drinks, is one of the key uses of glucose oxidase in food industries (Singh et al., 2016). Bitterness in food is eliminated using sodium gluconate, which is manufactured from glucose oxidase. The enzyme is employed in the baking industry to produce bread with better quality and a larger loaf volume. In the food sector, removing glucose and oxygen from diabetic drinks and egg white enhances the flavour, aroma and stability of food products. Glucose oxidase improves the colour, texture, flavour and shelf life of food products and also slows down their deterioration (Hanft and Koehler, 2006).

3.3.10 Laccases

The phenol-oxidizing enzymes known as "laccases" are part of the family of enzymes known as phenol oxidases, which oxidize phenolic and aromatic compounds, including certain amines, ethers and esters. The enzyme is commonly called "blue oxidases" and is well known for its potential to oxidize phenolic compounds in industrial sectors (Giardina et al., 2010). Laccases are efficiently used in bioremediation, bleaching of paper pulp, chemical synthesis, textile finishing, wine stabilization and biosensing. These enzymes also catalyze the oxidation of ascorbate, aromatic amines and phenolics.

Rhusvernicifera, a tree, was used as the initial source of laccase isolation. These enzymes can also be produced from plants (mango, pines, maize, peaches, etc.), insects (Bombyx, Calliphora, Rhodnius, etc.) and microbes (most commonly from filamentous fungi) (Morozova et al., 2007). Among all, Basidiomycetes, Ascomycetes and Deuteromycetes are the major producers of laccases (Gochev and Krastanov, 2007). Funaliatrogii (white rot fungus) produces laccase by absorbent fermentation (Li et al., 2017). However, the bacterium *Bacillus licheniformis* produces laccases for industrial purposes. Laccases are used in beverages and food industries for colour modification. The enzymes are also used in place of chemical and physical absorbents for wine stabilization. In place of the traditional method, laccases are used in the brewing industry to avoid haze formation through polyphenol (Mathiasen, 1995). Commercial laccase "Flavourstar" is used in the brewing industry to remove the off-flavour formation (Pronk and Leclercq, 2004). Laccase is also used in the beer industry to remove oxygen in the final step, thus improving the storage life of the product. These enzymes are used in the baking industry to decrease stickiness and enhance the strength and stability of bread batter (Si, 1994).

3.3.11 Catalase

Catalase enzyme is involved in the decomposition of hydrogen peroxide by its reduction to oxygen and water. The enzyme is found in all aerobic living organisms exposed to oxygen and thus, it can be produced by several sources, including bovine liver, plants, animals and bacteria like *Micrococcus luteus* and *Aspergillus niger* (Puri et al., 2012). However, the highest level of enzyme activity (400 mg/g) is found in isolates of *Aspergillus niger.*

Catalases are generally used with other enzymes in the food industry. For food preservation, these are used with glucose oxidases. Catalase improves the shelf life of wine and reduces its alcohol content by removing oxygen from the product (Röcker et al., 2016). Catalases are used in the dairy industry to eliminate peroxide from milk products and prevent product rancidity (Abada, 2018). These enzymes are used in food wrappers to minimize perishability by preventing oxidation. For industrial purposes, catalases are mostly obtained from microbes like *Micrococcus luteus, Aspergillus niger, Bacillus maroccanus, Enterococcus faecalis, Rhizobium radiobacter, Pyrobaculum calidifontis, Bacteroides fragilis, Ureibacillus thermosphaericus* FZSF03, etc. (Zeng et al., 2011).

3.3.12 Peroxidase

Peroxidases belong to the family of oxidoreductases. These enzymes contain iron (III) protoporphyrin IX as the prosthetic group. Peroxidases are involved in the oxidation of various organic and inorganic compounds as well as in peroxide reduction. These peroxidases are obtained from animals, microorganisms and plants and are involved in defence mechanisms against infectious or damaged tissues and lignification in plants (Biles and Martyn, 1993). *Phanerochaete chrysosporium* (fungus) is found to be the best source of peroxidase. The use of fungal peroxidases in industries is limited due to post-translational protein modification (Conesa et al., 2002), whereas peroxidases obtained from bacterial sources are more stable, active, easy to produce and more suitable for industrial uses. Peroxidase can also be produced from bacteria such as *Actinobacteria*, *Firmicutes*, *Acidobacteria* and *Proteobacteria* (Brown and Chang, 2014). Peroxidase uses hydrogen peroxide to catalyze various substrates.

Peroxidases are used in food industry for texture, colour and flavour, as well as to improve the nutritional composition of food. These enzymes are also used in the management of environmental pollutants, in the synthesis of polymers and as biosensors (Bansal and Kanwar, 2013). Inspite of so many benefits, one drawback of using peroxidases is that they cause off-flavours in vegetables and lead to undesirable browning of fruits.

3.3.13 Alpha-Acetolactate Decarboxylase

Alpha-acetolactate decarboxylase can be produced from bacteria like *Saccharomyces cerevisiae* or *Brevibacillus brevis*. This enzyme is used in the beer industry for speeding up the maturation of beer (Godtfredsen and Ottesen, 1982). In the absence of enzymes, generally the maturation of beer takes two weeks to twelve weeks, but the use of alpha-acetolactate decarboxylase leads to maturation within a shorter time (around 24 hours); however, the time taken for maturation depends on the enzyme source. Along with speeding up the maturation, alpha-acetolactate also nullifies the off-taste in beer, which is due to the presence of diacetyl by utilizing a two-step reaction of decarboxylation and protonation and helps in the conversion of acetolactate to acetoin. The enzyme works efficiently in both bound (encapsulated) and free forms (Dulieu, 2000).

3.3.14 Naringinase (Debittering Enzymes)

Naringinase represents a group of enzymes also called "debittering enzymes." These enzymes degrade naringin (bitter taste-causing in citrus fruits) into glucose, rhamnose, prunin and naringenin (Puri and Banerjee, 2000). *Alpha-rhamnosidase* acts on naringin and breaks it down to rhamnose and prunin. Prunin is further broken down to glucose and naringenin by the action of β-glucosidase. Naringinase is mostly produced by fungi like *Rhizopus*, *Aspergillus niger*, *Trichoderma*, *Circinella*, *Penicillium*, *Fusarium*, *Eurotium*, etc., and bacteria like *Bacillus* sp., *Bacteroides distasonis*, *Burkholderia cenocepacia*, *Pseudomonas paucimobilis*, *Thermomicrobium roseum*, etc. As fungal sources provide increased yield of naringinases, they are preferred more than bacterial sources. Naringinase is added as a supplement to fruit juices. It is used in combination with arabinosidase and β-glucosidase to improve the aroma of wine. Additionally, naringinases are also used to prepare tomato pulp and treat leftover kinnow peels (Puri, 2012).

3.3.15 Xylanases

The enzymes that convert xylan (a major hemicellulose constituent) into xylose by hydrolyzing the β-1,4 glycosidic bonds are known as xylanases. Not a single enzyme but a group of xylanolytic enzymes including exoxylanase, endo-(1,4)-β-d-xylanase, xylan-1,4-β-xylosidase, α-glucuronidase, α-l-arabinofuranosidase, p-coumaric esterase, acetylxylan esterase and ferulic acid esterase are involved in the process. These enzymes depolymerize the polysaccharide into monosaccharides (Collins et al., 2005; Walia et al., 2017; Bhardwaj et al., 2019). Exoxylanases release xylo-oligosaccharides by breaking the β-1,4 bonds present in xylan. These enzymes act from the non-reducing ends. β-xylosidases release xylose by breaking xylo-oligosaccharides and xylobiose (Sukumaran, 2005).

Xylanases are mainly produced by bacteria, actinomycetes and fungi. *Bacillus* species, *Streptomyces* species and *Pseudomonas* species are the major bacterial and actinomycete species that produce xylanases (Sharma and Chand, 2012). The xylanases produced by actinomycetes and bacteria can effectively work at pH 5.0–9.0 and at temperatures between 35 and 60°C. Fungal xylanases are higher in activity level than those obtained from yeast and bacteria. Fungal species that produce xylanases are *Penicillium*, *Aspergillus* and *Fusarium* (Mandal, 2015).

Xylanolytic enzymes, along with other enzymes, are frequently utilized in the production of bread. Hydrolysis of polysaccharides (non-starch) by enzymes improves the rheological properties of dough (Martínez-Anaya and Jiménez, 1997). Xylanolytic enzymes increase the volume of bread and hence improve the bread quality. This enzyme breaks down the hemicellulose present in wheat, which in turn improves the water binding in dough. This water binding makes the dough softer and delays crumb formation. Xylanolytic enzymes are used to improve taste, palatability and texture of biscuits (Mandal, 2015) and in fruit juice production, to hydrolyse hemicellulose in juice, which enhances juice stabilization, extraction, clarification and total juice yield (Bhardwaj et al., 2019).

Xylanases are also used in beer making. These are employed to hydrolyze the barley cell wall, releasing lesser oligosaccharides and arabinoxylans and thus lowering the viscosity and turbidity of the brew (Dervilly et al., 2002). Xylanases can efficiently break arabinoxylan present in animal feed into a digestible form; therefore, they are used in animal feed processing industries (Bhat, 2000).

3.3.16 Pectinases

Pectinase enzymes can be categorized as pectate lyases, pectin lyases, pectin esterases and polygalacturonases. Further, the enzyme can also be categorized as basic or acidic on the basis of pH. When pectinases cleave randomly, these are categorized as endopectinases, whereas when they act on terminal ends, pectinases are grouped as exopectinases (Pasha et al., 2013). These enzymes are involved in the breakdown of glycosidic bonds of pectic polymers. Pectinases act on natural substrates like pectic substances in pineapple, tomato, orange, lemon pulp, apple, orange peel and other citrus fruits (Saadoun et al., 2013). Pectinases can be derived from both naturally occurring and recombinant microorganisms, which can act on either coarse or smooth pectin sections (Pedrolli et al., 2009).

Pectinases are used in industries like remediation, the food industry, paper bleaching, etc. Juices, when prepared with added pectinase, are filterable, clear in appearance, have better flavour and colour and are about nine times superior to conventional treatment (Tapre and Jain, 2014).

3.3.17 Lipoxygenases

The oxidoreductase enzyme lipoxygenase (LOX) is found in abundance in both plants and mammals. By controlling its production of volatiles, LOX can function as a natural flavouring agent in food manufacturing. A class of dioxygenase redox enzymes known as lipoxygenase OXs is found in large quantities in both plant and animal cells. They do not contain heme iron, sulphur iron, or manganese (Aanangi et al., 2016). Polyunsaturated fatty acids (PUFAs), esters with cis- and cis-pentadiene structure, and hydroperoxide derivatives with conjugated double bonds can all be selectively catalyzed by LOX. In plants, linoleic acid and linolenic acid and in animals, arachidonic acid, are the major substrates for LOX (Shi et al., 2020).

Lipoxygenases are used in bread making and for the generation of aroma in the food industry (Cumbee et al., 1997). This enzyme has a number of effects on wheat flour dough. It works well as a whitening agent, producing a whiter loaf and crumb, as well as improving dough rheology and increasing mixing tolerance. The oxidation of pigments and unsaturated fatty acids by ambient oxygen is assumed to be the cause of the bleaching activity. By reinforcing the gluten network, it renders the dough resistant to unintentional over-mixing. Through baking and proofing, this improved dough strength endures, ultimately increasing loaf volume.

3.3.18 Asparginases

Asparaginase, also known as an asparagine-depleting enzyme, is used in large amounts in nutraceutical and pharmaceutical industries as an anticancer agent in the field of medicine (Krishnapura et al., 2016). Asparaginase serves as a food processing additive that helps starchy food products produce less acrylamide, which is thought to be a neurotoxic and probable carcinogen. Acrylamide is a chemical compound generated with a combination of L-asparagine and reducing sugars when food containing carbohydrates is cooked to temperatures above 1,200°C. Examples of such items are bread, other baked products and fried or baked items (Croft et al., 2004). During the Maillard reaction, a process responsible for giving baked or fried food their brown colour, crust and toasted flavour, the amino acid asparagine is transformed into acrylamide (Vass et al., 2004). Addition of asparaginase before baking or frying of food, asparagine is transformed into aspartic acid and ammonium. As a result, asparagine is unable to participate in the Maillard process, which greatly reduces the amount of acrylamide production by about 97 per cent (Pariza and Johnson, 2001; Kumar, 2015).

3.4 CONCLUSION

Microbial enzymes are useful in a variety of industries, including food, pharmaceuticals, paper and detergent industries. Due to the eco-friendly nature, high yield, efficient process control, safety and low cost of production and hydrolysis of microbial enzymes, enzyme-based processes are preferred over chemical-based processes. Microbial enzymes can also be produced easily, rapidly and more efficiently through various fermentation techniques when compared with animal and plant-based enzymes. It is also found that microbial enzymes can be easily modified by biochemical and molecular techniques. Higher production of these specific microbial enzymes can be achieved by overexpression of their genes. In spite of so many enzymes identified so far and increased application of these enzymes in the food sector, many microbial enzymes are still unexplored, and there are still a lot of opportunities for wider application of these enzymes in industries. More effective measures need to be taken for the identification of more microbial enzymes and strengthening the food sector with their increased use.

REFERENCES

Aanangi, R., Kotapati, K. V., Palaka, B. K., Kedam, T., Kanika, N. D. & Ampasala, D. R. (2016). Purification and characterization of lipoxygenase from mung bean (*Vigna radiate* L.) germinating seed-lings. *3 Biotechnology*, 6(1): 113. https://doi.org/10.1007/s13205-016-0427-5.

Abada, E. A. (2018). Application of microbial enzymes in the dairy industry. In: M. Kuddus (Ed.), *Enzymes in food biotechnology: Production, applications and future prospects* (pp. 61–72). Cambridge, MA: Academic Press.

Aiyer, P. V. (2005). Amylases and their applications. *African Journal of Biotechnology*, 4(13): 1525–1529.

Anbu, P., Gopinath, S. C. B., Chaulagain, B. P. & Lakshmipriya. T. (2017). Microbial enzymes and their applications in industries and medicine. *Biomed Research International*. doi: 10.1155/2017/2195808.

Aravindan, R., Anbumathi, P. & Viruthagiri, T. (2007). Lipase applications in food industry. *Indian Journal of Biotechnology*, 6: 141–158.

Aruna, K., Shah, J. & Birmole, R. (2014). Production and partial characterization of alkaline protease from Bacillus tequilensis strains CSGAB 0139 isolated from spoilt cottage cheese. *International Journal of Applied Biology and Pharmaceutical Technology*, 5: 201–221.

Bansal, N. & Kanwar, S. S. (2013). Peroxidase(s) in environment protection. *The Scientific World Journal*, 714639. https://doi.org/10.1155/2013/714639.

Bhardwaj, N., Kumar, B. & Verma, P. (2019). A detailed overview of xylanases: An emerging biomolecule for current and future prospective. *Bioresources and Bioprocessing*, 6: 40. https://doi.org/10.1186/s40643-019-0276-26.

Bhat, M. K. (2000). Cellulases and related enzymes in biotechnology. *Biotechnology Advances*, 18: 355–383.

Biles, C. L. & Martyn, R. D. (1993). Peroxidase, polyphenoloxidase, and shikimate dehydrogenase isozymes in relation to tissue type, maturity and pathogen induction of watermelon seedlings. *Plant Physiology & Biochemistry*, 31(4): 499–506.

Blanco, C. A., Caballero, I., Barrios, R. & Rojas. A. (2014). Innovations in the brewing industry: Light beer. *International Journal of Food Science and Nutrition*, 6: 655–660.

Borrelli, G. M. & Trono, D. (2015). Recombinant lipases and phospholipases and their use as biocatalysts for industrial applications. *International Journal of Molecular Sciences*, 16(9): 20774–20840.

Brown, M. E. & Chang, M. C. Y. (2014). Exploring bacterial lignin degradation. *Current Opinion in Chemical Biology,* 19: 1–7.

Casado, V., Martin, D., Torres, C. F. & Reglero, G. (2012). Phospholipases in food industry: A review. *Methods in Molecular Biology*, 861: 495–523.

Collins, T., Gerday, C. & Feller, G. (2005). Xylanases, xylanase families and extremophilic xylanases. *FEMS Microbiology Reviews*, 29, 3–23.

Conesa, A., Punt, P. J. & Van den Hondel, C. A. (2002). Fungal peroxidases: Molecular aspects and applications. *Journal of Biotechnology,* 93(2): 143–158. doi: 10.1016/s0168-1656(01)00394-7.

Cong, S., Tian, K., Zhang, X., Lu, F., Singh, S., Prior, B. et al. (2019). Synthesis of flavor esters by a novel lipase from *Aspergillus niger* in a soybean-solvent system. *Biotechnology*, 9(6): 244. doi: 10.1007/s13205-019-1778-5

Croft, M., Tong, P., Fuentes, D. & Hambridge, T. (2004). Australian survey of acrylamide in carbohydrate-based foods. *Food Additives and Contaminants*, 21(8): 721–736.

Cuesta, S. M., Rahman, S. A. & Thornton, J. M. (2016). Exploring the chemistry and evolution of the isomerases. *Biochemistry*, 113(7): 1796–1801.

Cumbee, B., Hildebrand, D. F. & Addo, K. (1997). Soybean flour lipoxygenase isozymes effects on wheat flour dough rheological and bread making properties. *Journal of Food Science,* 62(2): 281–283. doi.org/10.1111/j.1365-2621.1997.tb03985.x

De Souza, P. M. & Magalhães, P. O. (2010). Application of microbial α-amylase in industry - A review. *Brazilian Journal of Microbiology,* 41(4): 850–861.

Deng, L., Wang, Z., Yang, S., Song, J., Que, F., Zhang, H. et al. (2016). Improvement of functional properties of wheat gluten using acid protease from *Aspergillus usamii. PLoS One,* 11(7): e0160101. https://doi.org/10.1371/journal.pone.0160101

Dervilly, G., Leclercq, C., Zimmerman, D., Roue, C., Thibault, J. F. & Sauliner, L. (2002). Isolation and characterization of high molar mass water-soluble arabinoxylans from barley and barley malt. *Carbohydrate Polymers,* 47(2): 143–149.

Dulieu, C., Moll, M., Boudrant, J. & Poncelet, D. (2000). Improved performances and control of beer fermentation using encapsulated α-acetolactate decarboxylase and modeling. *Biotechnology Progress*, 16: 958–965.

Faulds, C. B. (2010). What can feruloylesterases do for us? *Phytochemistry Reviews*, 9: 121–132. https://doi.org/10.1007/s11101-009-9156-2.

Gallage, N. J., Hansen, E. H., Kannangara, R., Olsen, C. E., Motawia, M. S., Jørgensen, K. et al. (2014). Vanillin formation from ferulic acid in *Vanilla planifolia* is catalysed by a single enzyme. *Nature Communications*, 5(4037). https://doi.org/10.1038/ncomms5037.

Garg, G., Singh, A., Kaur, A., Singh, R., Kaur, J. & Mahajan, R. (2016). Microbial pectinases: An ecofriendly tool of nature for industries. *3 Biotechnology*, 6(1), 47. doi: 10.1007/s13205-016-0371-4.

Gerits, L. R., Pareyt, B., Decamps, K. & Delcour, J. A. (2014). Lipases and their functionality in the production of wheat-based food systems. *Comprehensive Reviews in Food Sciences and Food Safety,* 13(5): 978–989.

Ghosh, P. K., Saxena, R. K., Gupta, R., Yadav, R. P. & Davidson, S. (1996). Microbial lipases: Production and applications. *Science Progress*, 79: 119–157.

Giardina, P., Faraco, V., Pezzella, C., Piscitelli, A., Vanhulle, S. & Sannia, G. (2010). Laccases: A never-ending story. *Cellular and Molecular Life Sciences*, 67: 369–385.

Gochev, V. K. & Krastanov, A. I. (2007). Isolation of laccase producing *Trichoderma* spp. *Bulgarian Journal of Agricultural Science*, 13: 171–176.

Godtfredsen, S. E. & Ottesen, M. (1982). Maturation of beer with α-acetolactate decarboxylase. *Carlsberg Res Com Carlsberg Research Communications*, 47(2): 93–102.

Guerrand, D. (2017). Lipases industrial applications: Focus on food and agroindustries. *OCL-Oilseeds and Fats, Crops and Lipids*, 24(4), 7. https://doi.org/10.1051/ocl/2017031.

Gupta, R., Beg, Q. K. & Lorenz, P. (2002). Bacterial alkaline proteases: Molecular approaches and industrial applications. *Applied Microbiology and Biotechnology*, 59: 15–32.

Gurung, N., Ray, S., Bose, S. & Rai, V. (2013). A broader view: Microbial enzymes and their relevance in industries, medicine and beyond. *BioMed Research International*, 329121. doi: 10.1155/2013/329121.

Hanft, F. & Koehler, P. (2006). Studies on the effect of glucose oxidase in bread making. *Journal of the Science of Food and Agriculture*, 86: 1699–1704.

James, J., Simpson, B. K. & Marshall, M. R. (1996). Application of enzymes in food processing. *Critical Reviews in Food Science and Nutrition*, 36: 437–463.

James, J. A. & Lee, B. (2007). Glucoamylases: Microbial sources, industrial applications and molecular biology - A review. *Journal of Food Biochemistry*, 21(6): 1–52.

Jooyendeh, H., Kaur, A. & Minhas, K. S. (2009). Lipases in dairy industry: A review. *Journal of Food Science & Technology,* 46(3): 181–189.

Krishnapura, P. R., Belur, P. D. & Subramanya, S. (2016). A critical review on properties and applications of microbial l-asparaginases. *Critical Reviews in Microbiology*, 42: 720–737.

Kulkarni, S., Patil, S. & Satpute, S. (2013). Microbial esterases: An overview. *International Journal of Current Microbiology and Applied Sciences*, 2(7): 135–146.

Kumar, S. (2015). Role of enzymes in fruit juice processing and its quality enhancement. *Advances in Applied Science Research*, 6: 114–124.

Li, G., Liu, X. & Yuan, L. (2017). Improved laccase production by *Funaliatrogii* in absorbent fermentation with nutrient carrier. *Journal of Bioscience and Bioengineering*, 124(4): 381–385.

Lilbaek, H. M., Broe, M. L., Hoier, E., Fatum, T. M., Ipsen, R. & Sorensen, N. K. (2006). Improving the yield of mozzarella cheese by phospholipase treatment of milk. *Journal of Dairy Science*, 89: 4114–4125.

Mandal, A. (2015). Review on microbial xylanases and their applications. *International Journal of Life Sciences*, 4: 178–187.

Manzum, A. A. & Al Mamun, A. (2018). Isolation of *Bacillus spp.* bacteria from soil for production of cellulase abstract introduction: Screening and isolation of *Bacillus spp. Nepal Journal of Biotechnology*, 6: 57–61.

Martarello, R. D., Cunha, L., Cardoso, S. L., Freitas, M. M., Silveira, D., Fonseca-Bazzo, M. F., et al. (2019). Optimization and partial purification of beta-galactosidase production by *Aspergillus niger* isolated from Brazilian soils using soybean residue. *AMB Express*, 9: 81. https://doi.org/10.1186/s13568-019-0805-6.

Martínez-Anaya, M. A. & Jiménez, T. (1997). Functionality of enzymes that hydrolyse starch and non-starch polysaccharide in bread making. *Zeitschrift fur Lebensmittel-Untersuchung und –Forschung,* 205(3), 209–214.

Mathiasen, T. E. (1995). *Laccase and beer storage.* WO patent 1995021240A2.

Melani, N. B., Tambourgi, E. B. & Silveira, E. (2020). Lipases: from production to applications. *Separation & Purification Reviews*, 49: 143–158.

Miguel, Â. S. M., Martins-Meyer, T. S., Verı́ssimo da Costa, Figueiredo, E., Lobo, B. W. P. & Dellamora-Ortiz, G. M. (2013). Enzymes in bakery: Current and future trends. In I. Muzzalupo (Ed.), *Food Industry*. Rijeka, Croatia, InTech.

Morozova, O. V., Shumakovich, G. P., Shleev, S. V. & Yaropolov, Y. I. (2007). Laccase-mediator systems and their applications: A review. *Applied Biochemistry and Microbiology,* 43(5): 523–535.

NC-IUBMB. (2021). *Enzyme nomenclature.* Nomenclature Committee of the International Union of Biochemistry and Molecular Biology. Patel Enzyme Nomenclature.

Nijpels, H. H. (1981). Lactases and their applications. In G. G. Birch, N. Blakebrough & K. J. Parker (Eds.), *Enzymes and food processing*. Dordrecht: Springer. Retrieved from https://doi.org/10.1007/978-94-011-6740-6_6.

Panda, T. & Gowrishankar, B. S. (2005). Production and applications of esterases. *Applied Microbiology and Biotechnology,* 67: 160–169.

Pandey, A., Nigam, P., Soccol, C. R, Soccol, V. T., Singh, D. & Mohan, R. (2000). Advances in microbial amylases. *Biotechnology and Applied Biochemistry,* 31(2): 135–152.

Pandey, P., Kuila, A. & Tuli, D. K. (2021). Cellulase: An overview. In D. K. Tuli& A. Kuila (Eds.), *Current status and future scope of microbial cellulases* (pp. 95–113). Amsterdam: Elsevier.

Panesar, P. S., Kumari, S. & Panesar, R. (2010). Potential applications of immobilized β-galactosidase in food processing industries. *Enzyme Research,* 473137. https://doi.org/10.4061/2010/473137 52.

Pariza, M. W. & Johnson, E. A. (2001). Evaluating the safety of microbial enzyme preparations used in food processing: Update for a new century. *Regulatory Toxicology and Pharmacology,* 33(2): 173–186.

Pasha, K. M., Anuradha, P. & Subbarao, D. (2013). Applications of pectinases in industrial sector. *Journal of Pure and Applied Science and Technology*, 16: 89–95.

Patel, A. K., Singhania, R. R. & Pandey, A. (2017). Production, purification and application of microbial enzymes. In G. Brahmachari (Eds.) *Biotechnology of microbial enzymes: Production, biocatalysis and industrial applications* (pp. 13–41). Cambrige, MA: Academic Press.

Pedrolli, D. B., Monteiro, A. C., Gomes, E. & Cano Carmona, E. (2009). Pectin and pectinases: Production, characterization and industrial application of microbial pectinolytic enzymes. *Open Biotechnology Journal*, 3: 9–18.

Pronk, I. M. E. J. & Leclercq, C. (2004). Laccase from *Myceliophthora thermophila* expressed in *Aspergillusoryzae.* In WHO Food additives series: 52. *International Programme on Chemical Safety (IPCS) INCHEM.* Retrieved from: http://www. Inchem.org/documents/jecfa/jecmono/v52je06.htm.

Puri, M. (2012). Updates on naringinase: Structural and biotechnological aspects. *Applied Microbiology and Biotechnology,* 93: 49–60.

Puri, M. & Banerjee, U. C. (2000). Production, purification, and characterization of the debittering enzyme naringinase. *Biotechnology Advances*, 18: 207–17.

Puri, M., Sharma, D., Barrow, J. C. & Tiwary, A. K. (2012). Optimization of novel method for the extraction of steviosides from *Stevia rebudiana* leaves. *Food Chemistry*, 132: 1113–1120.

Raveendran, S., Parameswaran, B., Ummalyma, S. B., Abraham, A., Mathew, A. K., Madhavan, A., Rebello, S. & Pandey, A. (2018). Applications of microbial enzymes in food industry. *Food Technology and Biotechnology*, 56(1): 16–30.

Research and Markets. (2021). *Food enzymes market, global forecast, impact of COVID-19, industry trends by types, growth, opportunity company analysis.* 125 pp. Retrieved from https://www.researchandmarkets.com/reports/5411844/food-enzymes-market-global-forecast-impact-of#rela3-4801548.

Robinson, P. K. (2015). Enzymes: principles and biotechnological applications. *Essays in Biocehmistry*, 15: 1–41.

Röcker, J., Schmitt, M., Pasch, L., Ebert, K. & Grossmann, M. (2016). The use of glucose oxidase and catalase for the enzymatic reduction of the potential ethanol content in wine. *Food Chemistry,* 210: 660–670.

Rodríguez C. S. & Ángeles S. M. (2006). Application of solid-state fermentation to food industry – A review. *Journal of Food Engineering*, 76(3): 291–302.

Saadoun, I., Dawagreh, A., Jaradat, Z. & Ababneh, Q. (2013). Influence of culture conditions on pectinase production by Streptomyces sp. (strain J9). *International Journal of Life Science and Medical Research,* 3: 148. https://doi.org/10.5963/LSMR0304002.

Sharma, P. K. & Chand, D. (2012). Production of cellulase free thermostable xylanase from Pseudomonas sp. XPB-6. *International Research Journal of Biological Sciences,* 1(5): 31–41.

Sharma, R., Chisti, Y. & Banerjee, U. C. (2001). Production, purification, characterization and applications of lipases. *Biotechnology Advances,* 19: 627–662.

Shi, Y., Mandal, R., Singh, A. & Singh, A. P. (2020). Legume lipoxygenase: Strategies for application in food industry. *Legume Science.* https://doi.org/10.1002/leg3.44.

Si, J. Q. (1994). *Use of laccase in baking industry.* US patent 6296883 B1.

Sindhu, R., Binod, P. & Pandey, A. (2016). α-Amylases. In A. Pandey, S. Negi, C. R. Soccol (Eds.) *Current developments in biotechnology and bioengineering – Production, isolation and purification of industrial products* (pp. 3–24). Amsterdam, the Netherlands: Elsevier Publications.

Singh, R., Mittal, A., Kumar, M. & Mehta, P. K. (2016). Microbial protease in commercial applications. *Journal of Pharmaceutical, Chemical and Biological Sciences,* 4(3): 365–374.

Singh, R. S., Singh, T. & Pandey, A. (2009). Microbial enzymes - An overview. In: R. S. Singh, R. R. Singhania, A. Pandey & C. Larroche (Eds.), *Advances in enzyme technology, biomass, biofuels, biochemicals* (pp. 1–40). Elsevier Publications.

Singh, S. & Bajaj, B. K. (2016). Bioprocess optimization for production of thermoalkali - stable protease from *Bacillus subtilis* K-1 under solid-state fermentation. *Preparative Biochemistry & Biotechnology,* 46: 717–724.

Su, L., Ji, D., Tao, X., Yu, L., Wu, J. & Xia, Y. (2017). Recombinant expression, characterization and application of a phospholipase B from *Fusarium oxysporum. Journal of Biotechnology,* 242: 92–100.

Sukumaran, R. K., Singhania, R. R. & Pandey, A. (2005). Microbial cellulases - Production, applications and challenges. *Journal of Scientific and Industrial Research,* 64: 832–844.

Tapre, A. R. & Jain, R. K. (2014). Pectinases: Enzymes for fruit processing industry. *International Food Research Journal,* 21: 447–453.

Van der Maarel, M. J., E.C. Van Der Veen, B., Uitdehaag, J. C. M., Leemhuis, H. & Dijkhuizen, L. (2002). Properties and applications of starch-converting enzymes of the α-amylase family. *Journal of Biotechnol*ogy, 94: 137–155.

Vass, M., Amrein, T. M., Schonbachler, B., Escher, F. & Amado, R. (2004). Ways to reduce the acrylamide formation in cracker products. *Czech Journal of Food Sciences,* 22: 19–21.

Virgen-Ortíz, J. J., dos Santos, J. C. S., Ortiz, C., Berenguer-Murcia, A., Barbosa, O., Rafael, C. et al. (2019). Lecitase ultra: A phospholipase with great potential in biocatalysis. *Molecular Catalysis,* 473(1), 110405. doi:10.1016/j.mcat.2019.110405.

Walia, A., Guleria, S., Mehta, P., Chauhan, A. & Parkash, J. (2017). Microbial xylanases and their industrial application in pulp and paper biobleaching: A review. *Biotechnology,* 7(1): 11. doi: 10.1007/s13205-016-0584-6.

Worthington, C. C., Worthington, V. & Worthington, A. (2019). *Introduction to Enzymes* (pp 1–16). Worthington-Biochem.com.

Xu, Z., He, H., Zhang, S., Guo, T. & Kong, J. (2017) Characterization of Feruloylesterases produced by the four *Lactobacillus s*pecies: *L. amylovorus, L. acidophilus, L. farciminis* and *L. fermentum*, isolated from ensiled corn stover. *Frontiers in Microbiology,* 8: 941. https://doi.org/10.3389/fmicb.2017.00941.

Yadav, A., Mahaboob Ali, A. A., Ingawale, M., Raychaudhuri, S., Gantayet, L. M. & Pandit, A. (2020). Enhanced co-production of pectinase, cellulase and xylanase enzymes from *Bacillus subtilis* ABDR01 upon ultrasonic irradiation. *Process Biochemistry*, 92: 197–201.

Zeng, H. W., Cai, Y. J., Liao, X. R., Zhang, F., Li, Y. L. & Zeng, X. K. (2011). Serratiamarcescens SYBC08 catalase isolated from sludge containing hydrogen peroxide shows increased catalase production by regulation of carbon metabolism. *Engineering in Life Sciences*, 11: 37–43.

Zou, H., Wang, H., Zhang, Z., Lin, H. & Li, Z. (2023). Immune regulation by fermented milk products: The role of the proteolytic system of lactic acid bacteria in the release of immunomodulatory peptides. *Critical Reviews in Food Science & Nutrition*, 21: 1–19.

4 The Use of Fermentation as a Preservative Technology

Palmei Gaibimei and Ningthoujam Manda Devi

4.1 INTRODUCTION

Fermentation is one of the ancient food preservation methods that have been handed down since time immemorial. The term has been derived from the Latin verb "fervere," which means "to boil," as it gives boiling appearance due to the production of carbon dioxide bubbles as a result of the breakdown of complex organic compounds present in food into simpler compounds by the action of enzymes and starter culture microorganisms, including bacteria, molds, and yeasts (Shokri and Terefe, 2019). In this metabolic process, the microorganisms use glucose, fructose, etc., for respiration and release carbon dioxide to make gas cells appear in the food to produce desired compounds. This process preserves fruits and vegetables by prolonging the shelf life of food, improving the sensory flavor and nutritional value of food, imparting texture to foods, reducing toxicity, and decreasing cooking time. In order to achieve fermented food products, the food has to undergo different processes such as acidification, maturation, ripening, flavoring, and preservation. Fermentation has also been widely utilized for the purposes of producing unique and novel foods (Rodgers, 2008) as well as for converting food wastes into high-value food ingredients, chemicals, and biochemicals (Terefe, 2015) such as food flavorings, organic acids (Zhang et al., 2013), and enzymes (Kiran et al., 2014).

It has been observed that abundant fruits and vegetables are largely wasted due to a lack of proper storage and processing methods, especially during their surfeit seasons. Food processing plays an important role in shrinking food waste and helps in boosting food security. In addition, it also makes the food more nutritious, easier to market and distribute, removes toxic substances, provides more convenience, allows mass production, and offers variety in the diet of consumers. Modern food processing techniques and processes also enhance the quality of life for people with allergies, diabetics, and others who cannot consume some common food elements.

DOI: 10.1201/9781032650616-4

4.2 HEALTH BENEFITS OF FERMENTED FOODS

I. Fermented foods consist of probiotics, which rupture hard-to-digest cellulose in food, making food digestible.
II. Fermented foods help in the detoxification of the body. Consumption of fermented foods can help people suffering from lactose intolerance, constipation, irritable bowel syndrome, yeast infections, allergies, and asthma to live a healthier lifestyle in their diet.
III. Enhance the nutritional value as the fermentation process actually enriches foods with higher amounts of nutrient sources.
IV. Consumption of probiotics fermented foods and drinks like Kefir and Kombucha introduced beneficial bacteria into the digestive system and helped to balance the bacteria in the digestive system. Thus, it helped to slow or reverse some diseases, improve bowel health, aid digestion, and improve immunity.
V. It helps in better absorption of food by balancing the gut bacteria and providing enough digestive enzymes that help to absorb more of the nutrients in the foods when we eat.
VI. It improves the shelf life of food and helps to preserve food for a longer period of time for future consumption.
VII. It enhances the overall flavor of the food by adding numerous varieties to the diet.
VIII. It improves the immune system of the body.
IX. They are budget-friendly by incorporating healthy foods into the diet.

4.3 FERMENTATION PLAYS AT LEAST FIVE ROLES IN FOOD PROCESSING (STEINKRAUS, 1997)

I. Enrichment of the human diet through the development of a wide diversity of flavors, aromas, and textures in food.
II. Preservation of substantial amounts of food through lactic acid, alcoholic, acetic acid, alkaline fermentations, and high salt fermentations.
III. Enrichment of food substrates biologically with vitamins, protein, essential amino acids, and essential fatty acids.
IV. Detoxification during food fermentation processing.
V. Decrease in cooking times and fuel requirements.

4.4 TYPES OF FERMENTATION

I. **Ethanol Fermentation (also called alcoholic fermentation)**: It is a biological process that converts sugars such as glucose, fructose, and sucrose into cellular energy, producing ethanol and carbon dioxide as by-products. For instance, the production of alcoholic beverages, making bread products rise, and the production of biofuel.

II. **Lactic Acid Bacteria Fermentation**: It is a metabolic process by which glucose or other six-carbon sugars (also, disaccharides of six-carbon sugars, e.g., sucrose or lactose) are converted into cellular energy and the metabolite lactate, which is lactic acid in solution. Lactic Acid Bacteria Fermentation is an anaerobic fermentation reaction giving rise to fermented foods such as yogurt, cheese, sauerkraut, and pickles.

III. **Acetic Acid Bacteria Fermentation**: It is a process of oxidation in which alcohol is converted into acetic acid by the action of bacteria of the genus *Acetobacter* leading to the production of vinegar, cheese, breads, and biodiesel.

4.5 APPLICATION OF EMERGING TECHNOLOGIES IN FOOD FERMENTATION PROCESS

Recent studies have shown many innovations in food fermentation technologies, such as emerging novel thermal and non-thermal food processing technologies, which bring significant advancements by improving sustainability, food security, human and environmental health, and the global economy. Application of such novel technologies can positively affect fermentation process variables by improving the efficiency of food fermentation processes and subsequently enable the production of valuable fermented products with better characteristics in terms of sensory attributes and nutritional quality. These novel heating/thermal and non-thermal technologies have certain advantages of improving the quality of processed food, enhancing the product yield, shortening fermentation time, extending the shelf life without using additives, and developing differentiated fermented products with novel sensorial and physicochemical properties. The difference between thermal and non-thermal effects lies in that an effect is regarded as thermal if it can be induced not only by a microwave field but also by heat generated in a different way; an effect is non-thermal if it is induced by radiation of such intensity that the temperature increase it causes is negligible or if the effect cannot be achieved by increasing the temperature in a different manner (Dębicki and Styłba, 2010). In recent times, it has been observed that there is an increase in consumers' interest toward natural and chemical-free fermented food products which can boost immune health. Therefore, it is becoming a responsible task for the industries to industrialize traditional fermented products by introducing these non-thermal novel technologies to the market.

4.6 NOVEL HEATING/THERMAL TECHNOLOGIES ON FERMENTATION PROCESSES

4.6.1 Microwave Heating (MWH)

It is a technique based on electromagnetic radiation (300–300,000 MHz), which has a direct effect on macromolecules of food products in terms of ion conduction and dipole rotation. When microwave heating is applied, the absorption of energy by the water molecules present in the capillary of food products facilitates the fermentation process by the production of heat when ions accelerate and collide or dipoles rotate and line up in the rapidly alternating electric field (Cunningham, 1978). The process efficiency in microwave treatment depends mainly on the type of material under treatment, its physical parameters, structure, conductivity, and dielectric properties. Microwave heating (Figure 4.1) treatment may shorten the fermentation process time and also act as an important pretreatment tool in the fermentation process to improve the texture and nutritional value. Hence, microwave heating is usually used for the fermentation of liquid fermentation products, such as juice, yogurt, etc. (Gholamhosseinpour et al., 2020).

4.6.2 Radiofrequency Heating (RFH)

An efficient heating process that forms a substitution for traditional methods is radiofrequency heating (RFH). This process utilizes a wavelength frequency of 0.003–300 MHz for heating purposes, allowing for a more uniform temperature distribution and heat transfer than conventional heating. The longer the wavelength of RFH, the deeper the heating penetrates into the material, producing more uniform heating, thereby delivering the electromagnetic energy directly to the whole volume

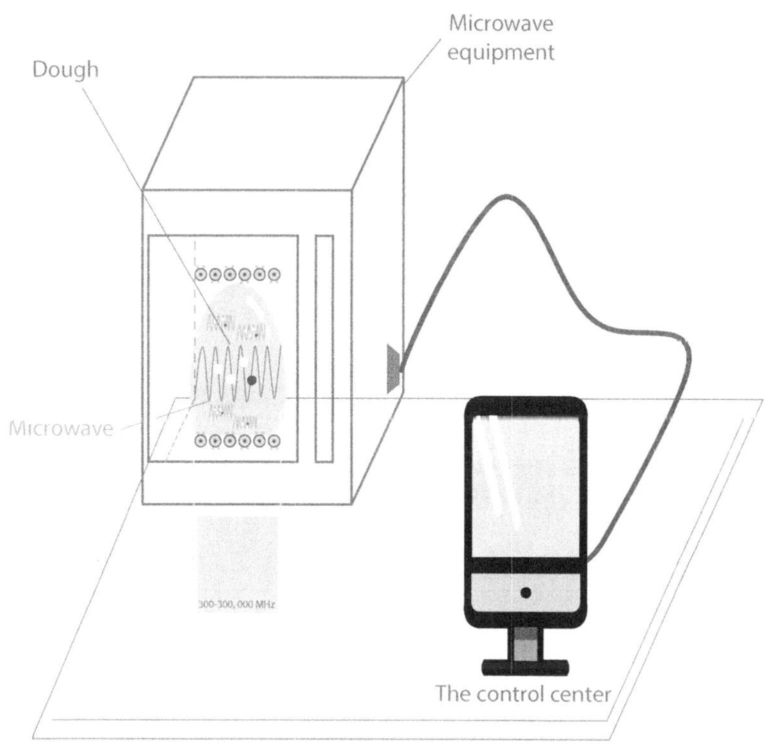

FIGURE 4.1 Schematic model of microwave heating (Wang et al., 2021)

of the substance being heated. In RF processing, its penetration depth depends on the function of the applied frequency and dielectric properties – dielectric constant and loss factor. RFH has been used for thermal processing of flour and powder-form products. It stands as a promising alternative with rapid volumetric heating ability and longer penetration depth compared to conventional approaches. In the fermentation process, the RFH process has shown potential for the production of fermented products by improving the shelf life, controlling harmful microorganisms as well as their sensory and nutritional properties. RFH equipment can stabilize the fermentation environment at about 35°C through the heat generated by the coil, so as to keep the environment similar to that of a traditional fermentation tank (Figure 4.2). RFH is utilized to process packaged food goods and food but is not yet feasible for bulk items (Siefarth et al., 2014).

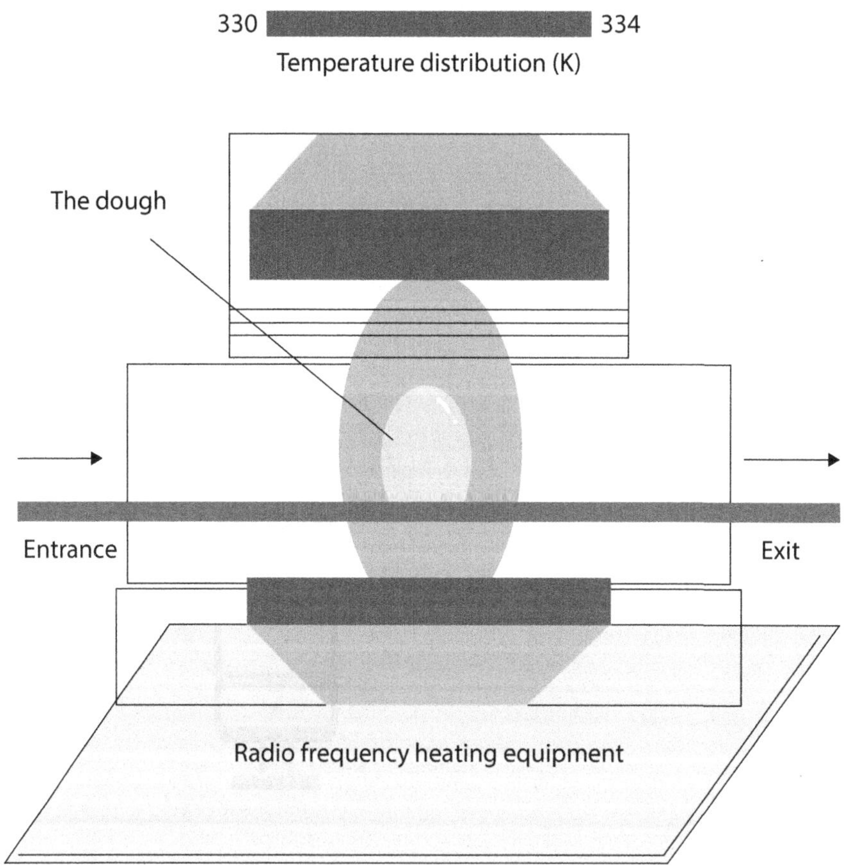

FIGURE 4.2 Schematic model of radiofrequency heating (Wang et al., 2021)

4.6.3 Ohmic Heating (OH)

Ohmic heating, one of the heat processing methods, is mostly known as joule heating, electrical resistance heating, or electro conductive heating. The OH experimental design usually consists of electrodes that contact the food liquids and solids, whereby electricity is passed through the substance using a variety of voltage and current combinations. The success of ohmic heating depends on the rate of heat generation in the system, the electrical conductivity of the food, electrical field strength, residence time, and the method by which the food flows through the system (Takhistov, 2007; Halden et al., 1990; Imai et al., 1995). The food acts as an electrical resistor, and the substance is heated by the dissipation of electrical energy. Electrical resistance heating allows particles and liquids to heat at the same rate and permits the rapid heating of mixtures with high solids fractions. When compared to conventional heating, where heat is conducted from the outside in using a hot surface, ohmic heating conducts heat throughout the entire mass of the food uniformly. Also, ohmic technology will minimize heating time by 90 percent and save energy by 82 to 97 percent (Castro et al., 2004). The shelf life of ohmically processed foods is comparable to that of canned and sterile, aseptically processed products. One of the important advantages of OH is that it is environmentally friendly technology, retaining better sensory and nutritional properties of products than traditional methods. For the fermented products, the irregular surfaces may have different contact areas after being squeezed separately by two electrode plates. After the products have been fermented, some of the sticky compounds may inevitably stick to the contact surface. In addition, the moisture content of the product seems to be another important factor since the moisture content and distribution determine the electrical conductivity of the material. This will create additional work for the next batch of products for fermentation and equipment maintenance (Figure 4.3) (Table 4.1).

4.7 NOVEL NON-THERMAL TECHNOLOGIES ON FERMENTATION PROCESSES

4.7.1 Ultrasound (US)

Ultrasound is generally defined as sound with a frequency above that to which the human ear can respond. US is also known as sonication and is of two types: probe and bath type (Figure 4.4). Traditionally, ultrasound has been used over the years in food processing for the purpose of inhibiting the actions and activities of microorganisms. Ultrasound-assisted fermentation is the application of an appropriate ultrasound treatment protocol to facilitate the growth and proliferation of microbial cells, effectively enhancing the overall efficiency of fermentation processes and increasing the yield and better quality of products. The application of US in the food industry is classified into two categories: high-frequency low-energy ultrasound (>100 kHz) or diagnostic ultrasound with intensities typically lower than 1 W cm^{-2} and low-frequency high-energy ultrasound (16–100 kHz) or high power ultrasound with intensities typically higher than 1 W cm^{-2}. High-frequency low-energy ultrasound

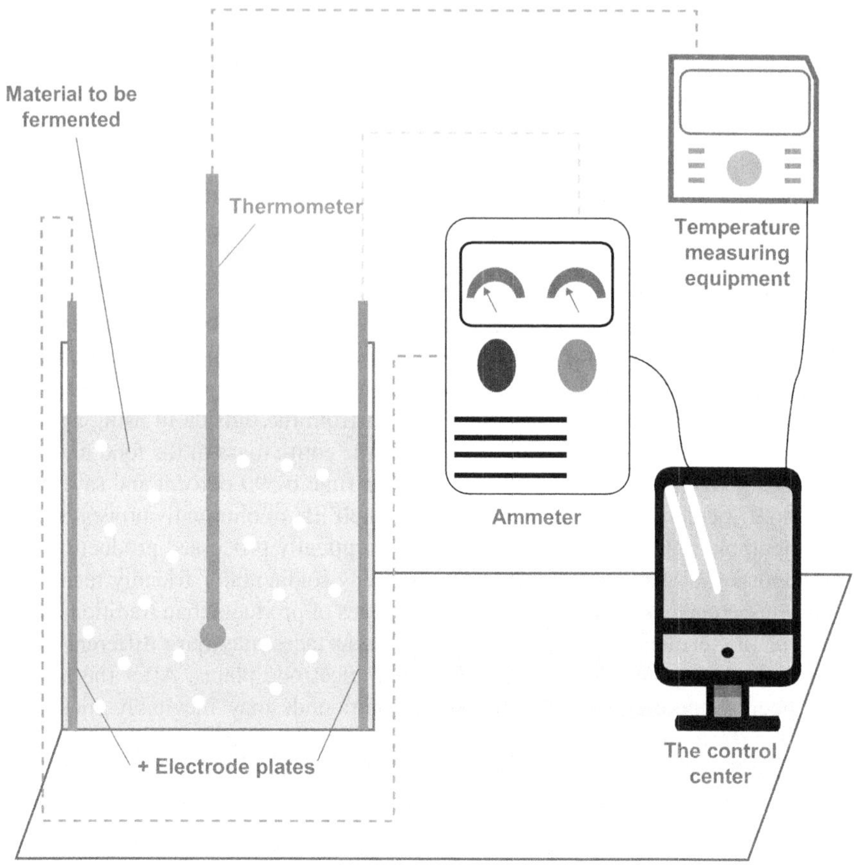

FIGURE 4.3 Schematic model of ohmic heating (Wang et al., 2021)

(HFLEUS) has been used for non-destructive measurement applications in the food industry, such as quality control and process monitoring (Awad et al., 2012). On the other hand, low-frequency high-energy ultrasound (LFHEUS) has been investigated as a method for enzyme inactivation and microbial disruption for sanitization and food preservation applications, as well as the release of internal components from the cells. High-frequency ultrasound employs very low power levels insufficient to cause acoustic cavitation, which therefore produces zero or minimal physical and chemical alterations in the material through which the wave passes. Hence, it can be employed for food analysis and quality control without affecting the product. In contrast, low-frequency ultrasound employs power levels high enough to generate cavitation and is capable of producing physical and chemical modifications in numerous applications. Ultrasound can be used in such processes to either monitor the progress of fermentation or to influence its progress. High-frequency ultrasound (>2 MHz) has been extensively reported as a tool for the measurement of the

TABLE 4.1
Application of thermal technologies in fermentation

Objective	Conditions	Results	References
(a) Microwave Heating (MWH)			
Impact of using MWH technology on fermentation time	Five treatments, viz. conductive, microwave, yeast, conductive + yeast, and microwave – yeast.	When the sample was pretreated with microwaves, the fermentation time was reduced by 40% as compared to the control sample.	Kapcsándi et al. (2013)
Effect of microwave pretreatment on enzymes during solid-state fermentation	Frequency of 2450 MHz, maximum power of 700W, treated for 10, 20, 30, 40, 50, and 60 s.	Results showed that MWH could increase the yield of hydrolase, as compared to untreated samples; the activity of glucose amylase and protease after microwave pretreatment (30s) was significantly increased.	Melikoglu M. (2012)
Production of xylooligosaccharide from wheat bran by microwave-assisted enzymatic hydrolysis	Microwave treatment 1400 W for 120 s	Wheat bran has almost 3–4 folds reducing sugar yield when pretreated by microwave irradiation.	Wang and Lu (2013)
Microwave treatments of milk: fermentation efficiency and biological activities	Microwaved at 720 W for 35 s.	After 24 h of fermentation, the average protease activity of the microwaved milk was 1.3–1.8 times more than the observed values in the other treated samples. Higher growth of lactic acid bacteria in fermented, microwaved, and sonicated samples.	Gholamhosseinpour et al. (2020)
(b) Radiofrequency Heating (RFH)			
Effect of radio frequency heating on yogurt	RF with 27.12 MHz operating frequency and 16 kW rated power. Applied electrode voltages during the yogurt heating trials were between 2.2 and 2.4 kV, with resulting field intensities in the water bath between 17.8 $kV{\cdot}m^{-1}$ and 19.2 $kV\ m^{-1}$.	Increased shelf life can be predicted as a reduction in microbial numbers was achieved. Hence, the potential to prolong yogurt shelf life is related not only to the inactivation of yeasts and molds, but also to the prevention of post-acidification due to reduced LAB numbers.	Siefarth et al. (2014)

(Continued)

Table 4.1 (Continued)

Objective	Conditions	Results	References
Novel RF-assisted thermal processing for reducing the come-up time of soft wheat flour and for microbiological safety	RF heating to 80°C, holding in a hot air oven at 80°C for 7 and 10 h; then to 90°C, holding for 2 and 3 h; and finally to 100°C, holding 0.75 and 1 h. Samples treated in a 6-kW parallel plate (electrodes) and RF heating system operating at 27.12 MHz.	Leads to the design of a reduction of a minimum of 7 log reductions of *Salmonella* spp. Without compromising the quality and functionality.	Boreddy et al. (2019)
(c) Ohmic Heating (OH)			
The effects of conventional pasteurization and OH under different process parameters on the physico-chemical properties, nutritional properties (bioactive compounds), and sensory profile of sweet whey.	Whey samples were subjected to both different OH parameters as 2, 4, 5, 7 and 9 $V \cdot cm^{-1}$ at 60 Hz, up to 72–75 °C/15 s and conventional processing at 72–75 °C/15 s.	The electric field intensity should be between 4 V and 5 V for practical and operational application of OH in dairy industries, providing a great release of bioactive peptides.	Costa et al. (2018)
Growth kinetics of *Lactobacillus acicfophilus* under ohmic heating	OH achieved by passing an alternating current of 60 Hz and a constant voltage (15 or 40 V).	Resulted in higher final pH and lower bacteriocin activity in the fermented medium. This shows that measurement of the electric current when ohmic heating is done at a constant voltage may be used in monitoring such fermentations.	Cho et al. (2000)

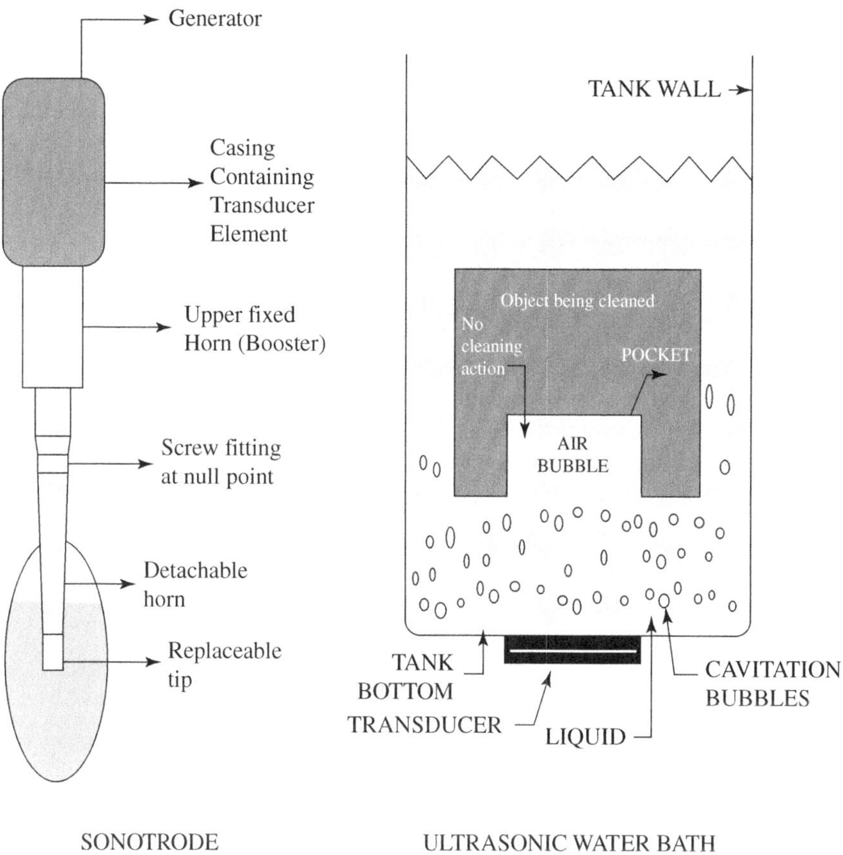

FIGURE 4.4 Schematic model of ultrasound (probe and bath type) (Bhargava et al., 2021)

changes in chemical composition during fermentation, providing real-time information on reaction progress. Low-frequency ultrasound (20–50 kHz) can influence the course of fermentation by improving mass transfer and cell permeability, leading to improved process efficiency and production rates.

4.7.2 High-Pressure Processing (HPP)

HPP, also known as pascalization, is a non-thermal emerging technique, consisting of treatments of pressure between 100 and 700 MPa for food processing and preservation by inactivation of vegetative cells of pathogenic and spoilage microorganisms, resulting in extending the shelf life and safety of treated food products. During HPP, the food is contained in a flexible package and subjected to hydrostatic pressure ranging from 100 to 900 MPa with a liquid (typically water) as a pressure transfer medium. Based on the isostatic principles, the pressure is transferred to the food in the container instantaneously and quickly with no gradient. As a result, all parts of the product obtain equal pressure irrespective of the size and external shape of the food product. HPP does not degrade pigments, aromatic compounds, and molecules (Figure 4.5). Thus no adverse effects are observed on the sensory and nutritional quality of treated foods (Considine et al., 2008; Yen and Lin, 1996). The efficacy of HPP treatments will be dependent on the pressure applied, the holding time and temperature, the characteristics of the food, and the target microorganism. HPP is applied mainly to pre-packed juices, sauces, dips, fishery products, meat products, and ready-to-eat meals (RTE)

4.7.3 Pulsed Electric Fields (PEFs)

Pulsed electric fields (PEFs) are a modern non-thermal technology that has been developed for food preservation and other applications such as extraction. PEF is based on the application of short pulses of high electric fields for less than 1 s, and an intensity of 10–80 kV cm^{-1} between the electrodes of a chamber containing liquid food. The electrical pulses can cause electroporation of the microbial cell

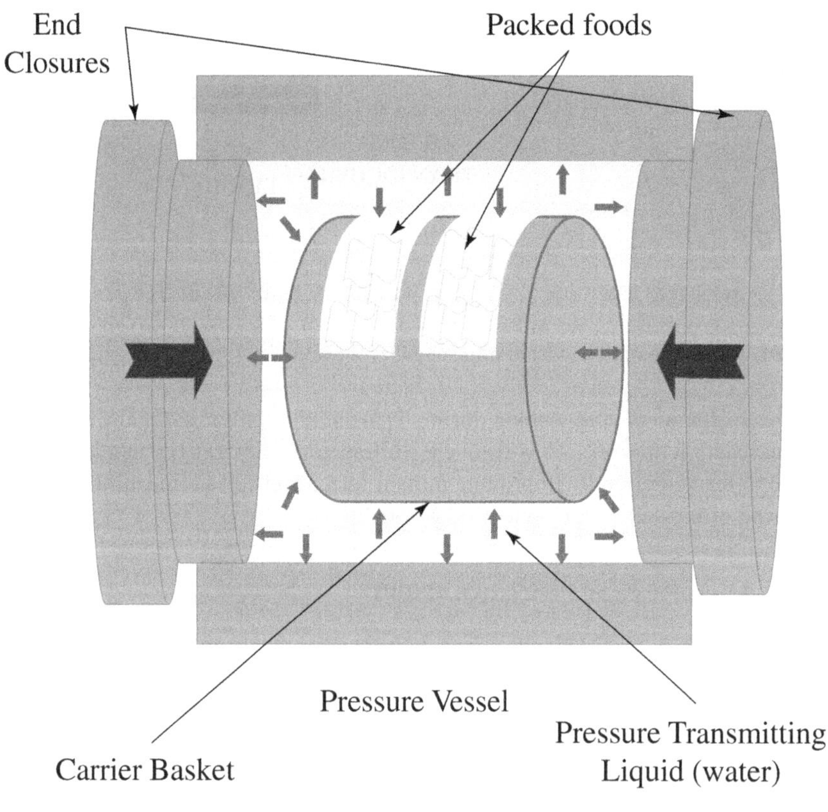

FIGURE 4.5 Schematic model of high-pressure processing (Balakrishna et al., 2020)

membranes and, under specific intensities, can cause irreversible cell membrane damage, intracellular leakage, and cell destruction (Chanos et al., 2020; Mohamed and Eissa, 2012). The efficacy of this technology for microbial deactivation depends on several factors, such as the characteristics of the microbe, the treatment media, the strength of the electric fields, the treatment time, the frequency of the pulses, the specificity of the applied energy, the polarity, and the geometry of the pulses (Chanos et al., 2020). PEF processing can be applied in a continuous flow and requires a very short time to complete, in contrast to other non-thermal food processing methods such as HPP (Puértolas et al., 2010). Application of sub-lethal levels of PEF can improve microbial growth and fermentation, by stimulation of metabolic cascades and control of metabolic pathways that can also lead to reversible electroporation, which corresponds to the formation of pores in the cell membrane (Chanos et al., 2020; Hunt et al., 2009; Mota et al., 2018; Shil et al., 2008). Application of PEF treatment (Figure 4.6) can also provide beneficial effects on the final characteristics of fermented products (Table 4.2).

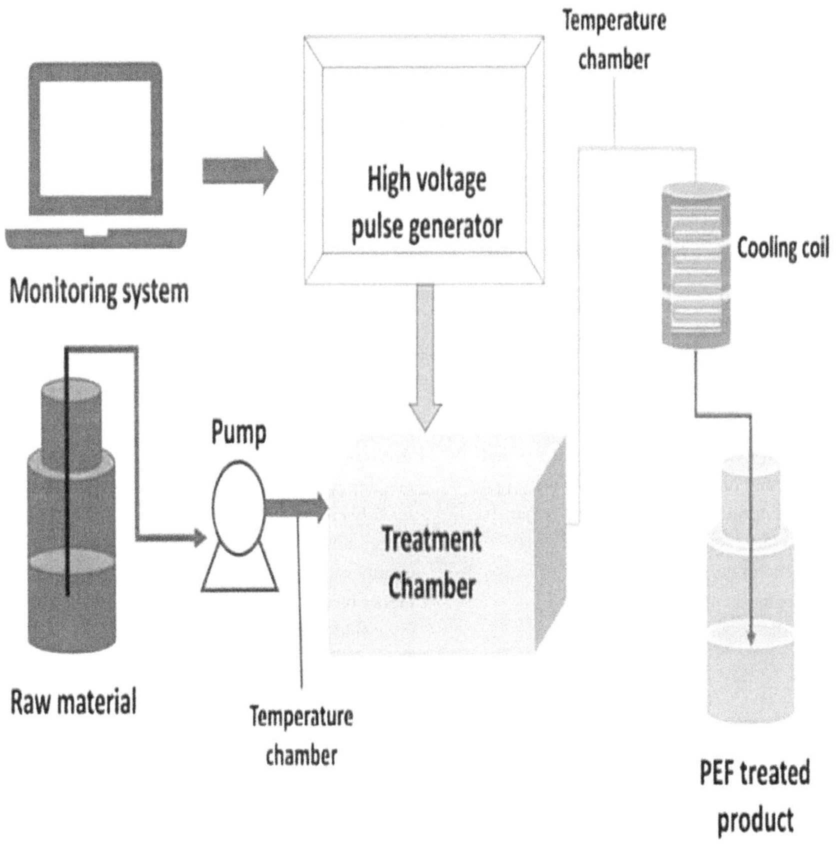

FIGURE 4.6 Schematic model of pulsed electric fields (Taha et al., 2022)

TABLE 4.2
Application of non-thermal technologies in fermentation

Objective	Conditions	Results	References
(a) Ultrasound (US)			
Stimulation of *Penicillium brevicompactum* to produce more of mycophenolic acid during fermentation	Pulsed sonication at 1.5 MHz with a 20% duty (200 ms ultrasound on and 800 ms ultrasound off)	Ultrasound treatment, eight times a day, each 10 min at the 0.2 W/cm^2 intensity, increased mycophenolic acid by 60%. Also results in an increase in biomass accumulation and the number of fungi hypha	Zhao et al. (2012)
Fermentation efficiency of *Lactobacillus casei subsp. casei*	24 kHz; amplitude 20%, 40% and 60%, durations 15, 30, 45 s	Increased lactic acid production, biomass production, cell proliferation, and protein utilization. An increase in cell membrane permeability	Dahroud et al. (2016)
Improvement in wine fermentation rate by *Saccharomyces cerevisiae*	43 kHz with 0.3 mW cm^{-2}	Results in 50%–60% reduction in the wine fermentation time and an increase in cell concentration from 5 g/L to a maximum of 10 g/L	Matsuura et al. (1994)
Improve milk homogenization prior to fermentation	20 kHz; power levels 90, 225, and 450 W; exposure times of 1, 6, and 10 min for homogenization of milk before yogurt fermentation and after inoculation with live yogurt cultures.	Results in better homogenization and 30 min shorter in yogurt end point fermentation. Increases water holding capacity and viscosity of the yogurt from ultrasound homogenized milk with increasing ultrasound amplitude, while reducing syneresis	Wu et al. (2000)
Effects of ultrasound treatment on isoflavone conversion during fermentation of soymilk	30 kHz; 20, 60, and 100 W power input; 1, 2, and 3 min; 10 mL	Results showed an enhanced growth rate and metabolic activity of *Lactobacilli* and *Bifidobacteria* starter cultures under ultrasound treatment, resulting in a notably higher bioconversion of isoflavone glycosides to isoflavone aglycones. These increases ranged from 4.3% to 69.9% compared to the control, dependent on ultrasonication power input and treatment time; i.e., the maximum increase of 69.9% was observed at 100 W power input for 3 min	Yeo and Liong (2011)

(*Continued*)

Table 4.2 (Continued)

Objective	Conditions	Results	References
(b) High-Pressure Processing (HPP)			
Metabolic selectivity and growth of *Clostridium thermocellum* in continuous culture under elevated hydrostatic pressure	Fermentation under elevated hydrostatic pressure (7.0 MPa, 17.3 MPa) was carried out at 333 K in a 100-mL stainless steel Parr mini reactor continuously stirred using an impeller.	Reported a 60-fold increase in ethanol: acetate ratio when sub-lethal pressures were applied to *C. thermocellum* fermentation in comparison to atmospheric pressure. This is due to the increase in the concentration of dissolved gases under HP conditions and the modification of the metabolic pathway leading to the production of ethanol	Bothun et al. (2004)
Reducing the ripening time of cheese via modifications in enzymatic activity or the substrate reactivity	Reggianito cheese was treated at 100 or 400 MPa and 20°C for 5 or 10 min to assess how HPP affects the ripening process at 12°C during 90 days.	HPP treatment did not affect gross composition and pH of the cheeses. The treated samples at 400 MPa for 10 min exhibited a higher plasmin activity compared to others HPP can provide a significant benefit to the cheese industry, such as enhanced accumulation of bioactive peptides and micronutrients in ripened cheeses and faster ripening, consequently lowering production costs of the final product	Costabel et al. (2016)
Effects of HPP on enological and quality properties of young white wine (Sauvignon blanc)	400 MPa, 450 MPa, and 500 MPa for 5 min, 10 min, and 15 min at ambient temperature	The °Brix, pH, tartaric acid, and reducing sugar contents showed no significant changes by HPP treatment, while the alcohol concentration (% v/v) increased at all pressure levels. Treatment at 500 MPa for 10 and 15 min increased turbidity, probably due to coagulation of residual pectin in the wine after HPP treatment	Briones-Labarca et al. (2017)
Application of HPP in yogurt to obtain an additive-free low-fat yogurt with improved performances	300, 400, 500, and 676 MPa for 5 min	Low-fat set yogurt made from milk subjected to a combinations of high hydrostatic pressure and thermal processing (85°C for 30 min) exhibited an improved elastic modulus, yield stress, resistance to normal penetration as well as reduced syneresis, particularly in the intermediate HHP range of 400–500 MPa compared to yogurts from thermally treated or raw milks, while pH and appearance were not affected	Harte et al. (2003)

(Continued)

Table 4.2 (Continued)

Objective	Conditions	Results	References
(c) Pulsed Electric Fields (PEFs)			
Impact of continuous PEF treatment on the growth of the non-Saccharomyces yeast cells *Hanseniaspora* sp.	Application of PEF treatment 72–285 V cm^{-1} during the different fermentation phases	The highest yeast growth attained when PEF intensity of 285 V cm^{-1} was used during Lag and early exponential and exponential phases This treatment reduces the fermentation time and increases biomass concentration	Al Daccache et al. (2020)
Survival rate of yeast cells after electroporation with a succession of bipolar pulses was investigated	Field strength values from 0 to 1.5 kV/cm, with 8 ms total pulse width	Growth of *Saccharomyces cerevisiae* under 0.85 kV cm^{-1} pulses increased by 100%	Fologea et al. (1998)
Application of low electrical field intensities on a recombinant *S. cerevisiae* yeast	Three different electric fields, ranging from 0.5 to 2.0 V cm^{-1}	Reduction of lag phase and an increase in biomass concentration (about 3 times increase)	Castro et al. (2005)
Effect of electroporation on the growth of *Lactobacilli* and *Bifidobacteria*, in mannitol-soymilk	PEF treatment from 2.5-7.5 kV cm^{-1} for 3.0-4.0 ms	Electroporation (PEF treatment) enhances bioactive aglycons yield due to the conversion of glucosides in mannitol-soy milk which is attributed to enhanced intracellular and extracellular b-glucosidase activities of the cells	Yeo and Liong (2013)
Changes in biosynthesis of exopolysaccharide in *Lactococcus lactis* subspecies *cremoris* treated by moderate pulsed electric field treatment	Pulse intensity set at 8 kV/cm, generating 10–12 mA of average electric current due to the conductivity of the ferment, which is approximately 680 mS/m	PEF treatment did not influence the biosynthesis of organic acids, amino acids, and nucleic acids, suggesting the applications potential of PEF in food fermentation process without changing the original taste and flavor of the fermented product	Ohba et al. (2017)

4.8 CLASSIFICATION OF FERMENTED FOOD PRODUCTS

There are numerous fermented food products. They are classified in a number of different ways. They may be grouped based on the microorganisms, the biochemistry, or the product type. The classification of different fermented food products is listed in Table 4.3.

4.9 SOME COMMON FERMENTED PRODUCTS

Fermented foods and their products are prepared and consumed worldwide. These traditional fermented food products are a rich source of various nutrients such as protein, vitamins, carbohydrates, fat, and minerals (Tamang et al., 2012). Fermentation of food not only preserves food by extending its shelf life but also enhances the sensory qualities of the final product (Chand et al.,2016). The rapidly increasing world population necessitates that the amount of food wasted due to spoilage is kept to a minimum. The fermentation process takes an important role in ensuring continuous, diverse, safe food supplies to meet the consumer demands. Some examples of the most common fermented foods that have wide geographical distributions include fermented fruits and vegetables, wine, beer, cider, sake, bread, yogurt, cheese, buttermilk, kefir, vinegar, tempeh, soy sauce, etc. Some of the fermented products are discussed below.

TABLE 4.3
Types of fermented food products

Sl. No	Types	References
1	**Based on Categories**	
	Alcoholic beverages fermented by yeasts, vinegars fermented with *Acetobacter*, milks fermented with *Lactobacilli*, pickles fermented with *Lactobacilli*, fish or meat fermented with *Lactobacilli*, and plant proteins fermented with molds with or without *Lactobacilli* and yeasts.	(Yokotsuka, 1982)
2	**Based on Classes**	
	Beverages, cereal products, dairy products, fish products, fruit and vegetable products, legumes, and meat products.	(Campbell-Patt, 1994)
3	**Based on Commodity**	
	Fermented starchy roots, fermented cereals, alcoholic beverages, fermented vegetable proteins, and fermented animal protein.	(Odunfa, 1988)
4	**Based on Commodity**	
	Cassava based, cereal, legumes, and beverages.	(Kuboye, 1985)
5	**Based on End Product Formed**	
	Lactic acid fermentation, alcohol fermentation, acetic acid fermentation, and butyric acid fermentation.	

4.9.1 Fermentation of Fruits and Vegetables

Fruits and vegetables are easily perishable commodities due to their high water activity. Proper storage methods are necessary for preserving fruits and vegetables. Fermentation of fruits and vegetables is the oldest technology practice for preservation by the action of lactic bacteria. Lactic acid (LA) fermentation of vegetables and fruits is a common practice that is done to preserve the nutritional and sensory features of food items (Karovicova and Kohajdova, 2005). A wide range of vegetables and fruits, including cabbage, olives, cucumbers, onions, peppers, green tomatoes, carrots, okra, celery, and cauliflower, have been preserved. Fermented fruits and vegetables can be utilized as a potential source of probiotics as they harbor a few lactic acid microorganisms, for example, *Lactobacillus plantarum*, *L. pentosus*, *L. brevis*, *L. acidophilus*, *L. fermentum*, *Leuconostoc fallax*, and *L. mesenteroides* (Sajjad et al., 2020). The general process for the fermentation of fruits and vegetables is shown in Figure 4.7. Fermented fruits and vegetables have many benefits;

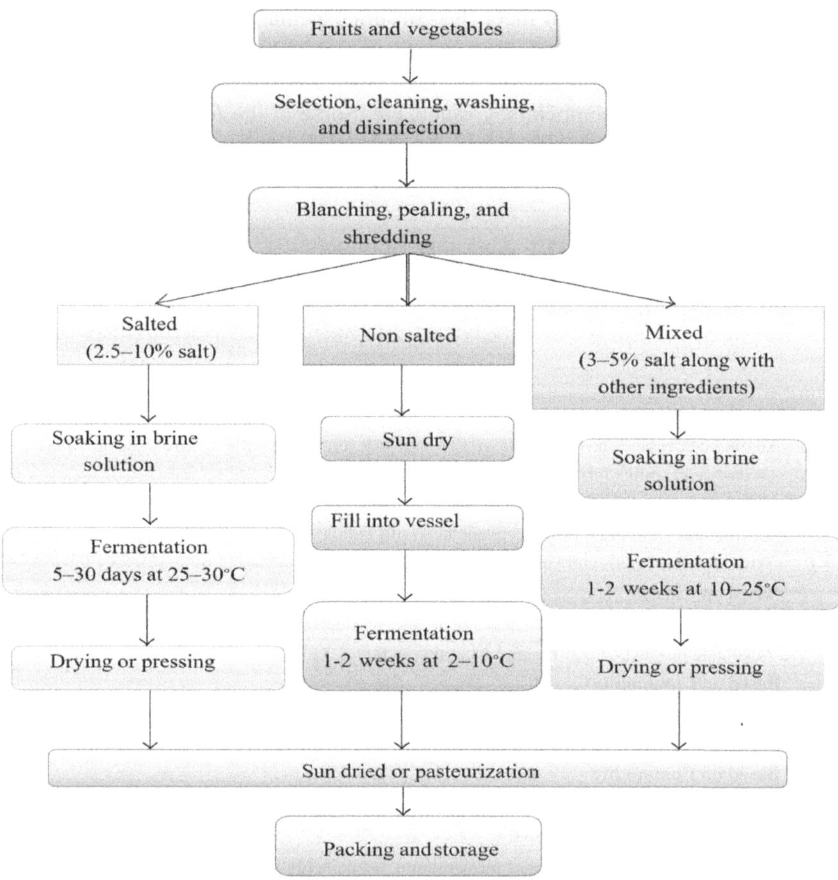

FIGURE 4.7 Overall fermentation process of fruits and vegetables (Swain et al., 2014)

they help in enhancing food quality and safety by increasing the nutritional quality, improving the digestibility and beneficial components, removing antinutrient compounds (oxalate, protease, and α-amylase inhibitors, lectins, condensed tannins, and phytic acid), improving the health benefits of humans, and also acting as biopreservatives (Swain et al., 2014). Some traditional products of fruits and vegetables in India include Khalpi, a fermented cucumber (*Cucumis sativus* L.) product, usually eaten by the Brahmin Nepalis in Sikkim; Sinki, an ethnic fermented radish tap root food, usually eaten in the Sikkim Himalayas; Gundruk, a nonsalted, fermented, and acidic vegetable product traditional to the Himalayas; Inziangsang, a traditional fermented leafy vegetable product prepared from mustard leaves, mostly consumed by the people of northeastern areas like Nagaland and Manipur; similar to gundruk; Goyang, an important fermented vegetable foodstuff consumed in Sikkim and Nepal; and Soidon, a fermented product prepared from the tip of mature bamboo shoots, mainly consumed by the people of Manipur (Yan et al., 2008; Sajjad et al., 2020). As a whole, the traditionally fermented fruits and vegetables not only serve as food supplements but also contribute to health benefits. The overall fermentation process of fruits and vegetables is given in Figure 4.7.

4.9.2 Wine

Wine is an undistilled alcoholic beverage usually having 5%–13% alcohol content. It is generally made from grapes or any other type of fruit with a good proportion of sugar content, such as peaches, plums, or apricots, bananas, elderberries, or black currant, etc. It is often named after the fruits used for making it. Fermentation involves a reaction that converts the sugars in the juice into alcohol and carbon dioxide. Yeasts utilize the sugars during the fermentation period. Table 4.4 gives the characteristics of different types of wine.

4.9.3 Beer

Beer is one of the most consumed alcoholic drinks in the world, which has numerous health benefits if it is consumed in reasonable amounts. It is produced by the brewing and fermentation of starches, mainly derived from cereals, usually malted barley, as well as maize, and flavored with hops and the like for a considerably bitter taste (Aroh, 2019). During the brewing process, fermentation of the starch sugars in the wort produces ethanol and carbonation in the resulting beer. Beer has a final ethanol content of about 3%–8%, and a vast variety of beers exists, such as ales, lagers, stouts, etc.

4.9.4 Yogurt

Yogurt is one of the most popular fermented milk products. It is prepared using either whole or skim milk, where the nonfat milk solids are increased to 12–15% by concentrating the milk or adding powdered skim milk or condensed milk. Their consistency, flavor, and aroma differ from one district to another. Yogurt can be produced in different forms, such as in the form of a highly viscous liquid or softer gel,

TABLE 4.4
Characteristics of different types of wine

Sl.No	Types of Wine	Common Aromas	Alcohol Content (%)	Acidity
1	Barbera (red wine)	Blackberry, plum, raspberry, black cherry	13	High
2	Cabernet Sauvignon (red wine)	Dark berries, cedar wood, cassis	>13.5	High
3	Malbec (red wine)	Black pepper, blackberry, Plum	13–14	Medium
4	Syrah/Shiraz (red wine)	Black pepper, blackberry	12.5–15	Medium to high
5	Zinfandel (red wine)	Blackberry, boysenberry, plum	13.5–17	Medium
6	Chardonnay (white wine)	Tropical fruit, pineapple, apple, lemon, oak	13.5–14.5	High
7	Gewurztraminer (white wine)	Tropical fruit and spice	11.5	Low to medium
8	Muscat/Moscato (white wine)	Peach, pear, citrus	5–7	High
9	Sémillon (white wine)	Ripe fruit, honey, nuts, orange peel	12–14	Medium
10	Viognier (white wine)	Apricot, fruit, floral, spice	13.5–15	Low

Adapted from Swami and Thakor (2014).

or in frozen form as a dessert or drink. It can be classified into the following groups according to the FAO/WHO code and principles (Table 4.5)

4.9.5 Tempeh

Tempeh is a traditional fermented soybean and a staple food in Indonesia, where it serves as a cheap basic foodstuff with high nutrition, and it is consumed by most Indonesian people. It is manufactured with dehulled, soaked, and cooked soybeans inoculated with a mold, normally of the genus *Rhizopus*. The cultured soybeans are bound together by newly grown thick white mycelia to form a cake (Handoyo and Morita, 2006).

4.9.6 Soy Sauce

Traditionally, soy sauce is made from a fermented paste of soybeans, roasted grain, brine, and *Aspergillus oryzae* or *Aspergillus sojae* molds. It is a liquid condiment

TABLE 4.5
Classification of yogurt

Sl. No.	Types	Fat Content (%)
1	Yogurt	>3
2	Partially skimmed yogurt	>0.5 and < 3
3	Skimmed yogurt	>0.5

known for its umami taste. It is a liquid seasoning that can be added directly to food and is used as a dip or salt flavor in cooking. It also promotes digestion by enhancing gastric juice secretion. There are two types of soy sauce, such as light soy sauce (lighter consistency) and dark soy sauce (thicker consistency), based on their fermentation time. Light soy sauce is more commonly used.

4.9.7 Fermented Sausages

A variety of procedures for producing stable, fermented meat sausages have developed around the world. In general, preservation of the meat is achieved by adding salts and the generation of lactic acid by bacteria, which leads to a rapid fall in the pH. *Micrococci*, *Staphylococci*, and yeasts are responsible for the development of color, taste, and flavor during the fermentation. In addition to fermentation, sausage processing may also include curing, smoking, drying, and aging to improve both the flavor and the shelf life.

4.9.8 Bread

Bread is a staple food prepared by the use of yeast to flour and ferment the carbohydrates, leaving behind carbon dioxide, which leavens the bread. The primary reason for their preparation is not for preservation but to improve the digestibility and eat ing appeal of grains; therefore, they have a relatively short shelf life. Bread is served in various forms with any meal of the day and is also an important component of healthy eating.

4.9.9 Vinegar

Vinegar means sour wine, which is named after the French term "vin aigre." It is one of the aged culinary products accidentally found in spoiled wine. It consists of an aqueous solution of acetic acid and trace compounds that may include flavorings. Vinegar usually contains from 5% to 8% acetic acid by volume.

4.9.10 Kefir

Kefir is produced by an acid/alcohol fermentation of pasteurized milk with a mixture of LAB (lactic acid bacteria), yeasts, and other bacteria. The final product is acidic, slightly alcoholic, liquid to semiliquid, and effervescent and is consumed as a beverage. Kefir has high nutritional value and contains a large number of probiotics.

4.9.11 Natto

It is one of the traditional Japanese foods made from soybeans fermented with *Bacillus subtilis* var. natto, having a powerful smell, strong flavor, and slimy texture. It is often served as a breakfast food with rice. Natto is a good source of probiotics, as fermented soybeans contain a lot of beneficial bacterial cultures that aid our digestive health and serve as a natural laxative. They are rich in proteins and vitamins and also promote heart health.

4.9.12 Kimchi

It is a traditional Korean dish consisting of pickled vegetables, which is mainly served as a side dish with every meal but can also be served as a main dish. Kimchi is most commonly made from napa cabbage and Korean radishes, with a wide selection of seasonings including chili powder, scallions, garlic, ginger, and jeotgal. Consumption of kimchi has various health benefits, including improved cardiovascular health and digestive health. The wealth of antioxidants in kimchi exercises healing effects in medical conditions like cancer, diabetes, obesity, atopic dermatitis, and gastric ulcers. The flavonoids and probiotics-rich kimchi help to combat aging, maintain healthy levels of cholesterol, and strengthen the immune system.

4.9.13 Idli

It is a traditional breakfast food in South Indian households, especially in Andhra Pradesh, Karnataka, Tamil Nadu, and Telangana, made from rice cakes. The cakes are made by steaming a batter consisting of fermented black lentils (de-husked) and rice and are usually served with chutneys (coconut based), sambar, and Medu vada. It is also one of the healthy and highly nutritious foods and easily digestible.

4.10 SAFETY MEASURES TO BE CONSIDERED IN FERMENTATION PROCESS

Some potential risks are associated with consuming fermented food products in addition to their positive aspects. Specific processes and safety measures should be followed to benefit from this product. In many parts of the world, especially in Africa and Asia, pathogens such as *enterotoxigenic* and *enterohemorrhagic Escherichia coli*, *Shigella* spp., *Salmonella* spp., *enterotoxigenic Staphylococcus aureus*, *Listeria monocytogenes*, and *Bacillus cereus* have been detected in fermented foods (Skowron et al., 2022). Some of the potential health risks in the fermentation process can be caused by the utilization of poorer quality raw materials, failure in maintaining the proper hygienic conditions while manufacturing, as well as in processing, storage, and distribution. Lack of appropriate safety standards and hygienic conditions while manufacturing these products may cause various illnesses and foodborne diseases. Using contaminated water in the fermentation process can also cause microbial contamination. Hence, it is very crucial to ensure that the mentioned points are avoided. It is essential to carry out the fermentation processes under good manufacturing and hygiene practices with an appropriate design of food processing plants to enjoy the benefits associated with fermented food consumption.

4.11 CURRENT AND FUTURE CHALLENGES IN FOOD PRESERVATION BY THE USE OF FERMENTATION TECHNOLOGY

The challenge of feeding the ever-rising human population cannot be met with an increase in food production alone. Even though advancements in technology lead to the invention of advanced techniques and machinery to increase food productivity, it also requires proper processing methods, storage, and preservation techniques. Therefore, the fermentation process of food is one of the methods of food preservation that improves the safety of foods by decreasing the risks of pathogens and toxins and extends the shelf life by inhibiting the growth of spoilage agents, which cause the sensory changes that make the food unacceptable to the consumer. The fermentation process can be widely used in various industries.

Worldwide, fermentation technology has an enormous scope and plays an exceptionally important part in the human diet due to its highly beneficial nutritional composition. It helps in bringing out the desirable changes in food. The fermentation technology-based products are marketed in different segments on the basis of the type of product, such as alcoholic beverages, pharmaceutical products, industrially important bulk products, food and feed products (Saran et al., 2019). There are various fermented food products. Some of the common fermented food products are fermented fruits and vegetables, wine, beer, yogurt, tempeh, soy sauce, fermented sausages, bread, vinegar, kefir, etc., which are described in detail above. Consumption of these food products gives various health benefits to our body. It also helps us to fight off harmful bacteria in our gut along with boosting our immune and digestive systems. They are also rich in nutrients and vitamins, and enhance energy in our body. Consumption of fermented food products also helps to reduce anxiety, moodiness, and depression. The demand for these fermented food products is rising globally due to their various health benefits, more palatability, and their importance in our diet. The probiotics are actively used in the food industry and have been incorporated in various food products, such as fermented dairy foods, non-dairy products, plant-based products, functional foods, etc. (Gupta et al., 2016).

Though fermentation processes have been used for generations as a requirement for sustainable production of materials and energy, today, it has become more demanding for the continuous formation and advancement of novel fermentation processes. An application of novel fermentation processes holds great potential for expanding the scope of fermentation to create novel foods, improve the sustainability of food production, and improve the quality of human life. Recently, the science of food fermentation has become increasingly popular and also assimilated with many recent non-thermal and thermal technologies. The various novel technologies, both for non-thermal, such as ultrasound, high-pressure processing, pulsed electric field, etc., and thermal technologies, such as microwave heating, radiofrequency heating, ohmic heating, etc., incorporated with the ancient fermentation process are described in detail above. The application of such novel, sophisticated technology helps to manufacture at an industrial scale pharmaceuticals, biochemicals, enzymes, foods, beverages, and food ingredients. Although the potential use of novel heating technologies seems obvious in the fermentation process, their use is still limited due to the lack of necessary theoretical and data-based support. With the increasing emphasis on sustainable processing and efficiency of energy and other natural resource utilization and consumers' demand for novel functional food products, the interest in improving the efficiency of fermentation processes and modulating product characteristics through adjunct technologies will continue. It also has a huge scope for the functional food and nutraceutical global market and has the potential to provide employment to enhance the overall socioeconomic status of the region. Though fermentation is one of the oldest technologies developed by humans, its potential for further improvement is far from limited. Therefore, further exploration and study are still necessary to investigate, which is an interesting and challenging task.

REFERENCES

Al Daccache M., Koubaa M., Maroun R. G., Salameh D., Louka N. and Vorobiev E. 2020. Pulsed electric field-assisted fermentation of *Hanseniaspora* sp. yeast isolated from Lebanese apples. *Food Res. Int.* 129: 108840.

Aroh K. 2019. Review: Beer production. *Electron J.* 1–16.

Awad T., Moharram H., Shaltout O., Asker D. and Youssef M. 2012. Applications of ultrasound in analysis, processing and quality control of food: A review. *Food Res. Int.* 48 (2): 410–427.

Balakrishna A. K., Abdul Wazed M. and Farid M. 2020. A review on the effect of high pressure processing (HPP) on gelatinization and infusion of nutrients. *Molecules.* 25: 2369.

Bhargava N., Mor R. S., Kumar K., Sharanagat V. S. 2021. Advances in application of ultrasound in food processing: A review. *Ultrason. Sonochem.* 70: 105293.

Boreddy S. R., Rose D. J. and Subbiah J. 2019. Radiofrequency-assisted thermal processing of soft wheat flour. *J. Food Sci.* 84: 2528–2536.

Bothun G., Knutson B., Berberich J., Strobel H. and Nokes S. 2004. Metabolic selectivity and growth of *Clostridium thermocellum* in continuous culture under elevated hydrostatic pressure. *Appl. Microbiol. Biotechnol.* 65 (2): 149–157.

Briones-Labarca V., Perez-Wom M., Habib G., Giovagnoli-Vicuña C., Cañas-Sarazua R., Tabilo-Munizaga G. and Salazar F. N. 2017. Oenological and quality characteristic on young white wines (sauvignon blanc): effects of high hydrostatic pressure processing. *J. Food Qual.* 16: 1–12.

Campbell-Platt G. 1994. Fermented foods-a world perspective. *Food Res. Int.* 27 (3): 253–257.

Castro I., Teixeira J. and Vicente A. 2005. The effect of the electric field on lag-phase, ethanol and β-galactosidase production of a recombinant Cerevisiae growing on lactose. Proceedings of the 2nd Mercosur Congress on Chemical Engineering – ENPROMER and 4th Mercosur Congress on Process Systems Engineering; 2005; Rio de Janeiro, pp. 2–4.

Castro I., Teixeira J. A., Salengke S., Sastry S. K. and Vicente A. A. 2004. Ohmic heating of strawberry products: Electrical conductivity measurements and ascorbic acid degradation kinetics. *Innov. Food Sci. Emerg. Technol.* 5: 27–36.
Chand K., Kumar S. and Saxena D. 2016. Fermented foods and their health benefits. *Eur. J. Biotechnol. Biosci.* 4 (9): 27–28.
Chanos P., Warncke M. C., Ehrmann M. A. and Hertel C. 2020. Application of mild pulsed electric fields on starter culture accelerates yogurt fermentation. *Eur. Food Res. Technol.* 246 (3): 621–630.
Cho H. Y., Yousef A. E. and Sastry S. K. 2000. Growth kinetics of *Lactobacillus acidophilus* under ohmic heating. *Biotechnol. Bioeng.* 49(3): 334–340.
Considine K. M., Kelly A. L., Fitzgerald G. F., Hill C. and Sleator R. D. 2008. High-pressure processing–effects on microbial food safety and food quality. *FEMS Microbiol. Lett.* 281 (1): 1–9.
Costabel L. M., Bergamini C., Vaudagna S. R., Cuatrin A. L., Audero G. and Hynes E. 2016. Effect of high-pressure treatment on hard cheese proteolysis. *J. Dairy Sci.* 99 (6): 4220–4232.
Costa N. R., Cappato L. P., Ferreira M. V. S., Pires R. P. S., Moraes J., Esmerino, E. A., … Cruz A. G. 2018. Ohmic Heating: A potential technology for sweet whey processing. *Food Res. Int.* 106: 771–779.
Cunningham F. E. 1978. The effect of brief microwave treatment on number of bacteria in fresh chicken patties. *Int. J. Poult. Sci.* 3 (3): 157–162.
Dahroud B. D., Mokarram R. R., Khiabani M. S., Hamishehkar H., Bialvaei A. Z., Yousefi M., Kafil H. S. 2016. Low intensity ultrasound increases the fermentation efficiency of *Lactobacillus casei subsp. casei* ATTC 39392. *Int. J. Biol. Macromol.* 86: 462–467.
Dębicki P. and Styłba S. 2010. *Environmental Impact of Electromagnetic Fields, Physical, Technical and Legal Aspects.* Bydgoszcz: Text Publishing House.
Fologea D., Vassu-Dimov T., Stoica I., Csutak O. and Radu M. 1998. Increase of Saccharomyces cerevisiae plating efficiency after treatment with bipolar electric pulses. *Bioelectrochem. Bioenerg.* 46 (2): 285–287.
Gholamhosseinpour A., Hashemi S. M. B., Jahromi L. R. and Sourki A. H. 2020. Conventional heating, ultrasound and microwave treatments of milk: Fermentation efficiency and biological activities. *Int. Dairy J.* 110: 104809.
Gupta M., Sharma S. and Sharma S. 2016. 'Probiotics' a new generation functional food-a review. *Biol. Forum – Int J.* 8(2): 363–375.
Halden K., de Alwis A. A. P. and Fryer P. J. 1990. Changes in the electrical conductivity of foods during ohmic heating. *Int J Food Sci Technol.* 25:9–25.
Handoyo T. and Morita N. 2006. Structural and functional properties of fermented soybean (Tempeh) by using *Rhizopus oligosporus. Int. J. Food Prop.* 9(2): 347–355.
Harte F., Luedecke L., Swanson B. and Barbosa-Canovas G. 2003. Low-fat set yogurt made from milk subjected to combinations of high hydrostatic pressure and thermal processing. *J. Dairy Sci.* 86 (4): 1074–1082.
Hunt R. W., Zavalin A., Bhatnagar A., Chinnasamy S. and Das K. C. 2009. Electromagnetic biostimulation of living cultures for biotechnology, biofuel and bioenergy applications. *Int. J. Mol. Sci.* 10 (10): 4515–4558.
Imai T., Uemura K., Ishida N., Yoshizaki S. and Noguchi A. 1995. Ohmic heating of Japanese white radish *Rhapanus sativus*. L. *Int J Food Sci Technol.* 30(4): 461–472.
Kapcsándi V., Neményi M. and Lakatos E. 2013. Effect of microwave treatment of the grape must fermentation process. Food Science Conference 2013- with Research for the Success Darenyi Program, Budapest, pp. 72–75.
Karovicova J. and Kohajdova Z. 2005. Lactic acid fermented vegetable juices-palatable and wholesome foods. *Chem. Pap.* 59: 143–148.

Kiran E. U., Trzcinski A. P., Ng W. J. and Liu Y. 2014. Enzyme production from food wastes using a biorefinery concept. *Waste Biomass Valorization.* 5 (6): 903–917.

Kuboye A. O. 1985. Traditional fermented foods and beverages of Nigeria. In Development of Indigenous Fermented Foods and Food Technology in Africa. Proceedings of the IFS/UNU Workshop, Douala, Cameroon, Oct. International Foundation for Science, Stockholm, Sweden.

Matsuura K., Hirotsune M., Nunokawa Y., Satoh M., Honda K. 1994. Acceleration of cell growth and ester formation by ultrasonic wave irradiation. *J. Ferment. Bioeng.* 77 (1): 36–40.

Melikoglu M. 2012. Solid-state fermentation of wheat pieces by *Aspergillus oryzae*: effects of microwave pretreatment on enzyme production in a biorefinery. *Int. J. Green Energy.* 9: 529–539.

Mohamed M. E. and Eissa A. H. A. 2012. Pulsed electric fields for food processing technology. *Struct. Funct. Food Eng.* 11: 275–306.

Mota M. J., Lopes R. P., Koubaa M., Roohinejad S., Barba F. J., Delgadillo I. and Saraiva J. A. 2018. Fermentation at non-conventional conditions in food-and bio-sciences by the application of advanced processing technologies. *Crit. Rev. Biotechnol.* 38 (1): 122–140.

Odunfa S. A. 1988. African fermented foods: From art to science. *MIRCEN J. Appl. Microbiol. Biotechnol.* 4(3): 259–273.

Ohba T., Uemura K. and Nabetani H. 2017. Changes in biosynthesis of exopolysaccharide in *Lactococcus lactis* subspecies cremoris treated by moderate pulsed electric field treatment. *Biosci. Biotechnol. Biochem.* 81 (4): 724–734.

Puértolas E., López N., Saldaña G., Álvarez I. and Raso J. 2010. Evaluation of phenolic extraction during fermentation of red grapes treated by a continuous pulsed electric fields process at pilot-plant scale. *J. Food Eng.* 98 (1): 120–125.

Rodgers S. 2008. Novel applications of live bacteria in food services: Probiotics and protective cultures. *Trends Food Sci. Technol.* 19 (4): 188–197.

Sajjad N., Rasool A., Fazili A. B. A. and Bhat E. A. 2020. Fermentation of fruits and vegetables. *Plant Archives.* 2 (20): 1338–1342.

Saran S., Malaviya A. and Chaubey A. 2019. Introduction, scope and significance of fermentation technology. In *High value fermentation products (Human health)*, pp. 1–25.

Shil P., Bidaye S. and Vidyasagar P. B. 2008. Analysing the effects of surface distribution of pores in cell electroporation for a cell membrane containing cholesterol. *J. Phys. Appl. Phys.* 41 (5): 055502.

Siefarth C., Tran T. B. T., Mittermaier P. Pfeiffer T. and Buettner A. 2014. Effect of radio frequency heating on yoghurt, I: technological applicability, shelf-life and sensorial quality. *Foods.* 3: 318–335.

Shokri S. and Terefe N. S. 2019. Ultrasonics for modulation of food fermentation processes. In *Elsevier's Food Science Module.*

Skowron K., Budzynska A., Grudlewska-Buda K., Wiktorczyk-Kapischke N., Andrzejewska M., Wałecka-Zacharska E. and Gospodarek-Komkowska E. 2022. Two faces of fermented foods- the benefits and threats of its consumption. *Front. Microbiol.* 13: 1–17.

Steinkraus K. H. 1997. Classification of fermented foods: Worldwide review of household fermentation techniques. *Food Control.* 8 (5/6): 311–317.

Swami S. B. and Thakor N. 2014. Fruit wine production. *J. Food Res. Technol.* 2 (3): 93–100.

Swain M. R., Anandharaj M., Ray R. C. and Rani R. P. 2014. Fermented fruits and vegetables of Asia: A potential source of probiotics. *Biotechnol. Res. Int*: 1–19.

Taha A., Casanova F., Šimonis P., Stankevič V., Gomaa M. A. E. and Stirkė A. 2022. Pulsed electric field: Fundamentals and effects on the structural and techno-functional properties of dairy and plant proteins. *Foods.* 11 (11): 1556. https://doi.org/10.3390/foods11111556

Takhistov P. 2007. Dimensionless analysis of the electric field based food processes for scale-up and validation. *J Food Eng.* 78: 746–754. https://doi.org/10.1016/j.jfoodeng.2005.11.015.

Tamang J. P., Tamang N., Thapa, S., Dewan S., Tamang B., Yonzan H., Rai A. P., Chettri R., Chakrabarty J. and Kharel N. 2012. Micro-organism and nutritive value of ethnic fermented foods and alcoholic beverages of North-East India. *Indian J. Tradit. Knowl.* 11: 7–25.

Terefe N. S. 2015. Emerging trends and opportunities in food fermentation. In *Elsevier's Food Science Module.*

Wang T. H. and Lu S. 2013. Production of xylooligosaccharide from wheat bran by microwave assisted enzymatic hydrolysis. *Food Chem.* 138: 1531–1535.

Wang Z., Huang J., Ma S., Wang X., Sun B., Wang F., Li L. and Bao Q. 2021. Novel heating technologies to improve fermentation efficiency and quality in wheat products: A short review. *Grain Oil Sci. Technol.* 4: 81–87.

Wu H., Hulbert G. J., Mount J. R. 2000. Effects of ultrasound on milk homogenization and fermentation with yogurt starter. *Innov. Food Sci. Emerg. Technol.* 1 (3): 211–218.

Yan P. M., Yan P. M., Xue W. T., Tan S. S., Zhang H. and Chang X. H. 2008. Effect of inoculating lactic acid bacteria starter cultures on the nitrite concentration of fermenting Chinese paocai. *Food Control.* 19 (1): 50–55.

Yen G. C. and Lin H. T. 1996. Comparison of high pressure treatment and thermal pasteurization effects on the quality and shelf life of guava puree. *Int. J. Food Sci. Technol.* 31 (2): 205–213.

Yeo S. K., Liong M. T., 2011. Effect of ultrasound on the growth of probiotics and bioconversion of isoflavones in prebiotic-supplemented soymilk. *J. Agric. Food Chem.* 59 (3): 885–897.

Yeo S. K. and Liong M. T. 2013. Effect of electroporation on viability and bioconversion of isoflavones in mannitol-soymilk fermented by lactobacilli and bifidobacteria. *J. Sci. Food Agric.* 93 (2): 396–409.

Yokotsuka T. 1982. Traditional fermented soybean foods. In: Rose A. H., eds. *Fermented Foods.* Academic Press, 180–188.

Zhang A. Y. Z., Sun Z., Leung C. C. J., Han W., Lau K. Y., Li M. and Lin C. S. K. 2013. Valorisation of bakery waste for succinic acid production. *Green Chem.* 15 (3): 690–695.

Zhao Y., Ang W. T., Xing J., Zhang J. and Chen J. 2012. Applications of ultrasound to enhance mycophenolic acid production. *Ultrasound Med. Biol.* 38 (9): 1582–1588.

5 Application of Membrane Technology in Upstream and Downstream Processing

Reena, Anupama Anand, and Harpreet Kaur Saini

5.1 INTRODUCTION

Membrane filtration is a physical separation technique that separates two phases from one another by rejecting a specific component and allowing other components to pass through it. Principally, a membrane (Latin: *membrana*, a thin layer of skin) is a semipermeable barrier that enables liquid-liquid separation based on differences between particle size and molecular weight of components present in a mixture (Obotey et al., 2020). The agro-food and bulk biotech industries have long used membrane filtration methods, but it wasn't until Sidney and Sourirajan's invention of the phase inversion membrane in the 1960s that these sectors began to experience the success that is the hallmark of modern membrane processes. Membrane separation has a long history, with applications in oil-water separation, metal rejection, biotechnology, pulp and paper, pharmaceuticals, food processing, desalination, and pulp and paper manufacturing (Asad et al., 2020). Although the idea of membrane processes is very straightforward, it is frequently not understood. A membrane's ability to function is influenced by a variety of driving variables, including pressure, temperature, electrical current, and concentration differences, among others. Since membrane technology has significant potential to reduce the energy usage associated with purification and separation, it is determined to be more effective for liquid separation. Nevertheless, these benefits that membranes have embraced also broaden the range of membrane filtration to liquid and gas phase separations (Liang et al., 2019).

Membranes are often categorized according to their composition, average pore size, and structure (Asad et al., 2020). Inorganic as well as organic substances can be used to make membranes. The most popular polymeric membranes for laboratory and commercial use are organic membranes like polysulfone, polyethersulfone, polyvinylidene fluoride, polyacrylonitrile, polytetrafluoroethylene, and polyamide-imide. In contrast, metallic, ceramic, and zeolite membranes are the most often

DOI: 10.1201/9781032650616-5

used inorganic membranes. Despite having better mechanical, chemical, and thermal qualities than polymeric membranes, inorganic membranes made of metals or ceramics are more expensive to produce and are therefore not favored (Lee et al., 2015). Regarding porosity, membranes can generally be divided into two categories: porous and nonporous. The former category is comparable to traditional filtration, with pressure acting as the driving force and size exclusion being the major separation mechanism (Cui et al., 2010). On the other hand, nonporous membranes do not rely on separation based on particle size, but transport through nonporous membranes usually follows the principle of solution diffusion. In this, the membrane surface is nonporous and homogeneous, and segregation is accomplished by variation in solute solubility and/or diffusivity (Scott, 1995). Porous membranes are favored for membrane processes such as microfiltration (MF), ultrafiltration (UF), and membrane distillation, while nonporous membranes find applicability in reverse osmosis (RO), nanofiltration (NF), forward osmosis (FO), pervaporation, and electrodialysis (Asad et al., 2020). According to their structural characteristics, porous membranes are often split into two categories: microporous membranes and asymmetric membranes. Membrane pores found throughout the membrane bodies are what distinguish microporous membranes. All the species above their ratings are intended to be rejected by microporous membranes, whilst those similar in size to the pores prefer to obstruct them. Asymmetric membranes, also called two-layer composite membranes, consist of a top layer that is a selectively loosely bound support layer and a dense active bottom layer. Unlike microporous membranes, which seldom block, asymmetric membranes block rarely (Cui et al., 2010). Composite membranes, integrated asymmetric membranes, as well as supported liquid membranes, are examples of anisotropic membranes, whereas microporous membranes, nonporous dense membranes, and electrically charged membranes are examples of isotropic membranes (Baker, 2000). Considering the separation relies on a physical mechanism, membrane filtration technologies can be employed for a variety of applications and can provide considerable advantages over traditional separation methods like distillation and adsorption. Membrane systems benefit from their large surface area per unit volume, selectivity, and ability to regulate the degree of contact and/or mixing between two phases. They function at very low temperatures and pressures, entail no phase shifts or chemical additives, and process biological molecules in a way that minimizes the degree of denaturation, deactivation, and/or degradation of the resulting products (Carole et al., 2020). The processing of food, beverages, and bio-products, where the processed products may be sensitive to temperature (in contrast to distillation) and solvents (in contrast to extraction), makes membrane separation particularly interesting. Both upstream and downstream processes are incorporating UF and MF to isolate macromolecules and uphold suspended colloids and particles. The concentration of proteins, the interaction of buffer systems, the clarification of suspensions for cell harvesting, and the sterilization of liquids to eradicate viruses and bacteria are just a few of the many applications for UF and MF that have been described (Van Reis and Zydney, 2001; Lee et al., 1995). Membrane bioreactors are another type of membrane process where enzymes, bacteria, or antibodies are suspended in solution and partitioned in a reaction vessel by a membrane

(Charcosset, 2006). Hence, this chapter presents an overview of membrane filtration by focusing more precisely on the application of this technology in upstream and downstream processing.

5.2 TYPES OF MEMBRANE FILTRATION

Membrane processes are a class of separation techniques where the properties of a membrane (porosity, selectivity, electric charge) are employed to separate the constituents of a solution or suspension. The feed stream is divided into two categories in these processes: the portion that penetrates through the membrane, known as permeate, and the portion that contains the components that have not been carried through the membrane, typically known as retentate (Asad et al., 2020). For the classification of membrane separation processes, as indicated in Tables 5.1 and 5.2, criteria are specified by the dimension of the constituent to be separated and driving force (such as pressure, concentration, or electrical potential gradients), which is dictated by the membrane's pore size. Pressure gradients are used to categorize the filtration process into five groups: MF, UF, NF, RO, and FO (Rezakazemi et al., 2018). Transmembrane pressure is essentially needed for the separation process in these approaches. Likewise, concentration-driven membrane processes include pervaporation and dialysis, while membrane distillation (MD) and electrodialysis (ED) are temperature-driven and electrical-driven membrane separation processes, respectively (Bera et al., 2022). All these processes are discussed in detail under the following heads and subheads:

5.2.1 Microfiltration

As the oldest membrane application currently used commercially, microfiltration membranes are second only to dialysis. After nitrocellulose was discovered in 1846,

TABLE 5.1
Classification of membrane filtration techniques based on driving force

Driving Force	Membrane Process
Pressure gradient	Microfiltration
	Ultrafiltration
	Reverse osmosis
	Nanofiltration
	Forward osmosis
Concentration gradient	Pervaporation
	Dialysis
Temperature driven	Micro-distillation
Electrical potential gradient	Electrodialysis

TABLE 5.2
Pore size, molecular weight cut-off, operating pressures, and rejection properties of different membrane filtration techniques

Membrane Filtration Technique	Membrane Material and Properties	Pore Size of Membrane	Molecular Weight Cut-Off	Operating Pressure	Rejects/Removes
Microfiltration	Membranes made up of cellulose acetate, regenerated cellulose, polycarbonate, polyvinylidene fluoride, polypropylene, aluminum, and zirconium oxide, etc.	0.10–5.00 μm	>10,00,000	1–3 atm	Latex, blood, paint, pigment, indigo dye, yeast, bacteria, plant gums, amylopectin
Ultrafiltration	Polymeric membranes made up of polyvinylidene fluoride, polyvinylidene fluoride-polyvinyl pyrrolidone, polysulfone, etc.	0.01–0.10 μm	4000–10,000	2–7 atm	Colloidal silica, virus, enzymes, protein, gelatin, amylose
Reverse osmosis	Flat sheet thin composite membrane made up of an active polyamide layer and a porous polysulfone support layer	0.1–0.6 nm	100	30–50 atm	Atoms, metal ions, fragrance, flavors, salts
Nanofiltration	Polymeric (cellulose acetate, polyamide, polyimide, polysulfone, polyethersulfone) or ceramic (silica-zirconia, titania, alumina, and zirconia) in nature	0.5–2.0 nm	200–500	5–15 atm	Synthetic dye, antibiotics, colorant, amino acids, sugars
Forward osmosis	Thin film composites consisting of a porous support layer and an active layer, made up of cellulose acetate, polyamide thin film composite	0.25–0.30 nm	300–500	Atmospheric pressure (osmotic pressure gradient)	Salts from seawater, large ions from wastewater, inorganic contaminants (Ca, Mg, silica), dissolved carbon

(Continued)

Table 5.2 (Continued)

Membrane Filtration Technique	Membrane Material and Properties	Pore Size of Membrane	Molecular Weight Cut-Off	Operating Pressure	Rejects/Removes
Pervaporation	Hydrophilic membranes made up of polyvinyl alcohol, polyimides, polymaleimides, nafion, and polyacrylamides Hydrophobic membranes made from polydimethylsiloxane or polytrimethylsilylpropyne	0.0001–0.001 μm	1,00,000	–	Water from ethanol-biofuels, alcohol from wines and beers
Dialysis	Made up of cellulose acetate, polymethylmethacrylate, ethylene vinyl alcohol, polyarylate, and polyethersulfone	5–10 nm	1000–3,00,000	–	Ions, labeling reagents, dyes
Micro-distillation	Hydrophobic microporous membranes made from polytetrafluoroethylene, polypropylene, or polyvinylidene fluoride	100 nm–1 μm	–	–	Non-volatile components, heavy metals, radioactive elements, minerals, and salts from water
Electrodialysis	Ion exchange membranes made up of hydrocarbon or partially hydrogenated hydrocarbon and perfluorocarbon polymers	10–100 Å	Does not affect membrane process significantly	–	Minerals, bacterial cells, salt ions, organic matter

Source: Cui et al., 2010; Abdullah et al., 2018; Conidi et al., 2020.

microfiltration was developed. Membrane research and development went on for many years, especially in Germany. Frick published the first description of cellulose nitrate MF membranes in 1855. The method was initially used in practice to cultivate microorganisms in drinking water. But even up until the middle of the 1960s, MF was only used in modest industrial applications (Anis et al., 2019). Pleated membrane cartridges were introduced in the 1970s, which encouraged widespread industrial applications of MF in the water, microelectronics, and pharmaceutical sectors (Baker, 2004). Numerous techniques for membrane modification have been identified, including the introduction of different filler compounds to the polymer casting solution, membrane post-treatment techniques such as cross-linking, and surface modification by the adsorption of active substances. There are spiral-wound and flat-sheet versions of MF membranes available. The membrane modules and filtration systems can also be altered to meet certain application requirements. This creates a wide range of opportunities for customized membrane systems with various uses (Anis et al., 2019).

In accordance with the concept of physical separation, MF is capable of eliminating micrometer-sized debris, such as suspended particles, significant pathogens, some bacteria, proteins, and yeast cells. The MF membrane process is a flexible one since it can reject a variety of large-scale pollutants. The pore sizes of MF membranes range from 0.1 to 5 μm and the membrane operates at pressures between 1 and 3 atm. A somewhat open membrane structure often separates particles with a diameter greater than 0.1 μm. These membranes necessitate the use of low hydrostatic pressures for optimal contaminant elimination and solvent flux since the hydrodynamic resistance is minimal (Anis et al., 2019). However, due to the vast range of pore sizes, MF can be used in a variety of industries, such as pharmaceuticals, food, and wastewater treatment (Hakami et al., 2020), microbiology (Charcosset, 2003), food (Almandoz et al., 2015), biotechnology (Davis, 2019), and pharmaceuticals (Davis, 2019). Further, to lessen the fouling, it is also used as a pretreatment for UF and a post-treatment for granular media filtration. Membrane components in MF might be either organic (polyvinylidene fluoride, polyamide, polysulfone, cellulose acetate, and so forth) or inorganic (porous alumina and metals). It is appropriate for isolating suspensions and emulsions and can hold up to 40% of organic material. While water, monovalent ions like $Na+$, Cl^-, dissolved or organic materials, tiny colloids, and viruses might pass through the filter, these filters mostly remove silt, algae, protozoa, and large bacteria (Anis et al., 2019).

5.2.2 Ultrafiltration

Ultrafiltration (UF) is a cutting-edge separation technology used in many different industry sectors. In the late 1960s, it was first developed as a fractionation method. Since then, UF membranes have undergone constant advancement, and their uses have spanned a wide range of industries, including chemical recovery, cell harvesting, dairy technology, medical use, wastewater restoration, water filtration, and juice concentration (Lee et al., 2017; Tarifa et al., 2018). These membranes are frequently employed as a pretreatment step to remove colloids, harmful microbes, macromolecule, and naturally occurring organic debris. They have pore sizes varying from 0.01 μm to 0.1 μm and significantly higher working pressures (from 2 to 7 atm) (Bera et al., 2022). However, some dissolved inorganic pollutants cannot be removed from water using this method, and it requires periodic cleaning to maintain a healthy pressure in the water stream. These membranes hold onto silica, endotoxins, proteins, and viruses. Also, in order to preserve RO membranes, UF is generally used as a pretreatment step for RO. It has a broad spectrum of applications but is best recognized as a clarifying and disinfection separation process. Nevertheless, the most significant obstacle to the widespread use of UF is the buildup of contaminants (physical, chemical, and biological elements) on or inside the membrane matrices, which makes the UF membrane susceptible to losing permeability when filtering source fluids. Under conditions of constant driving force (the trans-membrane pressure (TMP)), the membrane flux decreases over time below that of the equivalent pure solvent flow as a result of two significant conditions: concentration polarization and membrane fouling (Shi et al., 2014).

5.2.3 Concentration Polarization

Ultrafiltration is a state-of-the-art separation process utilized in diverse industries. In the late 1960s, it was first developed as a fractionation method. Since then, UF membranes have undergone constant advancement, and their uses have spanned a wide range of industries, including chemical recovery, cell harvesting, dairy technology, medical use, wastewater restoration, water filtration, and juice concentration (Lee et al., 2017; Tarifa et al., 2018). These membranes are frequently employed as a pretreatment step to remove colloids, harmful microbes, macromolecules, and naturally occurring organic debris. They have pore sizes varying from 0.01 μm to 0.1 μm and significantly higher working pressures (from 2 to 7 atm) (Bera et al., 2022). However, some dissolved inorganic pollutants cannot be removed from water using this method, and it requires periodic cleaning to maintain a healthy pressure in the water stream. These membranes hold onto silica, endotoxins, proteins, and viruses. Also, in order to preserve RO membranes, UF is generally used as a pretreatment step for RO. It has a broad spectrum of applications but is best recognized as a clarifying and disinfection separation process. Nevertheless, the most significant obstacle to the widespread use of UF is the buildup of contaminants (physical, chemical, and biological elements) on or inside the membrane matrices which makes the UF membrane susceptible to losing permeability when filtering source fluids. Under conditions of constant driving force (the trans-membrane pressure (TMP)), the membrane flux decreases over time below that of the equivalent pure solvent flow as a result of two significant conditions: concentration polarization and membrane fouling (Shi et al., 2014).

5.2.4 Membrane Fouling

When material in the feed solution exits the liquid phase and forms an accumulation on the membrane interface or within its porous structure, membrane fouling occurs. In UF, fouling is the accumulation of material on top of the membrane (external fouling) or inside its pore structure (internal fouling). Fouling may result in an irreversible decrease in a membrane's permeability, as opposed to the reversible nature of concentration polarization. Reversibility is an essential attribute of fouling. Based on their relative difficulty in cleaning, many studies differentiate between reversible and irreversible fouling. Reversible fouling is the kind that may be easily removed using specific cleaning techniques, but irreversible fouling does not disappear after cleaning (Shi et al., 2014). Membrane fouling causes a significant flux drop, a rise in operating and maintenance expenses for the plant, the requirement for membrane cleaning, a reduction in recoveries and feed water loss, and reduced membrane durability (Mohammad et al., 2012).

5.2.5 Nanofiltration

Since its initial recognition in the late 1980s, NF membranes have advanced significantly. NF membranes, which have characteristics halfway between RO and UF, have pores that are typically 1 nm in size, which is equivalent to a molecular weight cut-off (MWCO) of 300–500 Da. Due to the disintegration of surface functional groups or the adsorption of a charged solute, NF membranes in contact with solution are also mildly charged. For instance, ionizable groups in polymeric NF membranes, such as carboxylic and sulfonic acid groups, lead to charged surfaces when a feed solution is present. NF is a very intricate process that depends on micro-hydrodynamic and interfacial activities taking place at the membrane surface and inside its pores (Mohammad et al., 2015). NF membrane rejection may be a result of a confluence of steric, Donnan, dielectric, and transport processes. The typical Donnan process illustrates the equilibrium and membrane potential interactions between a charged group and the surface of the charged membrane. The membrane charge comes from the severance of ionizable groups at the membrane interface and from the inside of the membrane pore constitution. Depending on the particular materials employed during the production process, these groups may have an acidic, basic, or even a combination of both natures (Ernst et al., 2000). Along with having ionizable surface groups, NF membranes also have a weak capability for ion exchange, and occasionally, ions from the interacting solution can adhere to the membrane surface, slightly altering the membrane charge. Due to the aforementioned events, electrostatic repulsion or attraction occurs depending on the ion valence and fixed charge of the membrane, which may change depending on the localized ionic environment. There are two primary opposing theories regarding the precise nature of the relationship for the phenomenon of dielectric exclusion, which is still poorly understood. These are the "solvation energy barrier" process and the "image forces" phenomenon. These exclusion strategies are essentially charge-based exclusion phenomena and derive from the severe spatial containment and nano-length scales that occur in NF (Bowen et al., 2002; Oatley et al., 2012).

5.2.6 Reverse Osmosis

Osmosis is a natural process that occurs when water passes through a semipermeable membrane from one side of the membrane where there is a lower concentration of solutes to the other without the application of external force, creating a difference in osmotic pressure across the membrane. In the case of reverse osmosis, pressure greater than the osmotic pressure of the feed, usually around 30 to 50 atm, is applied across the membrane, which reverses the direction of water flow and causes separation of water from the solution. This is the basic principle of the working of RO membranes. The membranes currently employed in commercial RO construction are asymmetric, flat-sheet membranes made of cellulose acetate or cellulose triacetate and thin film composites, which consist of an incredibly thin layer of an extremely water-friendly polymer that has been applied to a microporous support, typically composed of polysulfone (Van der Bruggen, 2018). These membranes are equipped with very small pores of size 0.1–0.6 nm, which render them highly susceptible to membrane fouling and concentration polarization. These conditions result from phenomena such as bacterial growth, chemical interactions between solutes and membrane material, pore obstruction, and adsorption of solution molecules (Goosen et al., 2005). RO has been deemed most effective in the desalination of water by removing dissolved inorganic and minute organic substances. It was used for the first time in 1964 in Coalinga, California, USA, even though the method at the time was far from practical because of the extraordinarily high pressures that had to be applied and the meager water output. The performance of reverse osmosis in desalination, however, was significantly enhanced by the use of thin-film composite membranes, which have an extremely thin top layer created through interfacial polymerization. Reverse osmosis emerged as the most cost-effective desalination technique by the middle of the 1990s (Van Houte et al., 2013). Additionally, reverse osmosis has established itself as a routine procedure for ensuring the supply of ultrapure water and for water recycling from wastewater, either directly or indirectly by the infiltration of penetrates aquifers for later use as drinking water.

5.2.7 Forward Osmosis

FO uses the osmotic pressure differential generated across a semipermeable membrane, differentiating two feeds of differing concentrations to induce the permeation of water. An osmotic pressure difference will be created when two solutions with different concentrations are placed on opposite sides of a membrane, driving water to pass through the membrane from the region of low concentration (the feed) to high concentration (the draw). The solute movement is controlled by the chemical potential gradient, whereas the water transport is driven by the osmotic pressure differential. FO is a recommended energy-efficient method for desalinating saltwater with a wide range of input salinities (Nagy, 2019). Compared to other well-known pressure-based technologies, such as RO, this energy-efficient desalination method offers several advantages. FO employs natural osmosis, diluting the seawater feed stream with a draw solution at a greater pressure than the salt water so that water may pass through a semipermeable membrane. Additionally, it has been said that the lack of applied hydraulic pressure results in increased fouling reversibility and reduced fouling of FO membranes. Due to its lesser susceptibility to fouling, FO can recover clean water from difficult and highly fouling sources without requiring considerable preparation. Additionally, the reduced susceptibility to fouling makes cleaning simpler and less frequent, extending the membrane's durability (Van der Bruggen, 2018). Figure 5.1 shows the basic schematic of the working of pressure-driven membrane separation techniques.

5.2.8 Pervaporation

Kober used the phrase "pervaporation" in 1917 to describe the partial vaporization of the feed mixture. PV is a membrane filtration procedure that involves passing a liquid mixture across a substantial barrier that is typically held under a vacuum while only partially vaporizing it. This causes the permeating component to accumulate as vapor, which may be eliminated by either exerting low pressure on the permeant side or passing an inert liquid. The gradient in phase concentration on the opposing sides of the interfacial barrier corresponds to the disparity in chemical potential that drives the PV process. The principle of working of pervaporation is shown in Figure 5.2. The solution diffusion model commonly describes mass transport across PV membranes and takes three subsequent stages.

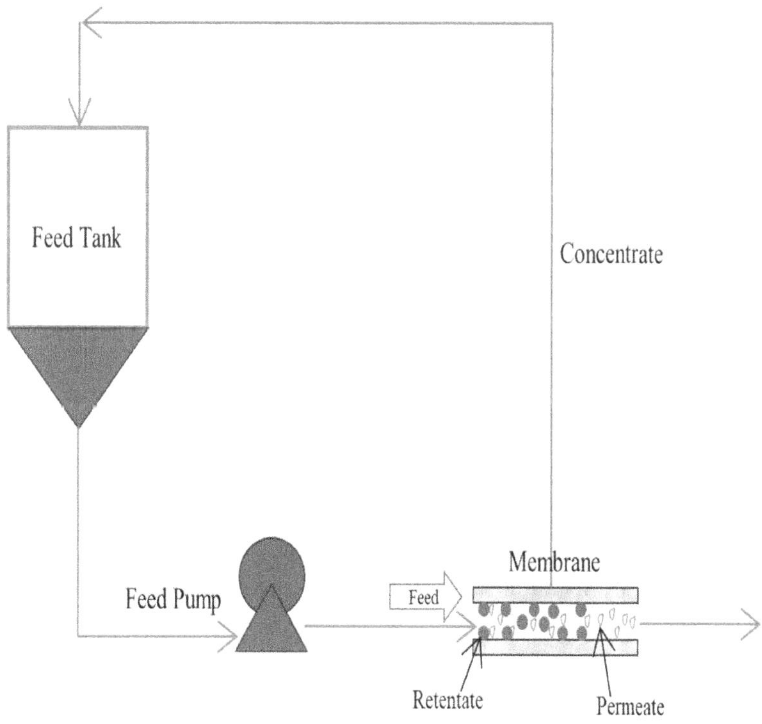

FIGURE 5.1 Basic schematic of pressure-driven process

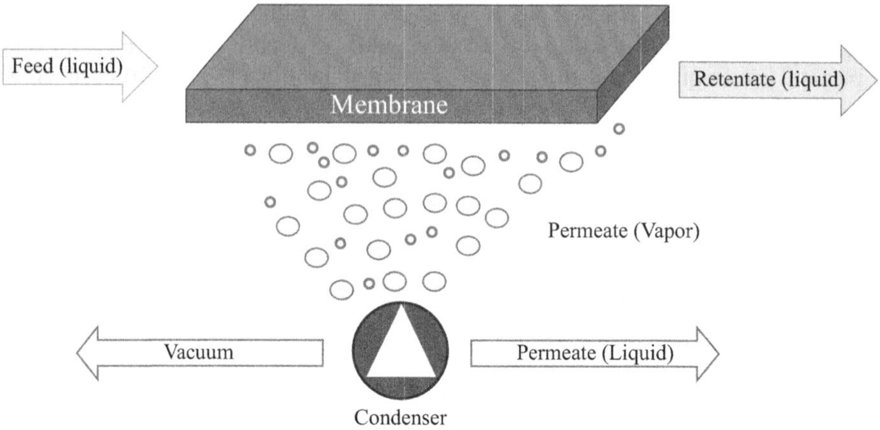

FIGURE 5.2 Principle of working of pervaporation

I. Partial dissolution (sorption) of the feed molecules in the membrane in contact with the liquid mixture. Additionally, it specifies the number of absorbed species present at the membrane's upstream interface, hence defining the chemical potential gradients for mass transfer.
II. Chemical potential gradient-based diffusion of components through the PV membrane.
III. Desorption of penetrated species at the downstream side of the membrane.

As PV results in significant organic fluxes and is very selective for hydrophobic species, these membranes find an application in the dehydration of aqueous organic mixtures, separation of solvent mixtures, and removal of volatile organic compounds from water (Smitha et al., 2004; Jonquières et al., 2013).

5.2.9 Membrane Distillation

Membrane distillation (MD) is a temperature-based filtration method that employs a microporous hydrophobic membrane connected with a condensing component (permeate or distillate) on one side and an aqueous heated solution (feed or retentate) on the other (Figure 5.3). This method falls under the category of membrane contactors, where a non-wetting membrane encourages mass and energy transfer between two opposing interfaces by following the laws of phase equilibrium rather than acting as a traditional barrier or filter.

Vapor pressure differences between the porous hydrophobic membrane surfaces are the major driving force for this separation operation. The feed solution need not be heated to the boiling point when using MD, which has numerous appealing properties, including reduced operating temperatures compared to those observed in conventional processes. Furthermore, hydrostatic pressure in MD is lower than

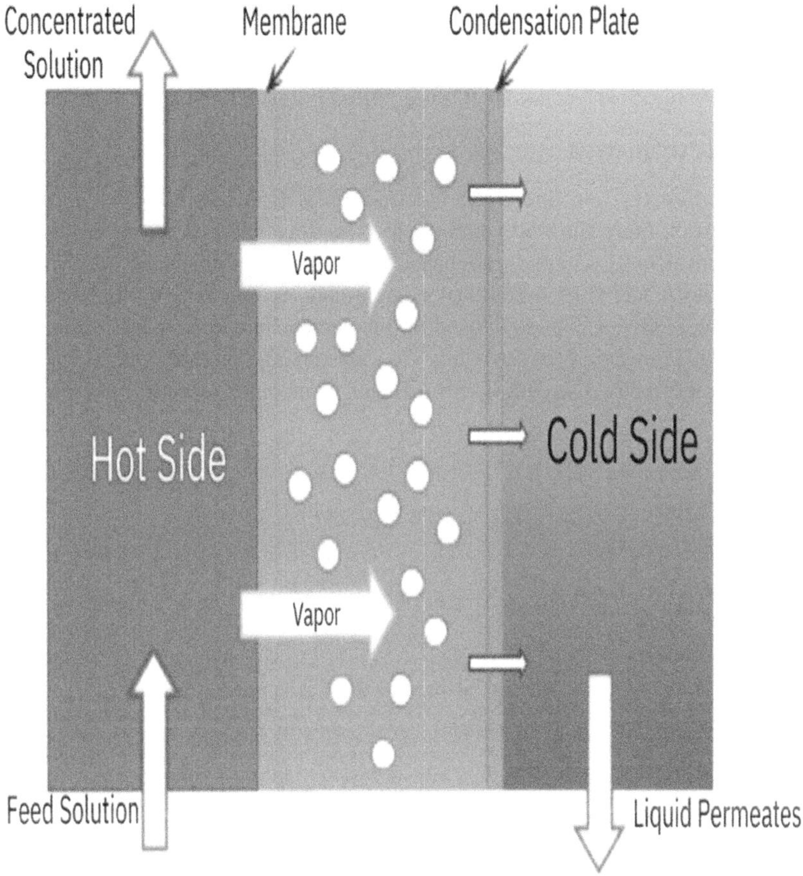

FIGURE 5.3 Principle of membrane distillation

that in pressure-driven membrane processes like RO. As a result, MD is anticipated to be a process with low operating costs and low demands on membrane properties (Alkhudhiri et al., 2012). Three main mechanisms, namely Knudsen diffusion, Poiseuille flow (viscous flow), and molecular diffusion, regulate the mass transport in MD. This results in a variety of forms of mass transfer resistance, including viscous resistance, molecular resistance, and resistance caused by the membrane itself (Knudsen resistance), which is caused by molecules colliding with one another. Low surface energy hydrophobic polymers with intrinsic or modified properties are used for the construction of these membranes. In the early stages of MD research, materials like silicone-coated glass fibers and nylon were examined, but their wetting resistance was found unsatisfactory. Polyvinylidene fluoride (PVDF), polypropylene (PP), and polytetrafluoroethylene (PTFE) continue to be the most widely employed polymers in MD membranes. These membranes have been successfully utilized for the removal of organic pollutants like ammonia, trichloroethane, etc., for the desalination of seawater to produce fresh water and also for the removal of some heavy metals (Wang et al., 2015).

5.2.10 Electrodialysis

Maigrot and Sabates introduced the electrodialysis (ED) desalination principle in 1890, and it took more than 50 years for it to be developed on a commercial level. The invention of ion exchange membranes, which yield significant water recovery and require no phase shift, reaction, or chemicals, is what drives the functioning of ED. A direct current voltage is used in the electrochemical separation procedure known as ED (Figure 5.4) to move ions via ion exchange membranes. More specifically, ED generates two streams with differing concentrations running in parallel

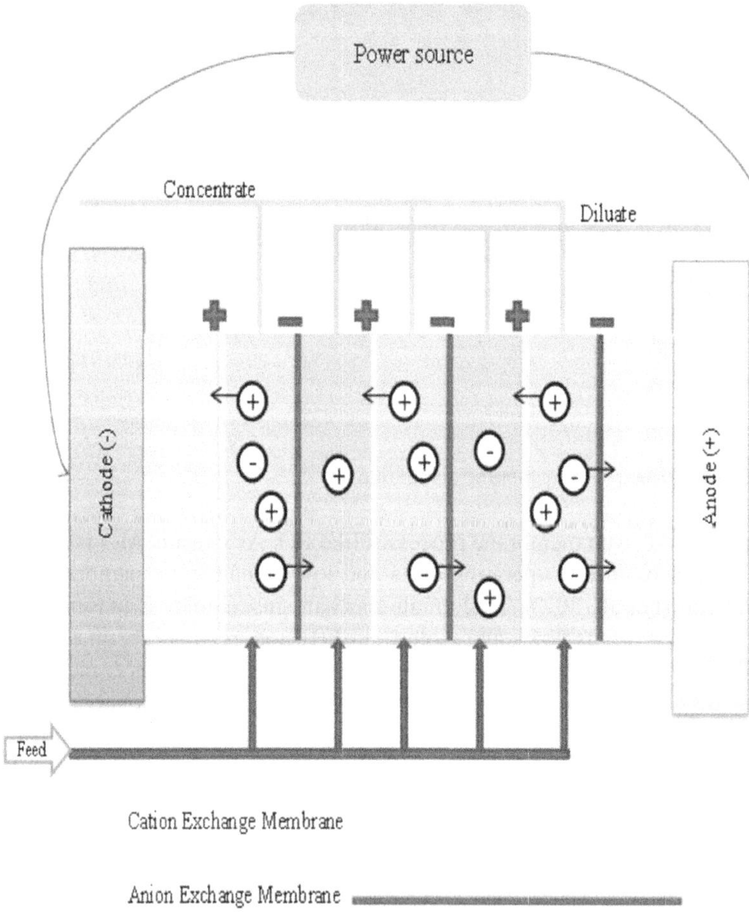

FIGURE 5.4 Principle of working of electrodialysis

compartments partitioned alternately by cation-exchange and anion-exchange membranes. Using a driving force, the process transfers positively and negatively charged ions from the origin water through cathodes and anodes to a concentrated wastewater stream, resulting in a more diluted stream (Gurreri et al., 2020).

These membranes are created as thin, foil-like structures with ion exchange groups attached by a polymer matrix. Membranes are strengthened with synthetic fiber so that they are impervious to water under pressure and to increase their mechanical vigor. The two distinct kinds of ion exchange membranes utilized in ED include cation-exchange membranes, which are electrically conductive membranes that only let positively charged ions flow through, and anion-exchange membranes, which are electrically conductive membranes that allow the flow of ions carrying a negative charge. Both these membranes exhibit reduced resistance to electricity, improved resistance to osmotic swelling, a change in pH and fouling, operate at temperatures greater than 46°C and have greater durability. The capacity of the membranes to distinguish between various ions in order to permit passage or permeation through the membrane is known as perm selectivity (or ion selection). In this respect, membranes may be designed to prevent the passage of divalent cations or anions like calcium, magnesium, or sulfates. At the industrial level, ED is mostly used to desalinate brackish water for the manufacture of drinking water. Additionally, certain ED facilities that use saltwater desalination to make table salt have been established. The employment of ED approaches in the biochemistry, food processing, and pharmaceutical sectors has, however, received a great deal of attention. This includes wastewater treatment, the recovery of chemicals or other useful products, and the elimination of harmful components (Valero et al., 2011).

5.3 UPSTREAM TECHNOLOGY

Upstream technology mainly refers to the initial stages or inputs before processing or preparation of raw material before manufacturing the product in food industries. While in microbiology and biotechnological applications, upstream generally refers to the selection of specific microbial strains for specified products having a desired commercial value and to create an environment for cell isolations to serve various purposes such as therapeutic proteins, antibiotics, hormones, enzymes, blood substitutes, amino acids, etc., respectively. The technology followed during upstream processing is of utmost importance as the initial stages of any manufacturing product directly influence the final quality of the product. Membrane filtration plays a significant role in various operations, such as isolation of proteins, keeping, recovering, and concentrating macro/micromolecules and suspended particles. Ultrafiltration or microfiltration is often used to remove impurities (such as bacteria and viruses); concentration of the feed via ultrafiltration to reduce overall volume; concentration by ultrafiltration to achieve desired product quality and buffer exchange via diafiltration to distribute the product in an appropriate buffer system are typical membrane filtration steps needed for upstream processes for a product (Charcosset, 2006; Asif et al., 2019) The current section highlights the importance of different membrane filtration techniques applied during upstream processing in various areas of industries.

5.4 APPLICATIONS OF MEMBRANE TECHNOLOGY IN UPSTREAM TECHNOLOGY

Membrane technologies such as ultrafiltration, microfiltration, nanofiltration, membrane distillation, etc. have been used significantly in various industrial uses for a long time. In order to produce molecules, emulsions, and particles, membrane processes are increasingly being used in reaction, clarification, and recovery operations. Membrane systems are advantageous because of their high surface-area-per-unit-volume, selectivity, and ability to regulate the level of contact and/or mixing between two phases. They operate at very low temperatures and pressures, lack changes in phase or chemical additives, and are particularly well adapted to processing biological molecules. This reduces the amount of denaturation, deactivation, and/or degradation of biological products (Baker, 2004).

Ultrafiltration and microfiltration, which are often used to recover macromolecules and retain suspended colloids and particles, are being used in upstream technology. Applications for ultrafiltration and microfiltration include sterilizing liquids in order to remove viruses and bacteria, exchanging buffer systems, clarifying suspensions for cell harvesting, and concentrating proteins. Membrane bioreactors are another type of membrane process in which enzymes, bacteria, or antibodies are immobilized within the membrane matrix itself or suspended in solution and compartmentalized by a membrane in a reaction vessel. Moreover, for a wide range of chromatographic purification techniques, including ion-exchange, hydrophobic, reversed-phase, and affinity chromatography, membrane chromatography is also being employed as an alternative to traditional resin-based chromatography columns in various initial steps of any field that requires separation (Baker, 2000; Asad et al., 2020) Finally, membrane contactors are being used to prepare emulsions and other sorts of particles, such as w/o emulsions, o/w emulsions, and polymeric particles, by applying pressure to drive the dispersed phase to permeate through a membrane into the continuous phase. A description of membrane filtration used in different aspects of industries is depicted in Figure 5.5.

5.5 APPLICATION IN BIOTECHNOLOGY

Membrane filtration is very well suited for both upstream and downstream processing in the field of biotechnology. The most common applications of microfiltration are in the first clarification of fermentation broths in order to remove the suspended cell mass and other particle debris, and in sterile filtration (bacterial elimination) prior to the final formation of numerous products. In order to completely remove *Brevundimonas diminuta*, "sterile filtering" is carried out in a dead-end arrangement employing 0.2 m pore size membranes (Kuriyel and Zydney, 2000). Moreover, microfiltration has widely contributed to virus removal from cell cultures in order to separate viruses having sizes ranging from 12 to 300 nm from proteins of 4 to 12 nm. Further, recombinant proteins with medicinal,

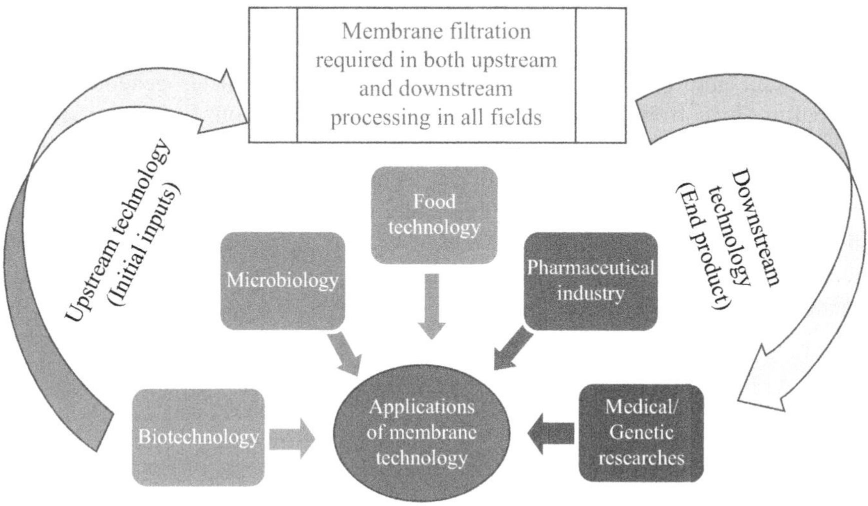

FIGURE 5.5 Application of membrane technology in various fields

preventive, or diagnostic uses are commonly produced using mammalian cell lines. Cell lines, however, can become infected with viruses or particles that resemble viruses. In addition to external pollutants, the addition of supplements and other ingredients into the fermentation process, as well as handling and other manipulations during processing, might introduce viruses. For the separation of viruses and proteins, membrane manufacturers are creating membrane filters with higher resolution. Because there could be instances of parvovirus contamination of cell cultures, which is crucial for the biotechnology sector, this is because they are both small (approximately 20 nm in diameter) and extremely resistant to various heat and chemical inactivation techniques; parvoviruses are exceptionally challenging to eradicate.

Membrane filtration techniques are also used in the production process of antibiotics. In particular, they are used in the initial clarification of fermentation broths in order to obtain a very high yield using a combined filtration and diafiltration process. If the nominal membrane molecular cut-off is higher than 20,000 g/mol, it is possible to achieve complete retention of the cells and particulate matter using membranes with pore sizes up to 0.2–0.45 nm and essentially complete passage of the antibiotics (Zeman and Zydney, 1996). In order to prevent emulsification and boost extraction effectiveness, emulsifiers in antibiotic broths may also be removed using ultrafiltration prior to solvent extraction (Li et al., 2004). The work of Nabais and Cardoso (1999) on the purification of benzyl penicillin filtered broths derived from fermented broths by ultrafiltration with diafiltration serves as a case study in this area.

Various other studies that highlight the importance and application of membrane filtration in biotechnology are the concentration and purification of recombinant Brain-Derived Neurotrophic Factor (rBDNF) inclusion bodies from *E. coli* cell suspensions by cross-flow microfiltration and diafiltration (Schutyser et al., 2002), the recovery of naturally glycosylated therapeutic proteins produced from animal cell cultures by microfiltration, and the recovery of heterogeneous immune globulins (IgG) from transgenic goat milk by nanofiltration. In another study by Baruah and Belfort (2005), membrane filtration was used in the upstream processing step of harvesting mammalian cells using the nanofiltration method to separate bacteria, yeast, mycoplasma, and endotoxins from the matrix. Hence, these filtration techniques play a crucial role in various upstream processes in the biotechnology field and can be further improved by better research and combination techniques for future times.

5.6 APPLICATION IN FOOD TECHNOLOGY

In various areas of food technology, separation of components is considered one of the most important initial steps that are followed in every process of formation wherein membrane filtration has its own significance in infinite manners. It has immense application in the dairy industry, enzyme manufacturing, fermentation, food and beverage industry, fisheries, and other areas (Tranchant C C and Selvamuthukumaran M, 2020). Utilization of these membrane processes in upstream stages not only helps in the separation of unwanted particles but also ensures better shelf life, safety, and quality of the developed product for downstream stages. In the dairy industry, these techniques are used for the separation of whey concentrates from milk, concentration of protein, removal of calcium (decalcification), protein fractionation, and extension of shelf life of the products by reducing microbial contamination (Darnon et al., 2003).

Microfiltration helps in the physical removal of bacteria, cells, and impurities in milk while it can also avoid the addition of natural inhibitors such as nitrates for achieving preservative-free cheese and whey. Moreover, it can be used in the concentration of whey protein and milk protein in the initial steps in order to prepare good quality products from the obtained masses. The amount of fat present is an important factor in the synthesis of protein isolate, such as milk protein isolate (MPI) or whey protein isolate (WPI), where a protein level of more than 90% in total solids is required. The final protein concentration must be achieved by removing the milk fat because the leftover fat has been concentrated to an extremely high degree, which can only be effectively removed by microfiltration (Casper et al., 1999). Further, ultrafiltration is commonly used in milk protein concentration and whey protein concentration. In some cases, dilution of water is done initially and then filtration is carried out, which is known as diafiltration, and is also being performed lately. Moreover, decalcification of pre-concentrated permeates is filtered by using reverse osmosis and nanofiltration for the collection of high-quality lactose from milk, which can be further processed downstream by evaporation and crystallization to produce lactose-free milk. In addition to this, nanofiltration and electrodialysis can also be used in the reduction of minerals like sodium and potassium chlorides (monovalent ions) prior to further processes such as concentration (Meyer and Kulozik, 2016). Furthermore, prior to the evaporation of milk or whey, water can be effectively removed using reverse osmosis. Hence, a reverse osmosis plant installed upstream of an existing evaporator can significantly improve the evaporator's capacity. Other than the dairy industry, protein separation has also been done by using advanced ultrafiltration techniques, also known as charged UF membrane separation, in which both size and charge-based separation of particles can be done at upstream stages. Other than this, pulse electric field ultrafiltration, nanofiltration, and UF in the presence of ultrasonic waves have also been found effective in the separation of protein from food matrices (O'Connell and Fox, 2000; Saxena et al., 2009).

Further, the wastewater produced during the manufacture of fishmeal contains a significant amount of potentially useful proteins. These proteins can be concentrated using UF technology and recycled into the production of fishmeal, boosting both the product's quality and its economic value from the raw material. Meanwhile, the treated water can either be released into the ocean or used again in the plant (Afonso et al., 2003). The recovery of proteins from fishing effluents by membrane separation methods has been researched by several authors. Despite making up only 1.5% of the overall wastewater, Lin et al. (1995) noted that the wastewater from the first washing in the manufacturing of surimi has the highest percentages of protein, nonprotein nitrogen, fat, and ash as well as a distinct fishy odor. The main uses of membranes in the field of fruit juices are deacidification by electrodialysis and clarification by MF or UF. Due to their efficiency and productivity, membrane separation techniques have been adopted by the apple juice industry as long-term investments. Further, membrane filtration enables the separation of pulp (retentate) and serum (permeate) in pulpy fruits and citrus (orange, lemon, grapefruit, tangerine) at upstream stages for the production of high-quality final products like ready-to-serve drinks, squashes, etc. The industrial UF of lemon juice, on the other hand, fills a growing market for highly concentrated liquids that may be used as an acidulous, bitter, or astringent ingredient in beverages, sauces, or household cleaning (Daufin et al., 2001).

Various combination techniques have also been used for high efficiency in food processing stages. Morin et al. (2006) used renetting and acid coagulation of buttermilk to remove caseins prior to fractionation by a mix of MF and UF to solve this issue. According to the authors, this method recovered 70–77% of the total phospholipids in buttermilk. Citrate was used to break up the casein micelles, and either UF or MF was then applied to polysulfone membranes. The findings of Corredig et al. (2003) suggest that this method appears to be effective for concentrating milk fat globule membrane (MFGM) proteins in the retentate. Additionally, new techniques combine the MF and RO membranes to produce fruit juice concentrates with organoleptic qualities that are comparable to fresh fruit drinks after dilution. Both clarified concentrates from turbid juices or pulps (lime, lemon) and stable pulp concentrates with high sugar content (65%) and suspended particles (30%) can be made using MF and RO or osmotic evaporation in combination with prior enzymatic treatment of raw material (Vaillant et al., 2009). Therefore, membrane filtration has an immense role in upstream processing as every step determines the final quality of the end product in the downstream processes; hence, it yield products of higher quality and better safety for human consumption.

5.7 APPLICATION IN OTHER INDUSTRIAL USES

Various other applications of membrane filtration have also been reported by various authors, such as in the upstream stages of fermentation (microbiology), pharmaceuticals, enzyme extraction, vaccine purification, wastewater treatment, etc. Pato et al. (2014) have reported the application of microfiltration in the clarification of vero cell culture, which was used for the preparation of a vaccine for yellow fever. Meanwhile, Li et al. (2004) purified benzylanaline antibiotic using a 50 kDa spiral wound ultrafiltration membrane at the initial stages by filtering fermentation broths containing bioemulsifiers. In addition to this, UF, MF, and RO are being used in wastewater treatments quite effectively. Different RO membrane materials, such as zeolite, polyvinylidene fluoride, polyethersulfone, and polyamide, have been evaluated for the treatment of generated water. Wei et al. (2010) carried out advanced treatment of complex pharmaceutical water at the initial step. It was found that the primary source of membrane fouling during the first stage of the NF process was the deposition of sulfate and calcium carbonate. A densely packed fouling layer was gradually generated on the membrane surface during the latter fouling stage, in addition to the deposition of calcium salts and complex organic foulants with carboxyl acid, amide, and alkyl halide functional groups. As a result, a cleaning agent's effectiveness in a cleaning procedure depends on its capacity to degrade the integrity of the compact fouling layer through chemical reactions with the foulants. The outcomes demonstrated that the agents' cleaning efficacies rose in the following order: NaOH (pH 11), HCl (pH 2), citric acid (pH 2), and EDTA (10 mM) for better purification and investigated membrane fouling and cleaning efficiency of the solvent.

5.8 LIMITATIONS OF MEMBRANE FILTRATION IN UPSTREAM TECHNOLOGY

Besides the advanced and effective applications of membrane filtration techniques in a myriad of fields, there are some limitations which are still to be worked upon for future times. These things should be taken into consideration for further research to make these techniques more effective and convenient for the next generations. These limitations, as reported by Saxena et al. (2009) and Mandenius (2021), are:

I. Cell retention systems need to communicate with media feed/ product removal/biomass removal.
II. Lack of In-line monitoring of process parameters at upstream stages.
III. A high degree of automation is required for integrated strategies at higher levels.
IV. Increased upstream costs can lead to increased downstream costs, resulting in non-economical applications of small-scale industries.
V. Lack of consistency at developmental stages.
VI. Lack of understanding of combined parameters during process development.
VII. Modulation of the external environment during the upstream stages.

VIII. Higher concentration and broader particle size at initial stages lead to higher chances of fouling.

5.9 DOWNSTREAM TECHNOLOGY

5.9.1 Downstream Processing

This procedure involves removing impurities and extracting a finished industrial product from a fermentation process. The recovery and purification of solutes/products are the most expensive processes in the pharmaceutical industry, and their cost is higher than the entire bioprocess (Degerman et al., 2008). These factors render downstream processing crucial in the pharmaceutical industry because an inefficient downstream process can significantly increase the cost of the entire manufacturing process. According to Warikoo et al. (2012), continuous operations are generally justified by increased productivity, a smaller environmental impact, and lower buffer usage. Development of DSP places equal emphasis on purity and process capabilities, as well as yield and productivity (Figure 5.6). It is possible to raise the separation efficiency of single-unit operations by expanding the current facilities and optimizing the current and existing processes to increase the separation efficiency of single-unit activities (Shukla et al., 2010).

Due to its physical principles for separation, downstream capacity always scales at least linearly with costs, whereas upstream capacity can be raised without increasing costs. Upstream manufacturing is more expensive than downstream manufacture at low product titers, but as titers increase, the main manufacturing expenses shift from USP to DSP (Chon et al., 2011; Kelley, 2009; Strube et al., 2012; Low et al., 2007). Gaining efficiency, flexibility, and the ability to adopt process-control methods that may be necessary for a parametric/real-time release are the primary factors in downstream processing that transition from a batchwise to a continuous operation (Jungbauer, 2013).

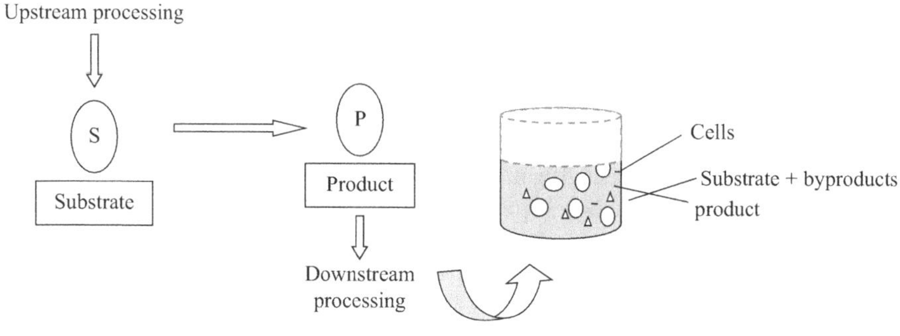

FIGURE 5.6 Downstream technology

Major Steps in Downstream Processing:

I. Isolation/purification of product
II. Concentration of product
III. Packaging of product

I. **Product purification**: Removing complete cell cultures, cell detritus, and colloidal particles during the clarification step lessens the workload on subsequent processes (Besnard et al., 2016; Chon and Zarbis-papastoitsis, 2011). In the companies that produce vaccines, centrifugation is employed to clarify the product (Besnard et al., 2016; Yavorsky et al., 2003). However, centrifuge maintenance and capital costs are significant. Due to its reliability and high throughput, membrane technology has recently attracted increased attention for the clarification of vaccines (Genzel et al., 2014). For upstream clarification, procedures utilizing tangential flow and regular flow/dead-end membrane filtering can both be used. For initial clarity, normal flow/dead-end membrane filtration is appropriate but is less common than tangential flow microfiltration (pore size: 1.40–0.65 m).

II. **Product concentration**: Centrifugation and filtration alone are insufficient when extracellular products are combined with all other mixes. In that situation, centrifugation and micro/ultrafiltration are used as the first steps in the purifying process (Besnard et al., 2016). Concentration is the next action. The residual liquid containing substrates, byproducts, and product is then used for that. Ion exchange chromatography will now be used to further clean up this procedure. Techniques for flocculation and sedimentation can also be utilized. Centrifugation or filtration is the first stage in the case of intracellular products. After receiving the cell, it must be cracked. Here, it is possible to destroy the cell through both physical and chemical means. Physical techniques such as homogenization and sonication are used. Osmolytic substances are used in chemical processes. The item will emerge from the cell once it has been broken.

Following the application of physical or chemical treatments to the cell, centrifugation will be carried out. The product is finally offered in two formats: (1) cellular waste and (2) fluid. The cell waste must be discarded, and the product-containing liquid must then be separated or purified using centrifugation and membrane filtration methods.

III. **Product packaging:** After undergoing procedures like centrifugation and homogenization, the product will finally be available in pure form and must be packaged carefully to prevent contamination. Packaging options include lyophilization, drying, and crystallization (Genzel et al., 2014).

5.10 APPLICATION OF MEMBRANE TECHNOLOGY IN DOWNSTREAM PROCESSING

Any generalization is challenging since it depends on the characteristics of the bioprocess how membranes and membrane processes may be used in downstream processing. But examining the separation issues of a typical downstream process might demonstrate the possibilities of membranes (Michaels, 1980).

5.10.1 Applications in Microbiology

In a bioreactor, several elements, including microorganisms, enzymes, nutrients, products, and byproducts, are typically present in a somewhat diluted aqueous solution (Figure 5.7). The products must be concentrated, purified, and segregated from the other reactor components. When boiled down to its bare minimum, the procedure entails several subsequent actions: (1) batch-culturing of the necessary microorganisms suspended in a sufficient nutrient solution that has been sterilized

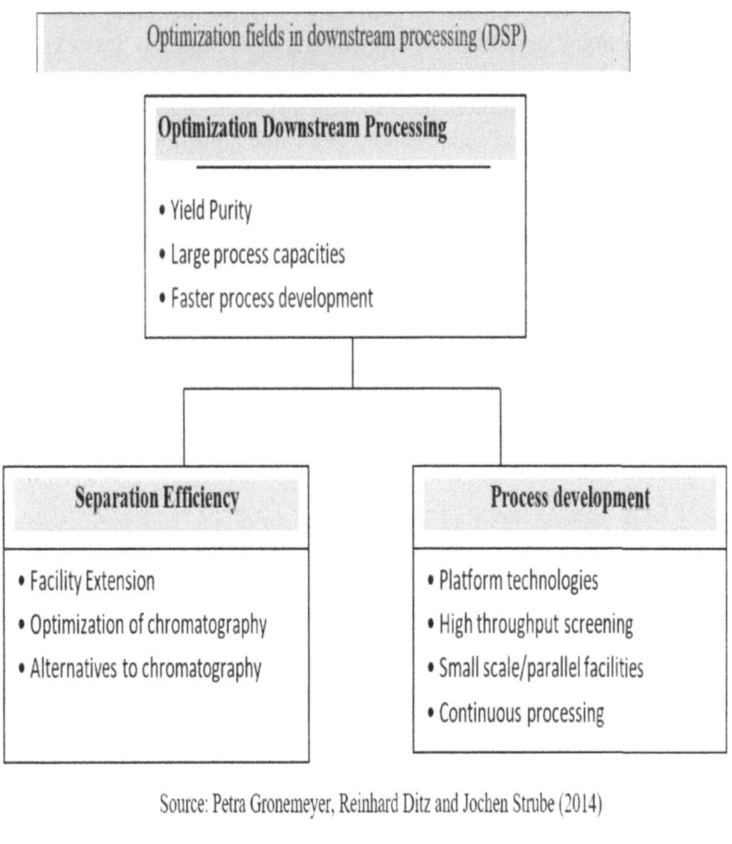

FIGURE 5.7 Downstream processing

while being subjected to the proper mechanical agitation and aeration; (2) separation of the product from the fermentation broth; and (3) the purification and concentration of the desired product in the clarified medium. Traditional separation techniques such as filtering, centrifugation, liquid-liquid extraction, ion exchange, fractional precipitation, crystallization, etc., may be used to recover the product. However, many of these procedures are wasteful, ineffective, and produce subpar goods. (Michaels, 1980).

The Application of Electrodialysis in Post-Processing: Other membrane processes, such as electrodialysis, are more challenging in their principal application and, as a result, are less frequently used today despite having obvious economic benefits. This is in contrast to the use of microfiltration, ultrafiltration, and reverse osmosis, which have almost no fundamental problems but many operational and practical ones. Anion- and cation-exchange membranes are positioned between two electrodes in an alternating sequence to generate separate cells in electrodialysis. When a direct current is provided to the system, ionic components of a feed solution that is going through the cell system will concentrate in alternating cells as the brine. Electrodialysis can be used in biotechnology to desalt solutions that contain proteins, dextrans, carbohydrates, etc. Although it frequently competes directly with ion exchange in many applications, electrodialysis is typically less expensive. The separation of various organic acids, particularly amino acids, which, due to their amphoteric nature, can be protonated when the pH value of the solution is lower than the isoelectric points can be separated by pH value adjustment and electrodialysis, is a rather interesting application of electrodialysis (Strathmann, 1985) (Figure 5.8).

5.10.2 Treatment of Oil Field and Refinery Waste

In a number of oil refineries, membrane technology is now used to purify produced water as well as provide process water. Instead of separating different types of produced water in a refinery and then sending the produced water to a central plant for membrane technology treatment, the possibility of using membrane technology in new applications is particularly high in the area of developing integrated membrane technologies to treat produced water at the source (P.Surinder, 2003). In this regard,

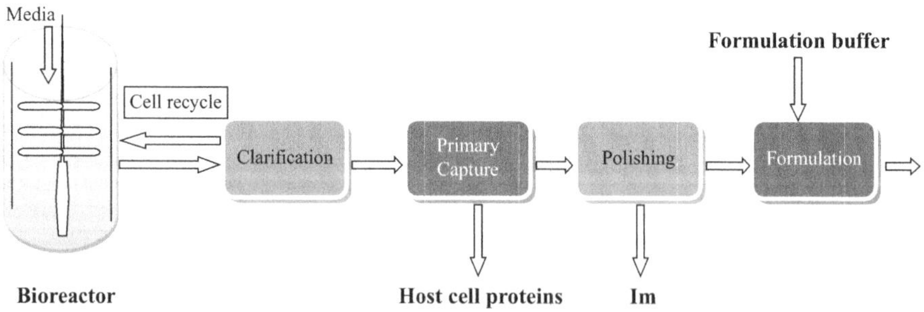

FIGURE 5.8 Schematic representation of continuous downstream process

membrane technology is not currently used in the segregation and pretreatment operations of certain types of produced water in refineries. According to Wong et al. (2004), in-plant source control, pretreatment, and end-of-pipe treatment are the main components of the treatment and management of generated water in petroleum refineries. Before transferring the pre-treated water to an end-of-pipe treatment system, some pollutant concentrations are reduced using in-plant source management. Prior to combining produced water with the main produced water stream, where some contaminants may be recovered, pretreatment involves removing specific pollutants from produced water. End-of-pipe treatment is the last step for meeting discharge standards (Wong and Hung, 2004). In refineries, many methods have been used to lessen the flow of contaminants and produced water (AECOM, 2010). For possible applications, it is crucial to include membrane technology into an in-plant source control or pretreatment process.

5.10.3 Applications in Pharmaceuticals

Downstream processing in pharmaceutical industries is very important because an ineffective downstream process can significantly increase the cost of the whole manufacturing process. Degerman et al. (2008) suggested that in the pharmaceutical industry, solute recovery and product purification are the most expensive processes—in certain cases, their cost exceeds that of the entire bioprocess. For the downstream recovery and purification of pharmaceuticals, a variety of methods, including crystallization, chromatography, membrane technology, and distillation, are available. (Siew et al., 2013; Szekely et al., 2013).

5.11 CONCLUSION AND FUTURE PERSPECTIVES

Applications of membrane technology for various upstream and downstream processes, such as clarification, recovery, and purification of industrial products (such as lactic acid, pharmaceuticals, and glycerol), have been reviewed in this chapter. Membrane technologies include microfiltration, ultrafiltration, and nanofiltration. For the extraction and purification of the finished product, membrane technology enables improved process control and high selectivity when compared to conventional upstream and downstream procedures (such as centrifugation and adsorption). These benefits have led to the integration of membranes with whole-cell and enzymatic bioreactors for the continuous operation of various biotechnology processes like fermentation and hydrolysis. Implementing membrane technologies is receiving more attention as a result of the requirement to strengthen downstream processing in order to complement improvements in upstream productivity. The stratified therapeutics will enable a shift in plant architecture in favor of highly adaptable, multi-functional spaces for modest production quantities. Such a development will aid in the advancement of single-use technologies as well as continuous, non-chromatographic, and flexible downstream processes. As a supporting tool for process optimization, these modifications in process development and the integration of USP and DSP development for impurity profiles would be beneficial and could be set up at a reasonable cost (Ma et al., 2009).

Future breakthroughs in membrane technology will probably be influenced by trends in the broader biotechnology sector, such as the need for improved productivity, reduced manufacturing costs, and faster development times. Additionally, disposable systems are becoming more popular, from bioreactors and buffer bags to single-use columns and membranes (Zhou et al., 2008). Not only could these technologies cut production costs, but they would also eliminate the need to create and test cleaning cycles. Additionally, second-generation biotherapeutics, such as protein-polymer conjugates (Molek and Zydney, 2007), plasmid DNA for gene therapy applications, and DNA-based vaccines (Latulippe et al., 2007), and virus particles (Grzenia et al., 2008), may be processed downstream using membrane systems, which may be appealing. The recent and future advancements in our understanding of the fundamental processes driving protein transport and fouling during membrane filtration will serve as the foundation for the successful creation of new membrane technologies aimed at these different applications.

REFERENCES

Abdullah, N., Rahman, M. A., Othman, M. H. D., Jaafar, J., & Ismail, A. F. (2018). Membranes and membrane processes: fundamentals. In *Current trends and future developments on (bio-) membranes* (pp. 45–70). Elsevier.

AECOM. (2010). Petroleum refining water/wastewater use and management. The International Petroleum Industry Environmental Conservation Association.

Afonso, M. D., & Bórquez, R. (2003). Nanofiltration of wastewaters from the fish meal industry. *Desalination*, *151*(2), 131–138.

Alkhudhiri, A., Darwish, N., & Hilal, N. (2012). Membrane distillation: A comprehensive review. *Desalination*, *287*, 2–18.

Almandoz, M. C., Pagliero, C. L., Ochoa, N. A., & Marchese, J. (2015). Composite ceramic membranes from natural aluminosilicates for microfiltration applications. *Ceramics International*, *41*(4), 5621–5633.

Anis, S. F., Hashaikeh, R., & Hilal, N. (2019). Microfiltration membrane processes: A review of research trends over the past decade. *Journal of Water Process Engineering*, *32*, 100941.

Asad, A., Sameoto, D., & Sadrzadeh, M. (2020). Overview of membrane technology. In *Nanocomposite membranes for water and gas separation* (pp. 1–28). Elsevier.

Asif, M. B., Hai, F. I., Jegatheesan, V., Price, W. E., Nghiem, L. D., & Yamamoto, K. (2019). Applications of membrane bioreactors in biotechnology processes. In *Current trends and future developments on (bio-) membranes* (pp. 223–257). Elsevier.

Baker, R. W. (2000). Updated by staff, membrane technology. In *Kirk-othmer encyclopedia of chemical technology* (pp. 44–76). John Wiley & Sons.

Baker, R. W. (2004). Overview of membrane science and technology. *Membrane Technology and Applications*, *3*, 1–14.

Baruah, G. L., Venkiteshwaran, A., & Belfort, G. (2005). Global model for optimizing crossflow microfiltration and ultrafiltration processes: A new predictive and design tool. *Biotechnology Progress*, *21*(4), 1013–1025.

Bera, S. P., Godhaniya, M., & Kothari, C. (2022). Emerging and advanced membrane technology for wastewater treatment: A review. *Journal of Basic Microbiology*, *62*(3–4), 245–259.

Besnard, L., Fabre, V., Fettig, M., Gousseinov, E., Kawakami, Y., Laroudie, N., ... Pattnaik, P. (2016). Clarification of vaccines: An overview of filter based technology trends and best practices. *Biotechnology Advances*, *34*(1), 1–13.

Bowen, W. R., & Welfoot, J. S. (2002). Modelling the performance of membrane nanofiltration—critical assessment and model development. *Chemical Engineering Science*, *57*(7), 1121–1137.

Carole Ayoub Moubareck,. (2020). Polymyxins and Bacterial Membranes: A Review of Antibacterial Activity and Mechanisms of Resistance. *Membranes*, 10(8), 181.

Casper, J. L., Wendorff, W. L., & Thomas, D. L. (1999). Functional properties of whey protein concentrates from caprine and ovine specialty cheese wheys. *Journal of Dairy Science*, *82*(2), 265–271.

Charcosset, C. (2006). Membrane processes in biotechnology: An overview. *Biotechnology Advances*, *24*(5), 482–492.

Chon, J. H., & Zarbis-Papastoitsis, G. (2011). Advances in the production and downstream processing of antibodies. New Biotechnology, 28, 458–463.

Conidi, C., Castro-Muñoz, R., & Cassano, A. (2020). Nanofiltration in beverage industry. In *Nanotechnology in the beverage industry* (pp. 525–548). Elsevier.

Corredig, M., Roesch, R. R., & Dalgleish, D. G. (2003). Production of a novel ingredient from buttermilk. *Journal of Dairy Science*, *86*(9), 2744–2750.

Cui, Z. F., Jiang, Y., & Field, R. W. (2010). Fundamentals of pressure-driven membrane separation processes. In Membrane technology (pp. 1–18). Butterworth-Heinemann.

Darnon, E., Morin, E., Belleville, M. P., & Rios, G. M. (2003). Ultrafiltration within downstream processing: Some process design considerations. *Chemical Engineering and Processing: Process Intensification*, *42*(4), 299–309.

Daufin, G., Escudier, J. P., Carrère, H., Bérot, S., Fillaudeau, L., & Decloux, M. (2001). Recent and emerging applications of membrane processes in the food and dairy industry. *Food and Bioproducts Processing*, 79(2), 89–102.

Davis, R. H. (2019). Microfiltration in pharmaceutics and biotechnology. In Current trends and future developments on (bio-) membranes (pp. 29–67). Elsevier.

Degerman, M., Jakobsson, N., & Nilsson, B. (2008). Designing robust preparative purification processes with high performance. *Chemical Engineering & Technology: Industrial Chemistry-Plant Equipment-Process Engineering-Biotechnology*, *31*(6), 875–882.

Ernst, M., Bismarck, A., Springer, J., & Jekel, M. (2000). Zeta-potential and rejection rates of a polyethersulfone nanofiltration membrane in single salt solutions. Journal of Membrane Science, 165(2), 251–259. https/doi.org/10.1016/S0376-7388(99)00238-0

Genzel, Y., Rödig, J., Rapp, E., & Reichl, U. (2014). Vaccine production: Upstream processing with adherent or suspension cell lines. In Animal cell biotechnology. Methods and protocols (pp. 371–393).

Goosen, M. F. A., Sablani, S. S., Al-Hinai, H., Al-Obeidani, S., Al-Belushi, R., & Jackson, A. (2005). Fouling of reverse osmosis and ultrafiltration membranes: A critical review. *Separation Science and Technology*, *39*(10), 2261–2297.

Grzenia, D. L., Carlson, J. O., & Wickramasinghe, S. R. (2008). Tangential flow filtration for virus purification. Journal of Membrane Science, 321, 373–380.

Gurreri, L., Tamburini, A., Cipollina, A., & Micale, G. (2020). Electrodialysis applications in wastewater treatment for environmental protection and resources recovery: A systematic review on progress and perspectives. *Membranes*, *10*(7), 146.

Hakami, M. W., Alkhudhiri, A., Al-Batty, S., Zacharof, M. P., Maddy, J., & Hilal, N. (2020). Ceramic microfiltration membranes in wastewater treatment: Filtration behavior, fouling and prevention. *Membranes*, *10*(9), 248.

Jonquières, A., Arnal-Herault, C., & Babin, J. (2013). Pervaporation. In *Encyclopedia of membrane science and technology*. (pp. 800–867). John Wiley & Sons.

Jungbauer, A. (2013). Continuous downstream processing of biopharmaceuticals. *Trends in Biotechnology, 31*(8), 479–492.

Kelley, B. (2009). Industrialization of mAb production technology: The bioprocessing industry at a crossroads. mAbs, 1, 443–452.

Kuriyel, R., & Zydney, A. L. (2000). Sterile filtration and virus filtration. In Downstream processing of proteins: Methods and protocols (pp. 185–194).

Latulippe, D. R., Ager, K., & Zydney, A. L. (2007). Flux dependent transmission of supercoiled plasmid DNA through ultrafiltration membranes. Journal of Membrane Science, 294, 169–177.

Lee, M., Wu, Z., & Li, K. (2015). Advances in ceramic membranes for water treatment. In *Advances in membrane technologies for water treatment* (pp. 43–82). Woodhead Publishing.

Lee, S., Ihara, M., Yamashita, N., & Tanaka, H. (2017). Improvement of virus removal by pilot-scale coagulation-ultrafiltration process for wastewater reclamation: Effect of optimization of pH in secondary effluent. *Water Research, 114*, 23–30.

Lee, S. S., Burt, A., Russotti, G., & Buckland, B. (1995). Microfiltration of recombinant yeast cells using a rotating disk dynamic filtration system. *Biotechnology and Bioengineering, 48*(4), 386–400.

Li, S. Z., Li, X. Y., Cui, Z. F., & Wang, D. Z. (2004). Application of ultrafiltration to improve the extraction of antibiotics. *Separation and Purification Technology, 34*(1–3), 115–123.

Liang, B., He, X., Hou, J., Li, L., & Tang, Z. (2019). Membrane separation in organic liquid: Technologies, achievements, and opportunities. *Advanced Materials, 31*(45), 1806090.

Lin, T. M., Park, J. W., & Morrissey, M. T. (1995). Recovered protein and reconditioned water from surimi processing waste. *Journal of Food Science, 60*(1), 4–9.

Low, D., O'Leary, R., & Pujar, N. S. (2007). Future of antibody purification. Journal of Chromatography B, 848, 48–63.

Mandenius, C. F. (2021). Measurement technologies for upstream and downstream bioprocessing. *Processes, 9*(1), 143.

Meyer, P., & Kulozik, U. (2016). Impact of protein removal by an upstream ultrafiltration on the reverse osmosis of skim milk and sweet whey. *Chemie Ingenieur Technik, 88*(5), 585–590.

Michaels, A. S. (1980). Membrane separation processes. In Encyclopedia of Chemical Technology (Kirk-Othmer), 3rd ed., John Wiley & Sons, New York.

Mohammad, A. W., Ng, C. Y., Lim, Y. P., & Ng, G. H. (2012). Ultrafiltration in food processing industry: Review on application, membrane fouling, and fouling control. *Food and Bioprocess Technology, 5*, 1143–1156.

Mohammad, A. W., Teow, Y. H., Ang, W. L., Chung, Y. T., Oatley-Radcliffe, D. L., & Hilal, N. (2015). Nanofiltration membranes review: Recent advances and future prospects. *Desalination, 356*, 226–254.

Molek, J. R., & Zydney, A. L. (2007). Separation of PEGylated α-lactalbumin from unreacted precursors and byproducts using ultrafiltration. Biotechnology Progress, 23, 1417–1424.

Morin, P., Pouliot, Y., & Jiménez-Flores, R. (2006). A comparative study of the fractionation of regular buttermilk and whey buttermilk by microfiltration. *Journal of Food Engineering, 77*(3), 521–528.

Nabais A.M.A and Cardoso J.P. (1999). Purification of benzylpenicillin filtered broths by ultrafiltration and effect on solvent extraction, Bioproc. Biosyst. Eng., 21, 157–163.

Nagy, E. (2019). Forward osmosis. In Basic equations of mass transport through a membrane layer (pp. 447–456).

O'Connell, J. E., & Fox, P. F. (2000). Heat stability of buttermilk. *Journal of Dairy Science, 83*(8), 1728–1732.

Oatley, D. L., Llenas, L., Pérez, R., Williams, P. M., Martínez-Lladó, X., & Rovira, M. (2012). Review of the dielectric properties of nanofiltration membranes and verification of the single oriented layer approximation. *Advances in Colloid and Interface Science, 173*, 1–11.

Obotey Ezugbe, E., & Rathilal, S. (2020). Membrane technologies in wastewater treatment: A review. *Membranes, 10*(5), 89.

Pato, T. P., Souza, M. C. O., Silva, A. N., Pereira, R. C., Silva, M. V., Caride, E., & Castilho, L. R. (2014). Development of a membrane adsorber based capture step for the purification of yellow fever virus. *Vaccine, 32*(24), 2789–2793.

Rezakazemi, M., Khajeh, A., & Mesbah, M. (2018). Membrane filtration of wastewater from gas and oil production. *Environmental Chemistry Letters, 16*, 367–388.

Saxena, A., Tripathi, B. P., Kumar, M., & Shahi, V. K. (2009). Membrane-based techniques for the separation and purification of proteins: An overview. *Advances in Colloid and Interface Science, 145*(1–2), 1–22.

Schutyser, M., Rupp, R., Wideman, J., & Belfort, G. (2002). Dean vortex membrane microfiltration and diafiltration of rBDNF E. coli inclusion bodies. Biotechnology Progress, 18, 322–329.

Scott, K. (1995). Membrane materials, preparation and characterisation. In *Handbook of industrial membranes* (pp. 187–269). Elsevier.

Shi, X., Tal, G., Hankins, N. P., & Gitis, V. (2014). Fouling and cleaning of ultrafiltration membranes: A review. *Journal of Water Process Engineering, 1*, 121–138.

Shukla, A. A., & Thömmes, J. (2010). Recent advances in large-scale production of monoclonal antibodies and related proteins. Trends in Biotechnology, 28, 253–261.

Smitha, B., Suhanya, D., Sridhar, S., & Ramakrishna, M. (2004). Separation of organic–organic mixtures by pervaporation—a review. *Journal of Membrane Science, 241*(1), 1–21.

Strathmann, H. (1985). Electrodialysis, a mature technology with a multitude of new applications. Desalination, 56, 279–295.

Strube, J., Grote, F., & Ditz, R. (2012). Bioprocess design and production technology for the future. In G. Subramanian (Ed.), Biopharmaceutical production technology (Vol. 2). Wiley-VCH.

Surinder, P. (2003). Refinery water systems. In Refining processes handbook (pp. 242–269). Gulf Professional Publishing.

Szekely, G., Bandarra, J., Heggie, W., Sellergren, B., & Livingston, A. G. (2013). Advanced membranes for organic solvent nanofiltration. *Chemical Engineering Research and Design*, 91(9), 1741–1753.

Tarifa, M. C., Lozano, J. E., & Brugnoni, L. I. (2018). Disinfection efficacy over yeast biofilms of juice processing industries. *Food Research International, 105*, 473–481.

Tranchant, C. C., & Selvamuthukumaran M. (2020). Introduction to membrane processing. In M. Selvamuthukumaran (Ed.), Applications of membrane technology for food processing industries (pp. 1–44). CRC Press.

Vaillant, F., Millan, P., Jariel, O., Dornier, M., & Reynes, M. (2009). Optimization of enzymatic preparation for passion fruit juice liquefaction by fractionation of fungal enzymes through metal chelate affinity chromatography. *Food Biotechnology, 13*(1), 33–50.

Valero, F., Barceló, A., & Arbós, R. (2011). Electrodialysis technology: Theory and applications. *Desalination, Trends and Technologies, 28*, 3–20.

Van der Bruggen, B. (2018). Microfiltration, ultrafiltration, nanofiltration, reverse osmosis, and forward osmosis. In Fundamental modelling of membrane systems (pp. 25–70).

Van Houtte, E., & Verbauwhede, J. (2013). Long-time membrane experience at Torreele's water re-use facility in Belgium. *Desalination and Water Treatment, 51*(22–24), 4253–4262.

Van Reis, R., & Zydney, A. (2001). Membrane separations in biotechnology. *Current Opinion in Biotechnology*, *12*(2), 208–211.

Wang, P., & Chung, T. S. (2015). Recent advances in membrane distillation processes: Membrane development, configuration design and application exploring. *Journal of Membrane Science*, *474*, 39–56.

Warikoo, V., Godawat, R., Brower, K., Jain, S., Cummings, D., Simons, E., ... Konstantinov, K. (2012). Integrated continuous production of recombinant therapeutic proteins. *Biotechnology and Bioengineering*, *109*(12), 3018–3029.

Wei, X., Wang, Z., Fan, F., Wang, J., & Wang, S. (2010). Advanced treatment of a complex pharmaceutical wastewater by nanofiltration: Membrane foulant identification and cleaning. *Desalination*, *251*(1–3), 167–175.

Wong, J. M., & Hung, Y. T. (2004). Treatment of oilfield and refinery wastes. In L. K. Wang, Y. T. Hung, H. H. Lo, & C. Yapijakis (Eds.), Handbook of industrial and hazardous wastes treatment (pp. 144–216). CRC Press.

Yavorsky, D., Blanck, R., Lambalot, C., & Brunkow, R. (2003). The clarification of bioreactor cell cultures for biopharmaceuticals. *Pharmaceutical Technology*, *27*(3), 62–77.

Ma, Z., Lan, Z., Matsuura, T., & Ramakrishna, S. (2009). Electrospun polyethersulfone affinity membrane: Membrane preparation and performance evaluation. Journal of Chromatography B: Analytical Technologies in the Biomedical and Life Sciences, 877, 3686.

Zeman, L. J., & Zydney, A. L. (1996). *Ultrafiltration and microfiltration: Principles and applications*.

Zhou, J. X., Solamo, F., Hong, T., Shearer, M., & Tressel, T. (2008). Viral clearance using disposable systems in monoclonal antibody commercial down-stream processing. *Biotechnology and Bioengineering*, *100*, 488–496.

6 Co-crystallization and Nanoencapsulation Techniques for Food Bioactive Components

Sadaf Nazir, Z. R. A. A. Azad, Sweta Joshi, Farah Naqash, Sadaf Parvez, and Shivam Yadav

6.1 INTRODUCTION

The presence of bioactive components in foods has attracted significant interest due to their potential health benefits, including antioxidant, antimicrobial, and anti-inflammatory properties. These compounds, such as polyphenols, carotenoids, and vitamins, have been linked to reducing the risk of chronic diseases and promoting overall well-being. Incorporating bioactive components into food matrices presents several challenges, including chemical instability, poor solubility, and susceptibility to degradation during processing and storage. These issues can significantly impact the bioavailability and functionality of these compounds, limiting their potential health benefits. To overcome these hurdles, co-crystallization and nanoencapsulation techniques have emerged as promising strategies to enhance the stability and functionality of food bioactive components. Co-crystallization and nanoencapsulation have emerged as promising techniques to address the challenges associated with bioactive component delivery. Co-crystallization can enhance the stability and solubility of these compounds by forming novel crystalline structures, while nano-encapsulation provides protection and controlled release through encapsulation in nanoscale carriers (Awuchi et al., 2022; Chezanoglou et al., 2021).

Co-crystallization, a fascinating and intricate process, is at the forefront of materials science, chemistry, and pharmaceutical research. This innovative technique involves the creation of crystalline structures containing two or more compounds in a precisely defined stoichiometric ratio. The intricate dance of molecular interactions, such as hydrogen bonding, van der Waals forces, and π-π stacking, underpins the formation of stable and unique crystal forms with properties that often surpass those of the individual components.

DOI: 10.1201/9781032650616-6

At its essence, co-crystallization is a harmonious symphony of molecules coming together in a well-defined ratio, akin to a chemical tango. The partners in this dance, referred to as co-formers, play a crucial role in determining the characteristics of the resulting co-crystal. These co-formers are carefully chosen to orchestrate the desired physicochemical properties, such as solubility, stability, and bioavailability, of the co-crystal. Each co-former contributes its unique molecular traits to the final structure, making the selection process akin to casting the right actors for a theatrical performance.

In the realm of pharmaceuticals, where co-crystallization has gained significant prominence, this process is akin to tailoring a bespoke suit for a client. Each client, in this case, is a bioactive compound with its distinct set of preferences and limitations. The co-former serves as the skilled tailor, ensuring that the co-crystal fits the compound perfectly, enhancing its attributes and mitigating its shortcomings.

The art of co-crystallization involves a deep understanding of the chemistry of interactions. It is akin to composing a symphony, with each molecule contributing its unique notes to create a harmonious whole. Hydrogen bonding, a fundamental force in molecular interactions, often takes center stage in co-crystallization. Like the notes in a musical score, hydrogen bonds create stability and structure in the co-crystal lattice. These bonds can form between the co-formers or between the co-formers and the bioactive compound, reinforcing the crystal's integrity and influencing its properties.

Van der Waals forces, another crucial component in this molecular choreography, contribute to the attraction between molecules. Much like the gravitational pull that keeps planets in orbit, van der Waals forces ensure that the co-formers and the bioactive compound are drawn together in a delicate equilibrium. They play a vital role in shaping the crystal lattice and influencing properties such as density and compressibility.

The intriguing phenomenon of π-π stacking adds an element of elegance to the co-crystallization dance. This interaction, akin to the graceful entwining of partners in a waltz, involves the stacking of aromatic rings. The resulting π-π interactions not only influence the crystal's structure but also impact its electronic properties, making them essential considerations in the design of co-crystals with specific optical or electronic characteristics.

One of the critical decisions in co-crystallization is the choice of suitable co-formers. This selection process resembles the casting of actors for a theatrical production. Just as actors must complement each other's talents and chemistry on stage, co-formers must interact harmoniously to create a stable and functional co-crystal. The compatibility between the bioactive compound and the co-formers is crucial, akin to selecting actors who can perform well together.

Each co-former brings its unique set of skills and attributes to the co-crystal. Some co-formers may excel in enhancing solubility, ensuring that the bioactive compound dissolves readily, while others may contribute to stability, extending the shelf life of the co-crystal. It is the synergy between these co-formers that determines the co-crystal's ultimate performance and utility.

Characterizing co-crystals is a multifaceted endeavor, akin to critically reviewing a masterpiece of literature or art. Researchers employ a combination of techniques to confirm the presence of co-crystals and assess their stability and purity. These methods are akin to the tools of an art historian or literary critic, each providing a unique perspective on the work.

X-ray diffraction, an essential technique in co-crystal characterization, serves as a powerful microscope that unveils the intricate lattice structure of the co-crystal. Much like examining brushstrokes on a canvas or deciphering the nuances of a script, X-ray diffraction allows researchers to visualize the arrangement of atoms within the co-crystal lattice. It provides the definitive evidence that a co-crystal has formed and offers insights into its crystallographic properties.

Fourier-transform infrared spectroscopy (FTIR) operates as the co-crystal's voice, revealing the vibrations and interactions between its molecular components. This technique is akin to analyzing the rhythm and cadence of a musical composition. FTIR spectra provide valuable information about the presence of specific functional groups and hydrogen bonding interactions within the co-crystal, aiding in its characterization.

Differential scanning calorimetry (DSC) is akin to assessing the co-crystal's response to temperature changes, much like examining an artist's work in different lighting conditions. This technique measures the heat flow associated with phase transitions, offering insights into the co-crystal's thermal behavior, stability, and purity.

Solid-state nuclear magnetic resonance (NMR) spectroscopy is akin to deciphering the co-crystal's hidden meanings and symbolism, much like a literary critic interpreting metaphors and allegories. NMR provides invaluable insights into the molecular dynamics, local environments, and interactions within the co-crystal, shedding light on its structural and functional attributes.

The culmination of these analytical techniques and approaches is akin to a comprehensive artistic or literary critique. Researchers meticulously examine every facet of the co-crystal, from its structural intricacies to its chemical composition and physical properties. This thorough examination ensures that the co-crystal's identity is fully understood and its potential fully realized.

In the grand tapestry of scientific research, co-crystallization stands as a testament to the intricate dance of molecules and the meticulous craftsmanship of scientists. It is an art form that merges the principles of chemistry, physics, and materials science to create structures with novel properties and enhanced functionality. As researchers continue to refine their techniques and deepen their understanding, co-crystallization holds the promise of unlocking new frontiers in drug development, materials engineering, and beyond. Much like a masterpiece that inspires generations, co-crystals have the potential to shape the future of science and technology.

Nanotechnology has revolutionized the food industry by enabling the design of nanoscale materials and structures with unique properties. Nanoencapsulation involves the confinement of bioactive components within nanocarriers, such as liposomes, polymeric nanoparticles, and nanoemulsions (Liao et al., 2023). Nanoencapsulation techniques can protect bioactive components from degradation, enhance their solubility, and provide controlled release profiles. Common methods include solvent evaporation, emulsification, and coacervation, each offering distinct advantages for specific applications (Pathakoti et al., 2017). Various materials, including lipids, proteins, and polysaccharides, have been employed as encapsulating agents for food bioactive components. The choice of material depends on factors such as compatibility, stability, and biocompatibility with the bioactive compound. Nanoencapsulation can improve the bioavailability of bioactive components by protecting them from adverse environmental conditions and enzymatic degradation. Furthermore, the controlled release properties of nanocarriers allow for targeted delivery, enhancing the efficacy of the bioactive compound. The characterization of nano-encapsulated bioactive components requires specialized techniques, including dynamic light scattering (DLS), transmission electron microscopy (TEM), and atomic force microscopy (AFM). These methods enable the determination of particle size, morphology, and encapsulation efficiency (Biswas et al., 2022).

6.2 PRINCIPLES AND MECHANISMS OF CO-CRYSTALLIZATION AND NANOENCAPSULATION

Principles and mechanisms of co-crystallization delves into a compelling exploration of co-crystallization processes, shedding light on their underlying principles and intricate mechanisms. Co-crystallization, the formation of crystalline structures composed of two or more distinct molecular or ionic compounds, has garnered significant attention due to its potential in diverse fields ranging from pharmaceuticals, nutraceuticals, and materials science (Queiroz et al., 2022). The fundamental principles that govern co-crystallization are the role of intermolecular interactions, such as hydrogen bonding, π-π stacking, and van der Waals forces. These interactions play a pivotal role in dictating the arrangement of different molecular entities in the co-crystal lattice, ultimately influencing the physicochemical properties of the resulting material. Furthermore, the mechanisms of co-crystallization are elucidated, revealing a multi-step process that involves nucleation, growth, and crystalline transformation. The researchers propose a comprehensive model to explain the kinetics and thermodynamics of co-crystal formation, considering factors such as solvent effects, temperature, and concentration of the reactants (Wang et al., 2022).

The co-crystallization has significance in pharmaceutical development. Co-crystals can enhance drug solubility, stability, and bioavailability, offering a potential solution to challenges posed by poorly water-soluble drugs. By tailoring co-crystal structures, researchers can manipulate target or controlled release profiles, revolutionizing bioactive delivery systems. In the materials science domain, co-crystallization opens doors to the design of novel materials with customizable properties. The co-crystals exhibit enhanced mechanical, electronic, or optical characteristics compared to their individual components. The intricate interplay of intermolecular forces and the formation kinetics, the study contributes to a deeper understanding of co-crystal formation (Behnamnik et al., 2021). The potential applications underscore the relevance of this research, offering a pathway to harnessing co-crystallization for tailored functionalities in various industries.

A comprehensive investigation into the intriguing realm of nanoencapsulation elucidates its fundamental principles and intricate mechanisms. Nanoencapsulation involves entrapping active substances within nanoscale carriers, revolutionizing fields such as pharmaceuticals, cosmetics, and food science. The core principles governing nanoencapsulation highlight the critical role of carrier materials, which range from lipids and polymers to proteins and other biomolecules. These carriers not only protect the encapsulated substances but also facilitate controlled release and targeted delivery. The interactions between the encapsulant and carrier, often guided by surface modifications, play a pivotal role in achieving desired encapsulation efficiencies and stability. Mechanistically, the sequential stages of nanoencapsulation begin with the selection of appropriate carriers and encapsulants, followed by encapsulation methods such as emulsion, solvent evaporation, or supercritical fluid techniques. The kinetics of encapsulation is affected by factors like mixing intensity, solvent evaporation rate, and temperature that influence the encapsulation process (Wani et al., 2023).

Furthermore, the technique underscores the applications of nanoencapsulation in the nutraceutical domain, where it serves as a revolutionary approach to bioactive delivery. Nanoencapsulated bioactives can enhance bioavailability, reduce toxicity, and offer sustained release profiles, thereby improving therapeutic efficacy and patient compliance (Peters et al., 2016). The principles of nanoencapsulation find applications in the food industry, where encapsulated active ingredients enable targeted penetration and prolonged release. In the food industry, nanoencapsulation preserves flavors, stabilizes sensitive nutrients, and even enables controlled release of additives, revolutionizing food product design (Rashidinejad et al., 2020). This provides insights into these applications, emphasizing how nanoencapsulation's mechanisms align with the specific needs of each domain.

In summary, the principles and mechanisms that underpin nanoencapsulation offer a comprehensive understanding of this transformative technology. By dissecting the interplay between carrier materials, encapsulants, and encapsulation techniques, the study contributes to advancing nanoencapsulation's applicability across pharmaceuticals and food science (Plati et al., 2022). As research in this field continues to evolve, the chapter serves as a cornerstone for researchers and industries aiming to harness the potential of nanoencapsulation for innovative and efficient delivery systems.

6.3 METHODS OF CO-CRYSTAL FORMATION

Methods of co-crystal formation represent a diverse and multifaceted toolbox of techniques, each a unique brushstroke on the canvas of materials science. These methods serve as the means through which scientists orchestrate the assembly of crystalline structures composed of two or more distinct compounds, producing co-crystals with properties meticulously tailored to meet the demands of diverse applications. The artistry lies in the careful selection of the method, each akin to a different artistic medium, to craft co-crystals that exhibit the desired attributes and characteristics.

6.3.1 Solvent Evaporation Method

This method involves dissolving the compounds of interest in a common solvent and allowing the solvent to evaporate, inducing crystallization of the co-crystal. The slow evaporation rate influences nucleation and crystal growth, facilitating the formation of well-defined structures.

6.3.2 Solvent-Drop Grinding Method

Mechanochemical methods involve grinding the individual compounds together. The mechanical force breaks intermolecular bonds and promotes new interactions, leading to co-crystal formation. This technique is advantageous for its simplicity and the absence of solvent-related issues.

6.3.3 Antisolvent Methods

Here, a compound is precipitated from a solution by adding an antisolvent, inducing supersaturation and co-crystal formation. The sudden change in solubility prompts nucleation and growth, yielding co-crystals with controlled properties.

6.3.4 Solvate-Assisted Co-crystallization

Solvates, where solvent molecules are part of the co-crystal lattice, can be used to facilitate co-crystal formation. The solvent molecules act as bridges between the components, aiding in their assembly.

6.3.5 Inclusion Complexes

This method involves using a host compound to encapsulate guest molecules in a co-crystalline lattice. Host-guest interactions, such as hydrogen bonding or π-π stacking, stabilize the complex.

6.3.6 Co-crystallization from the Melt

Co-crystals can be formed by melting the compounds together, followed by controlled cooling. The melted mixture promotes molecular diffusion and reorganization, leading to co-crystal growth upon solidification (Luján-Torres et al., 2023; Sardar et al., 2013).

Each method offers distinct advantages depending on factors like compound properties, desired co-crystal characteristics, and scalability. The choice of method influences the resulting co-crystal's structure, stability, and properties, making it a critical consideration in co-crystal design for applications ranging from pharmaceuticals to materials science.

6.4 METHODS OF NANOENCAPSULATION

Methods of nanoencapsulation are diverse techniques used to entrap active substances within nanoscale carriers, enhancing their stability, bioavailability, and controlled release. These methods play a pivotal role in industries like pharmaceuticals, cosmetics, and food, offering innovative solutions for effective delivery and targeted action (Figure 6.1).

6.4.1 Emulsion Techniques

Emulsion-based methods involve dispersing the active substance within an immiscible solvent phase, forming droplets that are stabilized by surfactants. These droplets can be solidified to create nanoparticles with the active substance encapsulated (Chen et al., 2017).

6.4.2 Solvent Evaporation

This method entails dissolving the active compound and carrier material in a common solvent, followed by solvent removal through evaporation. The process leads to the formation of nanoparticles containing the encapsulated substance (Jafari et al., 2017).

6.4.3 Supercritical Fluid Technology

Utilizing supercritical fluids as solvents or antisolvents, this technique allows precise control over particle size and morphology. The supercritical fluid enhances solute diffusion and can be easily removed after encapsulation (Campardelli et al., 2013).

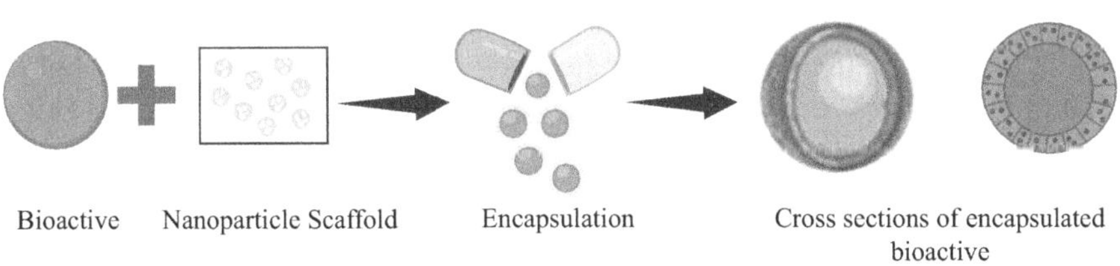

FIGURE 6.1 Encapsulation of bioactive

6.4.4 Spray Drying

In this method, a solution or suspension of the active substance and carrier is atomized into fine droplets and dried using a stream of hot air, resulting in encapsulated particles. The method of atomization and spray drying is akin to the intricate craft of creating delicate sculptures from molten glass. It is a technique employed to encapsulate active substances within carriers, and it plays a pivotal role in various industries, including pharmaceuticals, food, and materials science. This process involves the transformation of a solution or suspension containing the active substance and carrier into finely dispersed droplets, which are subsequently dried using a stream of hot air. The result is the creation of encapsulated particles, each with a protective shell that shields the active substance within. At its core, atomization and spray drying are akin to the artistry of controlled transformation. Much like a glassblower who skillfully shapes molten glass into intricate forms, this method takes a liquid solution or suspension, containing the active substance and carrier, and transforms it into a different state. It is this transformation that endows the active substance with new properties, making it more stable, bioavailable, or suitable for various applications.

6.4.5 Layer-by-Layer Assembly

This technique involves depositing alternating layers of oppositely charged materials onto a template core, leading to the creation of multilayered capsules with the active substance trapped within.

6.4.6 Nanoprecipitation

Nanoparticles are formed by rapidly mixing a solution containing the active compound and carrier with a nonsolvent, inducing precipitation and encapsulation of the active substance. The process of nanoprecipitation is akin to the art of sculpting, where the controlled mixing of elements results in the creation of intricate and finely detailed structures. In the realm of nanotechnology, nanoprecipitation is a technique used to form nanoparticles by rapidly mixing a solution containing the active compound and carrier with a nonsolvent. This carefully orchestrated dance of molecules induces precipitation, leading to the encapsulation of the active substance within the protective shell of nanoparticles.

> Much like a sculptor who meticulously molds clay into a desired shape, researchers employ nanoprecipitation to engineer nanoparticles with specific properties and functionalities. This process is particularly valuable in various fields, including pharmaceuticals, materials science, and nanomedicine, where the precise control of particle size, composition, and encapsulation efficiency is paramount.
>
> **(Jiang et al., 2022)**

These methods offer flexibility in designing nanoparticles with precise properties, such as size, shape, and release kinetics. The choice of method depends on the nature of the encapsulated substance, the carrier material, and the intended application (Jafari et al., 2017). By harnessing these techniques, nanoencapsulation facilitates the development of advanced bioactive delivery systems and improved food additives, contributing to advancements across multiple industries.

6.5 FACTORS INFLUENCING CO-CRYSTAL FORMATION

Co-crystal formation, the process of creating crystalline structures composed of two or more distinct compounds, is influenced by a multitude of factors that intricately dictate the outcome of the process. These factors play a crucial role in determining the success of co-crystal design and its properties.

I. **Molecular Interactions**: The nature and strength of intermolecular interactions between the co-formers significantly impact co-crystal formation. Hydrogen bonding, π-π stacking, van der Waals forces, and electrostatic interactions guide the assembly of the co-crystal lattice.
II. **Stoichiometry**: The ratio of co-formers in the mixture affects co-crystal stability and composition. Achieving the appropriate stoichiometry is vital to ensure balanced interactions and desired properties.
III. **Solubility**: Solubility of the co-formers in various solvents influences the supersaturation levels during co-crystal formation. Controlled solvent conditions can promote nucleation and growth of co-crystals.
IV. **Temperature and Pressure**: Temperature and pressure alterations impact molecular mobility and crystallization kinetics. Adjusting these parameters can lead to different polymorphs or co-crystal structures.
V. **Solvent and Solvent Evaporation Rate**: The choice of solvent and its evaporation rate can influence nucleation and crystal growth, affecting co-crystal size and morphology.
VI. **Mechanical Mixing**: Mixing intensity and duration during co-crystal preparation, especially in grinding or milling methods, can induce molecular interactions that facilitate co-crystal formation.
VII. **Additives and Modifiers**: The addition of third-party compounds, often referred to as co-formers or additives, can alter co-crystal properties by introducing additional interactions or altering crystallization pathways.
VIII. **pH and Ionic Strength**: For ionizable compounds, pH and ionic strength of the solution can influence molecular charge and solubility, impacting co-crystal formation.
IX. **Crystal Habit Modifiers**: Certain compounds or conditions can influence crystal habit, promoting specific crystal faces and morphologies during co-crystal growth.
X. **Seeding**: Introducing pre-formed crystals of the co-crystal can serve as nuclei for crystal growth, aiding in controlling the outcome of co-crystal formation (Karangutkar et al., 2020).

Understanding and manipulating these factors is paramount in designing co-crystals with desired properties. Researchers and industries consider these variables to engineer co-crystals tailored for applications ranging from pharmaceuticals to materials science.

6.6 FACTORS INFLUENCING NANOENCAPSULATION

Nanoencapsulation, a technique involving the encapsulation of active substances within nanoscale carriers, is governed by a myriad of factors that intricately shape the properties and performance of the encapsulated materials. These factors collectively influence the efficiency, stability, and controlled release capabilities of the nanoencapsulation process.

6.6.1 Carrier Material

The choice of carrier material, such as lipids, polymers, or proteins, determines the stability and compatibility of the encapsulated substance within the carrier matrix. Each type of carrier material has its own set of advantages and limitations:

6.6.1.1 Lipid-Based Carriers

Lipid-based carriers like liposomes or solid lipid nanoparticles are excellent for encapsulating lipophilic compounds due to their natural affinity for lipids. They provide protection against oxidation and can mimic cell membranes, making them ideal for drug delivery systems.

6.6.1.2 Polymeric Nanoparticles

Polymeric materials, such as biodegradable polymers like poly(lactic-co-glycolic acid) (PLGA), offer a wide range of options for encapsulation. They can be tailored to release encapsulated substances over extended periods, making them suitable for sustained drug delivery.

6.6.1.3 Protein-Based Carriers

Proteins like albumin or zein can be used as carriers, especially in the food and pharmaceutical industries. They offer biocompatibility and can encapsulate bioactive compounds like peptides or enzymes.

6.6.2 Solvent System

The choice of solvents or solvent mixtures is a critical aspect of the nanoencapsulation process that profoundly influences the outcome. The solvent system affects various aspects of nanoencapsulation, including dissolution, particle formation, and encapsulation efficiency. Here are some key considerations related to the solvent system:

6.6.2.1 Solvent Compatibility

It's crucial to select solvents that are compatible with both the active substance and the chosen carrier material. Compatibility ensures that the active substance dissolves effectively and that the carrier remains stable during the encapsulation process.

6.6.2.2 Solubility Enhancement

Solvent selection can be used strategically to enhance the solubility of poorly soluble active substances. By choosing solvents in which the active substance is highly soluble, you can improve loading efficiency.

Phase Separation: The choice of solvents can lead to phase separation during the encapsulation process. This can result in the formation of nanocarriers with unique structures, such as core-shell nanoparticles or Janus nanoparticles, which can impact release kinetics and stability.

6.6.2.3 Solvent Removal

After nanoparticle formation, the solvent or solvents used in the process must be removed. The method of solvent removal, such as evaporation, freeze-drying, or spray drying, can impact the size, shape, and properties of the resulting nanoparticles.

Environmental Impact: Considerations related to the environmental impact of solvent choice are increasingly important. Green solvent options or solvent-free methods are being explored to reduce the ecological footprint of nanoencapsulation processes.

6.6.2.4 Scale-Up Challenges

When scaling up nanoencapsulation processes for commercial production, solvent choice becomes even more critical. Factors like safety, cost, and regulatory compliance must be carefully evaluated when selecting solvents for large-scale manufacturing.

6.6.2.5 Solvent Mixing

In some cases, a mixture of solvents may be used to achieve the desired encapsulation outcome. This approach can offer advantages in terms of solubility and control over particle formation.

6.6.2.6 Residual Solvents

The presence of residual solvents in the final product can have regulatory implications, particularly in pharmaceuticals. Ensuring that residual solvents are within acceptable limits is crucial for product safety and compliance.

In summary, the solvent system is a key factor in nanoencapsulation that significantly influences the dissolution, particle formation, and overall success of the encapsulation process. Careful consideration of solvent properties and their compatibility with both the active substance and carrier material is essential to achieve optimal results in nanoencapsulation applications across various industries.

6.6.3 Emulsification Technique

The emulsification technique chosen for nanoencapsulation is a pivotal factor that directly impacts the characteristics and performance of the resulting nanoparticles.

Different emulsification methods offer unique advantages and challenges, and their selection can significantly influence the properties of the nanoencapsulated materials. Here's a closer look at the role of emulsification techniques:

I. **High-Shear Homogenization**: This method involves the use of mechanical shear forces to break down the active substance and carrier into smaller droplets, which then self-assemble into nanoparticles. High-shear homogenization is often used in the pharmaceutical and food industries due to its scalability and versatility. The intensity and duration of homogenization can be adjusted to control particle size and distribution.

II. **Ultrasonication**: Ultrasonication employs high-frequency sound waves to create cavitation bubbles in the emulsion, leading to the breakup of larger droplets into smaller ones. It is a gentle and precise method that is often preferred for sensitive compounds. The choice of sonication parameters, such as frequency and amplitude, influences the particle size and stability of the resulting nanoparticles.

III. **Microfluidization**: Microfluidization involves forcing the emulsion through microchannels or homogenization chambers at high pressure. This method offers precise control over particle size and distribution and is especially useful for achieving nanoscale encapsulation. It is commonly used in pharmaceutical and biotechnology applications.

IV. **Coacervation**: In coacervation, phase separation is induced by adding a coacervating agent, causing the encapsulation material to separate from the continuous phase and form droplets. This method is particularly effective for encapsulating hydrophobic substances and can produce core-shell nanoparticles.

V. **Double Emulsion (W/O/W or O/W/O)**: Double emulsion techniques involve multiple emulsification steps to encapsulate substances within multiple layers. Water-in-oil-in-water (W/O/W) and oil-in-water-in-oil (O/W/O) emulsions allow for the creation of complex nanostructures with precise control over core-shell arrangements and release profiles.

VI. **Implications on Particle Characteristics**: The choice of emulsification technique directly affects the size, distribution, and morphology of the nanoparticles. Smaller droplets often lead to finer, more uniform nanoparticles, while the use of multiple emulsification steps can create more complex structures.

VII. **Scalability and Production Considerations**: When selecting an emulsification method, scalability and production efficiency should be considered. Some techniques are more suitable for laboratory-scale research, while others can be adapted for large-scale manufacturing.

VIII. **Energy Input**: Different emulsification techniques require varying levels of energy input. This can influence the stability and potential damage to sensitive compounds during the encapsulation process.

In summary, the emulsification technique employed in nanoencapsulation is a critical factor that dictates the size, uniformity, and complexity of the resulting nanoparticles. Understanding the strengths and limitations of each method is essential for tailoring the encapsulation process to the specific needs of diverse applications, ranging from drug delivery to food encapsulation and beyond.

6.7 SURFACTANTS

Surfactants play a crucial role in nanoencapsulation by stabilizing emulsions and influencing the properties of the resulting nanoparticles. These amphiphilic molecules have both hydrophobic and hydrophilic regions, which enable them to reduce the interfacial tension between immiscible phases, such as oil and water. Here are several key aspects of surfactant selection and their impact on nanoencapsulation:

6.7.1 Stabilization of Emulsions

Surfactants are essential for preventing the coalescence or aggregation of emulsion droplets during the nanoencapsulation process. They form a protective layer at the oil-water interface, reducing the tendency of droplets to merge and ensuring the formation of stable nanostructures.

6.7.2 Emulsifier Type

The choice of surfactant can significantly affect the size and stability of the nanoparticles. Surfactants can be classified into ionic (anionic, cationic, or zwitterionic) or non-ionic types, each with distinct properties. Ionic surfactants, for instance, may influence the zeta potential of nanoparticles and their interaction with biological systems.

6.7.3 Surfactant Concentration

The concentration of surfactants in the emulsion system is critical. Too little surfactant may result in unstable emulsions, while excessive surfactant can affect the properties of the encapsulated material and the final nanoparticles. Optimization of surfactant concentration is necessary for achieving the desired nanoparticle characteristics.

6.7.4 Biocompatibility

In applications involving the encapsulation of substances intended for consumption or administration, such as in the pharmaceutical or food industries, the biocompatibility of surfactants is a critical consideration. Biocompatible surfactants are essential to ensure the safety and efficacy of the nanoencapsulated product.

6.7.5 Release Control

The choice of surfactant can influence the release kinetics of the encapsulated substance. Some surfactants can form a more tightly packed layer at the interface, slowing down the release of the active compound. This can be advantageous for achieving sustained release profiles.

6.7.6 Surface Charge

Surfactants can also impart surface charge to the nanoparticles, influencing their stability and interaction with biological systems. Zeta potential measurements can help assess the surface charge of nanoencapsulated materials and their colloidal stability.

6.7.7 Compatibility with Active Substance

Surfactants must be compatible with both the active substance and the chosen carrier material. Incompatibility can lead to changes in the stability and release properties of the nanoencapsulated material.

6.7.8 Environmental Impact

Considerations related to the environmental impact of surfactant use are increasingly important. Choosing surfactants with low environmental toxicity or exploring green surfactant alternatives aligns with sustainability goals.

In summary, surfactants are vital components in nanoencapsulation, serving to stabilize emulsions, control particle size, influence release kinetics, and impact the biocompatibility of the final nanoencapsulated product. The careful selection and optimization of surfactants are essential steps in tailoring the encapsulation process to meet the specific requirements of diverse applications across various industries.

6.8 PH AND IONIC STRENGTH

The pH and ionic strength of the formulation are critical factors that exert a substantial influence on the nanoencapsulation process. These factors affect the interactions between the active substance, carrier, and the surrounding environment, ultimately impacting nanoparticle stability and encapsulation efficiency. Here are some key considerations:

I. **pH**: The pH level of the formulation plays a pivotal role in controlling the ionization and solubility of both the active substance and carrier materials. Depending on the chemical nature of these components, pH can have a profound effect on their behavior during encapsulation. For instance:
   - pH can influence the degree of ionization of acidic or basic active substances, impacting their solubility and partitioning into the carrier.

- pH can alter the charge on the carrier or surfactant molecules, affecting their electrostatic interactions and potential for aggregation.
- pH can lead to changes in the stability of nanoparticles, as alterations in pH can promote flocculation or aggregation if not carefully controlled.

II. **Ionic Strength**: The ionic strength of the formulation, which is determined by the concentration of ions (e.g., salts) in the solution, can significantly impact nanoencapsulation. Key considerations include:

Ionic strength can influence the stability of emulsions or dispersions. Higher ionic strength may promote aggregation or precipitation of particles due to changes in electrostatic repulsion or attraction forces.

Salts can affect the solubility and conformation of polymers used as carriers, which in turn can impact encapsulation efficiency and release kinetics.

Ionic strength can also influence the osmotic pressure within nanoparticles, affecting their swelling or collapse behavior.

III. **Buffer Systems**: In many formulations, buffer systems are employed to maintain a stable pH. These buffers help control pH changes that may occur during the encapsulation process or when the nanoparticles are exposed to different physiological environments (e.g., in drug delivery applications).

IV. **Biological Relevance**: In biomedical applications, the pH and ionic strength of the formulation must align with the physiological conditions of the target site. For instance, in oral drug delivery, the pH of the gastrointestinal tract can vary significantly, and formulations must be designed to withstand these variations.

V. **Compatibility with Active Substances**: The pH and ionic strength should be selected to ensure the stability and efficacy of the active substance. Some compounds are sensitive to pH changes, and their activity or stability may be compromised under certain conditions.

In summary, pH and ionic strength are essential parameters in nanoencapsulation, and their careful control is necessary to achieve stable, efficient encapsulation processes. The selection of pH and ionic strength should align with the characteristics of the active substance, carrier, and intended application to optimize the performance and stability of the nanoencapsulated materials.

VI. **Temperature and Mixing Speed**: Controlling temperature and mixing speed during the encapsulation process is crucial, as they have a significant influence on the rate of nanoparticle formation and the final particle size distribution. These factors play a central role in determining the quality and efficiency of nanoencapsulation. Here's a closer look at their impact:

VII. **Rate of Nanoparticle Formation**: Temperature and mixing speed affect the kinetics of the encapsulation process. Higher temperatures generally lead to faster dissolution and diffusion of the active substance and carrier materials, promoting rapid nanoparticle formation. Similarly, increased mixing speed enhances the collision frequency between particles, facilitating more efficient encapsulation.

VIII. **Particle Size Distribution**: Both temperature and mixing speed can influence the particle size distribution of the resulting nanoparticles. Higher temperatures and faster mixing speeds tend to produce smaller particles, but the degree of influence depends on the specific formulation and the encapsulation method employed. Careful optimization is needed to achieve the desired particle size distribution.
IX. **Emulsion Stability**: The stability of the emulsion or dispersion is sensitive to temperature and mixing speed. Inadequate mixing can result in poor emulsion stability, leading to coalescence or aggregation of droplets. Conversely, excessively high mixing speeds can cause excessive shear forces, potentially damaging the nanoparticles or compromising stability.
X. **Phase Transitions**: Temperature variations can trigger phase transitions in certain carrier materials, such as lipids or polymers. These transitions can affect the properties and structure of nanoparticles, including their size, morphology, and release kinetics. Knowledge of phase behavior is crucial for achieving desired nanoparticle characteristics.
XI. **Encapsulation Efficiency**: Temperature and mixing speed can influence the encapsulation efficiency by affecting the interaction between the active substance and the carrier. Optimal conditions must be established to ensure efficient encapsulation while avoiding losses due to aggregation or degradation.
XII. **Scale-Up Challenges**: Scaling up nanoencapsulation processes for industrial production introduces additional challenges in controlling temperature and mixing. Ensuring uniform temperature distribution and consistent mixing across larger volumes is critical for maintaining product quality and reproducibility.
XIII. **Energy Consumption**: Higher mixing speeds and elevated temperatures can increase energy consumption in the encapsulation process. Balancing the need for efficient encapsulation with energy efficiency is essential for cost-effective manufacturing.
XIV. **Regulatory Compliance**: In regulated industries like pharmaceuticals and food, temperature and mixing speed must be documented, controlled, and validated as part of good manufacturing practices (GMP) to ensure product consistency and compliance with regulatory requirements.

In summary, temperature and mixing speed are key process parameters that impact the rate of nanoparticle formation, particle size distribution, emulsion stability, and encapsulation efficiency in nanoencapsulation. Their optimization must be tailored to the specific formulation and encapsulation method to achieve the desired nanoparticle characteristics for diverse applications across various industries.

6.9 ENCAPSULATION METHOD

The choice of encapsulation method is a critical factor in nanoencapsulation, as it determines the degree of control over particle size, structure, and loading efficiency. Different methods offer unique advantages and limitations, making their selection crucial for achieving the desired nanoparticle characteristics. Here are some key considerations related to different encapsulation methods:

6.9.1 Solvent Evaporation

Solvent evaporation is a commonly used method in nanoencapsulation. It involves dissolving the active substance and carrier in a volatile solvent and then removing the solvent to form nanoparticles. This method offers good control over particle size and loading efficiency, making it suitable for a wide range of applications. However, it may not be ideal for thermally sensitive compounds.

6.9.2 Supercritical Fluid Technology

Supercritical fluid technology, particularly supercritical CO_2, is well-suited for encapsulating lipophilic compounds. It allows for precise control over particle size and can produce nanoparticles with a narrow size distribution. This method is preferred for applications where the use of organic solvents is undesirable.

6.9.3 Layer-by-Layer Assembly

Layer-by-layer assembly is a versatile technique used to create multilayered nanoparticles with fine-tuned properties. It involves sequentially depositing layers of oppositely charged polymers, surfactants, or materials onto a core particle. This method is advantageous for achieving controlled release and can encapsulate a wide range of substances, including biomolecules.

6.9.4 Coacervation

Coacervation is a phase separation technique that is particularly effective for encapsulating hydrophobic substances. It involves the addition of a coacervating agent that induces the carrier material to separate and form droplets. This method can produce core-shell nanoparticles and offers control over release kinetics.

6.9.5 Electrostatic Assembly

Electrostatic assembly relies on the electrostatic interactions between charged components to encapsulate active substances. This method is often used for encapsulating biomolecules, where precise control over the encapsulation process is required.

6.9.6 Spray Drying

Spray drying involves atomizing a solution or suspension of the active substance and carrier into fine droplets that are rapidly dried to form solid nanoparticles. This method is suitable for heat-sensitive compounds and can produce nanoparticles with tailored properties.

6.9.7 Emulsion Techniques

Various emulsion techniques, such as oil-in-water (O/W) or water-in-oil (W/O) emulsions, offer flexibility in controlling particle size and structure. Double emulsion methods (W/O/W or O/W/O) allow for the encapsulation of hydrophilic and lipophilic substances in distinct layers.

6.9.8 Implications for Release and Stability

The choice of encapsulation method can significantly impact the release kinetics and stability of nanoparticles. Some methods are better suited for achieving sustained release profiles, while others may be more suitable for rapid release.

6.9.9 Scale-Up Challenges

When transitioning from laboratory-scale to industrial-scale production, the scalability of the chosen encapsulation method becomes critical. Factors like production rate, equipment, and cost-effectiveness need to be considered.

In summary, the encapsulation method is a key determinant of the nanoparticle characteristics and the success of nanoencapsulation. Careful consideration of the advantages and limitations of each method, along with their compatibility with the specific active substance and application, is essential for achieving the desired encapsulation outcome.

6.10 TARGETED RELEASE

The design of nanoparticles for specific release profiles is a crucial aspect of nanoencapsulation. Achieving targeted release can involve tailoring the carrier characteristics or incorporating additional stimuli-responsive components. Here are some key considerations related to targeted release in nanoencapsulation:

6.10.1 Carrier Modification

Altering the properties of the carrier material can be a fundamental strategy for achieving targeted release. This may involve modifying the carrier's surface charge, hydrophobicity, or composition to control the release kinetics of the encapsulated substance. For instance, adding functional groups to the carrier surface can enable pH-responsive release, where the nanoparticles release their cargo in response to changes in pH levels, such as in the acidic environment of tumors.

6.10.2 Stimuli-Responsive Nanoparticles

Designing nanoparticles that respond to specific external stimuli, such as temperature, light, magnetic fields, or ultrasound, is an advanced approach for achieving targeted release. Stimuli-responsive nanoparticles can be engineered to release their cargo when exposed to the appropriate trigger, enabling precise control over the release profile.

6.10.3 Layered or Multicompartmental Nanoparticles

Layered or multicompartmental nanoparticles consist of multiple layers or compartments with different release characteristics. These nanoparticles can be designed to release substances sequentially or in response to specific environmental conditions. For example, a nanoparticle might have an outer layer that releases rapidly and an inner layer with sustained release properties.

6.10.4 Encapsulation with Ligands or Antibodies

Incorporating ligands or antibodies on the nanoparticle surface can enable targeted delivery to specific cells or tissues. This approach, often used in drug delivery systems, enhances the selectivity of the release by ensuring that the nanoparticles bind to target receptors or antigens before releasing their cargo.

6.10.5 Biodegradable Components

Choosing biodegradable carrier materials allows for the gradual degradation of nanoparticles in the body, releasing their cargo over time. This approach is particularly useful for achieving sustained release and minimizing the potential accumulation of carrier materials in tissues.

6.10.6 Incorporation of Triggerable Components

Some nanoparticles are designed with triggerable components, such as gates or caps, that can be opened or removed upon exposure to certain conditions. For example, a temperature-sensitive polymer may be used to form a "gate" that opens when the temperature reaches a specific threshold.

6.10.7 Hydrogels and Matrices

Hydrogels or solid matrices can be used as carriers to provide controlled release of encapsulated substances. These materials can be engineered to respond to environmental factors, such as changes in pH or the presence of enzymes, leading to controlled release.

6.10.8 Regulatory Considerations

In pharmaceutical and medical applications, achieving targeted release while ensuring safety and efficacy is subject to regulatory scrutiny. It's essential to carefully document and validate the design and performance of nanoparticles intended for targeted release.

In summary, achieving targeted release in nanoencapsulation involves a combination of strategies, including carrier modification, stimuli-responsive components, ligand-based targeting, and thoughtful consideration of the environmental conditions where the nanoparticles will be deployed. These approaches enable precise control over release profiles, making nanoencapsulation a versatile technology with applications in drug delivery, diagnostics, and other fields where controlled release is essential.

6.11 SCALE-UP CONSIDERATIONS

Scaling up nanoencapsulation processes from laboratory-scale to larger production scales introduces a range of engineering and processing constraints that must be carefully considered. Factors that may influence nanoencapsulation at different scales include

6.11.1 Production Rate

The production rate required at an industrial scale is typically much higher than what can be achieved in a laboratory setting. To meet production demands, it may be necessary to optimize equipment and processes for higher throughput, such as using larger mixing vessels or increasing the flow rate of materials.

6.11.2 Equipment and Infrastructure

Transitioning to larger scales often requires the acquisition or modification of specialized equipment that can handle larger volumes. This includes larger reactors, pumps, and filtration systems, as well as facilities with sufficient space and utilities to support production.

6.11.3 Consistency and Reproducibility

Achieving consistent and reproducible results becomes more challenging as production scales increase. Variations in raw materials, mixing efficiency, and environmental conditions can have a greater impact on product quality. Robust process control and quality assurance measures are essential to ensure uniformity across batches.

6.11.4 Heat Transfer

Efficient heat transfer becomes increasingly important at larger scales. Scaling up may require the implementation of more advanced heating and cooling systems to maintain precise temperature control during the encapsulation process.

6.11.5 Safety

Safety considerations become even more critical during large-scale production. The handling of larger quantities of materials, potential chemical hazards, and the use of high-pressure equipment all necessitate comprehensive safety protocols and training.

6.11.6 Economic Considerations

Scaling up nanoencapsulation processes can bring about changes in cost structures. It's essential to perform cost analyses, including factors like raw material costs, labor, energy consumption, and equipment depreciation, to assess the economic viability of large-scale production.

6.11.7 Regulatory Compliance

Compliance with regulatory requirements becomes more complex as production scales increase, particularly in industries like pharmaceuticals and food. Large-scale production must adhere to good manufacturing practices (GMP) and meet stringent quality control standards.

6.11.8 Validation and Testing

Validation of the encapsulation process at larger scales is crucial to ensure that the product consistently meets quality specifications. Testing methods may need to be adapted to accommodate larger sample sizes and higher throughput.

6.11.9 Environmental Impact

As production scales increase, so does the potential environmental impact. Strategies to reduce waste, energy consumption, and environmental emissions should be considered in line with sustainability goals.

6.11.10 Downstream Processing

Large-scale production often involves more extensive downstream processing steps, such as purification, drying, and packaging. These steps need to be carefully integrated into the production process to maintain product quality and efficiency.

6.11.11 Technology Transfer

The transfer of technology from laboratory-scale research to industrial-scale production is a complex process that involves knowledge transfer, training, and close collaboration between research and development teams and production teams.

In conclusion, scaling up nanoencapsulation processes presents a range of challenges that require careful planning, investment in equipment and infrastructure, and adherence to rigorous quality control and safety standards. Successful scale-up is essential for bringing nanoencapsulated products to market in a cost-effective and commercially viable manner (Haris et al., 2023; Ünlüsayin et al., 2016; Gutiérrez et al., 2013).

By carefully considering these factors, researchers and industries can optimize the nanoencapsulation process to achieve desired characteristics for drug delivery, formulation, food enhancement, and other applications.

6.12 ADVANTAGES AND APPLICATIONS OF CO-CRYSTALLIZATION

Co-crystallization, the formation of crystalline structures containing two or more different molecules, offers several advantages and finds diverse applications across various industries due to its ability to modify and enhance the properties of the constituent compounds. Co-crystallization can lead to improved properties such as solubility, stability, bioavailability, and mechanical strength of the resulting material compared to individual components. Co-crystallization can improve the stability and release of flavors, nutrients, and additives in food products, enhancing taste, nutritional value, and shelf life. By selecting specific co-formers and adjusting their ratios, co-crystals can be engineered with precise properties suitable for particular applications. Co-crystallization can provide a new intellectual property strategy, offering patent protection for a new crystalline form of a known compound. Co-crystallization can provide a new intellectual property strategy, offering patent protection for a new crystalline form of a known compound. Co-crystals can be applied to a wide range of compounds, from pharmaceuticals and agrochemicals to materials science, electronics, and even food science (Chezanoglou et al., 2023; Bajaj et al., 2021; Irigoiti et al., 2021).

Nanotechnology enables the manipulation and control of materials at the nanoscale, allowing for the precise design and engineering of novel materials and devices with tailored properties. Nanomaterials often exhibit unique physical, chemical, and biological properties due to their small size and high surface-to-volume ratio. This can lead to improved strength, conductivity, reactivity, and other advantageous characteristics. Nanotechnology facilitates the creation of smaller and more efficient devices, leading to advancements in fields such as electronics, medicine, and energy storage. Nanoparticles can be designed to carry drugs, nutraceuticals, bioactives, or other therapeutic agents directly to specific cells or tissues, minimizing side effects and improving treatment efficacy. Nanotechnology and co-crystallization enable the development of bioactive delivery systems that enhance solubility, bioavailability, and controlled release, revolutionizing nutraceuticals. Nanotechnology is used in food packaging to improve shelf life, and nanoparticles can also be added to food to enhance nutritional value and safety (Azeredo et al., 2009; Liu et al., 2023).

6.13 IMPLICATIONS OF CO-CRYSTALLIZATION AND NANOTECHNOLOGY FOR THE FOOD INDUSTRY

The amalgamation of co-crystallization and nanotechnology holds immense implications for the food industry, revolutionizing various aspects from food safety and preservation to nutritional enhancement and consumer experience (Rai et al., 2022). Co-crystallization can lead to the creation of new food additives with improved solubility and stability, contributing to better texture, flavor, and appearance of food products. By forming co-crystals of certain compounds, the industry can also reduce the need for excessive additives, promoting cleaner and healthier ingredient lists (Kaur et al., 2021). Nanotechnology's application in the food sector is equally transformative. Nanoscale materials can be used for enhanced packaging, extending shelf life and ensuring food safety (Jadhav et al., 2022; Lopez-Polo et al., 2021). Nanosensors could detect contaminants and spoilage, thereby enhancing quality control and traceability. Nanoencapsulation can protect sensitive nutrients, vitamins, and flavor compounds, preserving them through processing and storage. Moreover, nanotechnology enables controlled release of additives, optimizing flavor and texture over time (Gökmen et al., 2011). However, ethical considerations, safety concerns, and regulatory challenges must be navigated. The introduction of novel materials and technologies demands rigorous testing to ensure they are safe for consumption. Transparent labeling and clear communication about the use of co-crystallization and nanotechnology are crucial for consumer acceptance (Saini et al., 2023).

In essence, the convergence of co-crystallization and nanotechnology in the food industry holds the promise of safer, more nutritious, and better-tasting food products, while also necessitating responsible innovation and comprehensive safety assessments.

6.14 NEED FOR CONTINUED RESEARCH AND DEVELOPMENT

The ongoing research and development of co-crystallization and nanotechnology are crucial to harness the full potential of these cutting-edge fields and address the complex challenges and opportunities they present. Research in co-crystallization and nanotechnology continuously expands our understanding of materials at the atomic and molecular levels. By manipulating matter at this scale, researchers can unlock entirely new materials with unprecedented properties. These materials may possess exceptional electrical conductivity, extraordinary strength, or remarkable thermal stability, enabling innovations in electronics, materials science, and beyond. In the realm of pharmaceuticals and healthcare, ongoing research into co-crystallization and nanotechnology holds the promise of personalized medicine and targeted therapies. These fields enable the design of drug formulations tailored to individual patient needs. Through precise control of drug release and bioavailability, co-crystals and nanoparticles offer the potential to maximize treatment efficacy while minimizing side effects. Energy storage and conversion are among the most pressing global challenges. Nanotechnology research is pivotal in developing advanced materials for energy storage systems like batteries and supercapacitors. The quest for more efficient and sustainable energy solutions relies on innovations in nanomaterials, allowing for the development of high-capacity and fast-charging energy storage devices.

In the case of co-crystallization, further research is needed to explore the vast landscape of possible combinations and conditions for forming co-crystals. Understanding the underlying mechanisms, thermodynamics, and kinetics of co-crystal formation will enable researchers to predict and control outcomes more effectively, ensuring reproducibility and scalability. This will be especially important in industries like pharmaceuticals, where co-crystals can offer tailored drug formulations with enhanced properties.

Similarly, nanotechnology demands continuous exploration due to its broad applications and potential to reshape industries. Ongoing research is necessary to delve into the safety aspects of nanomaterials, ensuring they do not pose risks to human health or the environment. Additionally, optimizing the synthesis methods, enhancing the precision of manipulation, and designing nanomaterials with specific functions are areas that require further investigation.

Both fields require collaboration across disciplines, as well as the alignment of research efforts with ethical, regulatory, and sustainability considerations. As co-crystallization and nanotechnology continue to advance, they hold the promise of addressing pressing challenges in healthcare, energy, materials science, and more. Continued research and development will enable us to unlock new possibilities, innovate responsibly, and drive positive impacts across various sectors. Co-crystallization and nanotechnology hold immense promise for addressing some of the most pressing challenges in science, technology, and society. However, to unlock their full potential and navigate the complexities they present, continuous research and development are essential. These fields offer opportunities for innovation, improved healthcare, sustainable energy solutions, and advanced materials, but they also demand responsible and ethical stewardship to ensure positive impacts on humanity and the environment.

6.15 CONCLUSION

In conclusion, the integration of co-crystallization and nanoencapsulation techniques into the realm of food bioactive components offers a compelling avenue for revolutionizing the food industry. The synergy between these advanced technologies presents unprecedented opportunities to enhance food quality, safety, nutritional value, and consumer experience.

Co-crystallization brings a fresh perspective to food formulation by enabling the creation of novel structures that enhance the solubility, stability, and bioavailability of bioactive compounds. This, in turn, leads to improved taste, texture, and visual appeal of food products. The precision of co-crystallization allows for the reduction of undesirable additives and the promotion of cleaner labels, aligning with the demands of health-conscious consumers (Acevedo and Franchetti, 2016).

Nanoencapsulation, on the other hand, empowers the food industry to encapsulate sensitive bioactive components within protective nanoscale carriers, safeguarding their integrity during processing and storage. This technology contributes to the controlled release of flavors, nutrients, and additives, enabling sustained benefits and prolonged shelf life (Nejatian et al., 2022). Additionally, nanoencapsulation enhances the potential for targeted delivery of bioactive compounds, providing health-promoting effects while maintaining the sensory attributes of the food (Karim et al., 2022).

However, the adoption of these techniques in the food industry necessitates a holistic approach that encompasses scientific rigor, regulatory compliance, and ethical considerations. Rigorous research is imperative to ensure the safety of these novel structures and the mitigation of potential risks. Transparency in labeling and effective communication will be paramount to building consumer trust and acceptance of these innovative approaches.

As co-crystallization and nanoencapsulation techniques continue to evolve, their impact on the food landscape will undoubtedly be profound. By addressing challenges related to nutrient degradation, flavor instability, and health-conscious consumer preferences, these techniques pave the way for a new era of functional foods that are both delicious and nutritious. The amalgamation of these cutting-edge technologies with culinary creativity holds the potential to redefine how we perceive and interact with food, offering a more personalized and health-enhancing dining experience for consumers around the globe.

REFERENCES

Acevedo, Nuria C., and Danielle Franchetti. 'Analysis of Co-Crystallized Free Phytosterols with Triacylglycerols as a Functional Food Ingredient'. *Food Research International* 85 (2016): 104–12. https://doi.org/https://doi.org/10.1016/j.foodres.2016.04.012.

Awuchi, Chinaza Godswill, Sonia Morya, Terwase Abraham Dendegh, Charles Odilichukwu R. Okpala, and Małgorzata Korzeniowska. 'Nanoencapsulation of Food Bioactive Constituents and Its Associated Processes: A Revisit'. *Bioresource Technology Reports* 19 (2022): 101088. https://doi.org/https://doi.org/10.1016/j.biteb.2022.101088.

Azeredo, Henriette M. C. De. 'Nanocomposites for Food Packaging Applications'. *Food Research International* 42, no. 9 (2009): 1240–53. https://doi.org/10.1016/j.foodres.2009.03.019.

Bajaj, Seema R., and Rekha S. Singhal. 'Enhancement of Stability of Vitamin B12 by Co-Crystallization: A Convenient and Palatable Form of Fortification'. *Journal of Food Engineering* 291 (2021): 110231. https://doi.org/https://doi.org/10.1016/j.jfoodeng.2020.110231.

Behnamnik, Ali, Mohsen Vazifedoost, Zohreh Didar, and Bahareh Hajirostamloo. 'Evaluation of Physicochemical, Structural, and Antioxidant Properties of Microencapsulated Seed Extract from Securigera Securidaca by Co-Crystallization Method during Storage Time'. *Biocatalysis and Agricultural Biotechnology* 35 (2021): 102090. https://doi.org/https://doi.org/10.1016/j.bcab.2021.102090.

Biswas, Rahul, Mahabub Alam, Animesh Sarkar, Md Ismail Haque, Md. Moinul Hasan, and Mominul Hoque. 'Application of Nanotechnology in Food: Processing, Preservation, Packaging and Safety Assessment'. *Heliyon* 8, no. 11 (2022): e11795. https://doi.org/https://doi.org/10.1016/j.heliyon.2022.e11795.

Campardelli, R., G. Della Porta, V. Gomez, S. Irusta, E. Reverchon, and J. Santamaria. 'Encapsulation of Titanium Dioxide Nanoparticles in PLA Microspheres Using Supercritical Emulsion Extraction to Produce Bactericidal Nanocomposites'. *Journal of Nanoparticle Research* 15, no. 10 (2013). https://doi.org/10.1007/s11051-013-1987-3.

Chen, Fang, Guang-Qin Fan, Zipei Zhang, Ruojie Zhang, Ze-Yuan Deng, and David Julian McClements. 'Encapsulation of Omega-3 Fatty Acids in Nanoemulsions and Microgels: Impact of Delivery System Type and Protein Addition on Gastrointestinal Fate'. *Food Research International* 100 (2017): 387–95. https://doi.org/https://doi.org/10.1016/j.foodres.2017.07.039.

Chezanoglou, Evangelos, and Athanasia M. Goula. 'Co-Crystallization in Sucrose: A Promising Method for Encapsulation of Food Bioactive Components'. *Trends in Food Science & Technology* 114 (2021): 262–74. https://doi.org/https://doi.org/10.1016/j.tifs.2021.05.036.

Chezanoglou, Evangelos, Nefeli Kenanidou, Chrysostomos Spyropoulos, Dimitra Xenitopoulou, Eleni Zlati, and Athanasia M. Goula. 'Encapsulation of Pomegranate Peel Extract in Sucrose Matrix by Co-Crystallization'. *Sustainable Chemistry and Pharmacy* 31 (2023): 100949. https://doi.org/https://doi.org/10.1016/j.scp.2022.100949.

Gökmen, Vural, Burçe Ataç Mogol, Roberta Barone Lumaga, Vincenzo Fogliano, Zoya Kaplun, and Eyal Shimoni. 'Development of Functional Bread Containing Nanoencapsulated Omega-3 Fatty Acids'. *Journal of Food Engineering* 105, no. 4 (2011): 585–91. https://doi.org/https://doi.org/10.1016/j.jfoodeng.2011.03.021.

Gutiérrez, Francisco J., Silvia M. Albillos, Elvira Casas-Sanz, Ziortza Cruz, Carlos García-Estrada, Ana García-Guerra, José García-Reverter, et al. 'Methods for the Nanoencapsulation of β-Carotene in the Food Sector'. *Trends in Food Science & Technology* 32, no. 2 (2013): 73–83. https://doi.org/https://doi.org/10.1016/j.tifs.2013.05.007.

Haris, Mohammad, Touseef Hussain, Heba I. Mohamed, Amir Khan, Moh. Sajid Ansari, Atirah Tauseef, Abrar Ahmad Khan, and Naseem Akhtar. 'Nanotechnology – A New Frontier of Nano-Farming in Agricultural and Food Production and Its Development'. *Science of The Total Environment* 857 (2023): 159639. https://doi.org/https://doi.org/10.1016/j.scitotenv.2022.159639.

Irigoiti, Yanet, Diego K. Yamul, and Alba S. Navarro. 'Co-Crystallized Sucrose with Propolis Extract as a Food Ingredient: Powder Characterization and Antioxidant Stability'. *LWT* 143 (2021): 111164. https://doi.org/https://doi.org/10.1016/j.lwt.2021.111164.

Jadhav, Rachana, Pallavi Pawar, Vishnu Choudhari, Niraj Topare, Sunita Raut-Jadhav, Shantini Bokil, and Anish Khan. 'An Overview of Antimicrobial Nanoparticles for Food Preservation'. *Materials Today: Proceedings* (2022). https://doi.org/https://doi.org/10.1016/j.matpr.2022.07.045.

Jafari, Seid Mahdi. 'Chapter 1 - An Introduction to Nanoencapsulation Techniques for the Food Bioactive Ingredients'. In *Nanoencapsulation of Food Bioactive Ingredients Jafari*, edited by Seid Mahdi Jafari, 1–62. Academic Press, 2017. https://doi.org/https://doi.org/10.1016/B978-0-12-809740-3.00001-5.

Jiang, Longwei, Ning Xia, Fenghui Wang, Cancan Xie, Rong Ye, Hongjie Tang, Huajiang Zhang, and Yingzhu Liu. 'Preparation and Characterization of Curcumin/β-Cyclodextrin Nanoparticles by Nanoprecipitation to Improve the Stability and Bioavailability of Curcumin'. *LWT* 171 (2022): 114149. https://doi.org/https://doi.org/10.1016/j.lwt.2022.114149.

Karangutkar, Amruta V., and Laxmi Ananthanarayan. 'Co-Crystallization of Basella Rubra Extract with Sucrose: Characterization of Co-Crystals and Evaluating the Storage Stability of Betacyanin Pigments'. *Journal of Food Engineering* 271 (2020): 109776. https://doi.org/https://doi.org/10.1016/j.jfoodeng.2019.109776.

Karim, Aiman, Abdur Rehman, Jianguo Feng, Asia Noreen, Elham Assadpour, Mohammad Saeed Kharazmi, Zhang Lianfu, and Seid Mahdi Jafari. 'Alginate-Based Nanocarriers for the Delivery and Controlled-Release of Bioactive Compounds'. *Advances in Colloid and Interface Science* 307 (2022): 102744. https://doi.org/https://doi.org/10.1016/j.cis.2022.102744.

Kaur, Prabhjot, Abdallah Elsayed, Jayasankar Subramanian, and Ashutosh Singh. 'Encapsulation of Carotenoids with Sucrose by Co-Crystallization: Physicochemical Properties, Characterization and Thermal Stability of Pigments'. *LWT* 140 (2021): 110810. https://doi.org/https://doi.org/10.1016/j.lwt.2020.110810.

Liao, Ziying, Lulu Dong, Muwen Lu, Songbai Zheng, Yong Cao, Michael Rogers, and Yaqi Lan. 'Construction of Interfacial Crystallized Oleogel Emulsion with Improved Thermal Stability'. *Food Chemistry* 420 (2023): 136029. https://doi.org/https://doi.org/10.1016/j.foodchem.2023.136029.

Liu, Lipeng, Bo Wang, Chao Li, Chang Wei, and Zhiyong Wei. 'Development of High Barrier Biodegradable Poly(Butylene Carbonate-Co-Terephthalate)-Based Blends with Balanced Mechanical Properties by Self-Reactive Compatibility and Co-Crystallization'. *Polymer* 281 (2023): 126130. https://doi.org/https://doi.org/10.1016/j.polymer.2023.126130.

Lopez-Polo, Johana, Angela Monasterio, Plinio Cantero-López, and Fernando A. Osorio. 'Combining Edible Coatings Technology and Nanoencapsulation for Food Application: A Brief Review with an Emphasis on Nanoliposomes'. *Food Research International* 145 (2021): 110402. https://doi.org/https://doi.org/10.1016/j.foodres.2021.110402.

Luján-Torres, Víctor E., Néstor Gutiérrez-Méndez, Diego E. Carballo-Carballo, Martha Y. Leal-Ramos, Iván Salmerón, Samuel B. Pérez-Vega, and Sergio Martínez-Monteagudo. 'Co-Crystallization of Lactose-Flavonoids Using Panela Cheese Whey'. *Journal of Food Engineering* 357 (2023): 111598. https://doi.org/https://doi.org/10.1016/j.jfoodeng.2023.111598.

Nejatian, Mohammad, Nazanin Darabzadeh, Samad Bodbodak, Hamed Saberian, Zahra Rafiee, Mohammad Saeed Kharazmi, and Seid Mahdi Jafari. 'Practical Application of Nanoencapsulated Nutraceuticals in Real Food Products; a Systematic Review'. *Advances in Colloid and Interface Science* 305 (2022): 102690. https://doi.org/https://doi.org/10.1016/j.cis.2022.102690.

Pathakoti, Kavitha, Manjunath Manubolu, and Huey-Min Hwang. 'Nanostructures: Current Uses and Future Applications in Food Science'. *Journal of Food and Drug Analysis* 25, no. 2 (2017): 245–53. https://doi.org/https://doi.org/10.1016/j.jfda.2017.02.004.

Peters, Ruud J. B., Hans Bouwmeester, Stefania Gottardo, Valeria Amenta, Maria Arena, Puck Brandhoff, Hans J. P. Marvin, et al. 'Nanomaterials for Products and Application in Agriculture, Feed and Food'. *Trends in Food Science & Technology* 54 (2016): 155–64. https://doi.org/https://doi.org/10.1016/j.tifs.2016.06.008.

Plati, Fotini, and Adamantini Paraskevopoulou. 'Micro- and Nano-Encapsulation as Tools for Essential Oils Advantages' Exploitation in Food Applications: The Case of Oregano Essential Oil'. *Food and Bioprocess Technology* 15, no. 5 (2022): 949–77. https://doi.org/10.1007/s11947-021-02746-4.

Queiroz, Marise Bonifácio, Felipe Resende Sousa, Lidiane Bataglia da Silva, Rosa Maria Vercelino Alves, and Izabela Dutra Alvim. 'Co-Crystallized Sucrose-Soluble Fiber Matrix: Physicochemical and Structural Characterization'. *LWT* 154 (2022): 112685. https://doi.org/https://doi.org/10.1016/j.lwt.2021.112685.

Rai, Akanksha, Vivek K. Sharma, Akansha Jain, Minaxi Sharma, Ashok Pandey, Harikesh B. Singh, Vijai K. Gupta, and Brahma N. Singh. 'Microbe-Fabricated Nanoparticles as Potent Biomaterials for Efficient Food Preservation'. *International Journal of Food Microbiology* 379 (2022): 109833. https://doi.org/https://doi.org/10.1016/j.ijfoodmicro.2022.109833.

Rashidinejad, Ali, and Seid Mahdi Jafari. '8 - Nanoencapsulation of Bioactive Food Ingredients'. In *Handbook of Food Nanotechnology Jafari*, edited by Seid Mahdi Jafari, 279–344. Academic Press, 2020. https://doi.org/https://doi.org/10.1016/B978-0-12-815866-1.00008-X.

Saini, Priyanshi, Kamalesu, Lalita, and Manikanika. 'Review on Nanotechnology "Impact on the Food Services Industry"'. *Materials Today: Proceedings* (2023). https://doi.org/https://doi.org/10.1016/j.matpr.2023.04.377.

Sardar, Bikash R., Kavita M. Tarade, and Rekha S. Singhal. 'Stability of Active Components of Cardamom Oleoresin in Co-Crystallized Sugar Cube during Storage'. *Journal of Food Engineering* 117, no. 4 (2013): 530–37. https://doi.org/https://doi.org/10.1016/j.jfoodeng.2013.03.035.

Ünlüsayin, Mustafa, Nicoleta G. Hădărugă, Gerlinde Rusu, Alexandra T. Gruia, Virgil Păunescu, and Daniel I. Hădărugă. 'Nano-Encapsulation Competitiveness of Omega-3 Fatty Acids and Correlations of Thermal Analysis and Karl Fischer Water Titration for European Anchovy (Engraulis Encrasicolus L.) Oil/β-Cyclodextrin Complexes'. *LWT - Food Science and Technology* 68 (2016): 135–44. https://doi.org/https://doi.org/10.1016/j.lwt.2015.12.017.

Villar, Mary Annilyn L., Mark Louis P. Vidallon, and Evelyn B. Rodriguez. 'Nanostructured Lipid Carrier for Bioactive Rice Bran Gamma-Oryzanol'. *Food Bioscience* 50 (2022): 102064. https://doi.org/https://doi.org/10.1016/j.fbio.2022.102064.

Wang, Hengpeng, Sumin Gao, Di Zhang, Yinlan Wang, Yu Zhang, Songsong Jiang, Bing Li, et al. 'Encapsulation of Catechin or Curcumin in Co-Crystallized Sucrose: Fabrication, Characterization and Application in Beef Meatballs'. *LWT* 168 (2022): 113911. https://doi.org/https://doi.org/10.1016/j.lwt.2022.113911.

Wani, M., M. Younus, N. A. Ganie, K. A. Dar, S. Q. Dar, Afzal Husain Khan, Nadeem A. Khan, Sasan Zahmatkesh, Mohammad Saood Manzar, and Ratul Banerjee. 'Nanotechnology Future in Food Using Carbohydrate Macromolecules: A State-of-the-Art Review'. *International Journal of Biological Macromolecules* 239 (2023): 124350. https://doi.org/10.1016/j.ijbiomac.2023.124350.

7 DNA Recombination and PCR Technology for Food Quality Determination

Monika Mahajan and Prabhjot Singla

7.1 INTRODUCTION

Food safety has emerged today as a very important scientific discipline to prevent foodborne illness caused during food handling, processing, storage and packaging. This has taken into consideration various additives, chemical residues, contaminants and biotechnological products, as well as pathogens, which are knowingly or unknowingly added to the food products. Food pathogens include bacteria, viruses, fungi and other parasites. Assessing food quality and safety for human health is of utmost importance because pathogens found in foods are easily transmitted to humans and produce significant damage by causing infections. Traditional methods for detecting pathogens in food include multiple subculturing and biotype identification steps that are time-consuming, costly, and laborious (Swaminathan and Feng, 1994). However, a rapid identification of the pathogen that causes food poisoning is essential for overcoming the pathogenic effects. Moreover, it has been found that a very short DNA fragment of genetically modified organisms (GMOs) is identified in animals, depicting the transfer of genes from forage to animal tissue (Boldura et al., 2015). Nowadays, the incidence of food fraud is so high that it is deeply impacting governments as well as industry and consumers. An increase in adulteration with cheap ingredients or unauthenticated components raises health, safety and religious concerns. Traditionally, the methods that are used for morphological characterization of food ingredients in terms of colour, shape, texture, odour, etc., are not fruitful for processed products as processing degrades the original characteristics. Therefore, there is a need for the development of advanced identification methods that constitute the first line of defence for both detecting foodborne or fungal pathogens, GMOs and deterring food fraud. The most frequently used techniques were chromatography or spectroscopy-based methods. Nevertheless, the presence of secondary metabolites in food products affects the chromatographic and spectral fingerprints, thus causing difficulty in analysing various food types. Moreover, various methods that rely on

DOI: 10.1201/9781032650616-7

proteins and lipids are also limited because food processing degrades the protein biomarkers.

Consequently, DNA-based techniques have revolutionized the identification of pathogens and various EMA in processed food because their accuracy and sensitivity do not get affected by the manufacturing process, growing conditions and harvesting period. Polymerase chain reaction (PCR), known as the copying machine for DNA molecules, is used to exponentially amplify the single or a few copies of DNA. This is important for easy characterization and comparisons of genetic material obtained from different sources. The principle of PCR is based on thermal cycling that utilizes the thermodynamics of nucleic-acid interactions, where alternate heating and cooling of PCR samples denature the dsDNA into ssDNA at high temperatures and duplicate them at lower temperatures by using DNA polymerase. It is a common, simple and inexpensive tool having different applications. A second-generation PCR developed by Higuchi et al., 1992, 1993 can easily quantify PCR product during amplification and can detect multiple species at the time. It is also known as real-time or quantitative (q)PCR. It is a high-throughput method that is easy to perform and has significantly higher reliability, sensitivity and specificity as compared to conventional PCR, making it an advanced identification method to detect EMA and low levels of target pathogens in food matrices contaminated by other dominant microorganisms. Another PCR-based technique is multiplex PCR that allows amplification/detection of more than one target sequence in a single reaction simultaneously by using different sets of primers, thus saving considerable effort, time and money (Elnifro et al., 2000). However, some considerations should be taken before performing multiplex PCR, such as the lengths of target DNA should be similar because shorter targets are favourable over the longer ones, thus resulting in different yields of amplified products. Also, the presence of many PCR primers in a single tube can cause the amplification of nonspecific DNA fragments and the formation of primer dimers. But this shortcoming could be overcome by choosing the primers with the same annealing temperatures.

7.2 PCR TECHNOLOGY

7.2.1 Background

Assessment of food quality and safety has great importance for human health because pathogens found in foods are easily transmitted to humans and produce significant damage by causing serious infections (Sockett, 1991). Traditional methods for detecting pathogens in food include multiple subculturing and biotype identification steps that were time-consuming, costly and laborious (Swaminathan and Feng, 1994). However, a rapid identification of the pathogen that causes food poisoning is essential to overcome the pathogenic effects. Nowadays, the incidence of food fraud, which constitutes an illegal deception for economic gain using food, has a wide prevalence all over the world that deeply impacts governments, industry and consumers (Moore et al., 2012). An increase in economically motivated adulteration (EMA) of food that involves the substitution of the authentic components with a cheap ingredient, or the false manufacturing of an ingredient, has raised concerns for health and safety as well as religious beliefs (Moyer et al., 2017). Traditionally, detection methods that rely on morphological characteristics (colour, shape, texture and odour) of food or food ingredients rarely work for processed products because the original observable characteristics are degraded during processing. Therefore, the development of advanced identification methods constitutes an important first line of defence for both detecting foodborne or fungal pathogens, genetically modified organisms (GMOs) and deterring food fraud. The most frequently used techniques were chromatography or spectroscopy-based methods. Nevertheless, secondary metabolites present in processed food products render the chromatographic and spectral fingerprints, thus causing difficulty in analysing various food types (Ali et al., 2012). Moreover, various methods that rely on proteins and lipids are also limited because protein biomarkers are easily denatured by food processing.

Consequently, DNA-based techniques have revolutionized the identification of pathogens and various EMA in processed food because their accuracy and sensitivity do not get affected by the manufacturing process, growing conditions and harvesting period. Polymerase chain reaction (PCR) is a "copying machine for DNA molecules" used to exponentially amplify the single or a few copies of DNA that allows easier characterization and comparisons of genetic material from different organisms. The principle of PCR is based on thermal cycling that utilizes the thermodynamics of nucleic-acid interactions, where alternate heating and cooling of PCR samples denature the dsDNA into ssDNA at high temperatures and duplicate them at lower temperatures by using DNA polymerase. The schematic diagram representing conventional PCR has been shown in Figure 7.1.

It is a common, simple and inexpensive tool having different applications. Recently, Higuchi and collaborators developed a second generation of PCR that can detect multiple species from a mixture and quantify the amount of PCR products formed during the amplification process, also known as real-time or quantitative (q) PCR (Higuchi et al., 1993). It is a high-throughput method that is easy to perform and has significantly higher reliability, sensitivity and specificity as compared to conventional PCR, making it an advanced identification method to detect EMA and low levels of target pathogens in food matrices contaminated by other dominant microorganisms. Another PCR-based technique is multiplex PCR, which allows the amplification/detection of more than one target sequence in a single reaction simultaneously by using different sets of primers, thus saving considerable effort, time and money (Elnifro et al., 2000). However, some considerations should be taken before performing multiplex PCR, such as the lengths of target DNA should be similar because shorter targets are favourable over the longer ones, thus resulting in different yields of amplified products. Also, the presence of many PCR primers in a single tube can cause the amplification of unspecific DNA fragments and the formation of primer dimers (Rodriguez-Lazaro and Hernandez, 2013). But this shortcoming could be overcome by choosing the primers with the same annealing temperatures.

PCR analysis utilizes nuclear as well as mitochondrial DNA, but the most preferred are the mitochondrial D-loop region, NADH dehydrogenase subunit 5/6 genes, cytochrome oxidase subunit I (COI), 12S rRNA, cytochrome b and 16S rRNA in the

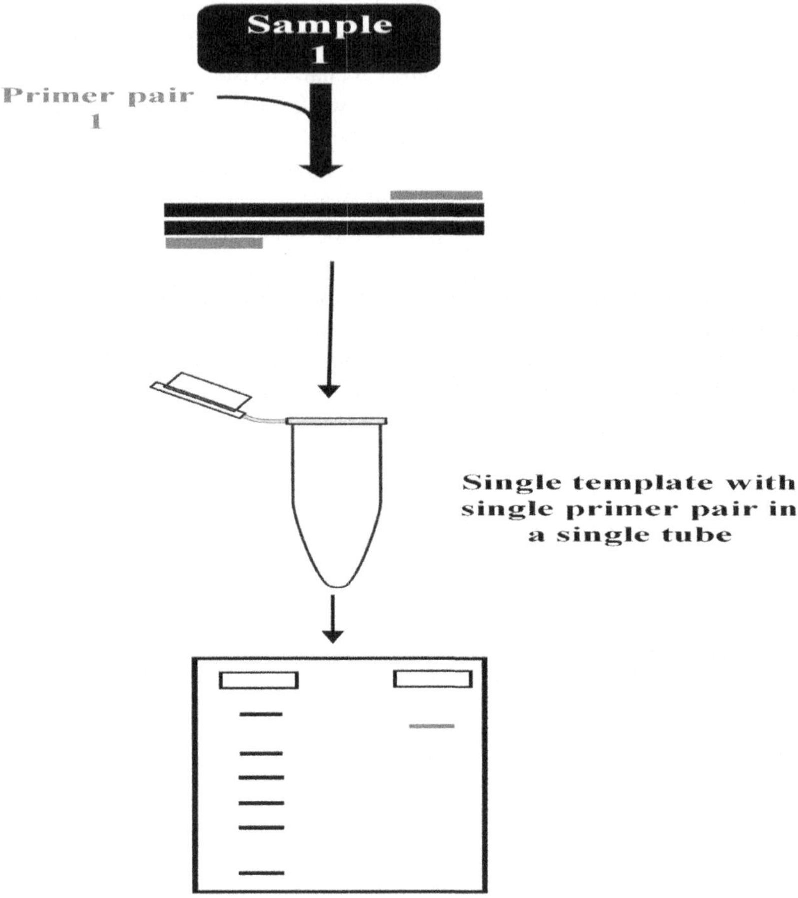

FIGURE 7.1 Schematic diagram representing conventional PCR

case of all animal species (Bartlett and Davidson, 1991). The important issues must be taken into consideration because each experiment has its specific particularities. The first and foremost is the adjustment of the DNA extraction protocol according to the amount of starting material/product and the matrix of the selected product. Second, is the selection of the biomarkers, which should be highly specific for the species and must be abundant in its genome to meet the requirements desired for the detection of a target in the food products.

7.2.2 Sample Preparation and DNA Extraction Methods

For successful real-time PCR results, the first key step is the isolation of genomic DNA from the suspected food items. The target can be a toxin-producing pathogen, any allergen, or it can be an adulteration of an unwanted component. However, various problems can be encountered during sample preparation but could be solved using suitable methods (Hill, 1996). Sample preparation involves the collection of food samples, separation of the target in the sample and isolation and purification of target DNA (Lantz et al., 1994). The homogeneity of food samples depends upon the kind of biological matrix from which they originate. The best option is to freeze-dry the samples, but air-drying under a contamination-free environment, to avoid the degradation of target DNA, can also be used. The samples are chopped into small pieces before drying, followed by grinding to obtain homogeneous samples for DNA extraction (Muhammed et al., 2015). The food samples, like milk, yoghurt, soybean milk, tofu, cheese and pâté, are shaken or mixed to form homogenized material before the collection of samples. For the extraction of DNA from foodborne or fungal pathogens, it is necessary to know whether the pathogen contaminates the foods at high levels or whether it is required to amplify the pathogen with an enrichment culture.

Template preparation from infected microorganisms requires cell lysis to release the DNA or RNA by cell wall digestion. Different factors like pathogen cell wall and treatment (chemical, thermal, solvents, osmotic shock, action of enzymes, detergents or mechanical) used for sample preparation affect the efficiency of nuclear extraction (Klancnik et al., 2003). Several sample preparation methods utilize multiple combinations of basic processes (centrifugation, filtration, hand massaging, etc.), which significantly exclude PCR inhibitors, thus increasing the PCR efficiency and specificity to increase the concentration of the target DNA (Radstrom et al., 2004). Moreover, fungal DNA degradation occurs in processed food materials, which makes its identification difficult. Additionally, vegetal samples infected with fungus have very low quantities of fungal DNA as compared to plant DNA, so to obtain precise results, it is necessary to inoculate the samples before morphological and molecular analysis.

Nowadays, a variety of commercial kits are developed that easily extract the DNA and give significantly higher qualitative and quantitative yields. However, the conventional method is always the best for DNA extraction as it uses CTAB reagent and has a thermal lysis step in the beginning, followed by the removal of protein and polysaccharide by various extraction steps (Demeke et al., 2017). Different biological matrices utilize different enzymatic steps. For instance, meat-derived products require longer treatment with proteinase-K that removes proteins and facilitates dissociation of tissue; the samples with higher starch content need to be treated with α-amylase; RNase treatment is given to remove RNA. For carrying out real-time PCR assays, commercial kits are generally preferred because of their convenience and effectiveness. To effectively remove PCR inhibitors from the processed food, hybrid methods are used that comprise a modified CTAB method along with the clean-up kits available commercially.

7.2.3 DNA Qualification and Quantification Methods

The efficiency of real-time PCR is critically affected by the concentration and purity of DNA extracts. The spectrophotometric or fluorometric methods are used for DNA quantification, with the former being the most used technique. DNA concentration is measured at 260 nm spectrophotometrically, and DNA purity is indicated by the ratios A260/230 and A260/280 (Costa et al., 2017). Organic compounds and aromatic amino acids present in proteins absorb at 230 nm and 280 nm, respectively; thus, the ratios A260/230 > 1.5 and A260/A280 = 1.7–2 depict the good quality of DNA. However, these values depend upon the biological matrix because the value lowers in the case of processed foods as DNA gets degraded in processed foods. In contrast, real-time PCR methods use highly sensitive fluorescent dye for the detection of DNA from processed food products.

In standard PCR protocols, the integrity of amplified products is detected by gel electrophoresis. The concentration of gel depends upon the size of the amplicon, thus ranging from 1.5– to 2.5%. The size of a PCR product is determined using a DNA 'ladder', a DNA molecule of different known lengths and molecular weights that act as known references to estimate the sizes of the unknown DNA molecules (Sambrook and Russell, 2001). To maintain accuracy, precision and lack of cross-contamination, the control along with target samples is analyzed on each gel.

7.2.4 PCR Inhibition, Amplification Controls and Approaches to Overcome Inhibition

The amplification efficiency and sensitivity of PCR can be dramatically reduced by the inhibitors that cause inhibition of the PCR amplification reaction. These inhibitors include a heterogeneous group of inorganic (salts and calcium ions) and organic (polysaccharides, detergents, proteins and polyphenols) compounds that can be introduced during food processing from different food ingredients (Schrader et al., 2012). They are found in cheese, milk, meat, spices, etc. For instance, nucleases, proteases, dead cells, fatty acids and chelators are present in milk (Bickley et al., 1996). In addition to this, liver and meat samples have bile salts, antibodies, collagen and fatty acids as PCR inhibitors (Lantz et al., 1997). Seafood like shrimps, fish and salmon have aldehydes, phenolics, cresol, fat and protein aldehydes as PCR inhibitors.

However, various approaches can be used to overcome inhibition. The simplest approach is to dilute the template that automatically dilutes the PCR inhibitor dead cells; fatty acids represent ins. The DNA polymerase that is tolerant to the effects of inhibitory compounds can be used. For instance, for preparing templates from cheese or meat samples, the use of Tfl and rTth is more authentic than Taq polymerase (Al-Soud and Radstrom, 2000). Moreover, in the presence of inhibitors, some facilitators (Tween 20, bovine serum albumin, Triton-X, betaine, dimethyl sulfoxide) can be used to improve the activity of DNA polymerases.

7.2.5 Selection of Target Genes and Design of Primers and Probes

The correct selection of the target, along with the design/selection of the primers and probes, is a key factor for the correct detection of foodborne pathogens, EMA and allergens. The multiple copy genes decide the lower limit of detection in real-time PCR compared to single-copy genes. For instance, for the consistent quantitative analysis of the food product and for the detection of potential food allergens, species-specific primers and probes are designed based on single-copy DNA targets (Laube et al., 2007, Druml et al., 2015). For instance, ara h1, 2S, Pru 1, Lup a4 and Jug r 3 genes for peanuts, Brazil nuts, almonds, lupine and walnut, respectively (López-Calleja et al., 2013a; 2013b; 2015b; 2014; Zhang et al., 2015; Villa et al., 2018).

The best way is to design your primer pairs using authentic software according to the requirements of the experiment. However, it is necessary to confirm the specificity of newly designed primers by aligning them with sequences from a suitable data set such as the NCBI database. The foremost issue for primers designed to be used in multiplex PCR reactions is their expected amplicon size that must be easily discriminated in an electrophoresis gel. The result of PCR analysis is seriously affected by fragment overlapping. The optimal length of an amplicon should be less than 150 bp; however, amplicons up to 300 bp amplify efficiently. Shorter amplicons are less susceptible to suboptimal reaction conditions and amplify more efficiently than longer ones.

7.3 APPLICATIONS OF PCR TECHNOLOGY

7.3.1 Detection of Genetically Modified Organisms

In 1946, scientists discovered that DNA can be transferred between organisms (Clive, 2011). There are several mechanisms by which DNA can be transferred from one organism to another. This occurs in nature on a large scale. For example, antibiotic-resistant bacteria are produced only by the natural DNA transfer. The first manmade genetically modified plant was produced in 1983 in tobacco by inserting an antibiotic-resistant gene into it. GM crops are developed mainly to improve the food quality and to solve the problems related to agriculture, herbicide and disease resistance, weed management, etc. China became the first country to introduce and commercialize the virus-resistant transgenic tobacco in 1990. The next transgenic crop was the Flavr Savr tomato with delayed ripening and enhanced shelf life. In 1994, these transgenic tomatoes were approved by the FDA to be marketed in the USA. In 1995, many transgenic crops were approved for marketing. These include canola, calgene with improved oil composition, herbicide roundup-tolerant GM soybean, Bt corn (Ciba-Geigy), Bt cotton (Monsanto), herbicide bromoxynil-resistant cotton (Calgene), herbicide glyphosate-resistant cotton (Monsanto), squash resistant to virus (Asgrow), etc. The USA leads in producing and commercializing GM crops as of 2011. GM soybean occupies 63% of the global area, while maize, cotton and canola occupy 19%, 13% and 5% of the global area, respectively. The dominant traits of GM crops are herbicide resistance, followed by insect resistance. Today, many genetically modified species of food exist, including soybean, cotton, canola, eggplant, potatoes, tomatoes, strawberries, corn, lettuce, carrots, cantaloupe, etc. (Johnson, 2008). New coming-ups are the genetically modified foods and food ingredients, vaccines, fibres, feeds, etc. Many GM crops are approved, but there is no GM animal which is approved as food. GM salmon is proposed, but it is under FDA approval. In most cases, these GM crops are sold as commodities, but they undergo further processing to make different food products. The Novel Food Regulation is concerned about novel foods and their ingredients. According to this, labelling of food or food products is required in order to ensure the consumer is informed of any changes in food ingredients, characteristics and properties such as composition, nutritional value, additives, flavours, etc. The regulation has been approved for the labelling of GMO crops.

According to the European Union regulations, all food items containing more than 0.9% of GMOs need to be labelled. For that, a highly reliable detection method is needed so that mislabelling can be prevented. The European Government has banned the cultivation of GM soybean, so it is imported and added as an important ingredient to the livestock feed. In that case, it is very important to detect the presence of GMOs in the product. Earlier, the molecular sequence specific to the plant transformation vector, such as promoter or terminator sequence, and later on, transgene-specific sequences were used to detect the GM organisms in food products. Food labelling policy has implemented detection methods to prevent food fraud in genetically modified organisms. Earlier, PCR was used to detect the presence of GMOs in food, but it is a single target method, so today it has been replaced with multiplex PCR, where multiple targets can be amplified in the same reaction. The schematic representation of multiplex PCR is shown in Figure 7.2.

This technique is highly reliable, flexible and cost-effective. Multiplex PCR was designed by Perming et al. to detect two transgenes, *CryIA(b)* and *pat*, in four different lines of GM maize. Matsuoka et al. designed multiplex PCR with five different lines of GM maize with primers specific to individual lines. Another multiplex PCR was developed by Demeke et al. for simultaneous detection of three GM lines using two as transgene-specific primers and one promoter or terminator-specific primer. The amplicon of varying sizes specific to different DNA sequences is generated through multiplex PCR. So, in this way, through multiplex PCR, with a single run, much information can be generated which not only saves the reagents but also saves time. The annealing temperature of the primers and amplicon size of the target are optimized in such a way that in a single reaction, many bands of different sizes should be obtained when visualized under gel electrophoresis. Multiplex PCR is used for the reliable detection of processed pasta made from Bt 176 and Bt 11 corn. GM maize and soybean were assessed with the primer sets specific to the transgenic construct, i.e., CaMV 35S promoter and NOS terminator. GM canola and GM maize could be distinguished from their non-GM counterparts using the various soybean primer set. The other primer set targeting soybean lectin genes lecMP1 and lecMP2 is not used for GM maize and GM canola. So, by using a single primer set for qualitative detection of GM maize, canola and soybean, not only reduces the testing cost but also saves time.

7.3.2 Detection of Allergens in Food Products

In the food industry, the most common method to detect food allergens in food products is based upon protein detection (Besler, 2001). The most common method is ELISA, where the species-specific protein is detected by specific antibodies either

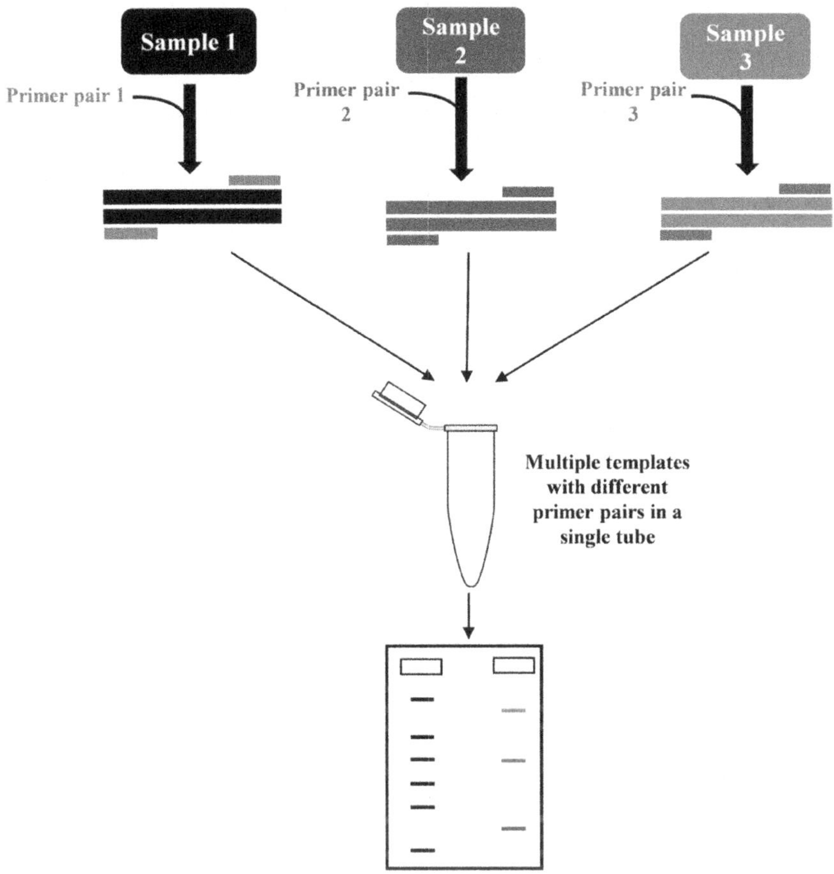

FIGURE 7.2 Schematic representation of multiplex PCR

raised in animals or produced through recombinant technologies (Scaravelli et al., 2009; de la Cruz et al., 2015, 2016). But when it comes to the detection of allergens in processed food items, it becomes difficult to use the immunochemical method for allergen detection. This is because processing, such as heat treatment or roasting, affects the integrity and quality of the protein, which results in false results (Scaravelli et al., 2009). The alternative to the protein detection method is the amplification of specific DNA sequences of the target through qualitative conventional end-point PCR or through quantitative real-time PCR using fluorescent probes. Real-time PCR is preferred over conventional PCR because of its high reliability, sensitivity and precision, its ability to detect very small quantities, reduced risk of contamination and the analysis of data in the exponential phase of the PCR reaction. In real-time PCR, the Cp value means cycle number. An increased Cp value means the fluorescence above the ground level, providing quantitative information about the target sequence.

Earlier in real-time PCR, dyes were used, but they led to unspecific amplification; therefore, the TaqMan probe, consisting of a short single-stranded oligonucleotide with a fluorescent probe as a reporter gene at the 5' end and a quencher dye at the 3' end, results in specific detection of the target sequence (Patel et al., 2006). By means of the FRET mechanism, the quencher dye reduces the fluorescence of the reporter dye. But the presence of exonuclease activity in Taq polymerase separates the donor fluorophore from the quencher, thereby increasing the expression of the fluorescent reporter gene, which is directly proportional to the amount of target DNA present in the sample (Jothikumar et al., 2009). The schematic representation of real-time qPCR is shown in Figure 7.3.

The TaqMan Real-Time PCR method has already been developed to identify or detect the peanut, pistachio, hazelnut, almond, cashew, walnut, pecan nuts as major allergens in food products (López-Calleja et al., 2013a, 2013b, 2014, 2015a). These allergens should be labelled according to EU Regulation 1169/2011.

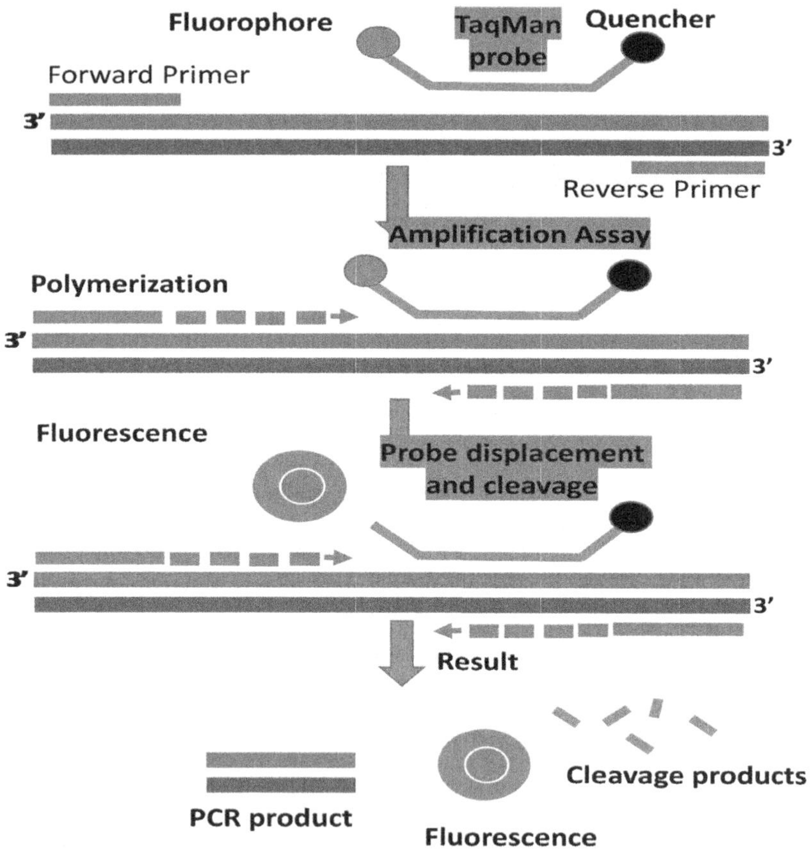

FIGURE 7.3 Schematic representation of real-time qPCR

7.3.3 Identification of Food Additives by DNA Barcoding

Plant- and animal-derived food additives are commonly used as preservatives, colourants, antimicrobials and nutritional additives for improving the quality, food safety and sensory properties. For instance, the use of herbal medicines as additives is very popular these days. But the major concern here is the increasing adulteration and food fraud, which has led to the mislabelling of products. This mislabelling may give rise to many health problems. Misidentification of food ingredients in food products can lead to many allergic reactions in allergic susceptible people. This food fraud or adulteration has become a very challenging issue today with respect to food quality and safety. Consumers are now demanding food authenticity to prevent food fraud. Food authenticity is assessed by various methods like morphological methods, ingredient analysis and some protein-based methods. But these methods are time-consuming and less efficient. A highly developed DNA-based method has come out to be more effective. But still, it has some drawbacks that include poor sequencing of DNA and degradation by inhibitors present in food. But now, DNA barcoding is receiving great attention as a new authentication method. It can easily track the raw material present in the food and prevent mislabelling. It is a highly accurate, rapid and cost-effective method. It is used to track the animal and plant species present a as an ingredient in food products.

DNA barcoding uses a small standardized DNA marker to amplify the gene at the target position at the 5′ end, sequenced and used as a marker key to identify the gene. The procedure for species identification through the DNA barcoding method is shown in Figure 7.4.

For example, COI (Mitochondrial cytochrome C oxidase) is used as a barcode marker to identify animal species. It has a high evolutionary rate and gives accurate species-specific distinction. But it has its own limitations as it is not applicable for amphibians because of the COI priming sites and mitochondrial variability. Therefore, chloroplast and nuclear genomes are used to identify the substituents in plant-based products. Consortium for the Barcode of Life (CBOL) and International Barcode of Life (iBOL) have recommended ribulose 1,5-bisphosphate carboxylase/oxygenase (*rbcL*) and maturase K (matK) (Mishra et al., 2016) as potential barcodes of plants. Therefore, DNA barcoding has emerged as a rapid, reliable and accurate DNA-based tool for detecting food fraud and food mislabelling.

7.3.4 Detection of Pathogenic Microorganisms in Food

Food safety is a very important concern for the public in order to ensure that the food is safe for consumption. It is also important for the economic sustainability of the agri-food sector. The incidence of food poisoning affects the goodwill of the company and makes the consumer less reliant on their products. In order to get assured that the food is safe for consumption, it needs to be tested for microbial pathogens. The pathogens which can cause foodborne illnesses include bacteria, viruses, fungi, parasites, etc. Viruses are mainly responsible for illness, but bacteria are responsible for most of the hospitalizations and even deaths of people. The

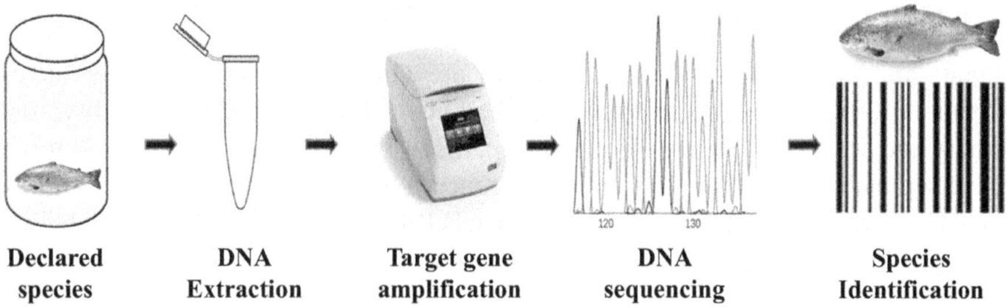

FIGURE 7.4 Flowchart diagram representing the procedure for DNA barcoding

most common pathogens responsible for foodborne diseases are *Listeria monocytogenes, Salmonella enteritidis, Escherichia coli, Staphylococcus aureus, Vibrio spp., Bacillus cereus, Shiga toxin-producing E. coli, Clostridium perfringens, Shigella flexneri, etc.* (Oliver et al., 2005; Scallan et al., 2011; Zhao et al., 2014).

These pathogens are present in very small amounts, i.e., < 100 cfu g^{-1}, either attached to the food surface or embedded in the matrix. These pathogens get injured when stored at low temperatures, low pH or after adding preservatives. Although it is difficult to revive them, they have the potential to cause illness, so it's important to detect their presence in the food used for consumption. Traditional methods for pathogen detection in food products rely on culture-based methods. In these methods, the contaminated food is cultured on agar plates, and the suspected colonies are further identified by morphological, biochemical or immunochemical methods. These traditional methods are time-consuming and usually take 5–7 days for the detection of the specific pathogen. Therefore, an alternative rapid qualitative or quantitative method is required, which can detect the specific pathogen and also show equivalence with the culture-based methods on validation. If any food item is found to be contaminated with a specific pathogen, it needs to be further validated by culture methods so that it can be fully characterized and tracked. One of the most reliable rapid methods that can easily distinguish positive from negative samples is based upon molecular techniques. This is because the genome of most food-borne pathogens has been annotated, and genetic information is available, so it becomes easy to identify and characterize them.

Molecular techniques involve the detection of DNA/RNA fragments specific to the target organism. The specificity is generally based on the target gene or it is at the level of genera, species, virulence or phenotype. PCR technology is used for specific pathogen detection that amplifies the target several fold for its easy detection. But the major hurdle comes when the target organism is present in very low numbers; in that case, liquid enrichment of the food sample for 24–48 hours is needed before the PCR amplification. This increases the amount of target DNA, which helps in the easy detection of the target organism.

The major concern with conventional PCR during pathogen detection is the possibility of getting false negative results during DNA extraction, degradation or inhibition of PCR. Therefore, both internal and external controls are added during the assay. External controls ensure that there is no DNA cross-contamination whereas internal controls amplify the target sequence and monitor that amplification and detection are working effectively. The main foodborne organisms detected by PCR are *Listeria monocytogenes*, *Salmonella*, *Escherichia coli*, *Staphylococcus aureus*, *Campylobacter jejuni*, *Shigella species*, etc. The main gene targets used for the detection of foodborne pathogens are given below (Alves et al., 2012; Chiang et al., 2012; Zhou et al., 2013)

But nowadays multiplex PCR is more preferred for the detection of foodborne pathogens as it can detect three to five pathogens simultaneously by using different primer sets in the same reaction. Chen et al. (2012) have carried out multiplex PCR for simultaneous detection of *Listeria monocytogenes*, *Salmonella enteritidis*, *Escherichia coli*, *Staphylococcus aureus and Shigella flexneri using* listeriolysin O (*hlyA*), invasion protein (*invA*), intimin (*eaeA*) gene, 16S rDNA and invasion plasmid antigen H (*ipaH*) gene, respectively. Ryu et al. (2013) have developed a modified multiplex PCR where six species of *Listeria*, i.e., *Listeria ivanovii*, *L. grayi*, *L. innocua*, *L. welshimeri*, *L. seelogeri* were distinguished with high accuracy in one reaction tube. The mixed genomic DNA limit used by Ryu et al. (2013) for multiplex PCR was 7.58×10^4 copies.

A more refined version of multiplex PCR has now come, which is the Genome Lab Gene Expression Profiler (GeXP-PCR), allowing the high-throughput detection of multiple pathogens in one reaction. This PCR involves primer designing, PCR amplification and capillary electrophoretic separation of the amplified product, instead of using agarose gel electrophoresis. GeXP-PCR makes use of chimeric primers containing the gene-specific sequence along with a universal tag at the 5' end. Then the universal primers complementary to the universal tag will be allowed to carry out the remaining PCR reaction. These universal primers have a fluorescent tag at the 5' end, which helps in detection during capillary electrophoresis (Zhou et al., 2013). This method is highly sensitive and is used for high throughput analysis. The detection limit of *Listeria monocytogenes*, *Salmonella enteritidis*, *Escherichia coli*, *Staphylococcus aureus and Shigella flexneri with* GeXP-PCR is 310, 420, 93, 270 and 85 CFU/ml, respectively.

Real-time qPCR is also used for the detection of pathogens in food samples. But this method does not need agarose gel electrophoresis to run, so it is better than conventional and multiplex PCR. qPCR makes use of fluorescent systems like TaqMan probes, SYBR green and molecular beacons. TaqMan probe has already been discussed earlier. SYBR green is a double-stranded DNA-binding fluorescent intercalating dye that binds to the minor groove of the DNA double helix, where it enhances the fluorescent signal after binding (Levin, 2005; Singh et al., 2009). Molecular beacons are another alternative to SYBR green. Molecular beacons are oligonucleotide probes having a stem and loop configuration. The loop portion has a sequence complementary to the target sequence, and the stem is produced after the annealing of two complementary arm sequences. The reporter dye is added to one end, and the quencher dye is added to the other end of the probe. When they are in close proximity due to the stem, there is no fluorescence. The fluorescence is produced when the probe gets hybridized with its complementary sequence in the amplicon. The conformational change during hybridization leads the two dyes to be apart from each other, which results in fluorescence (Liming and Bhagwat, 2004; Levin, 2005).

Liming and Bhagwat (2004) reported the *Salmonella* in fresh cut fruits and vegetables by molecular beacon qPCR by targeting the invasion-associated gene (iagA). The detection limit for Salmonella was 4 CFU/25 g of the commodity after enrichment. Later on, one sensitive and reproducible method of detection, i.e., SYBR green qPCR for *Vibrio parahaemolyticus* infection in tropical shellfish, was developed by Tyagi et al. (2009). Other methods, TaqMan and SYBR Green qPCR, were developed to detect the infection of *Staphylococcus aureus* in raw milk by targeting *the* enterotoxin gene cluster (egc). The detection limit for *Staphylococcus aureus* in raw milk was 10^4 cFU/ml and 10^3 CFU/mL for SYBR green qPCR and TaqMan qPCR, respectively (Fusco et al., 2011). Additionally multiplex PCR was developed for simultaneous detection of Vibrio cholerae, Vibrio parahaemolyticus and Vibrio vulnificus by targeting the zot, vmrA and vuuA genes, respectively. Fratamico et al. (2011) has also developed a multiplex PCR method to detect the *E. coli* serogroups O26, O45, O103, O111, O121 and O145 that produce Shiga toxins by targeting three genes: Shiga toxins (stx1 and stx2), intimin (eae) and wzx (Kim et al., 2012). molecular beacon-based multiplex qPCR was designed by Hu et al., 2014 where eight food-borne pathogens, i.e., *Listeria monocytogenes, Escherichia coli O157, Salmonella enterica subsp. enterica, Vibrio vulnificus, Vibrio parahaemolyticus, Campylobacter jejuni, Shigella spp., and Enterobacter sakazakii*, were detected simultaneously by targeting hlyA, rfbE, ssaR, vvh, toxR, ipaH, gyrA and 16S rRNA, respectively.

The advantage with SYBR green dye over other dyes, TaqMan and molecular beacons, is the DNA melting curve generated after PCR. It gives a Tm value for the amplified product, which is important to distinguish it from primer dimers, which are known to have a lower Tm value (Levin, 2005). The major disadvantage with SYBR green is that it lacks specificity and binds with non-specific sequences as well. In contrast, TaqMan and molecular beacons have high specificity and bind only to the target sequence, and no primer dimers are detected with these methods (Hein et al., 2001; Klerks et al., 2004). Nevertheless, the PCR method mainly depends on the primer sequence, its specificity and the annealing temperature, rather than on the type of detection probe used. But it is true that qPCR is always better than conventional PCR for the detection of foodborne pathogens, as it reduces the risk of cross-contamination (Omiccioli et al., 2009).

7.4 CONCLUSION

Universal demand of consumers for food quality and safety is of major concern in terms of health, lifestyle, religion, etc. This is due to the increasing adulteration in food by adding cheap ingredients, which can cause food poisoning, food allergies, and can also affect the religious sentiments of the consumer. PCR-based methods have a significant role in detecting any adulteration or food fraud and can also identify the potential risks that could arise after taking the adulterated commodities. These methods are highly sensitive and powerful and have vast practical applications in line with the current societal demands. This chapter has discussed all the techniques based on PCR which are being used today for food quality determination

REFERENCES

Al-Soud WA and Radstrom P (2000) Effects of amplification facilitators on diagnostic PCR in the presence of blood, feces, and meat. *J. Clin. Microbiol.* 38(12), 4463–4470.

Ali ME, Kashif M, Uddin K, Hashim U, Mustafa S and Che Man YB (2012) Species authentication methods in foods and feeds: The present, past, and future of halal forensics. *Food Anal. Methods* 5(5), 935–955.

Alves J, Marques VV, Pereira LFP, Hirooka EY, and Moreirade Oliveira TCR (2012) Multiplex PCR for the detection of *Campylobacter* spp. and *Salmonella* spp. In chicken meat. *J. Food Safety* 32, 345–350. doi:10.1111/j.1745- 4565.2012.00386.x

Bartlett SE and Davidson WS (1991) Identification of *Thunnus* tuna species by the polymerase chain reaction and direct sequence analysis of their mitochondrial cytochrome b genes. *Can. J. Fish. Aquat. Sci.* 48, 309–317.

Besler M (2001) Determination of allergens in food. *TrAc Trends Analyt. Chem.* 20, 662–672.

Bickley J, Short JK, McDowell DG and Parkes HC (1996) Polymerase chain reaction (PCR) detection of *Listeria monocytogenes* in diluted milk and reversal of PCR inhibition caused by calcium ions. *Lett. Appl. Microbiol.* 22(2), 153–158.

Boldura OM, Baltă C, Ahmadi M, Tulcan C, Huțu I, et al (2015) Traceability of transgenic soybean from forage trough animal tissue till the food product. *Lucrări Ştiinţifice Medicină Veterinară.* XLVIII(2), 5–13.

Chiang YC, Tsen HY, Chen HY, Chang YH, Lin CK, Chen CY, et al (2012) Multiplex PCR and a chromogenic DNA microarray for the detection of *Listeria monocytogenes, Staphylococcus aureus, Streptococcus agalactiae, Enterobacter sakazakii, Escherichia coli* O157:H7, *Vibrio parahaemolyti- cus, Salmonella* spp. and *Pseudomonas fluorescens* in milk and meat samples. *J. Microbiol. Meth.* 88, 110–116.

Clive J (2011) *Global status of commercialized Biotech/GM crops. ISAAA Briefs 43.* Ithaca: International Service for the Acquisition of Agri-biotech Applications.

Costa J, Amaral JS, Grazina L, Oliveira M and Mafra I (2017) Matrix-normalised real time PCR approach to quantify soybean as a potential food allergen as affected by thermal processing. *Food Chem.* 221, 1843–1850.

de la Cruz S, Alcocer M, Madrid R et al (2016) Production of *in vivo* biotinylated scFv specific to almond (*Prunus dulcis*) proteins by recombinant *Pichia pastoris. J. Biotechnol.* 227, 112–119.

de la Cruz S, Cubillos-Zapata C, López-Calleja IM et al (2015) Isolation of recombinant antibody fragments (scFv) by phage display technology for detection of almond allergens in food products. *Food Cont.* 54, 322–330.

Demeke T, Malabanan J, Holigroski M and Eng M (2017) Effect of source of DNA on the quantitative analysis of genetically engineered traits using digital PCR and real time PCR. *J. AOAC Int.* 100, 492–498.

Druml B, Hochegger R and Cichna-Markl M (2015) Duplex real-time PCR assay for the simultaneous determination of the roe deer (*Capreolus capreolus*) and deer (sum of fallow deer, red deer, and sika deer) content in game meat products. *Food Cont.* 57:370–376.

Elnifro EM, Ashshi AM, Cooper RJ and Klapper PE (2000) Multiplex PCR: Optimization and application in diagnostic virology. *Clin. Microbiol. Rev.* 13(4), 559–570.

Fratamico PM, Bagi LK, Cray Jr WC, Narang N, Yan X, Medina M, et al (2011) Detection by multiplex real-time polymerase chain reaction assays and isolation of Shigatoxin-producing *Escherichia coli* serogroups O26, O45, O103, O111, O121andO145 in ground beef. *Food Borne Pathog. Dis.* 8, 601–607. doi: 10.1089/fpd.2010.0773

Fusco V, Quero GM, Morea M, Blaiotta G, and Visconti A (2011) Rapid and reliable identification of *Staphylococcus aureus* harbouring the enterotoxin gene cluster (*egc*). and quantitative detection in raw milk by realtime PCR. *Int. J. Food Microbiol.* 144, 528–537.doi:10.1016/j.ijfoodmicro.2010.11.016

Hein I, Lehner A, Rieck P, Klein K, Brandl E, and Wagner M. (2001). Comparison of different approaches to quantify *Staphylococcus aureus* cells by real-time quantitative PCR and application of this technique for examination of cheese. *Appl. Environ. Microbiol.* 67, 3122–3126. doi:10.1128/AEM.67.7.3122-3126.2001

Higuchi R, Dollinger G, Walsh PS, and Griffith R (1992, April) Simultaneous amplification and detection of specific DNA sequences. *Biotechnol.* 10(4), 413–417.

Higuchi R, Fockler C, Dollinger G, and Watson R (1993) Kinetic PCR analysis: Real-time monitoring of DNA amplification reactions. *Bio/Technol.* 11(9):1026–1030.

Hill WE (1996) The polymerase chain reaction: Applications for the detection of foodborne pathogens. *Crit. Rev. Food Sci. Nutr.* 36(1–2), 123–173.

Hu Q, Lyu D, Shi X, Jiang Y, Lin Y, Li Y, et al (2014) A modified molecular beacons-based multiplex real-time PCR assay for simultaneous detection of eight food borne pathogens in a single reaction and its application. *Foodborne Pathog. Dis.* 11, 207–214. doi:10.1089/fpd.2013.1607

Johnson SR (2008) *Quantification of the Impacts on US Agriculture of Biotechnology-Derived Crops Planted in 2006.* Washington, DC: National Centre for Food and Agricultural Policy; 2008.

Jothikumar P, Hill V, and Narayanan J (2009) Design of FRET-TaqMan probes for multiplex real-time PCR using an internal positive control. *Biotechniques* 46, 519–524.

Kim HJ, Lee HJ, Lee KH, and Cho JC (2012) Simultaneous detection of pathogenic *Vibrio* species using multiplex real-time PCR. *Food Cont.* 23, 491–498.

Klancnik A, Smole Mozina S, and Jersek B (2003) The effect of DNA preparation procedure on sensitivity of PCR detection of *Listeria monocytogenes. Zbornik Biotehehniske fakultete* 81(1), 65–73.

Klerks MM, Zijlstra C, and van Bruggen AHC (2004) Comparison of real-time PCR methods for detection of *Salmonella enterica* and *Escherichia coli* O157: H7, and introduction of a general internal amplification control. *J. Microbiol. Meth.* 59, 337–349. doi:10.1016/j.mimet.2004.07.011

Lantz PG, Hahn-Hagerdal B, and Radstrom P (1994) Sample preparation methods in PCR based detection of food pathogens. *Trends Food Sci. Technol.* 5(12), 384–389.

Lantz PG, Matsson M, Wadstrom T, and Radstrom P (1997) Removal of PCR inhibitors from human faecal samples through the use of an aqueous two-phase system for sample preparation prior to PCR. *J. Microbiol. Methods* 28(3), 159–167.

Laube I, Zagon J, and Broll H (2007) Quantitative determination of commercially relevant species in foods by real-time PCR. *Int. J. Food Sci. Technol.* 42, 336–341.

Levin RE (2005) The application of real-time PCR to food and agricultural systems. A review. *Food Biotechnol.* 18, 97–133. doi:10.1081/FBT-120030386

Liming SH and Bhagwat AA (2004) Application of a molecular beacon-real-time PCR technology to detect *Salmonella* species contaminating fruits and vegetables. *Int. J. Food Microbiol.* 95, 177–187. doi: 10.1016/j.ijfoodmicro.2004.02.013

López-Calleja I, de la Cruz S, González I et al (2013) Survey of undeclared allergenis pistachio hydrolysis probe real-time PCR. *Food Cont.* 39, 49–55.

López-Calleja I, de la Cruz S, González I et al (2015a) Development of real-time PCR assay to detect cashew (*Anacardium occidentale*) and macadamia (*Macadamia intergriofolia*) residues in market analysis of processed food products. *Food Sci. Technol.* 62, 233–241.

López-Calleja I, de la Cruz S, González I et al (2015b) Market analysis of food products for detection of allergenic walnut (*Juglans regia*) and pecan (*Carya illinoinensis*).

López-Calleja I, de la Cruz S, Pegels N et al (2013a) Development of a real-time PCR assay for detection of allergenic trace amounts of peanut (*Arachis hypogaea*) in processed foods. *Food Cont.* 30, 480–490.

López-Calleja I, de la Cruz S, Pegels N et al (2013b) High resolution Taqman real-time PCR approach to detect hazelnut DNA encoding for ITS rDNA in foods. *Food Chem.* 141, 1872–1880.

López-Calleja I, de la Cruz S, Pegels N et al (2014) Sensitive and specific detection of allergenic almond (*Prunus dulcis*) in commercial food products by Taqman real-time polymerase chain reaction. *Food Sci. Technol.* 56, 31–39.

Mishra, P., Kumar, A., Nagireddy, A., Mani, D. N., Shukla, A. K., Tiwari, R., & Sundaresan, V. (2016). DNA barcoding: an efficient tool to overcome authentication challenges in the herbal market. *Plant Biotechnology Journal*, 14(1), 8–21.

Moore JC, Spink J, and Lipp M (2012) Development and application of a database of food ingredient fraud and economically motivated adulteration from 1980 to 2010. *J. Food Sci.* 77(4), 118–126.

Moyer DC, DeVries JW, and Spink J (2017) The economics of a food fraud incident - Case studies and examples including melamine in wheat gluten. *Food Cont.* 71, 358–364.

Muhammed MA, Charan Bindu BS, Sri B, Jini R, Harish Prashanth KV, and Bhaska N (2015) Evaluation of different DNA extraction methods for the detection of adulteration in raw and processed meat through polymerase chain reaction-restriction fragment length polymorphism (PCR-RFLP). *J. Food Sci. Technol.* 52(1), 514–520.

Oliver, S. P., Jayarao, B. M., & Almeida, R. A. (2005). Foodborne pathogens in milk and the dairy farm environment: food safety and public health implications. Foodborne Pathogens and Disease, 2(2), 115–129.

Omiccioli, E., Amagliani, G., Brandi, G., & Magnani, M. (2009). A new platform for Real-Time PCR detection of Salmonella spp., Listeria monocytogenes and Escherichia coli O157 in milk. *Food Microbiology*, 26(6), 615–622.

Patel, P., & Markey, M. (2006). Real-time PCR in clinical microbiology. Clinical Microbiology Newsletter, 28(13), 97–102.

Radstrom P, Knutsson R, Wolffs P, Lovenklev M, and Lofstrom C (2004) Pre-PCR processing strategies to generate PCR-compatible samples. *Mol. Biotechnol.* 26(2), 133–146.

Rodriguez-Lazaro D and Hernandez M (2013) Real-time PCR in food science: Introduction. *Curr. Issues Mol. Biol.* 15, 25–38.

Ryu J, Park SH, Yeom YS, Shrivastav A, Lee SH, Kim YR, et al. (2013). Simultaneous detection of *Listeria* species isolated from meat processed foods using multiplex PCR. *Food Cont.* 32, 659–664.

Sambrook J and Russell D (2001) *Molecular cloning: A laboratory manual* (3rd ed.). New York: Cold Spring Harbor Laboratory Press.

Scallan E, Hoekstra RM, Angulo FJ, Tauxe RV, Widdowson MA, Roy SL, et al. (2011) Food borne illness acquired in the United States-major pathogens. *Emerg. Infect. Dis.* 17, 7–15.

Scaravelli E, Brohee M, Marchelli R, et al (2009) The effect of heat treatment on the detection of peanut allergens as determined by ELISA and real-time PCR. *Anal. Bioanal. Chem.* 395:127–137

Schrader C, Schielke A, Ellerbroek L, and Johne R (2012) PCR inhibitors-Occurrence, properties, and removal. *J. Appl. Microbiol.* 113, 1014–1026.

Singh J, Batish VK, and Grover S. (2009) A molecular beacon-based duplex real-time polymerase chain reaction assay for simultaneous detection of *Escherichia coli* O157:H7 and *Listeria monocytogenes* in milk and milk products. *Food Borne Pathog. Dis.* 6, 1195–1201. doi:10.1089/fpd.2009.0310

Sockett PN (1991) The economic implications of human Salmonella infection. *J. Appl. Bacteriol.* 71, 289–295.

Swaminathan B and Feng P (1994) Rapid detection of food-borne pathogenic bacteria. *Annu. Rev. Microbiol.* 48, 401–426.

Tyagi A, Saravanan V, Karunasagar I, and Karunasagar I. (2009) Detection of *Vibrio parahaemolyticus* in tropical shell fish by SYBR green real-time PCR and evaluation of three enrichment media. *Int. J. Food Microbiol.* 129, 124–130. doi: 10.1016/j.ijfoodmicro.2008.11.006

Villa C, Costa J, Gondar C, Oliveira M, and Mafra I (2018) Effect of food matrix and thermal processing on the performance of a normalised quantitative real-time PCR approach for lupine (*Lupinus albus*) detection as a potential allergenic food. *Food Chem.* 262, 251–259.

Zhang WJ, Cai Q, Guan X, and Chen Q (2015) Detection of peanut (*Arachis hypogaea*) allergen by real-time PCR method with internal amplification control. *Food Chem.* 174, 547–552.

Zhao X, Lin CW, Wang J, and Oh DH (2014) Advances in rapid detection methods for food-borne pathogens. *J. Microbiol. Biotechn.* 24, 297–312. doi: 10.4014/jmb.1310.10013

Zhou B, Xiao J, Liu S, Yang J, Wang Y, Nie F, et al. (2013) Simultaneous detection of six food-borne pathogens by multiplex PCR with GEXP analyzer. *Food Cont.* 32, 198–204.

8 Fermenters and Bioreactor Systems

An Overview for Practical Applications and Current Trends

Shabnum Showkat, Rishi Richa, Anwesa Sarkar, and Kshirod Kumar Dash

8.1 INTRODUCTION

The first large-scale (over 20-liter capacity) fermenter for yeast production was utilized by De Beeze and Liebmann in 1944. However, a British scientist by the name of Chaim Weizmann (1914–1918) created a fermenter for the synthesis of acetone during the First World War. Since it was realized how important aseptic conditions were, efforts were made to design and build pipes, couplings, and valves that would allow for the production of sterile conditions when needed. In the 1930s, compressed yeast was produced in central Europe using large-scale aerobic fermentors for the first time (de Becze & Leibmann, 1944). The fermentor was a large cylindrical tank with an air-allowing network of perforated pipes at the base. Later iterations included mechanical impellers to accelerate mixing and disperse, and break up air bubbles.

The need for pressurized air resulted from this process. A vortex in the liquid was kept from developing by baffles on the vessel walls. A method that used steam and water in aeration tubes for sterilization and cleaning was invented in 1934 by Strauch and Schmidt. The choice to produce penicillin using the submerged culture technique, which requires aseptic conditions, enough aeration, and agitation, most likely had a major influence in driving the creation of meticulously planned and specially constructed fermentation containers. None of the fermentation plants were initially appropriate for deep fermentation when the British government concluded in 1943 that surface culture was insufficient. At Hindustan Antibiotic Ltd. at Pimpri, Pune, the first pilot fermenter was built in India in 1950. The pharmaceutical industry advancements in the 1940s necessitated the use of large-scale bioreactors for the synthesis of streptomycin and penicillin, which gave rise to the area of biochemical

DOI: 10.1201/9781032650616-8

TABLE 8.1
Important historical bioreactor events

Year	Event
4000–3000 BC	Brewer, baking
2000 BC	Distillation (China) and ethanol production
1923	Commercial generation of citric acid (Pfizer, USA)
1940s	Fermentation-based penicillin production (USA)
1950s	Design and large-scale aerated fermenter scaling
1970s	More than a hundred novel medications and vaccines have been made via bioprocessing
1980s	Control of fed-batch bioreactors

Source: Heinzle et al. (2006) and Dochain (2008).

engineering. Research on oxygen transmission, air and medium sterilization, and pH control led to advancements in bioreactor design and control. The creation of bioreactors capable of achieving and sustaining the chemical and physical environment for the organism that the biochemist/microbiologist advised was the primary focus of the early biochemical engineers. Instrumentation like the sterilizable oxygen electrode had to be developed in order to scale up from lab bioreactors to big fermenters. As illustrated in Table 8.1, important historical bioreactor events are as follows.

8.2 FERMENTER AND BIOREACTOR

A bioreactor is the vessel that facilitates various types of biochemical reactions, whereas the fermenter is the vessel that facilitates fermentation. Therefore, the fermentor is a type of bioreactor. Furthermore, bioreactors are used in the production of medicines, pharmaceutical products, antibodies, or vaccines, while fermenters are used to produce lactic acid or ethanol. A closed system used to process a biochemical reaction is called a bioreactor. It facilitates the proliferation of cells in a culture, such as those of mammals or insects, or it promotes the synthesis of secondary metabolites, such as those found in medicines, vaccines, or antibodies. A bioreactor is a container where, given specific, perfect conditions, microorganisms use a substrate, nutritional medium, and carbon source to carry out the fermentation process.

As a result, the fermentation process proceeds with the help of microorganisms, a medium containing nutrients, gas sparging, and ideal processing conditions. Since the fermentation process involves a lot of variables, the vessel that contains them must create a friendly environment for the living organisms to grow, multiply, and proliferate in order for them to do their job, which is to consume the substrate in the presence of nutrients under ideal process conditions (Chandrashekhar H and Rao J V, 2010). The vessel, referred to as a bioreactor, is anticipated to meet the necessary requirements for sufficient mass transfer, heat transfer, fluid flow characteristics,

FIGURE 8.1 Biochemical process

and process conditions, including temperature, agitation, pH, dissolved oxygen, gas flow rate, and fluid-fluid mixing. This requires a certain amount of systematic design with reference to various bioreactor configurations, flow characteristics, mixing and blending methodologies for different phases, techniques to maintain and control the process conditions, and methods to maintain asepsis. This task cannot be completed by a simple vessel that is chosen at random. The generalized outline of the biochemical process is illustrated in Figure 8.1 (Moo–Young and Blanch, 1981; Hochuli, 1992).

The bioreactor's dimensions can range over many orders of magnitude. The microbial cell (few mm), shaking flask (100–1000 mm^3), lab fermenter (1–50 L), and pilot scale (0.3–10 m^3) to plant scale (2–500 m^3) are all forms of bioreactors (Jagani et al., 2010).

8.2.1 Design and Construction

The flow rates, aeration, temperature, pH, foam control, and agitation rate are just a few of the factors that these bioreactors are made to regulate. The quantity of sensors and control components built into a particular bioreactor determines how many parameters may be tracked and managed. A schematic representation of the bioreactors is listed below and illustrated in Figure 8.2.

The following special features of biological processes must be considered when building a bioreactor (Jagani et al., 2010).

1. The reaction mixture usually contains modest quantities of both products and starting ingredients, or substrates, which have the potential to impede the process. The cells' dietary requirements (salts, oxygen), as well as the preservation of ideal biological circumstances (temperature, reactant concentration, pH) within certain bounds, determine cell development, the structure of intracellular enzymes, and the creation of products.
2. A few compounds affect the speed and mechanism of processes as well as intracellular control. These include inhibitors, effectors, precursors, and metabolic products.
3. Microorganisms can metabolize unconventional or even contaminated raw materials (cellulose, molasses, mineral oil, starch, wastewater, exhaust air, biogenic waste), a process that is frequently carried out in highly viscous, non-Newtonian media.
4. In contrast to isolated enzymes or chemical catalysts, microorganisms adapt the structure and activity of their enzymes to the process conditions,

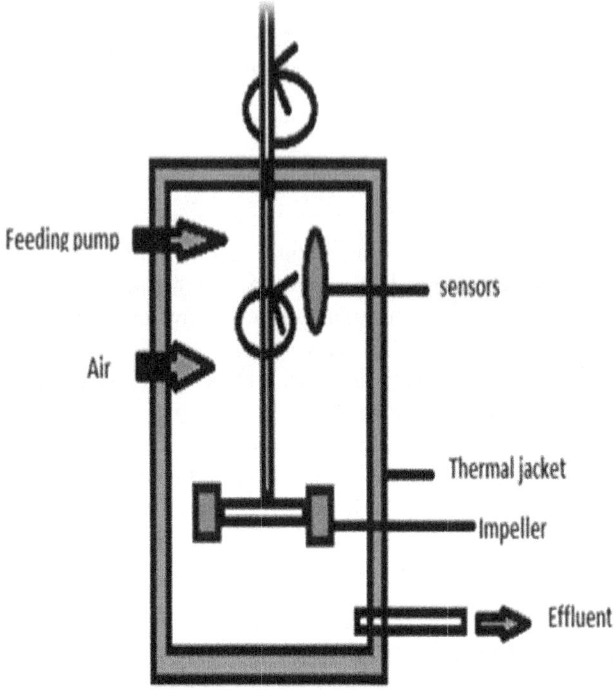

FIGURE 8.2 Schematic representation of bioreactor

whereby selectivity and productivity can change. Mutations of the microorganisms can occur under suboptimal biological conditions.

5. Microorganisms are frequently sensitive to strong shear stress and to thermal and chemical influences.
6. Reactions generally occur in gas-liquid-solid systems, the liquid phase usually being aqueous.
7. As biochemical conversion proceeds, the microbial bulk may rise. Microorganisms may develop on the walls, flocculate, or undergo autolysis during the process.
8. Complex dynamic behavior is a common feature of continuous bioreactors.

8.3 CLASSIFICATION

8.3.1 On the Basis of Flow Parameters and the Condition of Operation

As illustrated in Figure 8.3, three types of flow regimes may be found in bioreactors based on the flow parameters and the condition of operation (steady state or unsteady state). The concentration of reactants and products varies over time in an unsteady state process, whereas it stays constant in a steady state operation. Nonetheless, depending on where you are in the reactor, the concentration could or might not vary. Reactants are supplied into a batch reactor, which is the optimal unsteady state operated reactor. The reactor's intake and output are then sealed (Figure 8.3a). The reactor is under fully mixed circumstances, meaning that while the concentration is constant throughout the reactor, it varies over time.

Unsteady state operation is characterized by a buildup of product concentration and a decrease in reactant concentration over time. There are two scenarios under steady state settings. Reactants may flow through the reactor in a piston-like fashion with no mixing at all (referred to as plug flow); alternatively, they may fully mix as soon as they enter the reactor and the concentration is constant throughout (similar to a batch reactor); this latter scenario is known as a continuous stirred tank reactor (CSTR). Under the plug flow circumstances depicted in Figure 8.3b, the reactants enter the reactor from the other side. Inside the reactor, the reaction progresses progressively in the axial direction. There will not be any change in concentration in the radial direction. The reaction take place in an orderly direction as if they are a

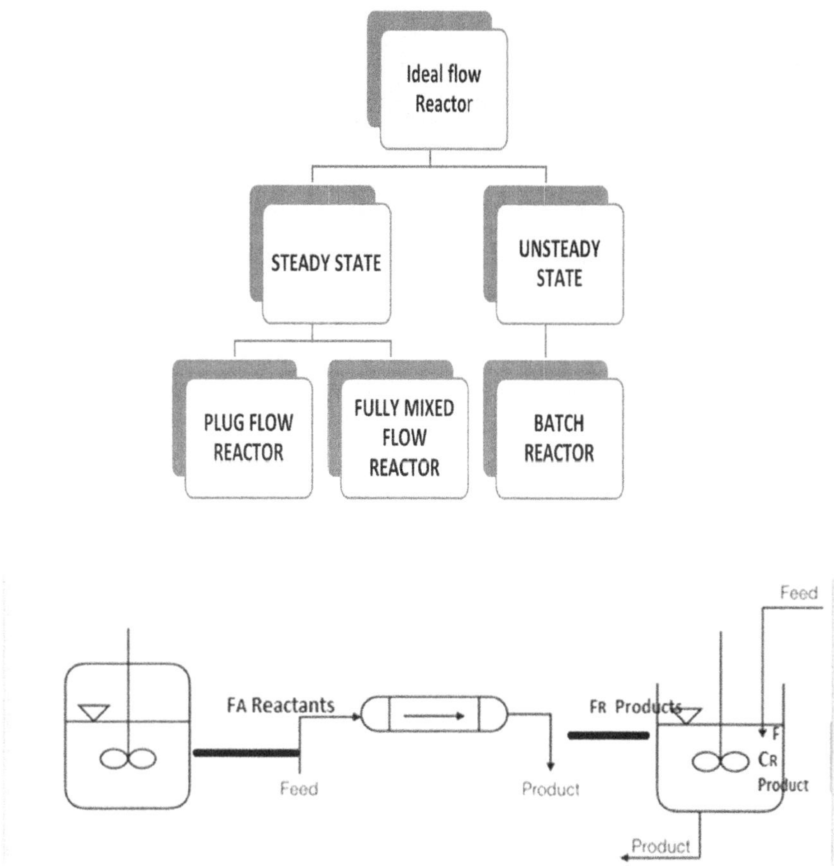

FIGURE 8.3 Classification of various types of ideal flow reactors. (a) Batch reactor, (b) plug flow reactor, and (c) CSTR

plug/piston along the axis, and hence the concentration of the reactant falls and that of products increases in the axial direction (i.e., in the x-axis); however, there will not be any change in the radial direction (i.e., in y-axis). There also will not be any accumulation of reactant or products, which is a typical characteristic of steady state flow conditions.

The third type of ideal reactor flow conditions is a fully mixed flow condition Figure 8.3c. In this, the reactant with initial concentration CAo enters the reactor continuously at the rate of FA. The contents in the reactor are perfectly mixed, hence the concentration of reactant falls the moment it enters the reactor to CA instantaneously, which is the concentration prevailing in the reactor. The contents are allowed to react to CR continuously; these conditions prevail in the reactor as long as the external and internal conditions do not change in the reactor; the volume remains constant and there will not be any accumulation of reactants or products in the reactor. Hence, the reactor is typically called a continuous stirred tank reactor, which is a case of the existence of perfect mixing conditions to prevail in the reactor. Thus, the two ideal continuous flow conditions are (1) total plug flow (non-mixing) conditions and (2) total mixing conditions. However, in reality, the realization of either is impossible. Most reactors operate in between these two extreme conditions. Mathematical equations are developed to model these two ideal flow regimes, which are helpful in the design of reactors/bioreactors.

8.3.2 Classification Based on Feeding Mechanism

1. **Batch Reactor**: All reactants are fed into the batch reactor, and its contents are sealed (Figure 8.3a). No further materials may enter or exit the reactor. Once the reaction time has elapsed, the goods are withdrawn. A batch reactor is the best option for small-scale operations, R&D stage process testing, producing some expensive products, producing a variety of products on a shift basis with ease and flexibility, and processing difficult-to-transform processes into continuous operations. Nevertheless, there are drawbacks to batch reactors. For instance, large-scale manufacturing in batch procedures is not cost-effective due to the high operational costs.
2. **Continuous Reactor**: The contents of the two types of continuous previously stated (Figure 8.3b and c) are well mixed using a stirrer, allowing for good mixing conditions and the preservation of a consistent temperature. In biochemical terms, chemostats are the overall term for such a setup used for cell culture. One of the usual drawbacks of CSTR is that it has the lowest conversion per unit reactor volume, requiring larger reactor volumes to achieve the same target conversion/yield.
3. **Fed-batch Reactors**, also known as semi-batch reactors, are a typical feature of biochemical operations. They are also often called fed-batch reactors. They operate in a mode halfway between batch and continuous processes. They have better yields and selectivity, gradual addition or removal assists in controlling temperature, particularly when the net reaction is highly exothermic. Thus, the use of a semi-batch reactor intrinsically permits more

stable and safer operation than in a batch operation. The semi-batch reactor is probably used in the chemical industry and in biotechnological processes. As with the continuous reactor, it is possible to achieve high productivities due to the fact that controlling the flow rate of the feed entering the reactor can optimize the growth rate of the cells.

8.4 VARIOUS OTHER TYPES OF REACTORS

Bioreactor science has experienced rapid progress and in recent years, different bioreactors have received great interest. With the tremendous progress in functional genomics, metabolic engineering, and systems biology, there is a great potential for a single cell working as a super bioreactor. It is also very exciting to see more and more achievements using plants and animals as integrated bioreactor systems.

8.4.1 Stirred Tank Reactors (STRs)

The best possible interactions between the culture media and the cells are necessary for bioreactors intended to exhibit the biological characteristics of living cells as efficiently as possible. They must offer effective ways to transport mass, heat, and mixing between the various phases in a well-regulated setting. In order to maximize and raise productivity, new kinds of bioreactors and current reactor technologies are always being created. The industry standard for mechanically stirred tank reactors (STRs) is their adaptability and flexibility. Najafpour (2008) stated that three primary types of fermenters are utilized on an industrial scale:

1. Non-mixed and non-aerated systems: approximately 70%.
2. Non-mixed and aerated systems: approximately 10%.
3. Mixed and aerated systems: approximately 20%.

Conventional items like wine, beer, and cheese are produced in non-mixed, non-aerated tanks. Microorganisms are cultivated in mixed and aerated tanks, where they produce the majority of the new products (Najafpour, 2008). Providing a sufficient and regulated environment for cell growth and product production is the bioreactor's primary shortcoming. Thus, when building a bioreactor, a number of aspects need to be taken into account. These include sterility, appropriate size and material, low energy consumption, geometry, temperature and pH control, aeration and mixing systems (where needed), and sterility (Stanbury, 1995). Because of its cheap operating costs, the standard STR bioreactor is the most significant bioreactor for industrial applications. The tanks' capacities range from a few dm^3 to hundreds of m^3. Glass is used to make laboratory-scale tanks with a maximum capacity of 20 L. They are usually constructed of stainless steel for larger quantities. They may be made using a variety of materials and combinations of those materials. Depending on the amount of heat to be extracted, the relationship between height and diameter can range from 2:1 to 6:1 (Najafpour, 2008). Since the ratios 1:1 require less material and offer a smaller surface area, it is likely the most economical.

Mechanical agitation in the bioreactor is used to ensure homogeneity and bubble dispersion; this demands a comparatively high energy use per volume unit (Doran, 1995). Many types of impellers that result in various flow patterns within the tank are detailed. Typically, liquid fills between 70% and 80% of the STR total capacity. A foam breaker may regulate the headspace for gas depletion and foam production in this situation. For both free and immobilized cells and enzymes, STRs are utilized. Every situation calls for a thorough examination and analysis of the biocatalyzer sensitivity (Doran, 1995).

8.4.2 CSTRs or Continuous Stirred Tank Reactors

The term "continuous STR" refers to STRs that are capable of feeding and removing mass and energy. Without taking into account changes in temperature, concentration, fluid characteristics, or reaction rate, an ideal CSTR may be modeled with perfect mixing. This indicates that the bioreactor medium characteristics are the same in the intake and outflow streams. The CSTR attributes are all precisely the same as those of the batch bioreactor. The instrumentation and control systems in this instance are more advanced and distinct (Laska & Cooney, 1999). CSTR operation tactics come in two varieties. One of them is the Chemostat, which is used to cultivate cells and allows for the addition of all nutrients while maintaining a consistent liquid volume by setting.

8.4.3 Microbioreactor

Microbioreactor miniature bioreactors, also known as low-cost microbioreactors, are used in high-throughput bioprocessing for a variety of purposes, including process optimization, medium creation, and strain enhancement. Continuous measurements of pH, dissolved oxygen, and optical density were made in a microbioreactor with a working volume of 2 mL using an optical sensor device (Kostov et al., 2001). The microbioreactor pH, dissolved oxygen, and optical density profiles, when utilized for *Escherichia coli* fermentation, were comparable to those of a standard 1-L bioreactor. This work served as the foundation for the development of a multiple-bioreactor system for high-throughput bioprocess optimization. The system included a 24-well plate miniature bioreactor for microbial cultivations assessment (Isett et al., 2007), a low-cost noninvasive optical CO_2 sensing system for fermentation and cell culture, and a scale-down model for the development of cell culture processes. A microbioreactor with a microliter volume was created by Zanzotto et al. (2004) using glass and polydimethylsiloxane (PDMS).

Through a gas-permeable PDMS membrane, aeration was carried out. Integrated sensors were used to detect pH, dissolved oxygen (DO), and optical density (OD) online using optical techniques. It was discovered that bacterial fermentations conducted in the microbioreactor under precisely specified conditions were comparable to those conducted in a 500 mL bench-scale bioreactor. The microbioreactor bacterial population behaved similarly to that of the bigger bioreactor. Additionally, it was shown that the microbioreactor system's sensitivity and repeatability allowed for the statistically significant distinction between various physiological states based on variations in the temporal development of the OD, DO, and pH.

8.4.4 Cell as a Super Bioreactor

The cell factory produces a wide range of commercially significant goods, and metabolic engineering offers a comprehensive method for designing new cell factories by offering logical design processes as well as flexible experimental and mathematical tools (Patil et al., 2004). For instance, diacetyl, alanine, and exopolysaccharides were produced by metabolically modifying lactic acid bacteria (Kleerebezem et al., 2000). Large-scale sequencing initiatives have made the whole genome sequence accessible for a growing number of species. This has led to significant research efforts in functional genomics, the field of assigning roles to all found open reading frames. There is a tendency in both metabolic engineering and functional genomics toward applying a macroscopic perspective to cell function, which will increase the role of the classical method in microbial physiology. It will be feasible to characterize the interaction between every component in a biological system, the cell, at the quantitative level, thanks to the growing understanding of molecular processes. Systems biology aims to do this, which would greatly aid research on microbial physiology and metabolic engineering Nielsen et al. (2002). Because plants synthesize a wide variety of secondary metabolites that may be utilized as medications, colors, flavors, and scents, it is particularly attractive to design plant cells to produce these compounds. Frequently, these plant metabolites have extremely intricate structures.

8.4.5 Wave Bioreactors

Wave bioreactors have been shown to have the potential to be used for growing grape, apple, and tobacco suspension cells. With a doubling time of two days, Eibl and Eibl (2006) discovered high plant cell (*V. vinifera*) biomass productivities of 40 g/(L day). There have been a number of small-scale photobioreactor proposals; however, most of them have drawbacks or challenges when it comes to scaling up. A sufficient supply of light is essential for the successful scaling up of photosynthetic organism culture. Nonetheless, a suitable technique for collecting solar energy and dispersing light within the photobioreactors is necessary. Internal optical fibers for light distribution have been reported for scale-up photobioreactors, according to Ogbonna et al. (1996), despite the fact that these fibers are costly and might cause technical issues. The morphology of the cells did not significantly alter in comparison to cultivations in stirred tank bioreactors. Wave bioreactors also save time by eliminating the need for cleaning and sterilization procedures, reducing foaming, being simple to operate, and having a minimal risk of contamination.

8.4.6 Photobioreactors

Only a small number of the photobioreactor models that have been developed can really employ biomass from algae. Photobioreactors are mostly used in photosynthetic activities, such as the controlled development of microalgae or vegetable biomass. Ugwu et al. (2005) state that mass transfer, which is essential to the smooth functioning of mass algal cultures, is one of the main obstacles to their practical deployment in algal mass cultures. Algal biomass is widely utilized in aquaculture, water treatment, the synthesis of fine compounds, as beneficial supplements for people and animals, heavy metal biosorption, and CO_2 fixation. According to Ugwu et al. (2005), there are several types of photobioreactors, including laboratory and industrial size versions. They are further divided into closed photobioreactors, open ponds, flat-plate, horizontal/serpentine tubular airlift, and inclined tubular photobioreactors (Rasoul et al., 2011). The development of increasingly advanced techniques for growing microalgae with improved productivities and resistance to contamination was made possible by various kinds of closed photobioreactors used outdoors (Bergeijk et al,.2010). The benefit of closed ponds or tanks, according to the same author, is that the light path length is significantly shorter than in open ones, resulting in higher cell densities, which lowers the possibility of contamination and makes harvesting easier. Additionally, it makes temperature regulation easier and decreases evaporative losses (Tredici, 2004). Fluorescent or other light bulb distributors are typically used to artificially light laboratory-scale photobioreactors.

8.4.7 Membrane Bioreactors

Membrane bioreactors use specialized membranes for product separation, nutrient delivery, or in situ aeration that have a defined molecular weight cutoff. They are made to hold onto recombinant protein products and cell biomass inside a certain cell compartment. Research has shown that mass transfer is crucial, even for relatively slow-growing plant cells, even while membrane bioreactors reduce shear stress on cultured cells (Huang & MacDonald, 2009). Using transgenic rice cell culture with a rice alpha amylase promoter, McDonald et al. (2005) used a membrane bioreactor to produce recombinant human alpha-1-antitrypsin (AAT), yielding an extracellular product titer close to 250 mg/L and 4–10% of the extracellular total soluble protein. However, the membrane bioreactor is primarily used for small-scale processes and is difficult to scale up for large-scale.

8.4.8 Disposable Bioreactors

The large-scale plant cell culture industry uses disposable bioreactors in an effort to save production costs and simplify validation procedures in compliance with GMP standards. FDA-approved biocompatible polymers like polyethylene, polystyrene, polytetrafluoroethylene, polypropylene, and ethylene vinyl acetate are commonly used to construct disposable bioreactors (Eibl et al., 2010). There are no clean-up or re-sterilization processes required because the bioreactor vessel is pre-sterilized and disposed of after harvest, which lowers the possibility of culture contamination (Xu et al., 2011).

8.5 PRACTICAL APPLICATIONS

Fermenters and bioreactor systems are vital tools in the field of biotechnology, facilitating the growth and manipulation of microorganisms or cells for various applications. Here's an overview of their practical applications and current trends:

8.5.1 Food and Beverage Industries

In the food and beverage business, bioreactors are now a necessary piece of equipment, especially for producing fermented goods. For ages, people have utilized the fermentation process to create a wide range of food and drink items, including bread, wine, yogurt, cheese, and beer. However, the method by which fermentation is carried out has changed as a result of the usage of bioreactors (Figure 8.4).

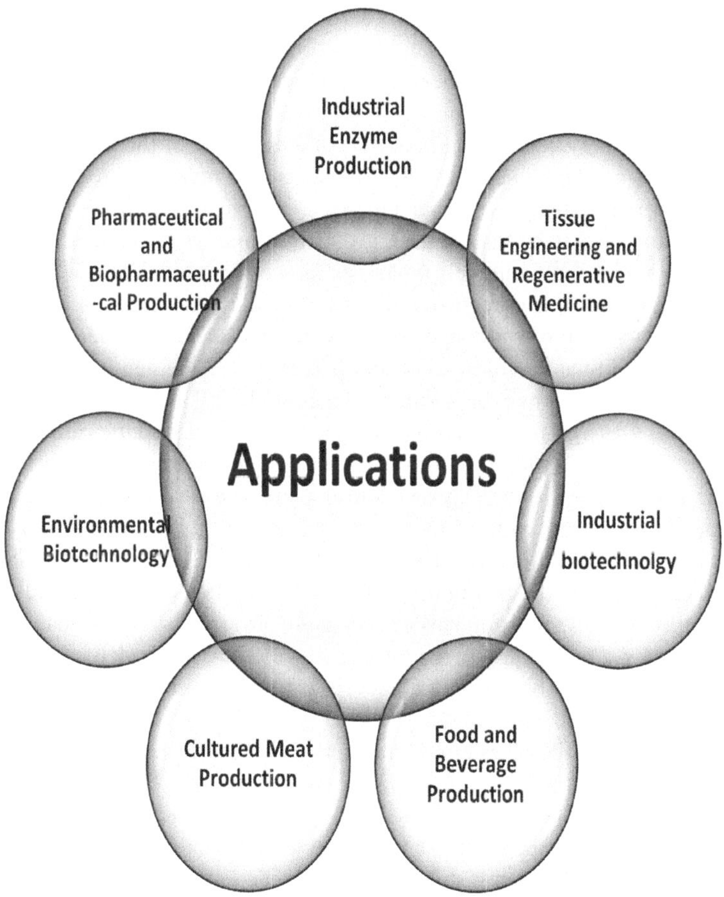

FIGURE 8.4 Practical application of various bioreactors

For the particular microbes required to develop and create the desired product, cheese, yoghurt, etc., bioreactors offer a more exact and regulated environment.

In addition to providing control over crucial pH, oxygen levels, and temperatures for microorganism growth, the use of bioreactors greatly increases the consistency and efficiency of the fermentation process, leading to increased yields of high-quality products. It is possible to modify bioreactors to accommodate various kinds of microbes and fermentation procedures. As an illustration, whereas some bioreactors are made for anaerobic fermentation, others are made for aerobic fermentation. Producing novel and inventive items that were previously unattainable has become feasible for food and beverage producers because of the customization capabilities of bioreactors.

8.5.2 Cultivated Meat

Standard cell culture dishes and stacked flasks have been used to make grown meat products for taste testing; however, in order to develop cultivated meat at scale, bioreactors with sizes of several thousand liters or more will be needed. The usual cell culture technician is not often concerned with issues like gas exchange, heat transfer, shear stress, mixing, and foaming; these issues are brought up by this fundamental difference in the way cells are cultivated. Although a lot of these methods may be taken from other sectors, such as cell treatments, the manufacture of recombinant proteins, or the production of other biologics, significant adjustments are needed to get cultured meat from taste testing events to a market-ready state. An interdisciplinary endeavor including cell biologists as well as chemical, mechanical, and bioprocessing engineers is required. To scale up the production of cultured meat beyond the bench and into pilot and commercial stages, a range of techniques can be chosen. Batch, fed-batch, continuous, and perfusion are the general categories into which these techniques may be divided (Meyer et al., 2017). When growing cells in batch culture, a set volume of medium is placed in a vessel, and the cells are allowed to develop to their maximum density before being harvested or moved to a bigger vessel. In fed-batch culture, the growth parameters, such as exponential cell growth or cell densities, are maximized by feeding the cells in a vessel with new media at varied rates from an in-line, independent feed vessel.

Depending on how the medium is combined and whether the cells are grown in suspension or affixed to a solid surface, there are several distinct bioreactor designs to select from. A continuous stirred tank reactor is the most widely used bioreactor for animal cell culture because, in comparison to airlift reactors, it provides better long-term sterility and less bubbling (Meyer et al., 2017). Continuous stirred tank reactors, in general, allow high mass transfer of oxygen to occur while allowing the development of cells in suspension by mechanical stirring. Though at lower maximum sizes, vertical wheel bioreactors and rocking platform bioreactors can provide outcomes similar to suspension growth. With attachment-dependent cells, suspension growth can also happen via the use of microcarriers.

The preference for stirred tank and rocking platform bioreactors in disposable, single-use systems up to 6000 L is evident in current developments in the cell therapy and biopharmaceutical industries. The benefits of single-use bioreactors include reduced turnaround times, cross-contamination, water, energy, and sensor costs, as well as the elimination of the need for hot sterilization. Consequently, single-use bioreactors could be a good choice for scale-out methods of production. However, since a batch of cultivated meat product is less valuable than a biologic drug or human cell therapy, single-use technologies are unlikely to be economically viable for cultivated meat unless current models can be made more affordable.

Notably, though, single-use bioreactor bags are made up of many material layers that must, at great expense, adhere to stringent regulatory requirements to guarantee that a valuable batch of product is not wasted. Therefore, there could be an opportunity to build commercially viable, food-safe disposable bags specifically for the production of farmed meat at lower batch values and food-grade regulatory criteria. It will be necessary to balance the waste generated by single-use bioreactors against the advantageous sustainability metrics of these systems, which demonstrate reductions in water and energy consumption during sterilization as compared to stainless steel bioreactors.

Stirred tank bioreactor vessels have been utilized in biopharma to produce animal cells in quantities as large as 20,000 L (Li et al., 2019). These amounts might not be enough for farmed meat, though. Due to the lack of moving parts in their mixing, which results in less inhomogeneity, less shear stress (discussed later), fewer nutrient or oxygen gradients, and a reduction in power requirements for mixing, airlift reactors also allow suspension growth and become advantageous at very large scales (e.g., >20,000 L) (Merchuk, 1990). An airlift reactor designed for microbial cell growth was the biggest bioreactor ever constructed, with a capacity of 1,500,000 L. Recently, the shape and internals of this airlift reactor design were optimized for animal cell growth using computational fluid dynamics models. The reactor was scaled down to 300,000 L. This led to a single reactor that could feed 75,000 people yearly with a potential cell density of 2×10^8 cells/mL (Li et al., 2019). Therefore, airlift reactors could provide a more manageable way to achieve very high sizes. Before such a scale is achieved, there are still a lot of unanswered questions about animal cell culture in airlift reactors. Airlift reactors will need to be used for a significant amount more modeling and empirical data collecting for animal fat, skeletal muscle, and other cell types. Because of this, a large portion of the discussion that follows will center on stirred tank bioreactors, where data is easier to translate.

8.5.3 Cell Culture: Growing and Culturing Cells

For the purpose of cultivating cells, including stem cells, the use of bioreactors has proven very advantageous. One kind of cell that can develop into many different kinds of specialized cells in the body is called a stem cell. Stem cell culture benefits greatly from the use of bioreactors because of the special needs that these cells have for growth and differentiation. Because bioreactors offer the perfect conditions for stem cells to proliferate and thrive, scientists can generate vast amounts of these cells for use in the ability to manipulate the growing environment more is one benefit of cell culture in bioreactors. For example, it is possible to accurately monitor and alter the nutrition supply to guarantee the cells' ideal development circumstances. For example, it is possible to accurately monitor and alter the nutrition supply to guarantee the cells' ideal development circumstances. Furthermore, bioreactors offer a sterile environment that minimizes the possibility of contamination, a crucial factor in the culture of stem cells.

8.5.4 Tissue Engineering in Medical Applications

In the past several years, tissue engineering has drawn a lot of interest from the medical community. It includes using cells, biomaterials, and biological components to create functioning tissues and organs. In medicine and drug testing, the use of bioreactors has created new opportunities, especially in the area of regenerative medicine such as transplantation. Tissues, including skin, bone, cartilage, and organs like the liver and heart, are being developed in bioreactors.

Comparing bioreactors to conventional tissue culture techniques reveals a number of benefits. They offer a more regulated environment, which makes it possible to better control how cells divide and develop. Additionally, bioreactors enable the constant delivery of nutrients and oxygen to cells, both of which are essential for their survival and development. Beyond what was possible even a century ago, bioreactors enable the production of tissues with more functionality and superior quality by offering a more natural environment for cell development.

8.5.5 Industrial Biotechnology

In recent times, industrial biotechnology has experienced tremendous growth, with bioreactors playing a crucial role in this development. Bioreactors are used to produce bioproducts, including enzymes, antibiotics, and biofuels. Large-scale growth and multiplication of microbes, plants, or animal cells are made possible by these devices because they offer a regulated environment. Thus, bioreactors have completely changed the manner in which several items that were previously difficult to manufacture on an industrial scale are produced. Bioreactors are employed in a variety of industrial biotechnology applications. They may be used, for instance, to create biofuels from renewable resources like algae, maize, and sugarcane. Additionally, they may be utilized to produce the enzymes needed for paper, textile, and food processing. Furthermore, antibiotics and other pharmaceuticals needed in medicine are made in bioreactors. The industrial manufacture of these items has improved in efficiency and economy as a result of bioreactors.

8.5.6 Pharmaceutical Industry

The pharmaceutical industry is one of the biggest in the world; it is responsible for discovering and developing new drugs that might save millions of lives. One of its most crucial components is the bioreactor, which is required for this company to produce vaccinations, antibodies, and other medications. Bioreactor containers offer sufficient volume to facilitate the growth of bacteria, cells, and tissues while preserving environmental equilibrium. Scientists in the pharmaceutical industry use these bioreactors to produce massive volumes of active compounds that are subsequently used to make life-saving antibiotics and immunizations. One application where bioreactors are extremely important is the manufacture of vaccinations.

8.5.7 Environmental Biotechnology

Using biotechnology to address environmental issues, the discipline of environmental biotechnology is expanding quickly. The main goal of environmental biotechnology is to address environmental problems in a sustainable manner. The treatment of industrial waste is one of environmental biotechnology's primary goals. Waste is treated in bioreactors by employing microorganisms to decompose contaminants. Industrial trash may be costly and challenging to properly dispose of, as it is a major cause of pollution. The use of bioreactors in environmental biotechnology to remediate hazardous waste provides a long-term solution to this issue. The waste contaminants are broken down and transformed into safe compounds by the bioreactor microorganisms. Compared to more conventional trash disposal techniques like incineration or landfilling, this approach is far more environmentally friendly. Environmental remediation, or cleaning up polluted sites, is another application of biotechnology. This might involve clearing spills and leaks as well as eliminating toxins from groundwater and soil. Microorganisms are used in bioremediation, a kind of environmental remediation, to break down contaminants. Oil, insecticide, and other polluted places can be cleaned up via bioremediation.

Environmental biotechnology has many more uses besides cleaning up dirty areas and handling industrial waste. It may be used, for instance, to produce biofuels from plant waste or algae, two renewable sources (Putt et al., 2011). Additionally, it may be utilized to make bioplastics, which are biodegradable polymers consisting of renewable resources. The application of bioreactors in environmental biotechnology will only grow and progress as people become more conscious of environmental issues.

8.5.8 Industrial Enzyme Production

Process factors and bioreactor designs for the synthesis of industrial enzymes. Selecting the right microorganisms and fermentation conditions, Actinomycetes, yeast, fungus, bacteria, and other kinds of microbes may all produce enzymes either intracellularly or extracellularly. According to Liu and Kokare (2023), the majority of microbial enzymes are generated extracellularly in industry, with fungi accounting for 50% of synthesis, bacteria (both gram-positive and gram-negative), and plants for the remaining 35%. Because it is easier to purify and retrieve enzymes from extracellular matter, extracellular generation of enzymes has several advantages over intracellular production. *Aspergillus*, *Bacillus*, *Kluyveromyces*, *Streptomyces*, and *Trichoderma* species are the most common bacteria and may produce around 50 g/L of extracellular proteins (Cheng-Wu et al., 2001; Patel et al., 2016). A good industrial enzyme manufacturing factory must pick an appropriate strain, but even this does not ensure maximal enzyme output until the fermentation process is sufficiently optimized. The kind of fermentation technique (SSF), the temperature during the process, the media pH, the availability of nutrients, and the length of the fermentation period are some other variables that affect how adequate a fermentation process is. To produce the necessary amount of enzymes, these parameters must be properly optimized and experimented with. The ideal circumstances for enzymes that are of high biotechnological interest because of their potential uses in the chemical, pharmaceutical, and food and beverage industries.

8.6 USE OF BIOREACTOR FOR DIFFERENT APPLICATIONS

8.6.1 Bioreactors for Fermentation

Various bioreactors are used for the following range of fermentation processes:

1. For the production of microbial biomass.
2. For the production of microbial enzymes.
3. For the bioconversion of organic feed using microorganisms.
4. For the production of secondary metabolites.

Mostly, a semi-batch/fed batch or a batch fermenter is used for the production of a large number of fermentation products. Ethanol may be produced by batch, fed-batch, or continuous fermentation processes (Thangprompan et al., 2013). Brazil normally employs fed-batch systems with high cell density and continuous fermentation processes to produce sugarcane ethanol. The remaining manufacturing facilities employ continuous fermentation, while around 85% use fed-batch fermentation. The volume of fermentation tanks varies from 300 to 3000 m^3. To maintain a low sugar concentration in a fed-batch operation, the fermenter is progressively supplied with the sugar feed medium. The feeding process takes around four hours. By doing this, the fermentation is not inhibited by a high sugar content. Over time, the fermenter volume grows. Due to a high concentration of yeast (10–14 g (fresh weight) mL^{-1}), the entire fermentation process takes 6–9 hours (Domínguez, 2001).

Because the sugar concentration is always maintained at a low value, a fed-batch process converts sugar to ethanol more effectively in a given amount of time than a batch operation. In order to maintain the same volume of broth in the fermentation vessel throughout a continuous fermentation, the volume flow rate of the sugar medium given to the fermenter must be maintained constant with the rate of broth harvest from the fermenter. Six to nine hours are spent living in the fermenter. Because it enables extended periods of continuous operation with occasional cleaning, continuous fermentation is becoming more and more popular. One potential issue is bacterial contamination. If contamination is not reduced, bacteria will eat the sugar that will be lost during fermentation. *S. cerevisiae* is the yeast that is usually used for fermentation. In the fermenter, the yeast is combined with the prepared feed medium, often known as wort. Temperatures between 32 and 34°C and pH values between 3.8 and 4.5 are typical for fermentation. Upon initiating the fermentation process, the yeast is combined with a medium containing 180–200 g L^{-1} of total reduced sugars, resulting in a fresh weight concentration of the yeast of around 300 g L^{-1}. The fermenter is filled to a quarter to a third of its capacity by this suspension. As the fermenter is fed, the remaining capacity is filled. To reduce product inhibition, feeding is done so that the broth's ethanol content stays at 7.5% w/v (g per 100 mL). About 7.5% ethanol and 100–140 g (fresh weight) of yeast biomass per liter are present in the fermented broth. The broth's residual sugar content is limited to less than 1.0 g L^{-1}.

Following fermentation, the clarified broth (or wine) is sent to distillation after the yeast is extracted from it using centrifugation. Thirty to eighty percent (g fresh weight per 100 mL) of yeast is present in the yeast slurry that exits the centrifuge. Sulfuric acid is added to this yeast cream after it has been diluted with an equivalent amount of water to bring the pH down to 1.8 to 2.5 (Hoekema et al.,2002). The pH is increased to around 3.8 after nearly two hours. The bacterial burden is lowered by this therapy. The fermentation vessel is refilled with the treated yeast. In the USA, the process of producing dry-milled corn ethanol typically involves the staggered employment of numerous batch fermenters (Ezemba et al., 2022). It's possible that one fermenter is filling in advance of fermentation, another is fermenting, and a third is draining. In this manner, it is possible to operate the processes continuously, both upstream and downstream of the fermenters. SSF, or simultaneous saccharification and fermentation, is used in many facilities. The fermentation mash frequently has a nitrogen source (urea or ammonium sulfate) added to it. The most often utilized microbe is yeast (Engasser, 1988). The temperature during fermentation is 32°C. After 48 to 72 hours of fermentation, the mash has a final ethanol content of 10% to 12%.

8.6.2 Bioreactors for Solid State Fermentation (SSF)

Solid state fermentation (SSF), which is primarily utilized in laboratories these days, provides a number of biotechnological benefits over submerged fermentation. Greater fermentation productivity, greater end product concentration, higher product stability, reduced catabolic suppression, growing microorganisms specialized for water-insoluble substrates, or mixed fungal culture are some examples of factors that might be stated. Aside from this, the low water activity utilized in SSF makes medium sterilization both not essential and frequently unneeded (Singhania et al., 2009). The unique characteristics of the SSF, including the range of materials that can be used as growth media and their composition, size, strength, porosity, and water-holding capacity, as well as the substrate's low humidity, which causes issues with the heat transfer system, must be taken into consideration when building bioreactors. When designing and developing control techniques for a reactor that will run on solid-state technology, all of the following things must be taken into account. Since the process of solid state fermentation takes place without the presence of free water, filamentous fungi are microorganisms that are naturally suited to this kind of environment (Chen et al., 2009).

Solid-state fermentation bioreactors are employed in large-scale traditional processes like koji and sake, but commercial bioreactor designs are not available for application in other bioprocesses. The primary challenges to be addressed are the technical elements, namely the absence of standardized procedures and the restricted repeatability of the experimental outcomes. Regarding the construction, management, and expansion of SSF bioreactors, a substantial number of articles and patents are accessible. The primary challenges in scaling up the bioreactor are the high heat generation and system heterogeneity (Hung et al, 2005).

Temperature, humidity, and substrate concentration gradients that may occur throughout the process have an impact on both operations involving agitation of the substrate, as those carried out in revolving drums, and processes utilizing static solid beds. Important yet challenging to control factors in SSF are temperature, moisture content, and oxygen concentration. Temperature rises quickly due to the significant amount of heat produced by the aerobic microbial growth in the bioreactor. This result is frequently unwanted, particularly in some biotechnological processes where the fermentation process might yield heat-sensitive compounds or enzymes that can be heat-denatured at the conclusion of the process. It is challenging to remove the generated heat, for example, using the bioreactor's two walls, in the absence of a free aqueous phase. It is via evaporation that the process is cooled instead. But this necessitates extremely high aeration rates since high aeration intensity is needed to overcompensate for the increased heat output that comes with higher metabolic activity. In many circumstances, the water lost by evaporation needs to be replaced, which might cause an unfavorable local increase in water activity during static processing. This enhanced water activity can thus have negative consequences in semi-sterile processes by encouraging the growth of bacterial contaminants, while in sterile fermentations, it may result in a localized lack of oxygen supply for the microorganisms. Although it should be emphasized that many bacteria react quite strongly to the shear stress induced by substrate mixing, it may still be helpful. The generation of metabolic water by aerobic bacteria is another difficult-to-account-for component that can cause issues, particularly in the development of conidiophores.

8.6.3 Bioreactors for Plant Cell Cultures

The use of plant cell cultures to produce important natural goods, including fine chemicals, flavors and perfumes, and medications, has been gaining popularity. Nevertheless, a number of issues have restricted the use and economic exploitation of plant cell cultures. The primary issues include sluggish growth, genetic instability, low and unstable cell productivity, and an inability to sustain photoautotrophic development (Paek et al., 2005). Additionally, certain characteristics of plant cell cultures, such as their large size and complex morphology, propensity to aggregate, time-dependent rheological behavior, foaming and wall growth, shear sensitivity, and relatively low growth and oxygen uptake rate, must be taken into account when designing bioreactors (Xu et al., 2011). A thorough analysis of these element effects on bioreactor design and operation has been conducted (Huang & McDonald, 2009).

Although the needs for nourishment and control are different, the culture of plant cells in reactors bears many parallels to the development of mammalian cells. For plant cell cultures to be successfully used in large-scale industrial processes, bioreactor design and operation are essential (Ranjbar et al., 2008). Three classifications of culture systems are often used to characterize a range of bioreactors used for large-scale plant cell culture:

1. Bioreactors for producing metabolites and enzymes.
2. Bioreactors for producing biomass (the ultimate result being cells, organic, shoots, or roots).
3. Biotransformation of exogenously supplied metabolites is carried out in bioreactors.

To provide improved conditions for plant cell cultures, the traditional stirred tank reactor (STR) for microbial cultures has been adapted. In order to enable optimal cell development and produce high productivity, the aim of bioreactor design and operation is to create a sustained and sterile culture environment with effective mixing and oxygen exchange.

8.6.4 Bioreactors for the Culture of Mammalian Cells

Because human and animal cells are so sensitive to changes in their culture environment, batch culture's conventional depletion and buildup of hazardous byproducts can have a negative impact on them. Traditionally, tissue culture is conducted using batch or semi-batch culture techniques, in which the carbon source is added at the beginning of the culture. T-flasks and well plates are the most common 2D tissue culture techniques, where cells are cultured statically without any mixing. Because these static systems are non-homogeneous, they have some constraints. In the absence of mixing, the culture medium will have concentration gradients of gases, nutrients, toxins, pH, and growth factors (Collins et al., 1998). This will decrease culture reproducibility and result in varied cell behavior in various spatial areas within the culture. The behavior of cells in 2D culture has been shown to differ significantly from that of cells in 3D culture (Abbott, 2003), and this has significant ramifications for the in vitro production of human tissues. Tissue engineering is essentially a 4D problem as tissue development is a dynamic, non-steady state process where biological factors, the physical rate process, and the local mechanical environment are always changing with time (Oksman and Inze, 2004).

8.6.5 Membrane Bioreactors for the Production of Value-added Products

Membrane bioreactors for the generation of products with additional value. Membrane bioreactors have the ability to create high productivity value-added goods such as proteins, organic acids, lipids, biofuels, and biopolymers. Despite the method's encouraging outcomes in the lab, scale-up issues prevent this technology from being widely used in industry. The purpose of this study is to discuss these issues and provide a foundation for future studies aimed at using membrane bioreactors in the industrial setting to generate products with added value. By using membranes to boost the mass transfer rate of the limiting substrates, achieve high cell concentrations, and separate the inhibitory chemicals that can block the bioconversion reaction, this chapter outlines the present state of the art in such bioreactor systems. It also discusses the commercialization patterns of today, the difficulties associated with using membranes, high costs, membrane fouling, and potential future paths for the broader use of membrane bioreactors (Akkoyunlu et al., 2021).

8.6.6 Ionic Liquid in Bioreactor Technology

ILs, Ionic liquids, are a novel class of solvents made totally of ions. ILs are often regarded as environmentally friendly solvents because of their low vapor pressure and other characteristics, including their high thermostability and lack of flammability. Applications of ILs in biotechnology are now being developed. Enzymatic and whole-cell processes are the two primary biotechnological domains in which ILs may find use. The former are the most common in the literature, with a large range of enzymes having catalytic activity that are on par with or even higher than those seen in typical organic solvents (Yang & Pan, 2005). Therefore, one major obstacle to the broad application of IL in bioreactor technology is its toxicity. Conversely, the most encouraging feature of these solvents is most likely their flexibility. Another benefit of ILs over conventional solvents is the ability to adjust almost all of their physicochemical characteristics by changing the alkyl chain or the anion/cation identity while maintaining important characteristics like non-volatility, non-flammability, and thermal stability. Consequently, it was determined that IL tuning is an important area of study to maximize and develop IL's potential in bioreactor technology (Quijano et al., 2010).

8.6.7 Production of Polyhydroxyalkanoates (PHAs)

The significant expenses associated with PHA manufacture have limited its output. Because of this, researchers have long searched for inexpensive substrates to lower total production costs. To reduce manufacturing costs, a wide range of less expensive carbon sources, including vegetable oil, whey, waste frying oil, and organic substrate, have been researched recently. Using inexpensive carbon sources, a mutant strain of *P. aeruginosa* generated large amounts of PHAs (Hellwig et al.,2004). In a preliminary experiment, the strain demonstrated excellent PHA synthesis. Two criteria will determine how PHA manufacturing develops in the future: (a) reduced production costs with greater yields; and (b) broader uses in other industries.

It is anticipated that better yielding PHA strains with high growth densities under ideal fermentation conditions would be produced quickly by genetic engineering and synthetic biology approaches. The manufacturing cost may be decreased, and large-sized granules can be accumulated using purifying procedures with controlled lysis. In two single-stage continuous stirred tank reactor configurations and three single-stage continuous stirred tank reactor configurations, continuous fermentation with mixed microbial culture may be employed to reduce production costs while obtaining greater yields utilizing inexpensive substrates. To give PHAs distinct features, several functional groups are affixed to the side chain of their fundamental structure. The predicted structural result of short or medium chain PHAs should be produced using the molecular evolution approach. In the mechanical industry, as well as many other scientific and technical domains, downstream processing is necessary to achieve the required product purity for a particular application. When soybean oil is utilized with the mutant bacterial strain P. aeruginosa, it consistently yields mcl PHA similar to P (3 HB). When using vegetable oil as a growth substrate, *P. aeruginosa* synthase can effectively generate mcl PHA monomers with little byproduct formation. In addition, efforts must be made to produce desired PHA congeners by regulated fermentation and to clarify the physicochemical and biological characteristics of each unique congener (Reddy et al., 2022).

8.7 CONCLUSION

The purpose of a fermenter and bioreactor is to provide conditions that allow an organism to produce a desired product efficiently. By offering the ideal conditions for reactions, two types of vessels, bioreactors and fermenters, allow biological processes to occur. Factors affecting production flow rates, aeration, temperature, pH, foam control, and agitation rate are factors that these bioreactors are made to regulate. Hence, maintenance of these conditions is required for better quality products. Vegetable, algal, microbial, and mammalian cells may all be cultivated in bioreactors using different operating models in order to produce desired products (Park et al., 2011). Microbial technology is frequently employed in even more fermentation methods, such as submerged and solid-state fermentation. The goal of cells (fungi, yeast, and bacteria) is to produce biomolecules or metabolic products for the market. Depending on the application and the kind of cell culture, each of them has restrictions, even though the majority are designed to maximize process conditions. These trends indicate the dynamic nature of fermenters and bioreactor systems, with a focus on efficiency, sustainability, and the ability to meet the evolving demands of various industries. Advances in technology and a deeper understanding of biological processes continue to drive innovation in this field.

REFERENCES

Abbott, A. Biology´s new dimension. *Nature*, 2003, 424, 870-872.

Akkoyunlu, B; Daly, S; Casey, E. Membrane bioreactors for the production of value-added products: Recent developments, challenges and perspectives. Bioresource Technology, 2021, 341, 125793.

Bergeijk, SA; Salas-Leiton, E; Cañavate, JP. Low and variable productivity and low Efficiency of mass cultures of the haptophyte Isochrysis aff. Galbana (T-iso) in outdoor Tubular photo bioreactors. *Aquacultural Engineering*, 2010, 43(1), 14–23.

Blanch, H. W., & Moo-Young, M. (1981). Product recovery in bioprocess technology. Biotechnology and Bioengineering, 23, 1–18.

Chandrashekhar, H; Rao, J. V. An overview of fermenter and the design considerations to enhance its productivity. Pharmacologyonline, 2010, 1, 261–301.

Chen, A; Chitta, R; Chang, D; Amanullah, A. Biotechnology and Bioengineering, 2009, 102, 148 (2009).

Cheng-Wu, Z; Zmora, O; Kopel, R; Richmond, A. An industrial-size flat plate glass reactor for mass production of Nannochloropsis sp. *Eustigmatophyceae, Aquaculture*, 2001, 195, 35–49.
Collins, P. C; Miller, W. M; Papoutsakis, E. T. Stirred culture of peripheral and cord blood hematopoietic cells offers advantages over traditional static systems for clinically relevant applications. Biotechnology and Bioengineering, 1998, 59, 534–543.
De Becze, G. I., & Liebmann, A. J. (1944). Effect of heat on the destruction of ascorbic acid in vegetables. Food Research, 9, 261–268.
Dochain, D. *Automatic Control of Bioprocess*. London: John Wiley and Sons, 2008, 242
Domínguez, A; Rodríguez Couto, S; Sanromán, MA. Dye decolourization by Trameteshirsuta immobilised into alginate beads. *World Journal of Microbiology and Biotechnology*, 2005, 21, 405–409.
Doran PM. *Bioprocess Engineering Principles*. Elsevier Science & Technology Books, 1995.
Eibl, R; Eibl, D. Design and use of the wave bioreactor for plant cell culture. *Plant Tissue Culture Engineering*, 2006, 6, 203–227.
Eibl, R; Kaiser, S; Lombriser R; Eibl, D. Disposable bioreactors: The current state-of-the-art and recommended applications in biotechnology. *Applied Microbiology and Biotechnology*, 2010, 86, 41–49.
Engasser, J-M. Bioreactor engineering: The design and optimization of reactors with living cells. *Chemical Engineering Science*, 1988, 43, 1739–1748.
Ezemba, C; Ekwegbalu, E. A; Ezemba, A. S. Fermentation, types of fermenters, design & uses of fermenters and optimization of the fermentation process. 2022.
Ge, X.; Kostov, Y.; Rao, G. Biotechnology and Bioengineering, 2005, 89, 329.
Heinzle, E; Biwer, AP; Cooney, CL. *Development of Sustainable Bioprocesses: Modeling and Assessment*. John Wiley and Sons, 2006, 294 p.
Hellwig, S.; Drossard, J.; Twyman, R. M.; Fischer, R. Nature Biotechnology, 2004, 22, 1415.
Hochuli, E. Interrelations of chemistry and biotechnology-III. Purification techniques for biological products. Pure and Applied Chemistry, 1992, 64(1), 169–184.
Hoekema, S; Bijmans, M; Janssen, M; Tramper, J; Wijffels, RH A pneumatically agitated Flat-panel photobioreactor with gas re-circulation: anaerobic photoheterotrophic Cultivation of a purple non-sulfur bacterium. *International Journal of Hydrogen Energy*, 2002, 27(11), 1331–1338.
Huang, TK; McDonald, KA. Bioreactor engineering for recombinant protein production in plant cell suspension cultures *Biochemical Engineering Journal*, 2009, 45, 168–184.
Hung, P. J.; Lee, P. J.; Sabounchi, P.; Aghdam, N.; Lin, R.; Lee, L. P. *Lab on a Chip*, 2005, 5, 44.
Hung, P. J.; Lee, P. J.; Sabounchi, P.; Lin, R.; Lee, L. P. *Biotechnology and Bioengineering*, 2005, 89, 1.
Isett, K; George, H; Herber, W; Amanullah, A. Biotechnology and Bioengineering, 2007, 98, 1017.
Jagani, S., Chelikani, R., & Kim, D. S. (2010). Effects of nanotechnology in food science: A review. Food Technology and Biotechnology, 48(3), 283–288.
Kleerebezem, M; Hols, P; Hugenholtz, J. Enzyme and Microbial Technology, 2000, 26, 840.
Kostov, Y.; Harms, P.; Randers-Eichhorn, L.; Rao, G. Low-cost microbioreactor for high-throughput bioprocessing. *Biotechnology and Bioengineering*, 2001, 72, 346e352.
Laska, ME; Cooney, CL. *Encyclopedia of Bioprocess Technology: Fermentation, Biocatalysis, and Separation*. 1st ed. Wiley, 1999.
Li, J., Li, T., & Wang, Q. (2019). Bioreactor design for microalgae cultivation and industrial biotechnology. Bioresource Technology, 291, 121–131.
Liu, X; Kokare, C. Microbial enzymes of use in industry. In: Biotechnology of Microbial Enzymes. Academic Press, 2023, pp. 405–444.

McDonald, K. A; Hong, L. M; Trombly, D. M; Xie, Q; Jackman, A. P. Production of human alpha-1-antitrypsin from transgenic rice cell culture in a membrane bioreactor. Biotechnology Progress, 2005, 21, 728–734.

Merchuk, J. C. (1990). Air-lift reactors: applications in biotechnology. Journal of Chemical Technology and Biotechnology, 49, 105–120.

Meyer, H. P., Schmidhalter, D., & Lütz, S. (2017). Industrial biocatalysis and bioreactor design for large-scale production. Current Opinion in Biotechnology, 47, 1–8.

Nag, A. Biofuels: Refining and Performance. McGraw-Hill Professional, 2008, 312 p.

Najafpour, GD; Ghasem D. *Biochemical Engineering and Biotechnology*, 2008.

Nielsen, J; Olsson, L. *FEMS Yeast Research*, 2002, 2, 175.

Ogbonna, J. C; Yada, H; Masui, H; Tanaka, H. Journal of Fermentation and Bioengineering, 1996, 82(1), 61–67.

Oksman-Caldentey, K. M; Inze, D. Trends in Plant Science, 2004, 9, 433.

Paek, K. Y.; Chakrabarty, D.; Hahn, E. J. Application of bioreactor systems for large-scale production of horticultural and medicinal plants. Plant Cell, Tissue and Organ Culture, 2005, 81, 287–300.

Park, J. B. K.; Craggs, R. J.; Shilton, A. N. Wastewater treatment high rate algal ponds for biofuel production. Bioresource Technology, 2011, 102, 35–42.

Patel, A. K.; Singhania, R. R.; Pandey, A. Novel enzymatic processes applied to the food industry. Current Opinion in Food Science, 2016, 7, 64–72.

Patil, K. R; Akesson, M; Nielsen, J. Current Opinion in Biotechnology, 2004, 15, 64.

Putt, R.; Manjinder, S.; Chinnasamy, S.; Das, K. C. An efficient system for carbonation of high-rate algae pond water to enhance CO_2 mass transfer. Bioresource Technology, 2011, 102, 3240–3245.

Quijano, G.; Couvert, A.; Amrane, A. Ionic liquids: Applications and future trends in bioreactor technology. Bioresource Technology, 2010, 101(23), 8923–8930.

Ranjbar, R; Inoue, R; Katsuda, T; Yamaji, H; Katoh, S. High efficiency production of astaxanthin in an airlift photobioreactor. *Journal of Bioscience and Bioengineering*, 2008, 106(2), 204–207.

Rasoul-Amini, S; Montazeri-Najafabady, N; Mobasher, MA; Hoseini-Alhashemi, S; Ghasemi, Y. Chlorella sp.: A new strain with highly saturated fatty acids for biodiesel Production in bubble-column photobioreactor. *Applied Energy*, 2011, 8.

Reddy, V. U. N.; Ramanaiah, S. V.; Reddy, M. V.; Chang, Y. C. Review of the developments of bacterial medium-chain-length polyhydroxyalkanoates (mcl-PHAs). Bioengineering, 2022, 9(5), 225.

Singhania, RR; Patel, AK; Soccol, CR; Pandey, A. Recent advances in solid-state fermentation. *Biochemical Engineering Journal*, 2009, 44, 13–18.

Stanbury, PF. *Principles of Fermentation Technology*. Oxford: Pergamon Press; 1995.

Szabo, IJ; Johansson G; Pettersson, G. Optimized cellulase production by Phanerochaetechrysosporium: Control of catabolite repression by fed-batch cultivation. *Journal of Biotechnology*, 1996, 48, 221–230.

Thangprompan, P.; Thanapimmetha, A.; Saisriyoot, M.; Laopaiboon, L.; Srinophakun, P. Production of ethanol from sweet sorghum juice using VHG technology: A simulation case study. Applied Biochemistry and Biotechnology, 2013, 171, 294–314.

Tredici, M. R. (2004). Mass production of microalgae: photobioreactors. Applied Microbiology and Biotechnology, 65, 477–485.

Ugwu, CU; Ogbonna, JC; Tanaka, H. Characterization of light utilization and biomass yields of Chlorella sorokiniana in inclined outdoor tubular photobioreactors equipped with static mixers. *Process Biochemistry*, 2005, 40(11), 3406–3411.

Xu, J.; Ge, X.; Dolan, M. C. Towards high-yield production of pharmaceutical proteins with plant cell suspension cultures. Biotechnology Advances, 2011, 29, 278–299.

Yang, Z.; Pan, W. Ionic liquids: green solvents for nonaqueous biocatalysis. Enzyme and Microbial Technology, 2005, 37, 19–28.

Zanzotto, A.; Szita, N.; Boccazzi, P.; Lessard, P.; Sinskey, A. J.; Jensen, K. F. A membrane-aerated microbioreactor for high-throughput bioprocessing. Biotechnology and Bioengineering, 2004, 85, 376–381.

9 Scope of Genetically Modified Food in Food Industry and Social Concerns

Irfan Khan, Sweta Joshi, Ahmed Mobeen, Akanksha Soni, Sadaf Ahmad, Nazia Tabasuum, and Waseem Ahmed

9.1 INTRODUCTION

Genetically modified foods (GM foods) are obtained from crops with altered genetic material that does not occur naturally through mating or natural recombination. This technology assists in transferring selected genes from one organism to another. Tomato became the first genetically modified (GM) crop to be deregulated for commercialization by the US Food and Drug Administration (FDA) in 1992. The aim of genetically modifying crops is to introduce desirable traits, such as resistance to pests, diseases, or environmental conditions, as well as to improve crop yield and nutritional content. GM foods have been a topic of considerable debate since they were first introduced in the 1990s. Genetically modified foods are the foods that are not naturally occurring and not even produced by mating of two individual plants or animals; rather, it is the food produced by genetically modified plants or animals (FAO, 2012). In contrast to this definition, another definition was suggested, which says "GMO as biotechnologically modified organism" (Oliver, 2014).

Advocates argue that genetically modified crops can help address food security issues by increasing crop yields and reducing losses due to pests and diseases. They can also potentially enhance the nutritional content of foods, making them more suitable for addressing nutrient deficiencies in certain populations. Critics, on the other hand, raise various concerns about GM foods, including potential risks to human health, environmental impact, ethical and moral concerns, consolidation of seed and agricultural industries, and the business model of GM crops (Pavone and Martinelli, 2015). Some worry about the long-term effects of consuming genetically modified organisms (GMOs), although extensive scientific research has not provided conclusive evidence of harm to human and animal health (Heinemann, 2009). Regulation

DOI: 10.1201/9781032650616-9

and labeling of GM foods vary across countries. Some countries have strict regulations and require thorough testing and risk assessment (Goven & Morris, 2012) and approval processes before GM crops can be cultivated or sold. Others have more lenient regulations or do not require specific labeling, making it challenging for consumers to know whether the products they purchase contain GMOs. It's important to note that the debate around genetically modified foods continues to evolve as new research and technologies emerge, shaping the way these crops are developed, regulated, and integrated into agricultural practices. Currently, there are a number of varieties of crops that contain genetically modified genomes (Johnson, 2008).

9.2 HISTORY OF GM FOODS

The story of the possibility of DNA modification came into existence at the time of DNA discovery, and it can be transferred between different species around 1944 (Avery et al., 1944). Many other relevant discoveries, such as the double helical model of B-DNA in 1954 and the concept of the central dogma, opened the door to gene technology (Nirenberg et al., 1963). Some important events in the direction of DNA modification have been presented in chronological fashion in Table 9.1.

9.3 STATUS OF GENETICALLY MODIFIED FOODS AT GLOBAL LEVEL

Genetically modified (GM) foods are widely grown and consumed in many countries around the world. GM crops have been commercially grown since the 1990s and

TABLE 9.1
Important events in the history of DNA modification

S.No.	Year	Event
1	1865	Discovery related to heredity transmitted in units by Gregor John Mendel
2	1869	Discovery of isolated DNA by Frederick Miescher
3	1911	Discovery of chromosomes carrying genes by Thomas Hunt Morgan
4	1952	Discovery of genes, made of DNA by Alfred Hershey and Martha Chase
5	1953	Discovery of the double helix structure of DNA by Francis H. Crick and James D. Watson
6	1966	Discovery of genetic codes by Marshall Nirenberg et al.
7	1968	Discovery of the first restriction enzyme by Steward Linn and Werner Arber
8	1980	Invention of the first transgenic mice by Jon W. Gordon et al.
9	1985	Generate the first transgenic domestic animal, a pig
10	1993	Use of Bovine somatotropin for increased milk production in dairy cows approved by the FDA
11	1994	First GM food, the FLAVR SAVR tomato, approved by the FDA

Source: Zhang et al., 2016.

have become an important part of agriculture in several nations. The most commonly grown GM crops include soybeans, maize (corn), cotton, canola, etc. These crops are often genetically modified to be resistant to pests, diseases, or herbicides, or to have improved nutritional profiles.

However, the acceptance and regulation of GM foods vary significantly from one country to another, as per their regulations. Some countries, especially in North and South America, have embraced GM technology and grow a large portion of their crops from genetically modified seeds. The United States, Brazil, Argentina, Canada, and India are some of the leading producers of GM crops.

In contrast, many European countries have strict regulations and a cautious approach toward GM foods. Several European nations have banned the cultivation of GM crops or have imposed strict labeling requirements for food products containing genetically modified ingredients. It's important to note that the global status of GM foods may have changed. For the most current and specific information on the global status of GM foods, We recommend consulting reputable sources such as the Food and Agriculture Organization of the United Nations (FAO), the World Health Organization (WHO), or the respective agricultural and food safety regulatory authorities of individual countries. While the global status of GM foods remains complex and varies by region, there are a few trends and developments that have shaped the landscape of GM crops and foods. Recently, Golden Rice has moved toward field trials after 25 years of its preparation in the laboratory (Oliver, 2014). Continued advancements in biotechnology have led to the development of new genetically modified crops with enhanced traits. Researchers are working on crops that are more resilient to climate change, require fewer resources like water and pesticides, and have improved nutritional content.

Some developing countries, such as Bangladesh, the Philippines, and parts of Africa, have started to adopt GM crops to improve agricultural productivity and address issues like food security and malnutrition. For instance, Bt brinjal (genetically modified eggplant) has been approved for commercial cultivation in Bangladesh. Public awareness about GM foods and consumer preferences for non-GM or organic products have influenced the market. Some consumers actively seek out non-GM products due to concerns about health and environmental impacts (Martinez-Poveda et al., 2009). As a response, some companies and food producers label their products as non-GMO (genetically modified organisms) to cater to this demand. Regulatory frameworks governing GM foods continue to evolve in different countries. Some nations have updated their regulations to accommodate new technologies and ensure the safety of GM products. International organizations like the Codex Alimentarius Commission provide guidelines to harmonize standards related to food safety, including GM foods, on a global scale.

Ongoing research into the safety and impact of GM crops on the environment and human health remains a topic of debate. Scientists and organizations conduct studies to assess the long-term effects of GM crops on ecosystems and biodiversity, addressing concerns raised by various stakeholders. The development of gene-editing technologies, such as CRISPR-Cas9, has opened new avenues for crop improvement. These technologies allow for precise modification of genes without introducing foreign DNA, raising unique regulatory and ethical questions regarding their categorization as GM organisms.

It's crucial to keep in mind that the acceptance and regulation of GM foods are continually evolving. As new research findings emerge and public attitudes change, the global status of GM foods will likely continue to shift. For the most recent and region-specific information, it's advisable to refer to updated reports from relevant international organizations and regulatory agencies. Several genetically modified (GM) crops have been widely adopted globally. These crops are grown in numerous countries and have been commercially cultivated for years. Table 9.2 shows some of the most globally adopted GM crops.

GM soybeans are widely grown in countries like the United States, Brazil, Argentina, and Canada. They are often genetically modified to be resistant to herbicides, allowing for easier weed control. GM maize is extensively cultivated in the United States, Brazil, Argentina, Canada, and several other countries. Some varieties of GM corn are engineered to resist pests like insects, specifically the European corn borer and corn rootworm. Others are engineered for herbicide tolerance. GM cotton is grown in countries such as the United States, India, China, and several African nations. It is modified to resist certain pests, particularly cotton bollworms. GM canola, also known as rapeseed, is grown in Canada, the United States, and Australia. Canola plants are genetically modified to be tolerant to specific herbicides. GM papaya, specifically the Rainbow Papaya, has been commercially grown in Hawaii, the United States, and other countries. It is genetically engineered to resist the ringspot virus, which had previously devastated papaya crops. GM sugar beets are grown in the United States and Canada. They are engineered to be resistant to herbicides, allowing for more straightforward weed management. GM alfalfa is

TABLE 9.2
Category of GM foods obtained from different GM crops

S.No.	Type of GM Crop	Appear on Ingredient List As	Examples of Food
1	Soybean	Soy oil/vegetable oil, soy protein/vegetable protein, soy lecithin/emulsifier	Margarine-type spreads, bread, cakes, biscuits, snack foods, mayonnaise, chocolate, etc.
2	Canola	Canola oil	Margarine-type spreads, mayonnaise, cooking oil, biscuits, snacks, breads, cake, etc.
3	Maize	Dextrose, fructose, maltodextrin, starch modifier/thickener, glucose syrup, Glucose etc.	Muffins, breakfast cereals, simmer sauces, cakes, biscuits, snacks, peanut butter, etc.
4	Cotton	Cottonseed oil	Mayonnaise, snacks, simmer sauce, cooking oil, margarine-type spreads, etc.

Source: Moglia & Portis, 2016.

grown in the United States and Canada. It is engineered for herbicide tolerance and resistance to certain pests.

GM varieties of summer squash and zucchini are cultivated in the United States. They are engineered for resistance against specific viruses. Around 13% of zucchini grown in the USA was genetically modified as of 2005 to resist infection against viruses (Johnson, 2008).

It's important to note that the adoption of GM crops can vary significantly from one country to another due to regulatory policies, public acceptance, and agricultural practices. Additionally, new GM crops and varieties are continuously being developed and tested in various parts of the world, potentially leading to further adoption in the future. For the most current and specific information about globally adopted GM foods, it is advisable to consult updated reports from agricultural and regulatory authorities or international organizations like the Food and Agriculture Organization of the United Nations (FAO) and the International Service for the Acquisition of Agri-biotech Applications (ISAAA).

9.4 STATUS OF GM FOODS IN INDIA

Genetically modified (GM) crops are a topic of debate and regulation in India. India has allowed the commercial cultivation of GM cotton since 2002. Bt cotton, a genetically modified variety engineered to resist bollworm pests, has been widely adopted and cultivated across the country. Cotton is the only GM crop that has been approved for commercial cultivation in India. In 2009, India approved the commercial cultivation of Bt brinjal, a genetically modified variety of eggplant resistant to certain pests. However, following public outcry and concerns raised by environmental activists and some scientists, the cultivation was put on hold, and a moratorium was placed on Bt brinjal in 2010. As of my last update, Bt brinjal was not being commercially cultivated in India.

Apart from cotton and Bt brinjal, there were no other GM crops approved for commercial cultivation in India up to 2022. However, there have been field trials and research on various other GM crops, such as GM mustard, which has been under consideration for commercial approval. The Genetic Engineering Approval Committee (GEAC) in India is responsible for granting approvals for the environmental release of genetically modified organisms, including crops. The process involves rigorous testing and evaluation of the safety and efficacy of GM crops before they are approved for cultivation. The adoption of GM crops in India is a contentious issue. While proponents argue that GM crops can address issues like pest resistance and crop yield, opponents raise concerns about environmental impacts, biodiversity, and the long-term health effects of consuming GM foods. Public perception and concerns have influenced the government's approach to the commercialization of GM crops.

9.5 CATEGORY OF GM FOODS

Herbicide-tolerant and insecticide-producing soybean, maize, and canola (oilseed rape) are the main genetically modified food crops currently grown for commercial purposes. However, the technology has been applied to, or is currently being applied to, a number of other crop species in an effort to increase levels of disease resistance and abiotic stress tolerance as well as to improve the nutritional value of the final product (Moglia & Portis, 2016). Genetically modified (GM) foods are typically categorized based on the type of genetic modification or the traits that have been introduced into the organism. Here, the main GM crops are categorized as discussed below (Clive, 2011).

9.5.1 Herbicide-Tolerant Crops

These crops are genetically modified to withstand specific herbicides, allowing farmers to use herbicides to control weeds without harming the crop. Consequently, weeds will compete with crops for growing area, water, fertilizer, and sunlight. They will also either directly or indirectly transfer pests and illnesses (Délye et al., 2013; Quareshy et al., 2018). Common herbicide-tolerant crops include soybeans, maize, canola, and cotton. Growing maize, soybeans, and cotton has changed since 1996 due to the introduction of genetically modified herbicide-resistant (HR) crops, especially glyphosate-resistant (GR) crops, which have changed the way weeds are controlled. Several crops have gained resistance against glyphosate during the course of the period (Heap, 2010).

9.5.2 Insect-Resistant Crops

The introduction of genetically modified corn (maize), potato, and cotton plants expressing genes encoding the entomocidal δ-endotoxin from *Bacillus thuringiensis* (Bt; also known as Cry proteins) marked the beginning of the commercialization of insect-resistant transgenic crops in the mid-1990s. Insect-resistant GM crops are engineered to produce a protein toxic to certain insect pests. This reduces the need for chemical insecticides and helps protect the plants from insects. Bt (*Bacillus thuringiensis*) crops, such as Bt cotton and Bt corn, fall into this category.

9.5.3 Virus-Resistant Crops

Some GM crops are modified to be resistant to specific plant viruses. Global crop productivity is seriously threatened by the prevalence of phytophagous insects (Patil, 2021). An estimated 25% of crops are said to be destroyed by insects every year. In fact, the creation of insect-resistant crops is a significant advancement in agriculture that will help boost crop productivity and lessen the need for pesticides. For example, certain varieties of papaya have been engineered to resist the ringspot virus, a common disease affecting papaya plants.

9.5.4 Disease-Resistant Crops

GM crops can be modified to resist diseases that commonly affect plants. For instance, scientists are working on developing genetically modified varieties of bananas resistant to Panama disease, which is a fungal disease that affects banana plants. Viral infections cause a 3% reduction in crop yields, while bacterial and fungal pathogens cause a 15% reduction (Oerke & Dehne, 2004).

9.5.5 Nutritionally Improved Crops

Crops that have been nutritionally enhanced are intended to improve animal and poultry feed. For instance, higher levels of limiting amino acids are produced by animal feed crops, reducing the need for supplements. Simultaneously, GM foods can be engineered to have improved nutritional profiles. For example, "Golden Rice" is a genetically modified variety of rice that has been developed to produce higher levels of beta-carotene, a precursor of vitamin A. This can be particularly beneficial in regions where vitamin A deficiency is prevalent.

9.5.6 Delayed Ripening Crops

Certain fruits and vegetables are genetically modified to have a longer shelf life by delaying the ripening process. Fruit ripening is delayed by antisense technology through downregulation of the ACC oxidase gene, which suppresses ethylene production. Since ethylene indicates when fruit is about to ripen, some plants can have their ripening process delayed by altering their ethylene receptors. This can reduce food waste during transportation and storage. Certain crops have been approved in the recent past as delayed ripening crops such as Carnation in Australia and Norway, Melon and Pineapple in the USA while Tomato in China, Canada, Mexico, and the USA (ISAAA GM approval Database).

9.5.7 Crops with Altered Fatty Acid Profile

Some GM crops are modified to produce oils with altered fatty acid compositions. For example, soybeans have been engineered to produce oils with higher levels of oleic acid, which is considered a healthier fat. Recently, two novel bioactive omega-3 (also known as n-3) long-chain fatty acids, viz., eicosapentaenoic acid [EPA, 20:5n-3] and docosahexaenoic acid [DHA, 22:6n-3], have been produced by genetically engineering (GE) oilseed crops, and these fatty acids significantly increase the nutritional value of the seed (Colombo et al., 2018) (Table 9.3).

These categories represent the primary types of genetically modified crops that have been developed for agricultural and commercial purposes. Each category addresses specific agricultural challenges, such as pest infestations, diseases, and nutritional deficiencies, with the goal of improving crop yield, reducing the use of chemical pesticides, and enhancing food security.

9.6 IMPORTANCE OF GM FOODS FOR HUMAN WELFARE

For over 15 years, GM crops have been used to produce food for hundreds of millions of people worldwide. Even though many of these consumers are from the USA,

TABLE 9.3
List of genetically modified foods

S.No.	GM Food	GM Trait
1	Egg plant	Antibiotic resistance, insect resistance
2	Argentine canola	Fertility restoration, modified oil/fatty acid, glufosinate herbicide tolerance, male sterility, oxynil herbicide tolerance, antibiotic resistance, glyphosate herbicide tolerance, etc.
3	Tomato	Antibiotic resistance, Lepidopteran insect resistance, delayed fruit softening, viral disease resistance, delayed ripening/senescence
4	Apple	Non-browning phenotype, antibiotic resistance, etc.
5	Maize	Fertility restoration, visual marker, male sterility, modified alpha amylase, mannose metabolism, glufosinate herbicide tolerance, Lepidopteran insect resistance, glyphosate herbicide tolerance, etc.
6	Potato	Reduced acrylamide potential, black spot bruise tolerance, Coleopteran insect resistance, antibiotic resistance, modified starch/carbohydrate, etc.
7	Plum	Antibiotic resistance visual marker, viral disease resistance
8	Papaya	
9	Sugarcane	Drought stress tolerance, antibiotic resistance
10	Wheat	Glyphosate herbicide tolerance

Source: Zhang et al., 2016.

one of the most litigious nations in the world, there have been no documented negative effects or health-related legal cases involving these foods. Genetically modified (GM) foods have generated significant interest and debate due to their potential benefits and concerns. Here are some reasons why GM foods are considered important by proponents:

9.6.1 Enhanced Crop Yield

Increasing world population is the alarming call for the food producers and it poses some responsibility on the farmers' heads to increase their productivity. Higher crop yield is not possible with conventional seeds; rather, GM seeds can achieve this multiple times. So, the GM crops are playing an important role in filling the bellies of the world population. In some studies, agricultural economists have confirmed the benefits of GM crops in two significant reviews. According to the first, which reviewed 147 studies and was published in 2014, genetically modified crops have enabled an average increase in agricultural yield of 22% and a 68% increase in farmers' profits, with profit margins being even higher in developing nations.

9.6.2 Reduced Dependency on Pesticides

Insect-resistant GM crops, such as Bt cotton and Bt maize, produce a protein toxic to specific insect pests. By reducing the need for chemical insecticides, these crops promote more environmentally friendly farming practices, decreasing the environmental impact of agriculture. The introduction of genetically modified corn (maize), potato, and cotton plants expressing genes encoding the entomocidal δ-endotoxin from *B. thuringiensis* (Bt; also known as Cry proteins) marked the beginning of the commercialization of insect-resistant transgenic crops in the mid-1990s. But, simultaneously, it also claimed in some studies that insect-resistant crops are creating imbalance in the ecosystem by declining the population of other insects such as honey bees.

9.6.3 Improved Nutritional Profile

In the last ten years, significant advancements have been achieved in the use of biotechnology to produce food crops with enhanced nutritional value. Along with new crop varieties that can fight chronic disease, biofortified staple crops like rice, maize, and wheat that contain vital micronutrients to help the world's impoverished are being developed. New advances in agricultural biotechnology opened the path for producing food crops that are nutritionally enhanced, increasing the amount and bioavailability of vital nutrients like iron and vitamin A (Pérez-Massot et al., 2013; Farre et al., 2011; Pachón et al., 2009; Gilani et al., 2007).

Scientists can modify crops to enhance their nutritional profiles, addressing specific nutrient deficiencies in certain regions. For example, "Golden Rice" has been engineered to produce higher levels of beta-carotene, addressing vitamin A deficiency in populations where rice is a dietary staple.

9.6.4 Improved Shelf Life and Reduced Food Waste

GM crops can be modified to delay ripening and increase shelf life, reducing spoilage during transportation and storage. This characteristic can help minimize food waste, a significant global issue.

9.6.5 Drought and Salinity Tolerance

GM crops can be engineered to withstand drought and high salinity levels, allowing cultivation in areas affected by water scarcity and soil salinity. This trait enhances agricultural resilience in the face of climate change.

9.6.6 Lower Production Cost

By reducing the need for chemical inputs such as pesticides and herbicides, GM crops can lead to cost savings for farmers. This economic benefit can make farming more sustainable and financially viable for agricultural communities.

Environmental conservation: Reduced pesticide use and increased crop yields can lead to more efficient land use. This conservation of land can help prevent deforestation and preserve natural habitats, promoting biodiversity and ecological balance.

9.6.7 Support for Small-Scale Farmers

In some cases, GM crops have been designed to benefit small-scale farmers by increasing crop yield and reducing the need for expensive inputs, making farming more economically viable for these communities.

It's important to note that while GM foods offer potential benefits, there are also concerns related to environmental impact, human health, ethical considerations, and socioeconomic factors. Ongoing research and careful regulation are essential to addressing these concerns while harnessing the potential benefits of GM technology. Public dialogue and informed decision-making play a crucial role in shaping policies and practices related to GM foods.

9.7 GM Foods Alleviating Food Security Issues

Genetically modified crops have the potential to enhance food production and availability. Additionally, there might be effects on the nutritional makeup and quality of food. Lastly, the income that farmers receive from growing GM crops may affect their ability to afford food. When everyone has physical and financial access to enough safe, nourishing food, there is food security. Regretfully, a sizable segment of the global population lacks access to food security. Approximately 900 million individuals are undernourished, which means they consume insufficient amounts of calories (FAO, 2012). GM foods play a crucial role in alleviating global hunger significantly (Juma, 2011; Borlaug, 2007).

9.8 GM Foods Addressing the Nutritional Security Issues

Genetically modified (GM) foods can play a significant role in enhancing nutritional security by addressing specific nutrient deficiencies and improving the overall nutritional content of food crops. Here's how GM foods can contribute to nutritional security. GM technology allows scientists to enhance the nutrient content of crops. For example, "Golden Rice" has been genetically modified to produce higher levels of beta-carotene, a precursor to vitamin A. This biofortified rice can help combat vitamin A deficiency, a widespread health issue in many parts of the world, especially in developing countries where rice is a dietary staple. Furthermore, nearly 2 billion people worldwide experience "hidden hunger," a form of malnutrition brought on by an inadequate intake of vital micronutrients in their regular diet rather than a lack of calories (Grebmer et al., 2014).

GM crops, including staple foods like rice, wheat, and maize, can be engineered to contain higher levels of iron and zinc. Iron and zinc deficiency can lead to anemia and other health problems. By increasing the levels of these essential minerals in commonly consumed crops, GM technology can address these deficiencies and improve overall health. GM crops can be modified to enhance the quality of proteins they produce. For example, genetically modified soybeans have been developed to produce higher-quality protein with improved amino acid profiles. This can benefit populations relying on plant-based proteins as their primary protein source. Beyond vitamin A, iron, and zinc, GM technology can be used to enhance the content of other essential micronutrients such as folate and vitamin C in various crops. These improvements can help address specific nutrient deficiencies prevalent in certain regions. Malnourished individuals frequently eat meals that are centered around a staple crop, depriving them of access to the wide range of fruits and vegetables that are necessary for a balanced diet. Consequently, one in four children is stunted because of inadequate nutrition, and nearly one-third of all childhood deaths worldwide occur before the age of five (Maternal nutrition, 2015).

Some GM crops are engineered to have reduced levels of anti-nutritional factors, substances that interfere with the absorption of essential nutrients. By lowering anti-nutrient levels, the nutritional quality of the crop is improved, making it more beneficial for human consumption. GM technology can be applied to a diverse range of crops, not just major staples. By enhancing the nutritional content of various crops, including fruits, vegetables, and grains, GM technology contributes to a more diverse and balanced diet, reducing the risk of micronutrient deficiencies.

9.9 ENVIRONMENTAL SUSTAINABILITY

By increasing the nutritional content of crops, GM technology can reduce the need for dietary supplements, which often have environmental and economic costs associated with their production and distribution. This can promote more sustainable lifestyles.

It's important to note that while GM foods have the potential to address specific nutritional deficiencies, ethical considerations, safety assessments, and regulatory frameworks are crucial in ensuring that these crops are developed and utilized responsibly. Public awareness, education, and informed decision-making are essential components of integrating GM foods into efforts aimed at improving nutritional security.

9.10 CONCERNS RELATED TO ACCEPTANCE OF GM FOODS

Genetically engineered food technologies provide significant potential in addressing some of the biggest challenges of the twenty-first century. They also carry certain known and unknown hazards, much like any new technology. Concerns and debates about genetically modified foods and crops typically center on issues of consumer choice and labeling, human and environmental safety, ethics, food security, poverty alleviation, intellectual property rights, environmental preservation, and the consequences of "tampering with Mother Nature" using this new technology for manipulating genes.

9.11 CONSUMER PERCEPTION

Most often, the negative perceptions elicited by genetically modified food products are ascribed to understanding their merits and demerits that diverge from those supported by science and research (Scott et al., 2016; De Steur et al., 2014). Contrary to what scientists agree upon, the general public undervalues the possible advantages of eating GMO food while overestimating the risk (Sjöberg, 2008; Lusk et al., 2015). For instance, Bredahl et al. (1998) conducted consumer interviews in Europe to investigate the advantages and disadvantages of the genetic modification feature of beers and yogurts. Danish, German, Italian, and British customers linked genetic alteration more with risks than benefits.

Consumers believed genetically modified food to be dangerous, unethical, and unnatural (Dean & Shepherd, 2007). Previous studies have examined the relationship between attitudes regarding genetically modified foods and subjective hazards and advantages (Moon & Balasubramanian, 2001, 2004; Siegrist, 2000; Sjöberg, 2008). According to reports by Frewer and Shepherd (1995), perceived risks are the best indicators of customers' views. Similarly, Moon and Balasubramanian (2001, 2004) contended that attitudes are more accurately predicted by perceived risks than by advantages. Sjöberg (2008) reported that the perceived benefits were found to be just as predictive of genetically modified food views as perceived hazards. According to some studies, GM resistance may be predicted by a low or absent impression of the benefits. According to Gaskell et al. (2004), regardless of the hazards involved, consumers' acceptance or rejection of genetically modified foods is solely based on their perception of the products' lack of benefits. According to a recent meta-analysis report (Bearth & Siegrist, 2016), it is not evident which factor has a greater influence on people's adoption of food technologies: hazards or advantages. A key barrier to efforts to modify consumers' attitudes about biotechnology in food production is the widely held misperception of dangers and advantages (Lähteenmäki et al., 2002). This misconception could result in GM rejection altogether (Bredahl, 2001; Sleboda & Lagerkvist, 2022).

The public's apprehension over false information about genetically modified foods has reached a new level, to the point where food firms with a vested financial interest in opposing them have invested billions of dollars in marketing their products as "GM-free." This supports the false notion that people who eat food containing genetically modified elements would suffer consequences and strengthens anti-biotechnology attitudes (Casassus, 2013).

Opponents of genetically modified products have recently experienced a form of mass hysteria due to a highly disputed study that suggested eating GM food increased the number of tumors in mice (Herring, 2010). The paper was widely cited by opposition groups and contributed fuel to the GMO controversy, even though it was later retracted.

9.12 POLITICAL ISSUES

The main concerns expressed in opposition to genetically modified crops are biosafety and the susceptibility of farmers and food to corporate monopolies. Non-governmental organizations make a concerted effort to incite opposition to genetically modified organisms (GMOs). These organizations work by spreading misinformation and inciting fear among the general public and government officials.

Untrue tales like livestock dying from eating the leaves of Bt cotton crops, to Indian farmers killing themselves in large numbers due to GMO crops interfering with their livelihood (Rao, 2007; Herring, 2010). These organizations purposefully maintain a sufficient distance from their scientific counterparts to ensure that people who are knowledgeable about the "sound science" underlying this technology are unlikely to dispute their claims.

All genetically modified crops must go through a rigorous and comprehensive risk assessment process to determine their safety for the environment and human health. These rules center on a thorough examination of the crop's molecular characterization, an evaluation of the crop's potential for toxicity and/or allergenicity, and a nutritional analysis. Political resistance has hampered this process in Europe, and GM crops—which have enormous potential to aid the world's needy—are currently mired in red tape (FAO, 2016). The design of the EU's regulatory program has effectively put an end to genetically modified crops throughout the continent, preventing the advancement of nutritionally enhanced crop research and development for humanitarian purposes (Sutherlin, 2013). The greatest illustration of this moratorium in action is the Golden Rice biofortified with β-carotene, which has been ready for release for well over ten years but has still not been approved (Bawa & Anilakumar, 2013; Potrykus, 2010a; Beyer, 2010; Gartland et al., 2013; Hefferon, 2015).

The Parliamentary Committee on Agriculture in India demanded in 2012 that the Indian government halt all field trials and pursue a ban on GM food crops like Bt brinjal. The panel brought up the "ethical dimensions" of transgenics in agricultural crops and discussed studies showing long-term environmental and chronic toxicological effects. They also pointed out that farmers did not receive any appreciable socioeconomic benefits from transgenics (Bawa & Anilakumar, 2013).

The debate over GMO labeling centers on how much information should be disclosed to consumers and whether this information explains the contents well enough to enable consumers to make more informed decisions. There is a distinction between required and optional GMO labeling. Consumers are informed by voluntary labeling that bioengineered materials are not present in products. Because it mandates the inclusion of warning labels on all food products containing GMOs, mandatory labeling goes much further (Premanandh, 2011). Each of these labeling systems has advantages and disadvantages. Labeling needs to support consumers' decision-making so they can know what is in their food. But because consumers do not know much about new technology, like genetically engineered foods, they frequently cannot determine whether GM products are dangerous or how to weigh the pros and cons of a particular risk. Because of this, people rely on the experts they believe to be reliable to draw well-informed conclusions that they can then utilize to shape their own opinions. As a result, in layman's assessment of genetically modified foods, trust is crucial (Storey, 2021; Vega Rodríguez et al., 2022).

9.13 ENVIRONMENTAL AND AGRICULTURAL ISSUES

Another controversy related to GM foods is the emergence of resistant weeds and pests known as "superbugs" and "super weeds." There will be significant selective pressure in that habitat if these cultivars are planted commercially, which could lead to the emergence of resistant insects in a few years and negate the transgenic's effects. Likewise, surrounding weeds may become resistant to the herbicide that the crop is tolerant of if herbicide spraying becomes more frequent because of new cultivars. This would result in an increase in the quantity and variety of herbicides applied to crop plants, as well as an increase in the dose or modification of herbicides. Ironically, the companies that sell chemical pesticides are the ones that are funding this research.

Uncertainty about whether these crops' pest-resistant trait can spread to their weedy relatives, leading to more and more resistant weeds, is another problem. It is also possible that insect-resistant plants could reduce competition and encourage minor pests to become major issues if they increase the mortality rate of a specific pest. Furthermore, it might lead to the pest population moving to a different, previously unthreatening plant population. From a toxicological perspective, more research is needed to ascertain whether herbicide residues or pest-resistant plants could harm important groups of organisms like nematodes, fungi, bacteria, and other microorganisms that are present in the surrounding soil. Most of the possible hazards associated with disease-resistant plants are related to viral resistance. It is feasible that the emergence of new viruses and consequently new diseases can be attributed to viral resistance (Bawa & Anilakumar, 2013).

9.14 HEALTH IMPLICATIONS

It is well-known that toxicity, allergenicity, and the spread of antibiotic resistance are the three main health risks associated with genetically modified foods. From an allergy perspective, there are two problems. The known allergen transfer that may happen from a crop into a non-allergenic target crop, resulting in the development of a neo-allergen. The patients who were allergic to Brazil nuts but not soybeans later displayed an IgE-mediated reaction to genetically modified soy beans (Bawa & Anilakumar, 2013).

9.15 CONCLUSION

Global food safety, sustainability, and climate change are three new challenges that could be addressed by using GE techniques to develop GM crops. Even though there is an increasing number of genetically modified crops available, some consumers and the scientific community continue to oppose them.

One of the biggest issues is the possibility that eating GM crops could endanger the health of people, animals, or the environment, especially considering how difficult it is to determine the long-term risk of doing so. Historically, there have been several risks linked to genetically modified organisms (GMOs), such as toxicity issues, allergy development, tumor development, infertility, and the potential for transgene(s) to be horizontally transferred to other species or the environment. Nevertheless, there is not enough scientific data to conclude that they are harmful to health. The goal of giving crops more health benefits on a global scale will require a lot of work and interactions between several disciplines, such as nutritionists, plant breeders, molecular biologists, and even social scientists. Clinical trials must assess new crop varieties with improved nutritional qualities, and the populations most suited to benefit from them must be informed about how these benefits can improve the general health of their community. The world's rural poor will never again go hungry if governments, businesses, and non-profit organizations work together.

REFERENCES

Avery, O. T., MacLeod, C. M., & McCarty, M. (1944). Studies on the chemical nature of the substance inducing transformation of pneumococcal types: Induction of transformation by a deoxyribonucleic acid fraction isolated from pneumococcus type III. Journal of Experimental Medicine, 79(2), 137–158.

Bawa, A. S., & Anilakumar, K. R. (2013). Genetically modified foods: Safety, risks and public concerns—A review. Journal of Food Science and Technology, 50(6), 1035–1046.

Bearth, A., & Siegrist, M. (2016). Are risk or benefit perceptions more important for public acceptance of innovative food technologies? A meta-analysis. Trends in Food Science & Technology, 49, 14–23.

Beyer, P. (2010). Golden rice and "golden" crops for human nutrition. New Biotechnology, 27(5), 478–481.

Borlaug, N. (2007). Feeding a hungry world. Science, 318, 359.

Bredahl, L. (2001). Determinants of consumer attitudes and purchase intentions with regard to genetically modified food: Results of a cross-national survey. Journal of Consumer Policy, 24(1), 23–61.

Bredahl, L., Grunert, K. G., & Frewer, L. J. (1998). Consumer attitudes and decision-making with regard to genetically engineered food products: A review of the literature and a presentation of models for future research. Journal of Consumer Policy, 21(3), 251–277.

Casassus, B. (2013). Study linking GM maize to rat tumours is retracted. *Nature*, 28.

Clive, J. (2011). Global status of commercialized biotech/GM crops (ISAAA Brief No. 43). International Service for the Acquisition of Agri-biotech Applications, Ithaca.

Colombo, S. M., Campbell, L. G., Murphy, E. J., Martin, S. L., & Arts, M. T. (2018). Potential for novel production of omega-3 long-chain fatty acids by genetically engineered oilseed plants to alter terrestrial ecosystem dynamics. Agricultural Systems, 164, 31–37.

De Steur, H., Blancquaert, D., Lambert, W., Van Der Straeten, D., & Gellynck, X. (2014). Conceptual framework for ex-ante evaluation at the micro/macro level of GM crops with health benefits. Trends in Food Science & Technology, 39(2), 116–134.

Dean, M., & Shepherd, R. (2007). Effects of information from sources in conflict and in consensus on perceptions of genetically modified food. Food Quality and Preference, 18(2), 460–469.

Délye, C., Jasieniuk, M., & Le Corre, V. (2013). Deciphering the evolution of herbicide resistance in weeds. Trends in Genetics, 29, 649–658. doi: 10.1016/j.tig.2013.06.001.

FAO. (2012). *The state of food insecurity in the world 2012*. Food and Agriculture Organization of the United Nations.

Farre, G., Twyman, R. M., Zhu, C., Capell, T., & Christou, P. (2011). Nutritionally enhanced crops and food security: Scientific achievements versus political expediency. Current Opinion in Biotechnology, 22, 245–251. doi: 10.1016/j.copbio.2010.11.002.

Food and Agriculture Organization of the United Nations. (2016). The state of food and agriculture 2016: Climate change, agriculture and food security. Rome: FAO.

Frewer, L. J., & Shepherd, R. (1995). Ethical concerns and risk perceptions associated with different applications of genetic engineering: Interrelationships with the perceived need for regulation of the technology. Agriculture and Human Values, 12, 48–57.

Gartland, K. M., Bruschi, F., Dundar, M., Gahan, P. B., Magni, M. V., & Akbarova, Y. (2013). Progress towards the "Golden Age" of biotechnology. Current Opinion in Biotechnology, 24, S6–S13.

Gaskell, G., Allum, N., Wagner, W., Kronberger, N., Torgersen, H., Hampel, J., & Bardes, J. (2004). GM foods and the misperception of risk perception. Risk Analysis: An International Journal, 24(1), 185–194.

Gilani, G. S., & Nasim, A. (2007). Impact of foods nutritionally enhanced through biotechnology in alleviating malnutrition in developing countries. Journal of AOAC International, 90, 1440–1444.

Von Grebmer, K., Saltzman, A., Birol, E., Wiesmann, D., Prasai, N., Yin, S., Yohannes, Y., Menon, P., Thompson, J., & Sonntag, A. (2014). Global hunger index. International Food Policy Research Institute.

Goven, J., & Morris, C. M. (2012). Regulating biopharming: The prism of farmer knowledge. Science as Culture, 21, 497–527.

Heap, I. (2010). The international survey of herbicide resistant weeds. Retrieved April 15, 2010, from http://www.weedscience.com

Hefferon, K. L. (2015). Nutritionally enhanced food crops: Progress and perspectives. International Journal of Molecular Sciences, 16(2), 3895–3914.

Heinemann, J. A. (2009). *Hope not hype: The future of agriculture guided by the* international assessment *of* agricultural knowledge, science *and* technology *for development* (pp. 1–74). Third World Networ.

Herring, R. J. (2010). Epistemic brokerage in the bio-property narrative: Contributions to explaining opposition to transgenic technologies in agriculture. New Biotechnology, 27(5), 614–622.

Johnson, S. R. (2008). Quantification of the impacts on U.S. agriculture of biotechnology-derived crops planted in 2006. National Center for Food and Agriculture Policy.

Juma, C. (2011). Preventing hunger: Biotechnology is key. Nature, 479, 471–472.

Lähteenmäki, L., Grunert, K., Ueland, Ø., Åström, A., Arvola, A., & Bech-Larsen, T. (2002). Acceptability of genetically modified cheese presented as real product alternative. Food Quality and Preference, 13(7–8), 523–533.

Lusk, J. L., McFadden, B. R., & Rickard, B. J. (2015). Which biotech foods are most acceptable to the public?. *Biotechnology Journal*, *10*(1), 13–16.

Maternal and Child Nutrition. Retrieved February 11, 2015, from http://www.thelancet.com/series/maternal-and-child-nutrition

Martinez-Poveda, A., Molla Bauza, M. B., Gomis, F. J. C., & Martinez, L. M. C. (2009). Consumer perceived risk model for the introduction of genetically modified food in Spain. Food Policy, 34, 519–528.

Moglia, A., & Portis, E. (2016). *Genetically modified foods*. University of Torino.

Moon, W., & Balasubramanian, S. K. (2001). *Public perceptions and willingness-to-pay a premium for non-GM foods in the US and UK*.

Moon, W., & Balasubramanian, S. K. (2004). Public attitudes toward agrobiotechnology: The mediating role of risk perceptions on the impact of trust, awareness, and outrage. *Applied Economic Perspectives and Policy*, *26*(2), 186–208.

Nirenberg, M. W., Matthaei, J. H., Jones, O. W., Martin, R. G., & Barondes, S. H. (1963). Approximation of genetic code via cell-free protein synthesis directed by template RNA. Federation Proceedings, 22, 55–61.

Oerke, E. C., & Dehne, H. W. (2004). Safeguarding production: Losses in major crops and the role of crop protection. Crop Protection, 23, 275–285.

Oliver, M. J. (2014). Why we need GMO crops in agriculture. Missouri Medicine, 111(6), 492–507.

Pachón, H., Ortiz, D. A., Araujo, C., Blair, M. W., & Restrepo, J. (2009). Iron, zinc, and protein bioavailability proxy measures of meals prepared with nutritionally enhanced beans and maize. Journal of Food Science, 74, H147–H154. doi: 10.1111/j.1750-3841.2009.01181.x.

Patil, B. L. (2021). Plant viral diseases: Economic implications. In D. H. Bamford & M. Zuckerman (Eds.), Encyclopedia of virology (4th ed., pp. 81–97). Academic Press.

Pavone, V., & Martinelli, L. (2015). Cisgenics as emerging bio-objects: Bio-objectification and bio-identification in agrobiotech innovation. New Genetics and Society, 34, 1–20.

Pérez-Massot, E., Banakar, R., Gómez-Galera, S., Zorrilla-López, U., Sanahuja, G., Arjó, G., Miralpeix, B., Vamvaka, E., Farré, G., Rivera, S. M., et al. (2013). The contribution of transgenic plants to better health through improved nutrition: Opportunities and constraints. Genes & Nutrition, 8, 29–41. doi: 10.1007/s12263-012-0315-5.

Potrykus, I. (2010a). Lessons from the 'Humanitarian Golden Rice'project: Regulation prevents development of public good genetically engineered crop products. *New Biotechnology*, *27*(5), 466–472.

Premanandh, J. (2011). Global consensus–Need of the hour for genetically modified organisms (GMO) labeling. *Journal of Commercial Biotechnology*, *17*, 37–44.

Quareshy, M., Prusinska, J., & Li, J., Napier, R. (2018). A cheminformatics review of auxins as herbicides. Journal of Experimental Botany, 69, 265–275. doi: 10.1093/jxb/erx258.

Rao, C. K. (2007). Causes of death of cattle and sheep in the Telengana region of Andhra Pradesh in India. Bangalore, India: Foundation for Biotechnology Awareness and Education. http://www. plantbiotechnology. org. in/issues. html

SAAA GM. (n.d.). GM approval database. http://www.isaaa.org/gmap-provaldatabase.

Scott, S. E., Inbar, Y., & Rozin, P. (2016). Evidence for absolute moral opposition to genetically modified food in the United States. *Perspectives on Psychological Science*, *11*(3), 315–324.

Siegrist, M. (2000). The influence of trust and perceptions of risks and benefits on the acceptance of gene technology. *Risk analysis*, *20*(2), 195–204.

Sjöberg, L. (2008). Genetically modified food in the eyes of the public and experts. *Risk Management*, *10*(3), 168–193.

Sleboda, P., & Lagerkvist, C. J. (2022). Tailored communication changes consumers' attitudes and product preferences for genetically modified food. *Food Quality and Preference*, *96*, 104419.

Storey, D. (2021). *Mandatory GMO labeling: Pros and cons.* https://www.tracegains.com/blog/mandatory-gm

Sutherlin, E. (2013). Could congress finally end the GMO labeling war. *The Genetic Literacy Project.*

Vega Rodríguez, A., Rodríguez-Oramas, C., Sanjuán Velázquez, E., Hardisson de la Torre, A., Rubio Armendáriz, C., & Carrascosa Iruzubieta, C. (2022). Myths and realities about genetically modified food: A risk-benefit analysis. *Applied Sciences*, *12*(6), 2861.

Zhang, C., Wohlheuter, R., & Zhang, H. (2016). Genetically modified foods: A critical review of their promise and problem. Food Science and Human Wellness, 5(3), 116–123.

10 Potential Food Safety Problems Related to New Uses of Biotechnology

Baraskar Ruchi Prabhakar, Sumit Kumar, Megharaj Jayarajan, Ashraque Hussain, and Valekar Swapnil Vilas

10.1 INTRODUCTION

The development of biotechnology has emerged in the aspects "how food is produced" and "how agriculture must be practiced," which provide creative answers to urgent global issues. Genetically modified organisms (GMOs), cultured foods, and precision agriculture are a few biotechnological applications that have raised concerns about allergenicity, unintended genetic changes, nutritional changes, disruption of microbial ecosystems, long-term health effects, and ecological effects (Holst-Jensen et al., 2003). Despite the enormous potential of these new biotechnology applications, it is crucial to assess any potential dangers to food safety. In order to ensure and secure the integration of biotechnology into the food supply chain, the much-awaited attempts to give an in-depth investigation of these worries, highlighting the necessity for thorough assessments, stringent restrictions, and educated decision-making, are described in the present chapter. The recognition of a crisis in public confidence in modern society has been prompted by public worries about emerging technologies like genetically modified organisms (GMOs), etc. . However, several evidence and experiences have disproved the long-held theory by proponents of these technologies that the public's unwillingness to follow scientific recommendations is caused by public ignorance and media irresponsibility (which is not to say that these conditions do not exist). The crucial realization, however, is still routinely disregarded: skepticism among the public is not a response to (supposedly misperceived) risks per se or to media depictions of these, but rather to the actions and behaviors of the dominant institutions of science and policy, including how they portray the general public (Wynne, B., 2001).

Electroporation is a different method that makes it easier to incorporate genes into the host genome. Plant tissues without cell walls can be studied using this technique. Through tiny holes that are momentarily created by electric pulses, DNA

DOI: 10.1201/9781032650616-10

enters plant cells. Microscopic crystals may also be able to produce these holes. Microinjection, which is the direct delivery of DNA into the genome, is another modern technique. The inactivation of certain genes, such as those that prevent plant virus infections and soften fruits, is another advantageous use of antisense technology. The Codex Alimentarius Commission resolved to create an Ad Hoc Intergovernmental Task Force on Biotechnology Food, Safety and Nutrition in 1999 at its 23rd Session in response to growing concerns about the safety and nutritional value of biotechnology-derived foods among the global population. Biotechnology-derived foods to create standards, guidelines, or recommendations, as suitable, for foods created through biotechnology or for features added using biotechnology the Task Force's initiatives. The topic of the allergenicity of foods containing genetically modified ingredients was expressly addressed for the first time in the 1996 Joint FAO/WHO Consultation on Biotechnology and Food Safety. The IFBC/ILSI assessment approach was recommended, and it included the following evaluation factors: the origin of the genetic material that was transferred, molecular weight, sequence homology, heat and processing stability, the impact of pH and/or gastric juices (digestive stability), and prevalence in foods.

As part of the overall safety assessment method, the 1996 Consultation agreed that "a rational scientific approach to the assessment of the allergenicity of genetically modified organisms can and should be undertaken." In addition, the 1996 Consultation provided a number of suggestions regarding the allergenicity of genetically modified foods:

- It's important to prevent the transfer of genes from widely allergenic foods.
- Unless these products can be readily identifiable in the marketplace and this identity will not be lost through distribution and processing, foods found to have an allergy that was transmitted from the organism that gave the DNA should not be considered for marketing authorization. Additionally, labeling strategies may not be useful in these circumstances, and consumers who are illiterate or may not be given labels face unique difficulties.
- Despite the fact that there is no patient population known to have an allergy to this gene product, involved organizations should think about the appropriateness of, and/or steps to take, in relation to foods containing new proteins that are determined to have the characteristics of an allergen.
- The characterization of food allergies and their identification

The integration of biotechnology into agriculture and food production has ushered in a paradigm shift in how we address the challenges of global food security and sustainability. Biotechnology offers innovative solutions such as genetically modified organisms (GMOs), synthetic biology, and precision breeding, all with the potential to enhance crop yields, nutritional content, and disease resistance. However, these new applications come with a range of potential food safety concerns that warrant thorough investigation and mitigation (Najafi et al., 2014). Historically, the adoption of genetically modified crops has raised questions about allergenicity, gene flow, and unintended consequences. The possibility of introducing novel allergens through genetic modifications has prompted rigorous evaluation of proteins within GMOs. Additionally, the transfer of genes from GMOs to non-modified organisms, along with the potential for antibiotic resistance marker genes which has raised alarms within the scientific community (Holst-Jensen et al., 2012). Moreover, biotechnological modifications aimed at enhancing nutritional content can inadvertently disrupt the delicate balance of essential nutrients, potentially affecting human health. This concern calls for comprehensive nutritional profiling and assessment to ensure that any improvements do not inadvertently compromise the overall nutritional quality of the food (Chassy, 2002).

Disruptions in food-associated microbiomes are also a growing concern. Alterations to microbial communities could impact processes like fermentation, shelf life, and pathogen control, thereby influencing the safety and quality of the final product. Furthermore, the potential for long-term health effects arising from consuming biotech-derived foods requires in-depth investigation, given that such impacts might manifest over extended periods. Ecological implications are another dimension of concern (Zhang et al., 2016). The widespread adoption of genetically modified crops resistant to pests or herbicides can lead to shifts in pest populations and non-target species, with potential ripple effects across ecosystems. As biotechnology continues to evolve and expand into new realms, understanding and addressing these potential food safety problems is paramount to the responsible and sustainable integration of these technologies into our food systems. This chapter explores the potential problems with food safety brought on by the development of biotechnology in the food sector.

10.2 FOOD SAFETY AND QUALITY

Biotechnology in the fields of agriculture and processing of food may have an impact on food quality and safety in several ways: (1) by replacing or changing the genes that regulate the dietary components of food agricultural products and livestock; (2) by changing the mutations that regulate the quantity of toxins that naturally occur in food crops, animals, or seafood; and (3) by prolonging certain agricultural methods of production, such as applying pesticides, which lead to chemical residues in food and water (Bawa et al., 2013).

10.3 FOOD SAFETY PROBLEMS

10.3.1 Allergenicity Assessment and Novel Proteins

Specifically, when considering potential allergenicity, the safety of genetically modified organisms (GMOs) and their products is a serious concern. Sensitive people may be at risk from the unintended introduction of new proteins into the food supply. This chapter discusses methodologies, difficulties, and the value of thorough evaluation with a focus on allergenicity assessment in the field of biotechnology (Zhang et al., 2019). The concept of "allergenicity" describes a substance's capacity to cause an allergic reaction in those who are susceptible. From minor symptoms like itching and hives to severe anaphylaxis, a potentially fatal illness, allergic reactions can range widely. Having allergies to food—typically proteins—is a serious public health hazard. Novel proteins can be introduced or altered through biotechnological means, which could generate new allergies (Verhoeckx et al., 2016).

10.3.2 Methods of Allergenicity Assessment

- Sequence Homology Analysis

  This technique involves contrasting the novel protein's amino acid sequence to those of recognized allergens. If there is a lot of sequence similarity, the protein may be suspected of being allergenic. This approach has drawbacks because not all allergens have strong sequence similarities, and some proteins that are not allergens might, however, show homology.
- Protein Structure Comparison

  Three-dimensional structural analysis of the protein may shed light on its possible allergenicity. Some allergens have similar structural traits, such as persistent protein folds and digestion resistance. The use of structural analysis can be used to find proteins that share structural characteristics with recognized allergens.
- IgE Binding Assays

  The protein's capacity to bind to immunoglobulin E (IgE) antibodies, which are involved in allergic reactions, is tested in in vitro experiments. The new protein may have allergenic potential if it binds to IgE from allergic people. This technique calls for careful interpretation, though, as any binding may not always result in a clinical allergy.

10.3.3 Challenges and Considerations

- Species-Specificity

  The species with which a protein comes into contact can affect how allergic it is. In certain animals, a protein allergy may not cause the same reaction as in others. Assessment is made more difficult by cross-reactivity and variation in individual responses.
- Relationships between Thresholds and Dose-Response

  A particular threshold level must be reached before an allergic reaction takes place because allergic reactions are frequently dose-dependent. It is challenging to establish safe exposure levels for proteins that may cause allergies since doing so requires an understanding of the dose-response relationship.
- Lack of Comprehensive Databases

  For a reliable assessment of allergenicity, an extensive and current allergen database is required. Such databases, however, might not always be accessible or may not contain data on recently identified allergies.

10.4 GENE TRANSFER AND ANTIBIOTIC RESISTANCE

Biotechnology's application in agriculture and food production has unlocked tremendous potential for productivity and innovation. Along with the advantages, there are some serious drawbacks, such as the possibility of gene transfer and the spread of antibiotic resistance genes (Guo et al., 2019). In the context of biotechnology and food safety, this chapter examines the intricate interactions between genetic alteration, gene transfer methods, and the effects of antibiotic resistance.

10.4.1 Gene Transfer Mechanisms

- Transfer of Horizontal Genes

  The transfer of genetic material between creatures that are not parents and offspring is known as horizontal gene transfer, or HGT. Particularly, bacteria are well-known for their capacity for HGT via processes like conjugation, transformation, and transduction. These methods make it possible to transfer genetic material across species barriers, particularly genes associated with antibiotic resistance (Dennig et al., 2002).
- Transformation Mediated by Agrobacterium

  The bacterium *Agrobacterium tumefaciens* is frequently employed as a vector to introduce genes into plant cells. Although this method is useful for genetically altering crops, there is still a risk of inadvertent gene transfer to other organisms or the environment.
- Selection Markers and Genes for Antibiotic Resistance

  Genes that are resistant to antibiotics are frequently utilized in genetic alteration techniques as selectable markers. By conferring resistance to particular antibiotics, these genes enable researchers to recognize successfully altered cells. However, the use of these markers has sparked concern since they may transfer resistance traits to environmental bacteria or infections to humans.

10.4.2 Implications for Food Safety

- Impact on Microbial Communities

  Concerns about the spread of antibiotic-resistant bacteria are raised by the transfer of antibiotic resistance genes from genetically engineered organisms to environmental microorganisms. This might make antibiotics less effective when used in clinical settings.
- Implications for Human Health

  Antibiotics may become less effective at treating illnesses if antibiotic resistance genes spread from genetically modified foods to gut microbes. Given the rise of illnesses that are resistant to antibiotics globally, this is a serious worry.

- Regulatory Methods

  Global regulatory organizations are aware of the possible dangers brought on by the spread of antibiotic resistance genes. The use of antibiotic resistance markers in genetically modified organisms has been subject to limitations or guidelines in several jurisdictions.

10.5 UNINTENDED NUTRITIONAL ALTERATIONS

Enhancing crop nutrition has become more important as biotechnology continues to transform agriculture and food production. The pursuit of nutritional improvement via genetic modification, however, has a twin challenge: assuring desired benefits while avoiding unintentional changes that could jeopardize food safety and nutritional quality (Kleter et al., 2010). This chapter examines the techniques, difficulties, and importance of thorough assessment to understand the intricacies of unexpected nutritional changes brought on by biotechnological interventions. Nutritional enhancement and genetic engineering for crops are used to genetically modify them to increase their nutrient content, such as by boosting vitamin content, protein levels, or mineral intake. Although these changes offer hope for preventing diseases linked to deficiencies and malnutrition, they may also have unanticipated effects on nutritional profiles (Ying et al., 2019).

10.5.1 Unintended Nutritional Alterations: Challenges and Mechanisms

- Complex Interactions

  The intake, synthesis, and distribution of nutrients are controlled by a network of interconnected pathways that make up the plant metabolism. Due to complex relationships, genetic alterations that target one nutrient may unintentionally upset the balance of others.
- Nutrient Conflict

  Genetic changes may result in a different way that nutrients are distributed, potentially taking resources away from vital nutrients and lowering the overall nutritional quality.
- Antinutrients and Bioavailability

  Unintentional changes may affect how readily nutrients are absorbed and used by the body, affecting their bioavailability. Additionally, unanticipated increases in antinutrients, which are substances that obstruct the absorption of nutrients, might occur.
- Comprehensive Nutritional Profiling

  Assessing how genetic alterations affect the overall nutritional makeup of crops requires thorough nutritional profiling. Unintended changes must be detected using approaches like metabolomics, proteomics, and advanced analytical techniques.

10.6 MICROBIOME DISRUPTION IN BIOTECHNOLOGY-DERIVED FOODS

The implementation of biotechnology into food production has changed the agricultural landscape and provided creative answers to pressing global issues. The possible modification of food-associated microbiomes is one factor that requires cautious attention (Chellappan et al., 2019). This chapter explores the complex interactions between foods created from biotechnology and microbial ecosystems, looking at the ramifications, difficulties, and mitigation techniques for microbiome disruption. Production of food utilizing microbiomes in food production, microbial communities have a significant impact on procedures like fermentation, preservation, flavor creation, and pathogen control. These groups contribute to the wholesomeness, nutritional value, and organoleptic qualities of food (Kumawat et al., 2022).

10.6.1 Disruption Mechanisms

- Modified Substances

  The composition and accessibility of nutrients that enable microbial development in foods developed from biotechnology may change as a result of genetic alterations or processing techniques. These changes in microbial populations may have an effect on the quality of the final product.
- Antimicrobial Properties

  Some crops that have undergone genetic engineering are designed to express antibacterial peptides. These peptides may have unintended effects on foodborne microbes despite being designed to increase disease resistance.
- Relationships to the Environment

  Disruption of the microbiome can occur outside of the dietary matrix. Released GMOs or industrial byproducts may have an impact on microbial communities in the environment, which may cause ecosystem disruption.

10.6.2 Implications for Food Safety and Quality

Shelf life and fermentation disruptions to the microbiome can change the microbial communities in charge of fermentation processes, which can have an impact on the qualities, safety, and shelf life of products. Pathogen management of microbial communities supports pathogen defense mechanisms in nature (Odame et al., 2010). These defenses could be compromised by disruption, demanding stronger control methods. Texture, flavor, and aroma of fermented foods are influenced by microbes. Disruption may result in unanticipated modifications to the senses.

10.7 EFFECTS OF BIOTECHNOLOGY-DERIVED FOODS ON HUMAN HEALTH

Unprecedented prospects for addressing the world's food problems have been unlocked by the integration of biotechnology into agriculture and food production. However, consuming foods generated from biotechnology raises a number of complex and serious health concerns, as depicted in the Figure. 10.1. This chapter explores approaches, difficulties, and the value of thorough study as it digs into the complex landscape of evaluating and comprehending the potential health repercussions linked to the extended intake of genetically modified foods. Assessing the long-term effects due to the nature of long-term health impacts, extensive longitudinal research is required to detect any potential minor alterations (Hammond et al., 2011). From immunological reactions and allergenicity to chronic illnesses and metabolic abnormalities, these impacts may occur.

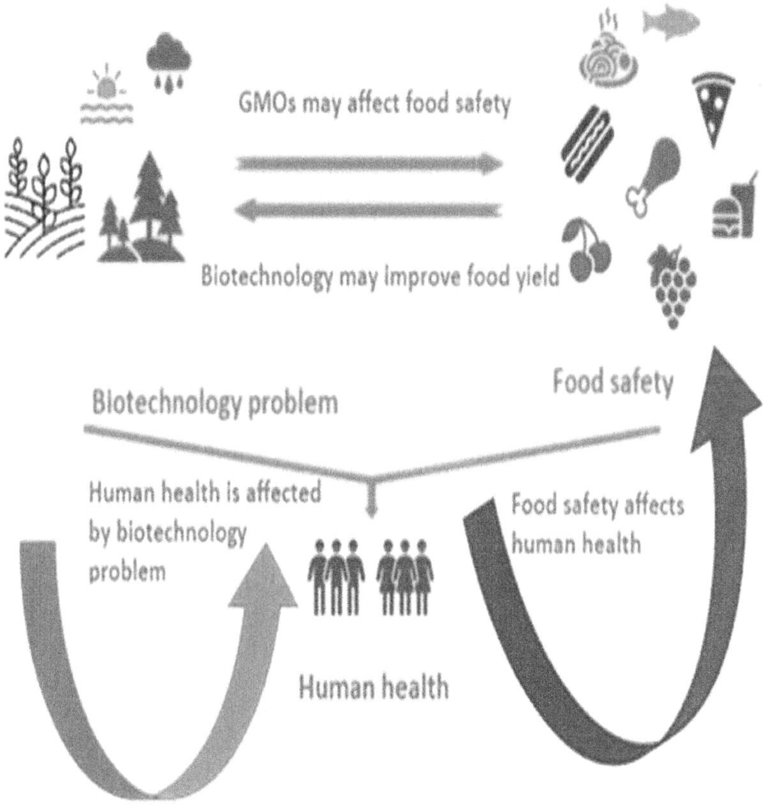

FIGURE 10.1 Human health and biotechnology problems

10.7.1 Methodologies for Assessment

- Epidemiological Research

  Large-scale epidemiological studies are frequently needed to discover trends and associations over protracted time periods for long-term health impacts. Such research is necessary to identify any connections between eating food developed from biotechnology and health effects.
- Animated Figures

  Although not directly applicable to people, animal research offers useful insights into potential health impacts. These models aid in the discovery of patterns, mechanisms, and biomarkers deserving of further study.
- Analysis of Molecules and Cells

  Researchers can examine modifications at the genetic, epigenetic, and metabolic levels thanks to molecular and cellular analysis. Understanding the mechanisms underlying potential health impacts may be aided by these discoveries.

10.8 ENVIRONMENTAL CONCERNS OF BIOTECHNOLOGY-DERIVED FOODS

The ability of humanity to meet the world's food demands has undergone profound changes as a result of the integration of biotechnology into agriculture and food production. However, a variety of environmental worries come along with this advancement and should be considered carefully (Wu et al., 2010). The complex interactions between foods developed from biotechnology and the environment, including their ecological effects, difficulties, and mitigation measures, are discussed in this section. Ecological interactions of foods developed from biotechnology have the potential to have both direct and indirect effects on ecosystem dynamics. In order to evaluate potential effects, it is essential to comprehend these relationships. Genetic pollution and gene flow concerns about gene flow between modified and unmodified species are brought up by the release of genetically modified organisms (GMOs) into the environment. If left unchecked, this gene flow may result in genetic contamination and unforeseen ecological effects.

Ecological balance disruption is the introduction of genetically modified crops with insect or herbicide resistance may upset ecological balances by modifying pest populations or non-target species. Weed evolution and herbicide-resistance crops may result in the establishment of herbicide-resistant weeds, necessitating harsher and environmentally harmful weed management measures. Effects of non-targets and biodiversity concerns have been raised about the unexpected consequences of genetically modified crops on non-target creatures and biodiversity. Changes in plant features, such as toxicity or nutritional value, may have unintended consequences for insects, birds, and other species. Microbial communities in soil where genetically modified crops grow may have an impact on soil microbes, which are important for the cycling of nutrients, the condition of the soil, and general ecosystem stability (Figure 10.2).

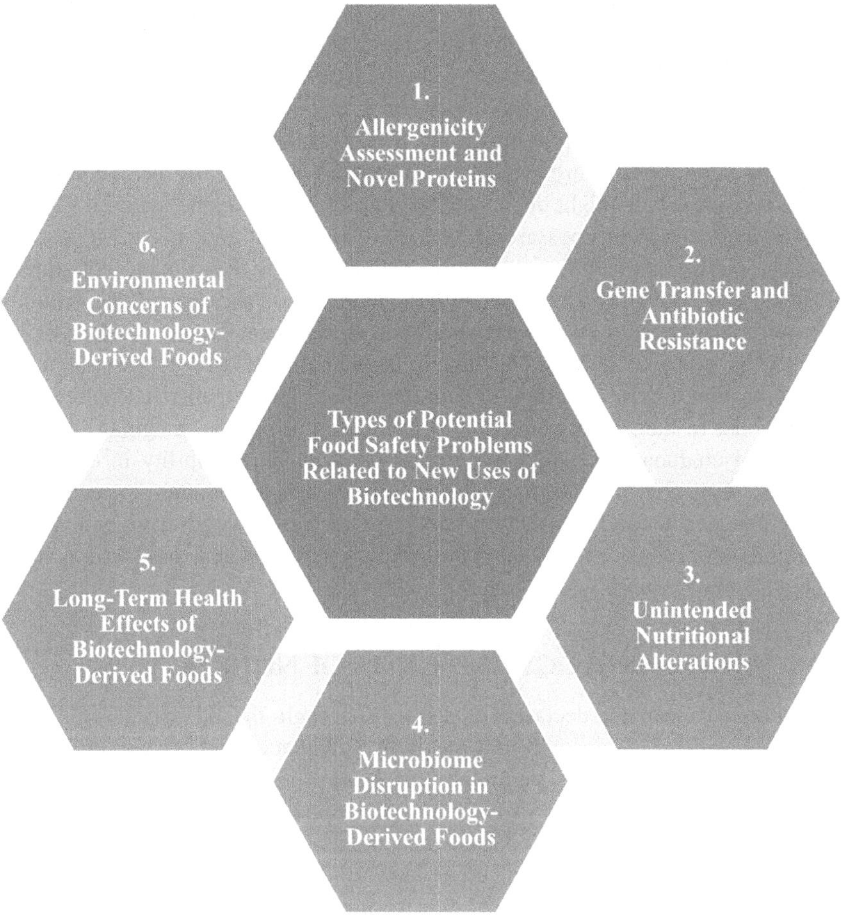

FIGURE 10.2 Types of potential food safety problems related to new uses of biotechnology

10.9 BIOTECHNOLOGY AND AGRICULTURE'S PRODUCTION COMMAND

Since it establishes the control of food production in the genes, biotechnology is a game-changer for agriculture and the food system. The genes have always been important in food production, yet humans have not been able to observe them, precisely choose them, or transport them beyond traditional species barriers (Saini et al., 2015). Although the researchers are gradually discovering which features in crops and cattle are controlled by individual genes, as well as how to switch those genes on and off, and understanding how to integrate genes into the organisms, as well as how to increase gene products. So, first and foremost, the amazing new technology that functions at the molecular level of the food system is available nowadays, which is perhaps the most essential level of food categorization. This means that throughout the food chain, production and quality directives begin with the genes and, more crucially, with people who control the genetics and utilize the new genetic engineering.

Second, along with new genetic technologies comes the legal right to possess genes. Those that are monitoring the changes in law in the scientific arena over the last six years are therefore aware that DNA, as well as some techniques used for genetic manipulation, are currently copyrighted. This means that a commercial interest or an inventor might possess ownership rights in genetic material. In terms of economics, this means possessing a genetic "invention" and specific genetic processes; a sole market right—a temporary patent for 17 years or more—is granted. Third, there are certainly a lot of genes in the field of food and agriculture. A number of genes control corn output, barley stalk strength, wheat protein levels, and soybean photosynthesis efficiency. There are genes in animals that control the amount of fat, breastfeeding rates, feed-to-meat processing rates, expansion, and resistance to infections. In fact, an organization of desired traits can be imagined, including agricultural characteristics such as greater yields or harvestability in cultivation; food-processing characteristics such as those regulating less water or greater solids in certain vegetables and fruits; nutritional traits such as those governing greater protein amounts in crops or lower fat material in animal products; and features implemented for their general well-being or ecological advantages.

10.10 BIOTECHNOLOGY AND GENES OF NUTRITION

The Kellogg Company declared in 1981, "Through the advances afforded by genetic engineering, grains will become more widely accepted because of improvements in taste, texture, form, and total nutritional profile". In a book titled *Genetic Engineering: A Natural Science*, the Monsanto Company includes "food plants with enhanced nutritional value" as one of the potential outcomes of current research in biotechnology. Monsanto executive Howard Schneiderman indicated in the November 1985 issue of *Science '85* that certain tropical root crops, such as cassava and taro, might be genetically modified for less cyanide and more protein. The Rockefeller Foundation, as part of its ongoing commitment to rice genetic enhancement, provided significant study grants to develop various kinds of yellow rice that can generate carotene within the kernel in order to fight a lack of vitamin A in nations with limited resources where rice is the primary source of nutrition. Several industrial and university laboratories in the United States are concentrating on nutrition genetics. For example, Phytogenic, a biotechnology subsidiary of J.G. Boswell Co., one of the nation's largest farms, is using recombinant DNA techniques to "increase the nutritional quality of the protein" in the russet Burbank potato. Researchers at the University of California, Los Angeles (UCLA) have been striving to uncover the genetic mechanisms influencing the seed retention protein in soybeans and other crops that are low in some necessary amino acids for consumption by humans. Several dietitians are against using genetic modification for nutritional betterment, claiming that meddling with nutrition genes may result in large fluctuations within levels of nutrients in industrial cultivars, causing nutritional havoc in the national availability of food.

10.11 APPLICATION OF BIOTECHNOLOGY TO FOOD SAFETY

Biotechnology integrated into the food industry has resulted in notable gains in agricultural output, nutritional value, and food manufacturing efficiency. However, as the sector evolves, the introduction of new biotechnology users poses significant food safety issues. Biotechnology can be used to solve these challenges by developing novel techniques and procedures that improve food safety protocols. Biotechnology provides tools and procedures for precise gene manipulation and microorganism modification. These tools can be used to enhance food safety by better detecting, preventing, and managing possible dangers. Biotechnological techniques can help identify contaminants, infections, and pollutants, eventually assuring the food supply chain's safety. However, the significance of biotechnology is as follows.

10.11.1 Pathogen Detection and Control

Foodborne pathogens are a serious hazard to public health, causing symptoms that range from mild gastroenteritis to serious infections. The introduction of new biotechnology users brings opportunities as well as difficulties in managing possible pathogen contamination problems in food. Biotechnology provides cutting-edge tools and strategies for identifying and managing foodborne pathogens, ultimately improving food supply chain safety. Foodborne infections like *Salmonella*, *E. coli*, *Listeria monocytogenes*, and *Campylobacter* can cause extensive outbreaks and serious health implications. Rapid and reliable identification of such pathogens is critical for preventing outbreaks, tracing contamination origins, and ensuring the quality of public food items.

10.11.2 Allergen Detection

Biotechnology helps to develop enhanced allergy detection technologies, ensuring that potentially allergic substances are correctly detected and labeled. Immunoassays and biosensors provide accurate and specific methods for detecting allergens in food products. The prevalence of food allergies is rising, and allergic reactions can range from minor annoyance to potentially fatal anaphylaxis. Food safety issues arise as new biotechnology users experiment with creative ways to improve food production. One such method is the possible introduction of novel allergens. But biotechnology offers sophisticated methods for allergy identification, making a huge contribution to consumer health protection.

10.11.3 Contaminant Monitoring

Mycotoxins, heavy metals, and pesticides are just a few of the contaminants that biotechnology helps detect and mitigate. Real-time monitoring of contamination levels is made possible by biosensors and molecular methods, strengthening food safety precautions. Consumers are at considerable risk for health problems from food pollutants, including mycotoxins, heavy metals, pesticides, and chemical additives.

There is a need to address possible issues with food safety brought on by the development of novel pollutants as new biotechnology users emerge. Improving food safety protocols dramatically, biotechnology provides innovative approaches to the identification, measurement, and reduction of numerous food pollutants.

10.11.4 Food Traceability and Authentication

The creation of traceability systems employing molecular indicators like DNA barcoding is made easier by biotechnology. These programs assist in tracking the provenance and integrity of food goods, reducing fraud, and fostering consumer trust. A comprehensive food safety system must include both food traceability and authenticity. The possibility of introducing innovative substances and processes as a result of the rise of new biotechnology users highlights the demand for precise traceability and authenticity methods. To improve food traceability and authenticity, ensure transparency, trust among consumers, and the avoidance of food fraud, biotechnology offers cutting-edge technologies and techniques. Food authentication guarantees that food products are correctly labeled and identified, whereas food traceability involves tracing the transportation of food products across the supply chain. Both elements are essential for avoiding fraud, assuring the integrity of the product, and successfully handling possible recalls.

10.11.5 Probiotics and Functional Foods

The development of probiotics and nutritional supplements that support gut health and strengthen the body's defenses against infectious agents is made possible by advances in biotechnology. By enhancing the human system's immune response, these products improve the safety of food overall. Due to their potential health benefits, which range from gut health improvement to immune system augmentation, probiotics and functional foods have attracted a lot of interest. Concerns about food safety must be addressed as new biotechnology users look for creative methods to create and use these goods. Probiotics and functional foods' safety and effectiveness are crucially dependent on the sophisticated tools and techniques that biotechnology provides. This article examines how biotechnology is used in the creation, manufacturing, and quality assurance of probiotics and functional foods, emphasizing its importance, techniques, and function in reducing the dangers to food safety brought on by novel biotechnology activities. When consumed in sufficient proportions, probiotics are live bacteria that have positive health effects. In contrast, functional foods have bioactive substances that improve bodily processes. These goods are essential for promoting general health and preventing chronic illnesses.

10.12 REGULATORY FRAMEWORK FOR BIOTECH FOOD SAFETY

Biotechnology has transformed the agriculture and food industries by providing novel approaches to increasing yields of crops, nutritional value, and total food production efficiency. Biotechnology, however, brings with it potential food safety risks in addition to its benefits. It is critical to ensure the quality of biotechnology-based foods because these products can have an influence on both individuals and the environment. Robust regulatory systems are essential for addressing these challenges. This article goes into the regulatory structure for biotechnological food safety, examining its importance in relation to possible food safety issues caused by new biotechnology users. The biotech food safety regulatory framework protects consumers from potential dangers connected with genetically modified organisms (GMOs), along with other biotechnology-based food items. It establishes the procedures for testing, assessing, approving, labeling, and monitoring these items to ensure that they fulfill safety requirements before entering the market. The framework aids in striking a balance between supporting technical progress and safeguarding human well-being and the ecosystem.

Labeling biotech foods correctly is critical for consumers in making educated decisions. Clear labeling of GMO-containing products is required by regulatory frameworks, allowing customers to distinguish between genetically modified and non-modified foods. The notion of "right to know" is preserved, increasing transparency. Due to a variety of genetic relationships, assessing the potential dangers of transgenic foods can be difficult. New technology users may introduce novel genetic changes that result in unforeseen results, necessitating adaptive regulatory techniques. During new biotechnology use, there may be insufficient historical data to conduct risk evaluation. This necessitates the development of solutions for dealing with circumstances in which thorough data is missing. Because regulatory frameworks differ between countries, safety assessment requirements are inconsistent. Harmonization initiatives are required to create a unified worldwide strategy for biotechnology safety in food.

10.13 BIOTECH IN FOOD SAFETY INCIDENTS

Several benefits have resulted from the quick adoption of biotechnology in the food sector, including improved nutrient quality and greater crop output. Yet, the complex nature of biotechnology also brings with it potential issues with food safety, especially for novice users who might not be familiar with the complexities of biotech-derived goods. For the purpose of preventing and resolving any biotechnology-related issues with food safety, learning from prior instances is essential. In this article, major biotech food safety incidents are looked at, their causes, effects, and lessons learned are examined, and their applicability to the concerns of new biotechnology consumers is discussed. In India (2005), due to ineffective pest resistance management and poor stewardship practices, the introduction of Bt cotton in India ran into difficulties. Cotton pests returned as a result of uncontrolled planting of Bt cotton varieties and poor resistance management.

Causes and Effects: Poor awareness of the value of establishing non-Bt refuge zones and the extensive use of Bt cotton cultivars without effective management of resistance to pests resulted in increased pest susceptibility and decreased crop yields.

Lessons Discovered: To prevent the emergence of insect resistance, it is crucial to conduct integrated pest management techniques correctly. Educating farmers on the significance of following stewardship guidelines and practices is essential for the sustainable use of GMOs. Although the taco shells were discovered to have Cry9C protein traces in 2000, the Bt toxin is not permitted for human consumption, concerns were raised by the event over the protein's propensity to cause allergies and the lack of thorough testing for genetically engineered ingredients. However, the taco shells' use of genetically modified corn, which contains the Cry9C protein, was blamed for the event. The discovery of proteins and genetic alterations that had not been authorized brought to light the requirement for more exacting testing and labeling procedures. For genetically modified crops in the food supply chain, thorough testing for the presence of unauthorized genetic changes and proteins is essential. Accurate information regarding genetically modified ingredients and transparent labeling are crucial for educating consumers and preventing inadvertent exposure.

A study on the ractopamine in pork (2011) was conducted, which concerns the safety of ractopamine for human ingestion that arose from its usage as a feed additive to encourage lean muscle growth in pigs. Imports of pork that had been ractopamine-treated were prohibited by some nations, including the European Union, China, and Russia. Consumer health issues were raised as a result of the discovery of ractopamine residues in pig products. Exporters of pork suffered from trade interruptions and import prohibitions. The inconsistencies in the safety requirements for food additives can impede trade and raise public health issues. It is crucial that safety laws are standardized globally. Using traceability systems can assist in locating the source of pollutants and guarantee the correct labeling of food products.

It is essential to learn from previous biotech food safety disasters in order to avoid such issues in the future, particularly for new consumers of biotechnology. The Bt cotton controversy, Cry9C corn incident, and ractopamine in pork are just a few examples of incidents that can help new users implement stringent safety protocols, responsible stewardship practices, transparent labeling, and effective communication strategies. These case studies highlight the significance of comprehending and addressing any issues with food safety linked to items generated from biotechnology.

10.14 FUTURE SCOPE: ENHANCING BIOTECH FOOD SAFETY

The food business has seen substantial innovation as a result of the quick development of biotechnology, which has improved agricultural yields, created innovative ingredients, and raised food quality. However, these possibilities may also present food safety issues, particularly for new biotechnology users. It is crucial to investigate potential future paths for improving biotech food safety in order to address these issues and guarantee the safety of food items generated from biotechnology. The complete product life cycle, including genetic alteration, potential allergenicity, and unanticipated effects, should be considered in thorough safety assessments. An integrated understanding of the product's safety profile can be obtained by integrating omics technologies (genomics, proteomics, and metabolomics). Pico et al. (2017). Future guidelines for food safety must incorporate quick and accurate detection techniques to find any pollutants or allergens created through biotechnology.

Real-time monitoring can benefit from high-quantities sequencing methods and biotechnology-enabledbiosensors. Future guidelines for food safety must incorporate quick and accurate detection techniques to find any pollutants or allergens created through biotechnology.

Real-time monitoring can benefit from high-quantities sequencing methods and biotechnology-enabled biosensors. Biotechnology and blockchain technology together can improve accountability and traceability. A distributed accounting system can offer precise details on the manufacturing, storage, and distribution of food products developed from biotechnology. The creation of individualized risk assessment frameworks based on unique genetic susceptibilities can modify assessments of food safety for various population subgroups. With this method, possible dangers related to particular genetic profiles can be identified.

10.14.1 Consumer Education and Engagement

1. Transparent Labeling: The precise and transparent labeling of goods developed from biotechnology can enable consumers to make knowledgeable decisions. Information about genetic engineering, allergenicity tests, and potential health impacts should be included on labels Lusk et al. (2004)
2. Initiatives in the Field of Education: By working together, regulatory agencies, business, and academic institutions may raise consumer understanding of the advantages and dangers of biotechnology. Campaigns aimed at educating the public can dispel myths and build trust.

10.15 RISK ASSESSMENT AND MANAGEMENT STRATEGIES IN BIOTECHNOLOGY FOOD SAFETY

The food sector is being transformed by biotechnology, which holds out the prospect of increased agricultural yields, greater nutrient quality, and novel food products. But the incorporation of biotechnology also raises potential issues with food safety, particularly for new users exploring this complex subject. Thorough risk assessment and efficient management techniques are crucial for ensuring the safety of foods developed from biotechnology. In this article, prospective food safety issues related to new biotechnology users are addressed along with the need for risk assessment, its essential phases, management solutions, and their applicability. The food sector is being transformed by biotechnology, which holds out the prospect of increased agricultural yields, greater nutrient quality, and novel food products. But the incorporation of biotechnology also raises potential issues with food safety, particularly for new users exploring this complex subject. Thorough risk assessment and efficient management techniques are crucial for ensuring the safety of foods developed from biotechnology. In this article, prospective food safety issues related to new biotechnology users are addressed along with the need for risk assessment, its essential phases, management solutions, and their applicability.

10.15.1 Steps in Risk Assessment

- Identification of Hazards: List any dangers that may be brought on by the manufacturing of food using biotechnology. These risks could include allergenicity, unexpected consequences on organisms other than the target, and the spread of antibiotic resistance genes.
- Exposure Assessment: Calculate the extent to which consumers may be exposed to the dangers. Consider elements like eating habits, quantity sizes, and frequency of consumption. Hazard classification analyze the dangers' features, such as their toxicity, allergic potential, and mechanism of operation. Identify any possible health effects of exposure to these risks.
- Risk Characterization: To determine the degree of risk involved with consuming foods generated from biotechnology, combine the data from hazard identification, exposure assessment, and hazard characterization.

10.15.2 Monitoring Techniques for Food Safety in Biotechnology

- Preventive Measures: Implement preventive actions to avert or lessen potential risks. To prevent cross-contamination, this entails using appropriate stewardship techniques, containment procedures, and segregating genetically modified crops.
- Monitoring and Surveillance: Keep an eye out for any pollutants or unwanted effects when using items generated from biotechnology. Utilize cutting-edge methods to detect compositional changes, such as metabolomics and molecular analyzes.
- Traceability and Labeling: Establish traceability mechanisms that track the production of items produced from biotechnology. Identification. Accurate labeling educates consumers and enables successful recall procedures, if necessary.
- Post-market Surveillance: After biotech-derived items hit the market, they should be continually monitored for safety. Evaluate any negative effects on human health or the environment in conjunction with regulatory organizations.

New biotechnology users must prioritize risk assessment and management to ensure food safety as they look for creative ways to improve food production. New users can make wise selections by methodically assessing potential dangers, comprehending exposure, and describing risks. Effective management techniques, from prevention to post-market monitoring, are essential for reducing possible issues with food safety. The Bt cotton case in India highlights how crucial it is to draw lessons from the past in order to steer clear of repeating the same mistakes. Ultimately, striking a delicate balance between innovation and consumer and environmental safety is necessary for the successful integration of biotechnology into the food industry.

10.16 CONCLUSION

In conclusion, addressing possible issues with food safety brought on by new biotechnology users necessitates a comprehensive strategy that incorporates legal compliance, scientific advancement, knowledge gained from past mistakes, and futuristic tactics. While the utilization of biotechnology improves the precision of identification and evaluation procedures, a solid regulatory basis protects the safety of products derived from biotechnology. New users can avoid mistakes and prioritize safety by learning from major instances and the lessons they taught. Future directions promise to further improve the safety of biotech food, including individualized risk assessment and integrated safety evaluations. In the end, maintaining a harmonious balance between the advantages of biotechnology and ensuring food safety for all depends on the cooperative efforts of regulators, researchers, industry, and consumers. The food sector has undergone radical change as a result of the quick development of biotechnology, which has made it possible to increase crop yields, boost nutrient content, and create novel ingredients. The potential for introducing novel ingredients, processes, and pollutants, however, needs a detailed awareness of possible issues with food safety as new users enter the biotechnology ecosystem. Regulatory frameworks, biotechnology application, prominent events, and future directions are just a few of the facets of potential food safety hazards associated with new users of biotechnology that have been covered in this article.

REFERENCES

Bawa, A. S., & Anilakumar, K. R. (2013). Genetically modified foods: Safety, risks and public concerns—a review. *Journal of Food Science and Technology, 50*(6), 1035–1046.

Chassy, B. M. (2002). Food safety evaluation of crops produced through biotechnology. *Journal of the American College of Nutrition, 21*(Suppl. 3), 166S–173S.

Chellappan, D. K., Ning, Q. L. S., Min, S. K. S., Bin, S. Y., Chern, P. J., Shi, T. P., & Dua, K. (2019). Interactions between microbiome and lungs: Paving new paths for microbiome based bio-engineered drug delivery systems in chronic respiratory diseases. *Chemico-biological Interactions, 310*, 108732.

Courvalin, P. (1994). Transfer of antibiotic resistance genes between gram-positive and gram-negative bacteria. *Antimicrobial Agents and Chemotherapy, 38*(7), 1447–1451.

Dennig, J., & Duncan, E. (2002). Gene transfer into eukaryotic cells using activated polyamidoamine dendrimers. *Reviews in Molecular Biotechnology, 90*(3–4), 339–347.

Guo, M., Ye, J., Gao, D., Xu, N., & Yang, J. (2019). Agrobacterium-mediated horizontal gene transfer: Mechanism, biotechnological application, potential risk and forestalling strategy. *Biotechnology Advances, 37*(1), 259–270.

Hammond, B. G., & Jez, J. M. (2011). Impact of food processing on the safety assessment for proteins introduced into biotechnology-derived soybean and corn crops. *Food and Chemical Toxicology, 49*(4), 711–721.

Holst-Jensen, A., Bertheau, Y., De Loose, M., Grohmann, L., Hamels, S., Hougs, L., ... Wulff, D. (2012). Detecting un-authorized genetically modified organisms (GMOs) and derived materials. *Biotechnology Advances, 30*(6), 1318–1335.

Holst-Jensen, A., Rønning, S. B., Løvseth, A., & Berdal, K. G. (2003). PCR technology for screening and quantification of genetically modified organisms (GMOs). *Analytical and Bioanalytical Chemistry, 375*, 985–993.

Kleter, G. A., & Kok, E. J. (2010). Safety assessment of biotechnology used in animal production, including genetically modified (GM) feed and GM animals-a review. *Animal Science Papers and Reports*, *28*(2), 105–114.

Kumawat, K. C., Razdan, N., & Saharan, K. (2022). Rhizospheric microbiome: Bio-based emerging strategies for sustainable agriculture development and future perspectives. *Microbiological Research*, *254*, 126901.

Lusk, J. L., House, L. O., Valli, C., Jaeger, S. R., Moore, M., Morrow, J. L., & Traill, W. B. (2004). Effect of information about benefits of biotechnology on consumer acceptance of genetically modified food: evidence from experimental auctions in the United States, England, and France. *European review of agricultural economics*, 31(2), 179-204.

Najafi, M. B., & Lee, B. H. (2014). Biotechnology and its impact on food security and safety. *Current Nutrition & Food Science*, *10*(2), 94–99.

Odame, H., Kameri-Mbote, P., & Wafula, D. (2010). *Globalisation and the international governance of modern biotechnology implications for food security in Kenya.*

Pico, T., Creveling, J. R., & Mitrovica, J. X. (2017). Sea-level records from the US mid-Atlantic constrain Laurentide Ice Sheet extent during Marine Isotope Stage 3. *Nature communications*, *8*(1), 15612.

Saini, J. K., Saini, R., & Tewari, L. (2015). Lignocellulosic agriculture wastes as biomass feedstocks for second-generation bioethanol production: Concepts and recent developments. *3 Biotech*, *5*, 337–353.

Verhoeckx, K., Broekman, H., Knulst, A., & Houben, G. (2016). Allergenicity assessment strategy for novel food proteins and protein sources. *Regulatory Toxicology and Pharmacology*, *79*, 118–124.

Wu, G., Kang, H., Zhang, X., Shao, H., Chu, L., & Ruan, C. (2010). A critical review on the bio-removal of hazardous heavy metals from contaminated soils: Issues, progress, eco-environmental concerns and opportunities. *Journal of Hazardous Materials*, *174*(1–3), 1–8.

Wynne, B. (2001). Creating public alienation: Expert cultures of risk and ethics on GMOs. *Science as Culture*, *10*(4), 445–481.

Ying, Y., Zhang, H., & Yu, P. (2019). Implications of recent research on microstructure modifications, through heat-related processing and trait alteration to bio-functions, molecular thermal stability and mobility, metabolic characteristics and nutrition in cool-climate cereal grains and other types of seeds with advanced molecular techniques. *Critical Reviews in Food Science and Nutrition*, *59*(14), 2214–2224.

Zhang, C., Wohlhueter, R., & Zhang, H. (2016). Genetically modified foods: A critical review of their promise and problems. *Food Science and Human Wellness*, *5*(3), 116–123.

Zhang, Y., Wu, Z., Li, K., Li, X., Yang, A., Tong, P., & Chen, H. (2019). Allergenicity assessment on thermally processed peanut influenced by extraction and assessment methods. *Food Chemistry*, *281*, 130–139.

11 Challenges and Considerations in Regulating Bioengineered Food

Shweta Parida

11.1 INTRODUCTION

Bioengineered foods, often referred to as genetically modified or genetically engineered (GE) foods, have been modified through genetic engineering techniques to possess specific traits or characteristics. These modifications involve the manipulation of an organism's genetic material and its proper deliberation. The development and regulation of bioengineered foods are influenced by various scientific, economic, ethical, and regulatory considerations. The science behind bioengineered foods has advanced significantly, enabling scientists to introduce desirable traits into plants, animals, and microorganisms. These traits can include enhanced resistance to pests, diseases, or environmental stressors, improved nutritional profiles, and extended shelf life. Biotechnology has revolutionized the precision and efficiency of genetic modifications, allowing more targeted changes to an organism's DNA.

Bioengineered foods offer numerous potential benefits, including increased crop yields, reduced reliance on chemical pesticides, improved nutritional content, and enhanced resistance to adverse environmental conditions. These traits have the potential to address food security challenges, mitigate losses from pests and diseases, and provide options for tailored nutrition to address specific health needs. The development and deployment of bioengineered foods raise ethical and societal questions. Discussions encompass topics such as environmental sustainability, potential effects on biodiversity, and concerns about corporate control over seed supply. Ethical considerations also include the potential impact on traditional farming practices and the potential to exacerbate socioeconomic disparities. International trade of bioengineered foods can be influenced by varying regulations and consumer perceptions in different regions. Some countries have embraced bioengineered crops, while others have placed restrictions or bans on their cultivation or import. Disparities in acceptance can impact trade relationships and contribute

DOI: 10.1201/9781032650616-11

to complex discussions in global food policy. As technological capabilities continue to advance, finding equilibrium in harnessing the benefits of biotechnology and addressing concerns about safety, transparency, and societal implications remains a critical challenge. Regulatory frameworks, informed by scientific evidence and societal discourse, play a pivotal role in guiding the development and responsible use of bioengineered foods.

Regulating bioengineered food presents several challenges and considerations due to the complex nature of the technology and its potential impact on human health, the environment, and socioeconomic factors. The science behind these modifications can be complex, making it challenging for regulators to fully understand and assess the potential risks and benefits associated with bioengineered foods. One of the primary considerations in regulating bioengineered food is ensuring its safety for consumption. Assessing the potential health risks of various fermented foods, probiotics, functional foods, and allergenicity of genetically modified organisms and plant-based meats requires comprehensive scientific studies and testing protocols. Regulators need to establish rigorous safety evaluation frameworks to determine whether a bioengineered food is substantially equivalent to its non-modified counterpart or if it requires additional scrutiny. Labeling and transparency for bioengineered foods can be a contentious issue. Balancing the need for transparency with potential impacts on consumer choice and market dynamics is a key consideration in regulatory decision-making. Bioengineered crops may have unintended environmental consequences. Regulatory issues help to evaluate the potential ecological risks associated with the cultivation and release of bioengineered organisms to ensure they do not harm biodiversity or ecosystem stability. The regulation of bioengineered food becomes more complex when considering international trade. Divergent regulatory frameworks and standards across countries can create trade barriers and complicate market access for bioengineered products. Harmonizing regulations, aligning safety assessment procedures, and establishing international standards can facilitate trade and address the challenges posed by global supply chains. These crops are often developed by private companies, which may hold intellectual property rights over the technology and the resulting products. Balancing the interests of innovators with broader public interests, accessibility, and affordability of bioengineered foods requires careful consideration. A balance between incentivizing innovation and ensuring fair access to the benefits of bioengineering must be carried out. Assessing the long-term impacts of bioengineered foods is challenging due to the evolving nature of the technology and the complexity of ecological and health systems. Evaluating potential risks, such as the development of resistance in pests or unintended consequences for human health, requires ongoing monitoring, research, and adaptive regulatory frameworks. Overall, regulating bioengineered food requires a multidisciplinary approach, involving scientific expertise, stakeholder engagement, and a thorough understanding of the potential risks and benefits. It is crucial to balance safety, transparency, environmental considerations, trade, intellectual property rights, and the long-term impacts of this technology to ensure responsible and effective regulation, as shown in Figure 11.1.

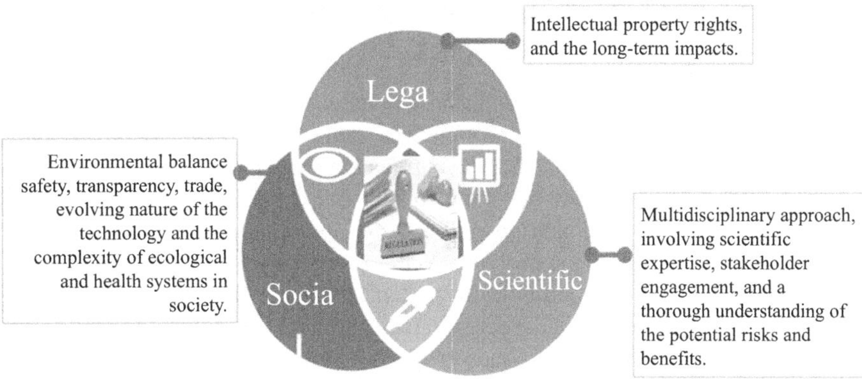

FIGURE 11.1 Various factors affecting regulation of bioengineered foods

The establishment of the first international food regulatory bodies marked significant milestones in the development of global food safety and trade standards. These organizations were founded to address the challenges posed by the increasingly interconnected world of food production, distribution, and consumption. Two key international bodies that played pivotal roles in shaping food regulations and standards are the Codex Alimentarius Commission (CAC) and the World Trade Organization (WTO). The CAC, often referred to as Codex, was established in 1963 by the Food and Agriculture Organization (FAO) and the World Health Organization (WHO), both United Nations agencies. Codex serves as the international food standards-setting body and operates as a joint FAO/WHO program. Its primary objective is to protect consumer health and ensure fair practices in food trade through the development of science-based standards, guidelines, and codes of practice. The WTO, established in 1995, aims to facilitate global trade and promote economic cooperation among its member countries. While not exclusively focused on food regulation, the WTO plays a significant role in shaping trade-related aspects of food safety and regulations through its agreements and mechanisms. The Agreement on the Application of Sanitary and Phytosanitary Measures (SPS Agreement) within the WTO provides guidelines for member countries to ensure food safety and protect human, animal, and plant health. It encourages the use of international standards, including those developed by Codex, as a basis for food safety measures. The SPS Agreement also emphasizes the importance of science-based decision-making in establishing food safety regulations. The establishment of these international bodies recognized the need for coordinated efforts to address food safety, quality, and trade issues that transcend national borders. By providing a platform for collaboration, information exchange, and the development of science-based standards, these bodies have played critical roles in harmonizing food regulations and ensuring the safety and quality of food products in the global market.

11.2 USE OF BIOTECHNOLOGY IN PRODUCTION OF BIOENGINEERED FOODS

Due to modernization and the increasing emphasis on well-being, individuals are placing greater importance on dietary choices, nutrition, and the safety of food to improve their overall quality of life. The outbreak of the SARS-CoV-2 pandemic has underscored the significance of nutritional health as a key determinant of vulnerability to severe illness, with conditions like obesity and malnutrition being identified as prominent risk factors. This evolving awareness among consumers has prompted various players within the food industry to depart from traditional methods of food production. Instead, they are adopting strategies to augment or incorporate health-enhancing attributes into food products, with the aim of maximizing their effectiveness in promoting better health.

Farmers, governments, and scientists are constantly looking for techniques to improve agricultural productivity. Crop quality and quantity have traditionally been improved through the application of breeding procedures. Pesticides and fertilizers have raised agricultural yields dramatically in the past decades. These chemicals, however, harm the ecosystem and may have an unfavorable footprint on the quality of food consumed. Crops of genetic engineering (GE), a more environmentally friendly alternative, provide a simple, non-toxic, and effective means of improving plant yields. Genetic engineering today is defined by its ability to combine sequences of precise DNA with additional desirable qualities. Guided nucleases can break DNA and be directed to a particular target site when linked with a desired DNA-binding domain. As a result, the host's DNA repair mechanisms are turned on, which causes a gene's desired mutation and, in certain cases, a complete change of function. Genome editing has already been done using meganucleases, zinc finger nucleases (ZFNs), transcription activator-like effector nucleases (TALENs), CRISPR-Cas9, and orthologs. All methods have been used to modify genomes, but the first three are more challenging to manipulate because of their closed architecture and high cost. Furthermore, rather than using protein-DNA complexes, CRISPR systems use complexes of ribonucleotide-DNA and guide-RNA (gRNA) for target recognition and subsequent partitions. This characteristic makes CRISPR simpler, more straightforward, and affordable to implement.

In human societies, fermented foods containing a variety of microbial populations first became common during the 15,000–20,000-year shift from hunter-gatherer to agricultural groups. Worldwide consumption of fermented foods is still prevalent; they may account for up to one-third of the average person's diet today. These foods differ in raw ingredients, cultural influences, geographic accessibility, and microbial diversity. The actual contribution of fermented foods' bacteria to health has improved host biomarkers and the gut microbiome. Notably, the rise in interest in traditional fermented foods from various locations around the world influences their potential to include new beneficial bacteria and/or have health advantages. New varieties of fermented foods have emerged as a result of these advances, and some of them are linked to positive health effects.

Recent studies by academics, government, and private research institutes worldwide are showing how functional foods and food ingredients can prevent chronic lifestyle diseases or optimize health, lowering healthcare costs and improving consumer quality of life. Nutrigenomics, which studies how nutrition affects genetically predisposed diseases, will greatly impact functional food research and development. This technology may allow nutrition to be tailored to a person's genetics (Fogg-Johnson and Meroli, 2000; Venter et al., 2001). Nutrigenomics has a significant impact on expected disease prevention efforts and the functional foods industry. Consumers must realize that functional foods are not a "magic bullet" for poor lifestyles. There are no such things as healthy or unhealthy "foods," but rather, there are healthy and bad practices of eating. Thus, they should be careful of many of these foods' touted or inferred benefits and mindful of functional food restrictions.

Probiotics, another bioengineered meal, include live bacteria that benefit the host when consumed in sufficient amounts. Probiotics include yeast (*Saccharomyces cerevisiae*), lactic acid bacteria (LAB; *Lactobacillus*, *Streptococcus*, and *Enterococcus* species), *Bifidobacterium*, *Propionibacterium*, *Bacillus*, and *Escherichia coli*. They may be naturally occurring (like all dietary microbes) or genetically modified for a specific effect. The genetically modified microorganisms were not directly addressed in the analysis (case-by-case examination is advisable); the non-recombinant microbe suggestions, even though insufficient; the microorganism administration technique, exposure level, users' health, and physiological functions must be considered while assessing safety, while considered as probiotics.

Another novelty in the area of lab-grown meat, which is also popularly known as cultured meat or cell-based meat, is an innovative food technology that involves growing animal muscle tissue from animal cells in a laboratory setting. It offers a potential solution to address some of the challenges associated with traditional animal agriculture, such as environmental impact, animal welfare concerns, and the increasing global demand for meat. Public perception, education, and awareness play a significant role in shaping the regulatory approach and market acceptance of these products. This sector is founded on a distinct ethical and moral claim, which is the promise of a future meat-free from the environmental and animal cruelty that is common in existing food chains headed by animal agriculture.

11.3 RISK AND REGULATORY ISSUES CONCERNING BIOENGINEERED FOOD

11.3.1 Regulatory Issues for the Impacts of GMOs on Biodiversity in Agriculture

Reduced chemical pesticide use among some genetically modified organisms (GMOs) is engineered to be resistant to pests or diseases, reducing the chemical pesticide applications. This can lead to a decrease in the overall use of pesticides and has positive effects on biodiversity by minimizing the negative impacts of pesticide exposure on non-target organisms. Whereas the negative impacts of GMOs on biodiversity in agriculture may include a reduction in genetic diversity and impacts on non-target organisms. The cultivation of genetically uniform crops, such as GMOs, can lead to a reduction in genetic diversity within agricultural systems. Genes from genetically modified crops can potentially flow into wild or non-GMO populations through cross-pollination or hybridization. This can result in the spread of genetically modified traits to wild relatives or related crops, potentially impacting the genetic integrity and diversity of native or non-GMO populations.

The problem of genetically modified (GM) crop production is a complex and multifaceted issue that is challenging to define. The testing of genetically modified crops is a crucial issue for the purposes of authenticating products, ensuring biosafety, and complying with regulations. The land that is being cultivated with genetically modified organisms is swiftly increasing, and various novel genes are being introduced into essential crops. In order to ensure the safety of humans, the environment, animals, and other forms of microflora, newly produced genetically modified organisms (GMOs) need to undergo thorough molecular testing before being released for commercial use. The regulatory procedures for the certification of genetically modified crops need a thorough risk assessment for each and every individual instance. According to Ives et al. (2017), the detection and identification of genetically modified organisms are incredibly important for assessing the authenticity of a sample, labeling food, and for business considerations. Therefore, it would be beneficial to integrate several testing procedures in order to conduct comprehensive analyses of genetically modified (GM) materials to evaluate their legitimacy and potential risks to human health. There have been some reports done on the potential risks that are associated with the use of genetically modified organisms, such as the potential for allergenicity, the potential for the development of herbicide-tolerant weeds and resistant insects, the potential for harm to non-target organisms, the possibility that antibiotic resistance could be caused by selection marker genes, which would reduce the effectiveness of antibiotics as a treatment for disease etc. (Mesnage et al., 2013). Using strong promoters to turn on specific genes may be detrimental to people, have an impact on nutritional profile, and allow transgenes to spread from unintended crops or weeds, among other things.

GM crops are engineered to possess specific traits, such as resistance to pests or tolerance to herbicides, through the introduction of foreign genes. While these modifications are intended to confer benefits like increased crop yields and reduced pesticide use, concerns have been raised about their impact on ecosystems and non-target organisms. One concern is the possibility of gene flow between GM crops and wild or non-GM cultivated relatives. This can lead to hybridization and the spread of modified genes to non-GM populations. Such gene flow may impact the genetic diversity and adaptation of natural plant populations, potentially leading to unintended ecological consequences. The expression of GM crop traits, such as insecticidal proteins, could affect non-target organisms like beneficial insects, birds, and soil organisms. For example, insect-resistant GM crops could inadvertently harm non-pest insects, impacting the broader ecosystem. Similarly, herbicide-tolerant crops may affect weed communities, soil health, and aquatic ecosystems if herbicides leach into water bodies.

The widespread cultivation of a single GM crop variety can lead to monocultures, where large areas are planted with genetically uniform crops. Monocultures can increase the vulnerability of crops to pests and diseases, disrupt local biodiversity, and lead to changes in ecosystem dynamics. Genetic modifications can lead to unintended effects due to the complex interactions within ecosystems. These effects might not be immediately apparent and may impact multiple trophic levels within food chains. Assessing the long-term effects of GM crops on the environment is challenging, as ecosystem changes can occur gradually over time. This emphasizes the importance of rigorous monitoring and post-release studies.

It's important to note that the potential environmental impacts of GM crops can vary based on the specific crop, trait, and local conditions. Regulatory frameworks, scientific research, and risk assessments play a crucial role in determining the safety of GM crops for the environment. Researchers continue to study the long-term effects of GM crops and refine risk assessment methodologies to ensure their environmental impact is thoroughly understood in Figure 11.2.

The potential negative impacts of genetically modified crops on human health have been a subject of scientific research. There is a possibility that the consumption of GM crops with newly introduced proteins could trigger allergic reactions in individuals with allergies to related proteins. Therefore, rigorous allergenicity assessments are crucial during the development and regulatory evaluation of GM crops. The process of genetic modification could potentially result in the synthesis of new compounds or metabolic changes that might have toxic effects on human health. Comprehensive toxicity assessments are necessary to identify any potential harmful effects of GM crops on human organs or systems. Alterations in the nutritional content of GM crops could impact human health; i.e., excessive consumption of that crop might lead to imbalances in overall nutrition. Genetic modification can lead to

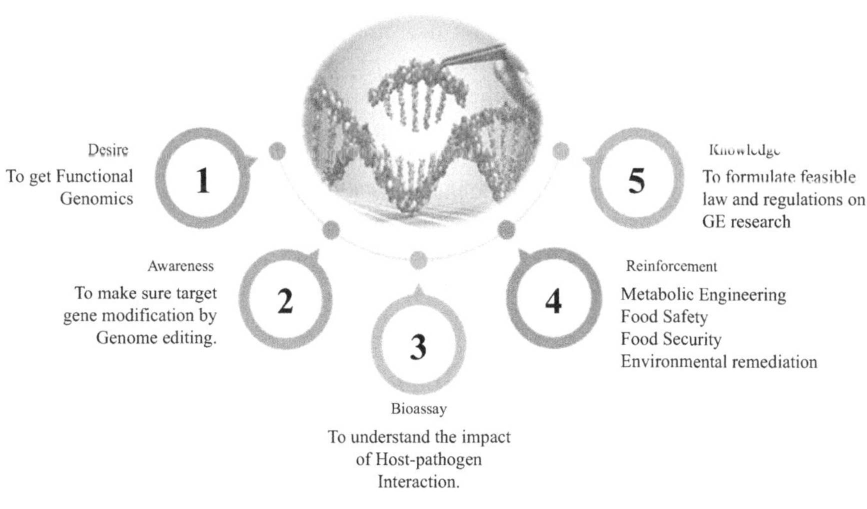

FIGURE 11.2 Risk assessment of GM crops

unintended effects due to the complex interactions within the plant's genome. These effects might not be immediately evident and could affect the nutritional profile or composition of the crop in ways that are not fully understood. Assessing the potential durable health effects of consuming GM crops can be challenging due to the complexity of diet-related health outcomes and the relatively short history of GM crop consumption. The assessments involve evaluating the potential allergenicity, toxicity, and nutritional changes associated with the genetic modifications. Research and scientific studies continue to assess a vital function in providing insights into the potential health impacts of GM crops. As the cultivation of GM crops continues, ongoing research, transparent evaluation processes, and comprehensive monitoring are essential to ensuring that any potential negative health impacts are identified and addressed appropriately.

Recent developments in biotechnology, such as genome editing, act as a tool for food products to reach the market more quickly. Revolutionary genome editing technologies are going to be absolutely necessary for the food production of the future. In comparison to conventional genetically modified crops, they are preferable because of resilience, process precision, and sensible regulation. The CRISPR/Cas9 technology has already resulted in the creation of items that do not fall under the purview of the GMO label. The development of the CRISPR/Cas9 system as a potent tool for editing genes has resulted in regulatory difficulties and challenges. According to statements made by the United States Department of Agriculture (USDA), crops that have been genetically modified through the use of the CRISPR/Cas9 platform are permitted to be farmed and sold without first undergoing the regulatory procedures and possible threat assessments mandated by GMO biosafety rules (Waltz, 2016). This decision will prevent expenses in millions of dollars in genetically modified agricultural research through field experiments and data collection and remove public concern about GM crop use. The United States Department of Agriculture (USDA) has decided not to impose any GMO-related regulatory measures on five cultivars that were produced utilizing the CRISPR/Cas9 system. These five cultivars include browning-resistant mushrooms by the deletion of the polyphenol oxidase gene, waxy maize seedlings with enhanced amylopectin, *Setaria viridis* with a delayed flowering time (Jaganathan et al., 2018), *Camelina* with higher oil content (Waltz, 2018), and soybean with modified Drb2a and Drb2b for drought tolerance (Cai et al., 2015; Kumar et al., 2020), were not subject to GMO regulatory measures.

Regulatory bodies must assess the potential risks associated with CRISPR/Cas9 modifications, including unintended off-target effects and potential long-term ecological impacts. Developing standardized safety assessment protocols specific to gene-edited organisms is necessary to evaluate and manage any potential risks. The regulation of CRISPR/Cas9 and gene-edited organisms is a global concern. Divergent regulatory approaches across countries can create barriers to research collaboration, hinder technology transfer, and complicate the global trade of gene-edited products. Harmonizing regulations and establishing international guidelines can facilitate the responsible and consistent oversight of

CRISPR/Cas9 technology. Regulatory agencies and policymakers are actively working to address these issues by developing guidelines, frameworks, and oversight mechanisms specific to gene-editing technologies like CRISPR/Cas9. The goal is to strike a balance between promoting innovation, ensuring safety, and addressing ethical and societal concerns associated with the use of this powerful gene-editing tool.

11.3.2 Regulatory Issues for Impacts of Fermented Foods

The Codex CAC, a part of the FAO and WHO, has established standards and guidelines for fermented foods. The CAC Standards cover fermented milks, such as yogurt, and provide guidelines for starter cultures, microbial viability, and composition. The Codex Standards also include guidelines for kimchi, describing its preparation steps and essential ingredients. Regional Codex Standards have been developed for fermented foods specific to certain regions, including fermented cooked cassava-based products from Africa and various fermented foods from Asia. These regional standards provide additional information on additives, contaminants, analytical methods, and labeling recommendations. However, there is currently no Codex Standard for kombucha. The Therapeutic Goods Administration (TGA) in Australia regulates complementary medicines, which include vitamins and herbal products. The TGA has guidelines for listed and registered complementary medicines, with listed medicines requiring certification by the sponsor and registered medicines undergoing a stricter assessment process. The TGA aligns its regulatory approach with international counterparts such as the European Food Safety Authority (EFSA) and the Food and Drug Administration (FDA). Australia and New Zealand have a joint agency, Food Standards Australia and New Zealand (FSANZ), which develops food standards and regulations. These standards cover food safety, fortification, and processing. The TGA and FSANZ separate the regulation of therapeutic goods and food products. Therapeutic goods are regulated by the TGA, while food products fall under the jurisdiction of FSANZ.

11.3.3 Regulatory Issues for Impacts of Functional Foods

Food regulations are changing to meet the growing demand for nutraceuticals and useful foods that are thought to be good for one's wellness. Nutraceuticals are products that blend the benefits of vitamins and minerals and are consumed as a form of supplement/drugs. In the United States, the Dietary Supplement Health and Education Act (DSHEA) lets dietary supplements, like herbs and plants, be put into different groups. The FDA is regulating food supplements and making sure they are safe. The USDA and FDA are the main organizations that regulate food, dietary supplements, and their ingredients. For quality control, manufacturers must follow the Current Good Manufacturing Practices, and must report any adverse side effects to the FDA. Additionally, it is the responsibility of the FDA to regulate product information, such as labeling and health claims. The Food and Drugs Act of 1906 was the first law about food in the United States, which was replaced by the Federal Food, Drug, and Cosmetic Act in 1938 and is still active today.

Claims for nutraceuticals

The regulation of nutraceutical claims differs among Asian countries, with no established harmonized definition or technical standards. In Asia, functional foods are mostly controlled as regular foods. Some claims that can be made are about the amount of nutrients and how they compare to other nutrients. However, claims that are based on international standards for reducing the risk of disease are usually not allowed. But the rules about health claims are always changing, and big changes are likely to happen in the next five years. When joining the Indian market, it's important to focus on making health and label claims that fit Indian regulatory rules. It should be thought about if the labeling and packing meet Indian requirements and if the product meets the requirements for specific product claims. The suitable label information and claims for the Indian market will be determined with the aid of a regulatory assessment of the product.

Health claims on label

When joining the Indian market, it is very important to focus on making health claims and labeling that are in line with Indian regulations. Clients from both outside and at home frequently have questions about the packaging and labeling regulations in India, the packing of consignments, the registration of samples, the composition of consignments, the content of the labels, and the claims made about the structure and function of the products. This will help to develop label content and claims that are specific to India. Newcomers should also learn about the health claims that are popular in India and make sure they meet the requirements to make certain product claims.

11.3.4 Regulatory Issues for Impacts of Probiotics on Human Health

The WHO defines probiotics as "live microorganisms that confer a health benefit on the host when given in adequate amounts." Food must have at least 10^6 CFU/mL to be labeled as probiotic. Most probiotics contain many strains of bacteria from different species. This is because many strains of probiotic products are thought to be more effective at providing health benefits and a wider spectrum of efficacy, as well as being additive or synergistic. However, in particular cases, probiotic organisms may have opposing effects. This seems rational and convincing; however, studies comparing single-species and multispecies probiotics found such claims are questionable and need further study.

Probiotics must be species and strains of bacteria that have been shown to benefit the host because many microorganisms can disrupt intestinal microbial homeostasis. Thus, *Lactobacillus*, *Bifidobacterium*, *Lactococcus*, and *Enterococcus* species dominate the global probiotic market. Probiotics like *Lactobacillus bulgaricus* and *Streptococcus thermophilus* are not normally part of the gut flora but are used as starter cultures in dairy products. Because they lack colonization characteristics, these bacteria may not affect gut microbial homeostasis.

There are currently three categories in which probiotics can be found and which are regulated: (i) foods (fermented foods), with *Lactobacillus*, *Bifidobacterium*, and *Lactococcus* claiming GRAS (generally recognized as safe) status; (ii) dietary supplements, which are frequently sold over-the-counter (OTC); and (3) drugs (pharmaceuticals). The classification of probiotics depends on their producers, their suggested uses, and different regulatory organizations' regulations. Human-consumed foods and medications, including probiotics, must be safe to avoid health hazards. Probiotics have been shown to be safe in a variety of populations, including healthy adults, women during late pregnancy and their infants during early infancy, infants aged 0–2 years and children, hospitalized children, critically ill children, and immune-compromised patients. Probiotics must not cause or aggravate disease in humans, whether they come from foods or over-the-counter supplements.

11.3.4.1 Probiotics Safety

Recent papers from prestigious scientific medical journals such as *The Lancet*, *Cell*, *The New England Journal of Medicine*, and *Nature* have raised questions about the safety of probiotics. Some large-scale clinical trials have reported unexpected results, leading to concerns about the methodology and scientific rigor of these trials. Cochrane reviews have recommended probiotics as safe and effective for certain conditions, but subsequent reviews have concluded that more research is needed to support these recommendations. Safety concerns and adverse effects of probiotic intake have been highlighted by researchers, suggesting that the true safety and efficacy of probiotics may be underestimated. Probiotics are typical food supplements rather than pharmaceuticals, which may lead to less stringent testing and a focus on health claims rather than safety. Regulatory frameworks for probiotics are still lacking, and global market growth has brought regulatory questions to the forefront. Regulatory bodies such as the European Food Safety Authority (EFSA) oversee health claims of probiotics, but the primary concern is the validity of claims rather than safety. Post-marketing regulatory vigilance is limited, and the term "health benefit" on labels is not clearly regulated. The strain-specific properties and differences among probiotics further complicate their evaluation and safety assessment. Thorough knowledge of microbial colonization properties is crucial for evaluating probiotic safety and health risks. Individual responses to probiotics can vary due to factors such as age, gender, and underlying health conditions. Dose-dependent toxicological and pharmacological concepts should be considered, and the manufacturing process and potential hazards should be thoroughly examined. Risks associated with systemic and local infections, metabolic activities, gene transfer, antibiotic resistance, modulation of the host's immune response, and drug interactions should be evaluated. Animal models have limitations, and human clinical trials are necessary to establish probiotic safety, despite the associated costs and challenges. Concerns about bacteremia and systemic inflammation have been reported in immunocompromised patients and animal models. Overall, the safety of probiotics warrants attention and further investigation, particularly when considering their use in critically ill patients.

11.3.4.2 Pathogenicity Studies of Probiotics and Clinical Cases

Reports and clinical instances have documented cases of invasive fungal infections linked to *Saccharomyces cerevisiae* and *Saccharomyces boulardii*. It is advisable to exercise some caution when considering these probiotics for patients with compromised immune systems or those in critical condition. Additionally, probiotics containing *Lactobacillus* strains have been linked to instances of bacteremia and endocarditis in individuals with weakened immune systems and cardiac abnormalities. The effects of probiotics on human health present a complex picture with mixed outcomes. Some research suggests potential drawbacks, such as an elevated risk of conditions like rhinitis, asthma attacks, atopic dermatitis, allergies, and sensitization. Prolonged usage of probiotics could potentially have adverse effects on human health, including the promotion of antibiotic resistance and increased virulence potential. Metabolites generated by probiotics, including D-lactate, have the potential to be toxic, potentially leading to D-lactic acidosis. The interaction of probiotic bacteria with amino acids and proteins can result in the production of substances that could be harmful. Unwanted effects related to probiotic use encompass symptoms such as constipation, excessive thirst, skin rashes, fatigue, and itching. However, the precise toxicological implications of these effects remain uncertain. The established contraindications for *Lactobacillus*-containing probiotics in individuals with hypersensitivity to lactose and dairy products have faced challenges. To fully comprehend the safety and toxicological implications of probiotics, further comprehensive research is imperative.

11.3.4.3 Drug Interactions of Probiotics

The composition of the gut microbiota is subject to dynamic shifts influenced by a range of factors, including nutritional status, diseases, medications, circadian rhythms, and environmental elements. It is now acknowledged that probiotics, as integral components of the host's microbiota, possess the capability to affect the availability, pharmacokinetics (PK), pharmacodynamics (PD), and toxicity of drugs and xenobiotics. Traditionally, it was assumed that drugs absorbed through the intestinal tract did not interact with the host's microbiota, unless considering sustained-release formulations or enterohepatic recirculation. Nevertheless, recent research has revealed that probiotic-drug interactions are indeed real, even in the absence of these specific circumstances.

Experiments conducted on rats have illustrated elevated drug availability in instances where probiotics were administered, as seen with drugs like amiodarone and gliclazide. This increase in drug availability could be attributed to alterations in intestinal pH, facilitating drug absorption, or the enhancement of drug transporters. Conversely, certain studies have indicated decreased drug availability, such as observed with tacrolimus. The gut microflora's impact on drug pharmacokinetics extends to numerous drugs, including digoxin, irinotecan, indomethacin, insulin, levodopa, ketoprofen, lovastatin, risperidone, and sulfasalazine. These interactions can hold clinical significance, particularly for drugs with a narrow therapeutic range, affecting their effectiveness, safety, and potential toxicity.

Bile acids and salts, which can be influenced by probiotics, also play a role in drug absorption. Bile salts possess the capacity to heighten membrane permeability and fluidity, influencing drug availability and delivery. They also form micelles, thereby improving drug solubility and dissolution. The impact of probiotics on drug pharmacokinetics/pharmacodynamics and its clinical implications in drug pharmacokinetics are subjects that necessitate further exploration. In terms of interactions between probiotics and antibiotics, it is advisable to administer antibiotics at least two hours before or after probiotics. Similarly, probiotics containing yeasts, like *S. boulardii*, can interact with antifungal medications, rendering their combination inadvisable.

The use of probiotics requires careful consideration in individuals with compromised immune systems or those employing chemotherapeutic or immunosuppressive drugs, as there is a potential for inducing pathogenic colonization and infections. However, contradictory reports exist within these vulnerable populations, and ongoing clinical trials seek to assess the safety of specific probiotic strains in cancer patients undergoing anticancer treatments. A number of factors are associated with the potential toxic nature of probiotics, and a universal standard cannot be applied. As the equilibrium of gut microbial populations can be easily disrupted, the safety assessment of probiotics remains subject to further exploration.

11.3.5 Regulatory Issues for Impacts of Plant-Based Meats on Biodiversity

Lab-grown meat, also known as cell-cultured meat, is a product of cellular agriculture where adult muscle stem cells are cultured in a collagen matrix. This technology has been in development for over a century but gained attention in 2013 when the first lab-grown hamburger was produced. The cost of production has significantly reduced over the years, but commercial production is still in its early stages. Lab-grown meat is produced by cultivating muscle tissue strips from stem cells, with fat cells co-cultured to enhance flavor, texture, and tenderness. The resulting tissue can be processed and packaged. Unlike traditional meat production, lab-grown meat only requires the development of muscle tissue, without the need for other biological structures. It has primarily been used in scientific and medical applications. The composition of the cell culture medium, which provides nutrients for cell growth, can be derived from animal materials or non-animal sources such as cyanobacteria and mushroom extract. Large-scale production of lab-grown meat faces challenges, including the development of suitable cells and growth media, the avoidance of animal components to prevent communicable diseases, and the creation of edible materials for thicker and more continuous meat pieces like steaks. Finding efficient methods for non-animal culture media production is considered a crucial step for the wider acceptance of lab-grown meat.

Lab-grown meat has the potential to be an excellent functional food due to its ability to be modified to meet specific nutritional needs. It can be enriched with essential amino acids, fats, vitamins, minerals, and bioactive compounds, surpassing the nutritional content of natural meat. Consumers are increasingly open to trying modified foods with functional and nutritional benefits. The strict hygiene conditions and sterile systems involved in lab-grown meat production contribute to improved safety by minimizing the risk of pathogens and diseases associated with conventional meat production. The use of pesticides, fungicides, growth factors, and antimicrobials could be reduced as lab-grown meat becomes more widely consumed. As production scales up and costs decrease, lab-grown meat could become more affordable than traditional meat, potentially addressing nutritional deficiencies in developing countries and supporting the growth and development of children.

The development and expansion of lab-grown meat production and consumption have the potential for various environmental benefits. Life cycle assessments suggest that replacing conventionally produced meat with cultured meat could help reduce greenhouse gas emissions, land use, water consumption, and energy consumption. Depending on the type of meat and the production methods, cultured meat can result in significant reductions in environmental impacts compared to conventional farming. Cultured meat production can also contribute to wildlife conservation by reducing the need to convert natural habitats into farmland and providing an alternative source of meat for endangered species. Additionally, the transportation of lab-grown meat is likely to have lower energy consumption compared to conventional meat due to reduced transportation of animals and non-edible parts. The reduction in slaughterhouse waste and associated environmental consequences is another potential benefit of lab-grown meat production.

Consumer perceptions and ethical concerns are significant factors that can impact the acceptance and commercial success of lab-grown meat. It is important for lab-grown meat to closely resemble natural meat in terms of sensory characteristics in order to be accepted by consumers. The complexity of tissue production and the ability to replicate muscle biology will determine the feasibility of producing whole muscle pieces. One of the potential benefits of lab-grown meat is the reduction in the population of animals raised for meat production, which could lead to improvements in animal welfare and the obsolescence of intensive livestock farming practices. However, there are ethical considerations regarding the collection of stem cells from animals and the use of animal-derived serum for cell cultures. The acceptance of lab-grown meat may also face obstacles from vegetarians, religious groups, and individuals who believe that consuming animal products is unnecessary for human health. Consumer knowledge about lab-grown meat is currently limited, but surveys suggest that while many consumers may be hesitant to try it, only a small minority outright reject the idea. Providing information about the viable benefits of lab-grown meat compared to conventional meat production is crucial for increasing consumer confidence and acceptance. The socioeconomic impact of lab-grown meat is another challenge, as it may disrupt the traditional livestock industry and have implications for stakeholders in terms of production, profits, and global meat consumption. These complex questions need to be addressed by governments and society as lab-grown meat technology advances.

11.4 REGULATION UNDER THE COORDINATED FRAMEWORK

11.4.1 Regulatory Framework for GMO

The USDA is not the sole US agency in charge of GMO regulation. The Environmental Protection Agency (EPA), in addition to the Food and Drug Administration (FDA), is a component of the US regulatory system. One to three agencies may need to get permission, depending on the GMO in question. For instance, the EPA monitors pesticide safety and regulates GMOs created to withstand particular pesticides. Most likely, both the USDA and the FDA will control goods intended for human consumption. The FDA and USDA may have authority over the insertion of a gene that confers enhanced drought tolerance, but the EPA does not. Regulations and policies are frequently technical in the United States and generally based on the product's intended purpose and engineered properties. Other considerations, such as ethical, cultural, and socioeconomic issues, barely or never affect how decisions are made. The FDA also concluded that there are no appreciable differences between cereals produced traditionally and those that have been genetically engineered. Therefore, the organization lacks the authority to require further labeling (FDA, 2001).

In order to conduct scientific evaluations on the safety and environmental impact of GMOs, the European Union (EU) relies on the European Food Safety Authority (EFSA). The EFSA normally works with the scientific agencies of the various member states to provide an expert opinion. The final decision-maker is the European Commission. It is difficult to have GM products approved in several of the member countries because of strong public opposition. As a result, it is doubtful that even a crop approved by the European Commission will be cultivated in all of the member nations. The EU demands accurate labeling of GM foods. The variations in GMO approvals are reflected in the varying rules and cultural viewpoints that govern various nations. Different systems are in place to control and monitor genetically modified (GM) crops and the foods made from them. Different approaches are used by the government to answer the general public and the scientific community. Regulations take into account various cultural contexts, environmental conditions, and frequently clout interests of numerous parties, like farmers, agricultural companies, environmentalists, or governmental organizations (Araki and Ishii, 2015).

The development of technology that allows gene manipulation without modifying the DNA sequence, as well as transgene-free GE plants, has increased the complexity of regulatory rules. In theory, employing CRISPR to knock off a gene rather than CRISPR-Cas9 should make it easier to circumvent regulatory barriers. Because their genomes were not altered, these plants should not be termed GMOs. The downside of such approaches in this context is that they often need the presence

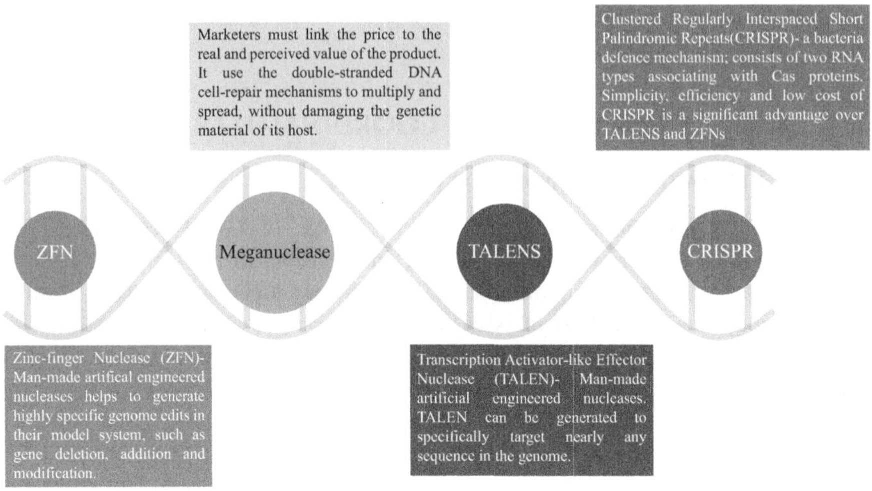

FIGURE 11.3 Various tools for genome editing

and consistent expression of CRISPR components. As a result of the technique, it is likely that they will be classified as GMOs, particularly because they contain exogenous DNA. Epigenome approaches might be an exception to this rule because they might theoretically be expressed briefly and then rejected in offspring. People advocate for a move from process-based to product-based regulation. Instead of assessing the procedure, which is often a lengthy and costly process, risks should be assessed based on the finished product, after off-target mutations have been investigated. Furthermore, USDA judgments show that certain mutations caused by ZFNs, meganucleases, TALENs, and CRISPR that involve the NHEJ pathway fall outside of its regulatory jurisdiction (as shown in Figure 11.3). This implies that CRISPR technology can be used to generate species that are not genetically modified (Wolt et al., 2016).

Here are some key ethical dilemmas associated with genome editing:

1. **Informed Consent and Autonomy**: One significant concern is ensuring that individuals undergoing genome editing fully understand the risks, benefits, and potential long-term consequences of the procedure.
2. **Unintended Consequences**: Genome editing might inadvertently introduce unintended genetic changes that could lead to unforeseen health issuesor other unintended consequences in the individual or future generations.
3. **Inequality and Access**: There is a risk that genome editing technologies could exacerbate existing social and economic inequalities. If these technologies become available only to the wealthy, it could lead to genetic privilege and inequality among different segments of society.

4. **Germline Editing and Future Generations**: Editing the germline (sperm, egg, or embryo) has the potential to impact not only the edited individual but also future generations. Deciding what genetic changes are appropriate to make and how they should be made involves making choices.
5. **Long-Term Safety**: The long-term effects of genome editing are largely unknown, and unforeseen consequences might arise. Implementing these technologies without a full understanding of their potential risks could lead to harm.
6. **Environmental Impact**: Genes are not isolated entities; they interact with the environment. Editing genes could have unforeseen effects on ecosystems and environments, especially if edited organisms are released into the wild.
7. **Ownership and Patents**: The development and commercialization of genome editing technologies raise questions about who owns the rights to edited genetic material and whether these technologies should be patented.
8. **Regulation and Oversight**: Striking a balance between allowing scientific progress and ensuring proper oversight and regulation to prevent reckless or harmful uses of genome editing technologies is a significant challenge.

Addressing these ethical dilemmas requires open dialogue among scientists, ethicists, policymakers, and the public to establish guidelines and frameworks that promote responsible and ethical use of genome editing technologies.

11.4.2 Regulatory Issues for Impacts of Fermented Foods

In Europe, fermented foods, including microorganisms used in food production, are considered ingredients and must adhere to food safety requirements. Food cultures utilized in fermentation generally don't face premarket regulations unless they're classified as novel foods. Novel foods go through a premarket authorization process, which involves risk assessment by the European Food Safety Authority (EFSA). Traditional fermented foods from non-EU countries can seek authorization through alternative routes, provided evidence of safe consumption for at least 25 years outside the EU is presented. For health claims associated with fermented foods, including those containing probiotics, prior approval from the EU Commission is necessary. A solitary health claim related to improved lactose digestion in yogurt has received authorization, while other claims have been declined due to insufficient proof of health benefits. Different interpretations of the term "probiotic" exist among EU member states, causing a lack of harmonization in the market. Some countries permit the use of "probiotic" as a general health claim, while others are considering its future usage. These regulatory variations have led to a fragmented marketplace for probiotic products and functional fermented foods in the EU.

In the United States, all foods are regulated by federal and state laws. The US Food and Drug Administration (FDA) oversees acidified foods, including yogurt, but generally doesn't regulate other fermented foods due to their history of safe consumption. Yogurt is defined by the FDA as a dairy product produced through milk fermentation with specific bacteria, allowing for the addition of probiotics and lactic acid bacteria. Minimum standards for cultured milks are also established, encompassing milk fat, milk solids-nonfat content, and titratable acidity. For acidified foods, including some fermented products, adherence to federal laws regarding production, storage, packaging, and quality control is required. While the FDA provides guidance on acidified foods, certain aspects have been revised or addressed in other documents. Furthermore, the United States employs Cottage Foods legislation across states, governing home-based food enterprises and permitting specific product sales at farmers' markets. Food Freedom Laws in some states allow the sale of homemade food, including fermented foods, without licensing or inspection requirements. Targeted regulations, like those for health claims such as "gluten-free," have been developed. Trade associations, such as Kombucha Brewers International, have influenced fermented food regulations by establishing codes of practice and certification programs for commercial products like kombucha. The National Environmental Health Association has issued guidance on kombucha fermentation, including practices for ensuring food safety.

Canada allows live bacterial cultures, including probiotics, to be added to food, and they are generally classified as fermented foods. However, if a food is claimed to have therapeutic use, it may be classified as a natural health product (NHP). Health claims for probiotics must be supported by scientific evidence and not mislead the public. The Safe Food for Canadians Regulations cover labeling requirements for certain fermented foods, but there are no federal regulations for fermented milk products apart from cheese. Canada's approach to regulating fermented foods is considered lenient to foster innovation. Safety guidelines for kombucha preparation are available through the British Columbia Centre for Disease Control.

The Australia New Zealand Food Standards Code (ANZFSC) governs the regulation of fermented foods (FFs) in Australia and New Zealand. It includes specific standards for dairy products like yogurt and cheese standards, including requirements for ripened or unripened milk products. The ANZFSC also addresses novel foods, which are non-traditional foods that require public health and safety considerations. Health claims for live microbes in Australia and New Zealand are limited to improved lactose digestion for yogurt cultures containing specific bacterial strains. Kombucha regulation in Australia is an example of the challenges and variations in FF legislation, with different states implementing different regulations based on alcohol content.

In Argentina, Argentina's Argentine Food Code (Codigo Alimentario Argentino or CAA) regulates fermented foods, including locally produced and imported foods. The CAA is enforced by Servicio Nacional de Sanidad y Calidad Agroalimentaria (National Service for Agri-Food Health and Quality, Argentina) (SENASA), Instituto Nacional de Vitivinicultura (National Institute of Viticulture, Argentina) (INV), and Administración Nacional de Medicamentos, Alimentos y Tecnología Médica (National Administration of Drugs, Foods and Medical Technology, Argentina) (ANMAT). It covers all fermented dairy products with specifications similar to Codex Alimentarius standards. Microbial loads, including a minimum of 107 CFU/g of lactic acid bacteria, are specified for fermented milks. Cheeses are regulated separately, with detailed specifications for labeling, classification, microbiological requirements, and nutrient content. In 2020, kombucha was added to the CAA, a carbonated non-alcoholic drink obtained through the fermentation of *Camellia sinensis* infusion and sugars. However, Argentina has no legal framework for foods with probiotics and functional dietary foods.

Brazil's regulatory framework for fermented foods, functional foods, and probiotics is governed by the Ministry of Agriculture, Livestock and Supply (MAPA) and the Agência Nacional de Vigilância Sanitária (ANVISA) under the Ministry of Health. These institutions enforce regulations related to food safety, registrations, imports, and good manufacturing practices. Microbiological standards for dairy products are protected through resolutions referring to Codex and Standard Methods for the examination of dairy products. Specific regulations exist for labeling, additives, and packaging of fermented dairy products. Brazil has detailed regulations for its diverse range of artisanal cheeses, including the Aseptic, Ready-To-Eat (ARTE) seal for interstate transport and sale of inspected products. Kombucha production has been regulated through the Identity and Quality Standard published by MAPA. Brazil has also developed regulations for novel foods and functional foods, requiring premarket clearance for probiotics and functional foods.

The Department of Agriculture, Forestry and Fisheries (DAFF), the National Department of Health, and the Department of Trade and Industry are the three government departments responsible for overseeing South Africa's food laws. As a result, laws governing many food groups, such as fermented foods like yogurt, overlap. The Foods, Cosmetics, and Disinfectants Act, as modified, forbids the term "probiotic" from appearing on yogurt labels but permits functional claims about the composition of the microbial community where the yogurt culture contains a particular strain. According to the Agricultural Products Standards Act, yogurt culture must include certain strains or comparable cultures that are suited for yogurt production. Other fermented milk products like kefir and maas are also covered by these laws, which outline starter cultures, viable lactic acid bacteria, and yeasts. To provide a consistent regulatory framework for yogurt in South Africa, these regulations must be harmonized.

The Russian Federation has legislation regulating fermented foods, including requirements for fermentations, enzyme preparations, and facilities for probiotic microorganism production. These regulations are outlined in the Federal Law on hygienic requirements for food manufacturing and trafficking. The Technical Regulations for Milk and Milk Products outline safety and compliance requirements for milk and dairy products, based on physicochemical, organoleptic, and microbiological criteria.

Japan's Food Sanitation Act regulates food safety and standards, including food additives and milk products. The Food for Specified Health Uses (FOSHU) guidelines were established in 1991 to regulate functional foods, including probiotic-containing fermented foods. FOSHU approval allows products to bear a special label indicating their health benefits. Japan does not require clinical trials for FOSHU approval and accepts scientific substantiation of active ingredients. The "Foods with Functional Claims" or New Functional Products (NFP) Regulation, introduced in 2015, provides a more flexible spectrum of acceptable health claims. The Japanese Ministry of Agriculture, Forestry and Fisheries is collaborating with the Codex to develop Asian Regional Standards for Soybean Products fermented using *Bacillus*, including foods like natto, cheonggukjang, douchi, thuanaosa, and kinema.

In China, the Food Safety Law is the main regulatory legislation for food, but it has no specific provisions for fermented foods. The New Food Raw Materials Regulation evaluates new food ingredients, including microbes used in fermentation, and has approved several microbial species for use in fermented food (FFs) products. China also has National Standards for certain FFs, such as fermented milk, soy sauce, fermented vinegars, and fermented alcoholic beverages. These standards outline technical requirements for various aspects, including organoleptic qualities, contaminants, mycotoxins, nutritional metrics, and microbial strains and loads. However, there are no specific standards for other types of fermented dairy products beyond fermented milk and yogurt. In light of a food poisoning incident caused by homemade fermented corn noodles, Chinese authorities have emphasized the need for standards and regulations to ensure the safety of FFs, given their traditional importance in Asian diets.

South Korea regulates fermented foods (FFs) through various laws and ministries. The Ministry of Food & Drug Safety (MFDS) sets food regulations, including standards for FFs. The Food Sanitation Act and Food Code outline specifications for FFs, including traditional Korean FFs like kimchi and fermented seafood. These specifications cover production, processing, packaging, storage, labeling, distribution, chemical and microbiological contaminant levels, and testing methods. Functional health foods, including FFs with probiotics, are regulated under the Functional Health Foods Act (FHFA) and the Health Functional Food Code (HFFC). Probiotics are recognized as functional ingredients, and specific species are listed for use in functional foods. The only approved health claim for probiotic functional ingredients is related to gut health, and the recommended intake and minimum viable count of live bacteria are specified.

The Food Safety and Standards Act in India regulates fermented foods (FFs), including yogurt and cheeses. The 2011 Regulations provide compositional standards for fermented milks, allowing for the addition of other lactic acid bacteria if declared on the label. The 2016 Regulations address FFs with health claims, including "food with added probiotic ingredients" and "novel foods." Probiotic labeling must be associated with a health claim supported by scientific evidence evaluated by the Food Safety and Standards Authority of India (FSSAI). Minimum viable microbe count requirements are specified, and specific labeling requirements are outlined. The regulation also provides a list of approved probiotic organisms, which are periodically updated. Novel foods are subject to production through innovative technologies; the regulation ensures that the food can't be imported without FSSAI approval.

11.4.3 Regulatory Issues for Impacts of Functional Foods

Europe established the definition of functional foods and established a regulatory commission, FUFOSE, in 1995. The European market for probiotics has experienced rapid growth, particularly in dairy products such as yogurts and fermented milks. European laws require that even microbial cultures in food meet legal requirements, and the Qualified Presumption of Safety (QPS) was introduced to ensure premarket safety assessment of foods and food supplements. However, probiotics do not have a specific legal definition and are classified under the category of functional foods. The regulatory framework for probiotics in Europe is not harmonized. If probiotics are marketed as food supplements, they are regulated under the Food Products Directive and Regulation (Regulation 178/2002/EC; Directive 2000/13/EU). If marketed as traditional herbal products, they fall under the Herbal Medicinal Products Directive (2004/24/EC). For probiotic products marketed as herbal medicinal products, it is recommended to obtain a drug registration, or they will be categorized as food supplements. Registered drugs are governed by the Drug Law (65/65/EC, amended).

In the USA, probiotics are regulated based on their intended usage, and the main regulatory bodies involved are the Dietary Supplement Health and Education Act (DSHEA) and the Food and Drug Administration (FDA). Probiotics used as dietary supplements are considered and fall under the regulation of DSHEA. Manufacturers of dietary supplements containing new ingredients are required to notify the FDA. If probiotics are intended for therapeutic purposes, they must undergo safety and efficacy testing before being marketed as probiotic drugs, which are regulated by the FDA. Biological products, including probiotics, may require approval through a Biological License Application (BLA). The determination of a biological product can be complex, as it encompasses various substances used for disease prevention, treatment, or cure. Overall, probiotics in the USA are regulated differently based on their classification as dietary supplements, drugs, or biological products, and they must meet the respective regulatory requirements established by the FDA and DSHEA.

In Canada, the regulatory framework for the functional food industry, including probiotics, is considered complex and unsatisfactory. Probiotics fall under the category of natural health products in Canada, and their approval and packaging are primarily governed by the Canadian Food and Drugs Act, specifically under the Natural Health Products Regulations.

Australia and New Zealand have a joint framework agency called Food Standards Australia and New Zealand (FSANZ). Functional foods in these countries are defined as foods that serve physiological roles beyond basic nutrient requirements, going beyond traditional foods in terms of their intended functions. The Therapeutic Goods Administration (TGA) in Australia has regulatory guidelines for complementary medicines. These products can be classified as therapeutic goods or medicines, and their classification determines whether they fall under the TGA or Food Standards Australia and New Zealand (FSANZ). Complementary medicines, which contain active ingredients like vitamins and herbs, are regulated as medicines by the TGA. They can be listed or registered, with listed medicines requiring certification by the sponsor and registered medicines undergoing a strict assessment process. The TGA aligns its regulatory approach with international counterparts such as the European Food Safety Authority (EFSA) and the Food and Drug Administration (FDA). Australia and New Zealand have established FSANZ to develop food standards, including safety, fortification, and processing guidelines. Most complementary medicines do not undergo significant scrutiny unless they are registered with the Australian Register for Therapeutic Goods (ARTG).

In Brazil, probiotics are considered functional foods, and they are regulated by the National Health Surveillance Agency of Brazil (ANVISA). Safety and efficacy demonstrations are also required for food products in Brazil, and they must be registered and approved by ANVISA.

Japan has been a global market leader in the probiotics industry, with a regulatory system in place since 1991. Foods for Specific Health Uses (FOSHU) claims are not allowed on the labeling of probiotic food products unless they obtain special permission from the Ministry of Health and Welfare (MHLW). FOSHU has different categories for food claims based on supporting evidence of health benefits. The government has designated specific subcategories for FOSHU health claims, including gastrointestinal health, cholesterol moderation, hypertension moderation, and more. New claims and combinations of claims are regularly approved by the MHLW. The registration process for FOSHU involves providing detailed information on safety, efficacy, processing, formulation data, and proposed labels, along with product samples, which must be submitted to a Japanese scientific journal.

China has a well-developed market for functional foods, regulated by the State Food and Drug Administration (SFDA), which oversees health foods, including functional foods and nutraceuticals. Functional foods in China are defined as foods with special health functions or the ability to supply vitamins or minerals. Safety and efficacy demonstration is required for these products, and they must be registered and approved by the SFDA.

Malaysia lacks an official definition for functional food products, which are regulated under food regulations. The Food Safety and Quality Division, Drug Control Authority, National Pharmaceutical Control Bureau, and Committee for the Classification of Food-Drug Interface Products are responsible for implementing and enforcing these regulations. Malaysian regulations are developed locally using Codex, which provides complex guidelines for functional foods.

The regulation of nutraceuticals in India is still evolving, and there may be conflicting or confusing content due to the recent implementation of the Food Safety and Standards Act (FSSA). To enter the Indian nutraceutical market, it is important to focus on areas such as product evaluation, actual product analysis, procuring licenses, and developing India-specific health and label claims. Product evaluation involves examining each active ingredient and additive to determine permissibility, standards, and dosage based on therapeutic, prophylactic, or recommended daily allowance (RDA) guidelines for Indians. Manufacturers may face uncertainty regarding the classification of their products as functional food, dietary supplements,

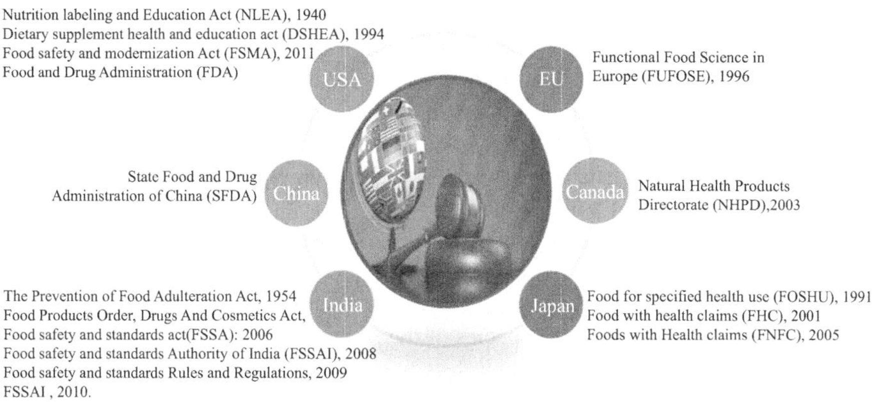

FIGURE 11.4 Assortment of regulations for functional foods across the world

or drugs under different acts and rules. The regulatory enforcement structure and procedures outlined in the Food Safety and Standards Rules 2011 guide the product analysis process, involving various steps from document authentication to sample collection, dispatch, and analysis. In India, probiotics are consumed both as food and pharmaceutical products, leading to different regulatory frameworks based on their intended use. Currently, probiotics are categorized as functional foods and regulated by food laws rather than pharmaceutical regulations. The FSSA was enacted to regulate various food products, including nutraceuticals and dietary supplements, but the specific categorization of food types such as nutraceuticals and biotherapeutic agents is still unclear. The Prevention of Food Adulteration Act (PFA) Rules provide minimum standards for the quality and content of food products, including labeling and packaging requirements. In the absence of specific regulatory guidelines, the Indian Council of Medical Research (ICMR) and the Department of Biotechnology (DBT) formed a Task Force to develop regulatory guidelines for evaluating the safety and efficacy of functional foods and assessing health claims, as well as labeling requirements to protect consumers' interests.

The regulations for functional foods across the world were summarized in Figure 11.4 with details of establishment.

11.4.4 Regulatory Issues for Impacts of Probiotics on Human Health

11.4.4.1 Assessment of Probiotic Safety

The FAO and WHO offer guidelines for probiotic safety assessment in food. Probiotic safety can be evaluated through various methods. Intrinsic characteristics of strains and species, like bile salt deconjugation or mucin degradation, can be studied in vitro for potential risks. In vivo research involves assessing probiotic survival and effects using techniques like fecal collection, intestinal intubation, or mucosal biopsies. The guidelines from FAO and WHO propose examining strain specificity, including tests for antibiotic resistance, metabolic activities, and unwanted outcomes such as bile salt deconjugation or D-lactate production. Monitoring consumer well-being through surveillance and epidemiological studies is advised, along with evaluating toxin production and hemolytic activity post-consumption. Animal models, particularly those that are immunocompromised, can be utilized to evaluate probiotic traits. Safety assessment should also encompass tests for properties like anti-mutagenicity, anticarcinogenicity, and non-pathogenicity.

The Joint Food and Agriculture Organization of the United Nations/World Health Organization Expert Consultation developed guidelines in 2001 to standardize the evaluation of probiotics in food and support health claims. These guidelines recommend identifying the genus and species of the probiotic strain using phenotypic and genotypic tests, conducting in vitro testing to understand the probiotic's mechanism of action, and conducting human trials to substantiate the clinical health benefits.

Safety assessment should include evaluating antimicrobial drug resistance patterns, metabolic activities, side effects observed in clinical trials and post-marketing, toxin production, and hemolytic potential. Animal studies should also confirm the lack of infectivity. Specific health claims can be made on probiotic food labels when sufficient scientific evidence is available, and manufacturers should ensure independent third-party review of the evidence. However, only a few manufacturers have conducted rigorous studies to demonstrate the efficacy and safety of their probiotic products. Until strict regulations are established, clinicians should consider the available evidence when assessing the therapeutic potential of probiotic products. Manufacturers, although not required by law, should provide guidance to consumers and clinicians regarding the safety assessments conducted on their products.

11.4.4.2 Current Regulation of Probiotics

Regulatory aspects of probiotics involve evaluating their efficacy, safety, quality control during manufacturing, and the regulation of health claims. The classification and regulations vary depending on the country and the claims made by the product.

In the European Union, probiotics and food supplements are regulated under the Food Products Directive and Regulation. Health claims for probiotics must be authorized by the European Food Safety Authority (EFSA). The EFSA has established a list of microbial cultures with a Qualified Presumption of Safety. However, there is little regulation of the manufacturing process and post-marketing follow-up.

In the United States, most probiotic products are classified as functional foods or dietary supplements. Dietary supplements are required to comply with Good Manufacturing Practice guidelines, but there is no requirement for testing quality or efficacy. Probiotic products can make structural or functional claims but not disease-specific claims. Medical foods are a category of probiotics intended for the dietary management of specific diseases and are not subject to drug regulations.

Both regions have regulatory gaps that do not adequately address the complexities of probiotic products, such as their dynamic nature, variation among species and strains, and interactions in multispecies or multistrain products. This can lead to issues of quality, safety, and claim validity, particularly in medical contexts among vulnerable populations.

The quality of probiotic products is influenced by the manufacturing process, which affects both safety and efficacy. Variations in manufacturing can impact the viability of live bacteria and the presence of dead bacteria in the product. Current labeling regulations do not provide complete information on the total number of bacteria consumed, including dead bacteria. Manufacturing processes also influence the biochemical and immunological profile of probiotic products, leading to variations in their therapeutic activities. In summary, there is a need for improved regulation and quality control of probiotics, considering their specific characteristics and potential impact on consumer health.

11.4.5 Regulatory Issues for Impacts of Plant-Based Meats on Biodiversity

Start-up enterprises both in the United States and various other countries are actively striving to bring lab-grown meat to the retail market, with a potential timeline of achieving this milestone by 2021. However, prior to the realization of this prospect, the establishment of a comprehensive legislative and regulatory framework is imperative to oversee this pioneering food production. In the context of the European Union (EU), lab-grown meat is classified as a novel food category and is subject to the provisions of Regulation EU 2015/2283, which specifically governs novel foods. To gain approval, producers are required to submit meticulous applications encompassing pertinent details on production methodologies, product safety assessments, ethical considerations, labeling strategies, and more. Following a rigorous scientific evaluation by the European Food Safety Authority (EFSA) to ensure safety, the Commission may grant endorsement through a formal regulation.

At present, no applications seeking the approval of lab-grown meat have been tendered within the EU, consequently prohibiting the introduction of such products into the market. Upon receiving approval, lab-grown meat will become subject to labeling prerequisites dictated by the Novel Food Regulation (EU) 2011/1169, which pertains to consumer information regarding food items. Nevertheless, challenges remain unresolved concerning the nomenclature for lab-grown meat, as it is imperative to deliver transparent and accurate details about its production methodology to consumers. The selection of an appropriate and fitting name represents a delicate matter that necessitates future deliberation.

In the United States, the current regulatory landscape lacks specificity when it comes to lab-grown meat, despite being home to numerous cultured meat companies. The prevailing definition of a "meat product" as outlined in the Federal Meat Inspection Act does not align with lab-grown meat, as it mandates that meat and its products originate from animal carcass.

11.5 NEW DIRECTIONS FOR FEDERAL REGULATION OF BIOENGINEERED FOOD

The process of continually updating federal regulations for foods is driven by a commitment to staying aligned with the ever-evolving landscape of scientific knowledge, emerging challenges, and the dynamic preferences of consumers. This ongoing refinement reflects the authorities' dedication to ensuring that the food supply remains secure, wholesome, and trustworthy. As new scientific discoveries emerge, federal regulatory agencies like the FDA and USDA assess their potential implications on food safety, nutritional standards, and overall consumer health. For instance, advancements in food testing methods, microbiology, and toxicology allow regulators to better identify potential hazards and develop more effective strategies to mitigate risks. These insights lead to the revision of regulations to reflect the most current understanding of food safety and quality.

The food industry is susceptible to novel challenges, ranging from the emergence of new foodborne pathogens to the adoption of innovative production techniques. Regulatory bodies are equipped to swiftly respond to these emerging issues by re-evaluating existing regulations or devising new ones. For example, the rise of plant-based meats and cell-based meats has prompted discussions about how best to categorize and regulate these products, resulting in potential adjustments to labeling and safety requirements. Consumer preferences and demands are dynamic and continuously evolving. Regulations are shaped to meet these changing needs, particularly in areas such as allergen labeling, nutritional information, and dietary trends. Regulatory agencies aim to provide consumers with accurate information to make informed choices about the foods they consume, as well as to ensure that foods catering to specific dietary needs are properly labeled and safe.

The overarching objective of these regulations is to guarantee a safe and transparent food supply for consumers. By setting and enforcing standards for food safety, labeling accuracy, and quality control, regulatory agencies establish a foundation of trust between consumers and the food industry. Consumers can have confidence that the foods they purchase meet established safety and quality standards. While the primary focus is on safeguarding consumer well-being, these regulations also play a crucial role in fostering innovation within the food industry. By setting clear guidelines and expectations, regulations enable food manufacturers and producers to explore new technologies, ingredients, and production methods while ensuring that innovation does not compromise safety or quality.

In essence, the continuous evolution of federal regulations for foods is a dynamic process that reflects a harmonious blend of scientific advancement, responsiveness to emerging challenges, and a commitment to meeting the ever-changing expectations of consumers. This multifaceted approach underscores the importance of maintaining a safe and resilient food supply while enabling the food industry to flourish and embrace innovation. As advancements in biotechnology and genetic engineering continue to shape our food system, it is crucial to update and adapt regulatory frameworks to ensure the safety, transparency, and responsible oversight of these products. The following are key elements for the new directions in federal regulation:

1. Science-Based Risk Assessment: Strengthen the risk assessment process by integrating the latest scientific research and methodologies. This approach should consider the specific traits and characteristics of bioengineered food products and assess their potential risks to human health and the environment. The risk assessment process should be transparent and based on sound scientific principles.
2. Proportional Regulation: Implement a risk-based regulatory approach that distinguishes between different types of bioengineered food products based on their potential risks. Tailor regulatory requirements to the specific traits, characteristics, and intended uses of these products. This approach will ensure that regulations are proportional, effective, and avoid unnecessary burdens on low-risk bioengineered food products.
3. Enhanced Transparency and Labeling: Improve transparency and labeling requirements for bioengineered food products. Enhance the clarity, consistency, and accessibility of labeling information to provide consumers with clear and accurate information about the presence of bioengineered ingredients. Explore the use of digital technologies, such as QR codes or smartphone apps, to provide additional product information beyond what can be included on packaging.
4. Post-Market Surveillance and Monitoring: Establish a robust post-market surveillance and monitoring system to continuously assess the safety and performance of bioengineered food products. This system should include active surveillance, data collection, and risk assessment to identify and address any potential long-term health or environmental impacts.
5. Public Engagement and Education: Foster public engagement and education initiatives to enhance consumer understanding of bioengineered food and the regulatory processes in place. Encourage open dialogue and provide opportunities for public input and stakeholder engagement to address concerns, build trust, and ensure that regulatory decisions reflect broader societal values.
6. International Collaboration: Collaborate with international regulatory bodies to harmonize standards and regulations for bioengineered food products. Promote information sharing, joint research efforts, and the development of globally recognized guidelines to facilitate trade, ensure consistency in safety assessments, and avoid unnecessary duplication of efforts.

By embracing these new directions, we can foster a regulatory environment that protects public health, encourages innovation, and supports the responsible development and use of bioengineered food products.

11.6 CONCLUSION

In conclusion, the safety considerations for probiotics are crucial to ensure consumer safety, including factors such as strain origin, absence of pathogenicity and virulence factors, lack of disease-promoting activities, and absence of transferable antibiotic resistance genes. The regulatory framework for probiotics varies across countries and regions, highlighting the need for harmonization and improved standards. Similarly, the regulation of genetically modified organisms (GMOs) remains a complex and controversial topic, with varying perspectives and regulatory approaches worldwide. Ongoing research, transparent communication, and robust risk assessment are important to address concerns and ensure the safety of GMOs in the context of food production. The regulatory frameworks for functional foods and nutraceuticals are rapidly evolving, but challenges remain in terms of clarity, consistency, and harmonization. Efforts are needed to improve regulations, promote consumer safety, and foster consumer and investor confidence. The nutraceutical industry is still in its early stages, and the adoption of technologies like blockchain can enhance traceability and safety in this sector. Lab-grown meat holds promise as an alternative to conventional meat production, addressing sustainability and dietary needs. However, it is still in the early stages and faces challenges in production, regulation, consumer acceptance, and ethical considerations. Stakeholder involvement and targeted marketing campaigns can help build acceptance and address these challenges. Further research and exploration are necessary to fully develop the market for lab-grown meat.

Overall, regulatory frameworks, safety considerations, and ongoing research are critical in ensuring the safety, efficacy, and acceptance of probiotics, GMOs, functional foods, nutraceuticals, and lab-grown meat. Continued efforts, collaboration, and informed decision-making by stakeholders are essential to address these complex and evolving areas in the food industry.

REFERENCES

Araki M, Ishii T. (2015). Towards social acceptance of plant breeding by genome editing. *Trends in Plant Science*, 20(3), 145–149.

Cai, Y., Chen, L., Liu, X., Sun, S., Wu, C., Jiang, B., et al. (2015). CRISPR/Cas9-mediated genome editing in soybean hairy roots. *PLoS One*, *10*, e0136064. doi: 10.1371/journal.pone.0136064

FDA (U.S. Department of Health and Human Services–Food and Drug Administration). (2001). Premarket notice concerning bio-engineered foods; Proposed rule. *Federal Register*, *66*, 4706–4738.

Fogg-Johnson, N., & Meroli, A. (2000). Nutrigenomics: The next wave in nutrition research. *Nutraceuticals World*, *3*, 86–95.

Ives, A. R., Paull, C., Hulthen, A., Downes, S., Andow, D. A., Haygood, R., &Schellhorn, N. A. (2017). Spatio-temporal variation in landscape composition may speed resistance evolution of pests to Bt crops. *PloS one*, *12*(1), e0169167. doi: 10.1371/journal.pone.01691675

Jaganathan, D., Ramasamy, K., Sellamuthu, G., Jayabalan, S., Venkataraman, G. (2018). CRISPR for crop improvement: An update review. *Frontiers in Plant Science*, *9*, 985. doi: 10.3389/fpls.2018.00985

Kumar, K., Gambhir, G., Dass, A., Tripathi, A., Singh, A., Jha, A., Yadava, P., Choudhary, M., & Rakshit, S. (2020). Genetically modified crops: Current status and future prospects. *Planta*, *251*(4), 91. doi: 10.1007/s00425-020-03372-8

Mesnage, R., Clair, E., Gress, S., Then, C., Székács, A., & Séralini, G. E. (2013). Cytotoxicity on human cells of Cry1Ab and Cry1Ac Bt insecticidal toxins alone or with a glyphosate-based herbicide. *Journal of Applied Toxicology*, *33*(7), 695–699. doi: 10.1002/jat.2712

Venter, J. C., Adams, M. D., Myers, E. W., Li, P. W., Mural, R. J., Sutton, G. G., & Kalush, F. (2001). The sequence of the human genome. *Science*, *291*(5507), 1304–1351.

Waltz, E. (2016). Gene-edited CRISPR mushroom escapes US regulation. *Nature*, *532*(7599), 293. doi: 10.1038/nature.2016.19754

Waltz, E. (2018). With a free pass, CRISPR-edited plants reach market in record time. *Nature Biotechnology, 36*, 6–7. doi: 10.1038/nbt0118-6b

Wolt, J. D., Wang, K., Yang, B. (2016). The regulatory status of genome-edited crops. Plant Biotechnology Journal, 14(3), 510–518.

12 Utilisation of Biomass for First-Generation Biofuel Production

Prospects, Challenges and Feedstock

Megharaj Jayarajan, Sumit Kumar, Baraskar Ruchi Prabhakar, Ashraque Hussain, and Valekar Swapnil Vilas

12.1 INTRODUCTION

The fuels generated from biological material are known as biofuels. In the 20th century, biofuels emerged to prominence, which was associated with the invention of the automobile (Songstad et al. 2009). The best-known biofuel is ethanol, which is widely used in America, even though many other biofuels, such as biodiesel, are commonly used in many parts of the world, such as Europe, Asia and increasingly in Brazil. Although the biomass generated cellulosic ethanol and dimethyl ether are developing and could be categorised as the second and third-generation biofuels, ethanol and biodiesel are still regarded as first-generation biofuels. In the early 2000s, both the federal and state governments supported business expansion for biomass utilisation. The subsidies encouraged farmers to start producing bioethanol in the 1980s. Further, the National Park Service had previously used biodiesel vehicles. In 2005, the Energy Policy Act mandated that government fleets contain vehicles which did not use imported petrol, opening up greater market space for biofuels. In the 2000s, certain US states started requiring the use of biodiesel additives in normal diesel. National legislation to boost the sale of biodiesel includes the Energy Independence Act of 2007, the National Renewable Fuel Standard of 2010 and the Volumetric Ethanol Excise Tax Credit of 2005–2011 (Bracmort et al. 2009).

Sugar, starch and vegetable oil are the primary ingredients of first-generation biofuels. The fermentation of sugars, starches or cellulose by the activity of microbes and enzymes results in the production of biologically produced fuels such as ethanol,

DOI: 10.1201/9781032650616-12

propanol and butanol. Although the first-generation biofuel production methods are helpful, there is a limit to the generation of biofuels without affecting food supply and biodiversity. The first-generation biofuels require government subsidies, which are not cost-competitive with oil and other existing fossil fuels, although some of them do not significantly reduce greenhouse gas emissions. First-generation biofuels often surpass conventional fossil fuels in terms of life cycle assessment when production and transportation emissions are considered (Dahman et al. 2019). Ethanol and biodiesel are examples of first-generation biofuels, which are connected directly to biomass. Ethanol is typically created by fermenting C6 carbohydrates, primarily glucose, with conventional or genetically modified yeast strains like Saccharomyces cerevisiae. First-generation bioethanol is only in fact produced using a few different feedstocks, primarily sugarcane or maize. Brazil is one of the top countries using sugarcane as a feedstock for the manufacture of biofuels.

Compared to fossil fuels, first-generation biofuels may be able to lower greenhouse gas emissions. By absorbing carbon dioxide during their growth, the plants used to make these biofuels balance emissions produced when they are consumed. By offering a substitute for fossil fuels, biofuels can help diversify the energy mix and possibly lessen reliance on imported oil. They also support local agriculture and foster employment in the bioenergy sector; the generation of biofuels can open up economic prospects (Gabrielle 2008). Although improving biomass conversion is important, creating a sustainable and economically viable biofuel production system also requires developing feedstock. The creation of bioenergy feedstocks comprises the identification of appropriate feedstock through molecular breeding and genetic modification. It is expected that different biofuel production models will be developed in various parts of the world depending on the climate, soil characteristics, social and economic structures and resource limitations. The right feedstock selection will be an essential component for these models and ultimately for global sustainable economic growth (Yuan et al. 2011).

12.2 BIOMASS AS FIRST-GENERATION BIOFUELS

12.2.1 Biomass

Organic material that can be transformed into energy is referred to as biomass. Food crops, energy crops (e.g., switchgrass or prairie perennials), agricultural leftovers (e.g., maize stover), wood waste and by-products (both mill wastes and traditionally non-commercial wood biomass) and animal manure are common forms of biomass, as shown in Figure 12.1. The concept of biomass has grown in popularity in recent years. Sources have expanded to include algae, building debris, municipal solid trash, and yard waste, as well as food waste. Some claim that the limited use of biomass as an energy source to date is due to its lack of year-round availability as a feedstock, the fact that it is frequently found at dispersed sites, the potential cost of transportation, the lack of long-term performance data, the need for expensive technology to convert it to energy and the possibility that it may not meet quality standards to be a reliable source of fuel for electric generators. Due to its extensive

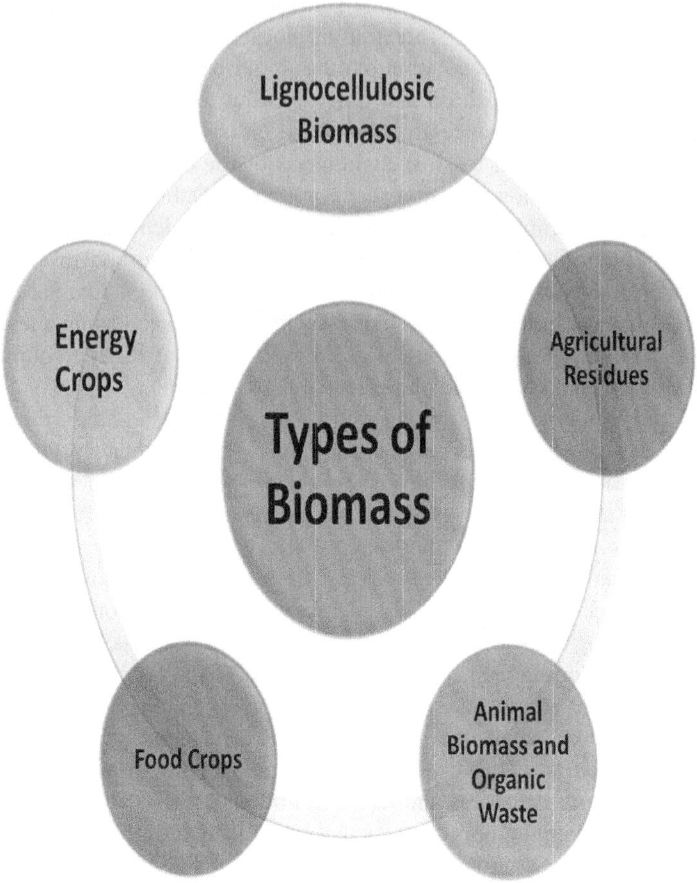

FIGURE 12.1 Cycle representing the types of biomass

availability, woody biomass has drawn special attention; however, aside from sawmill waste, it is still finding limited application in the production of energy. Wood can either be digested to produce liquid fuels or directly burned, which often produces heat, steam and electricity (a process known as combined heat and power or CHP) (Bracmort et al. 2009).

12.2.2 Types of Biomass

12.2.2.1 Lignocellulose Biomass

The plentiful and renewable resource known as lignocellulosic biomass (LB) is mostly made up of the polysaccharides cellulose and hemicellulose as well as the aromatic polymer lignin. The recalcitrance of LB to enzymatic hydrolysis brought on by the heterogeneous multi-scale structure of plant cell walls is one of the main barriers to LB valorisation. The causes of LB recalcitrance are complexly interconnected and challenging to separate. It also consists of resources like wood, agricultural waste, and special energy crops like switchgrass and miscanthus (Ashokkumar et al. 2022).

12.2.2.2 Agricultural Residues

By-products such as maize stover, rice husks, and sugarcane bagasse are examples of agricultural wastes that are left behind after crops are harvested. These leftovers can be used as biomass feedstock, giving farmers a new source of income and cutting waste. Agricultural crop wastes have been utilised for a long time for various purposes, including fuel and a supply of fibre for papermaking. However, there aren't many instances of crops other than trees being grown particularly for the aim of supplying raw materials or energy to industry (Bowyer et al. 2001).

12.2.2.3 Energy Crops

Plants like algae, jatropha, and camelina are examples of energy crops that are especially cultivated for their high energy content. They may be grown on marginal sites and are highly suited for biofuel production, reducing competition with food crops. Energy crops provide considerable potential for supplying the world's future energy needs. Additionally, agricultural fields provide energy farming, a substitute for traditional agriculture. According to studies on energy crops used in biofuel production, they are a relatively affordable and environmentally friendly method of producing sustainable energy. Today, the majority of developed nations use energy-producing mainstays including corn, sugar beets, soybeans, rapeseed, and wheat. In addition, the main sources of biodiesel and bioethanol, respectively, are oilseed plants and sugarcane. As a result, the production of food and feed is in competition with these created raw materials (Koçar et al. 2013).

12.2.2.4 Food Crops

Some food crops, such as maize and sugarcane, can also be used as biomass for energy generation, but their usage as a feedstock for biofuels raises questions about the food versus fuel conflict. Fermentation is a technique that can convert starch and sugar-rich food crops like maize and sugarcane into bioethanol. Transesterification can be used to turn oil-rich seeds from oilseeds like soybeans, canola, and palm oil into biodiesel. While providing prospects for the production of biofuels, the conversion of oilseeds into biodiesel also poses issues with land use and deforestation as shown in Figure 12.2.

12.2.2.5 Animal Biomass and Organic Waste

Animal biomass and organic waste products can be useful resources for energy production in addition to plant-based biomass. Animal biomass is any organic material that comes from animals and can be used to produce energy. This includes waste products from the meat and dairy industries, such as animal dung and corpses. Animal biomass is a viable resource for the generation of bioenergy, although it is frequently ignored. Biogas, a renewable energy source that can take the place of

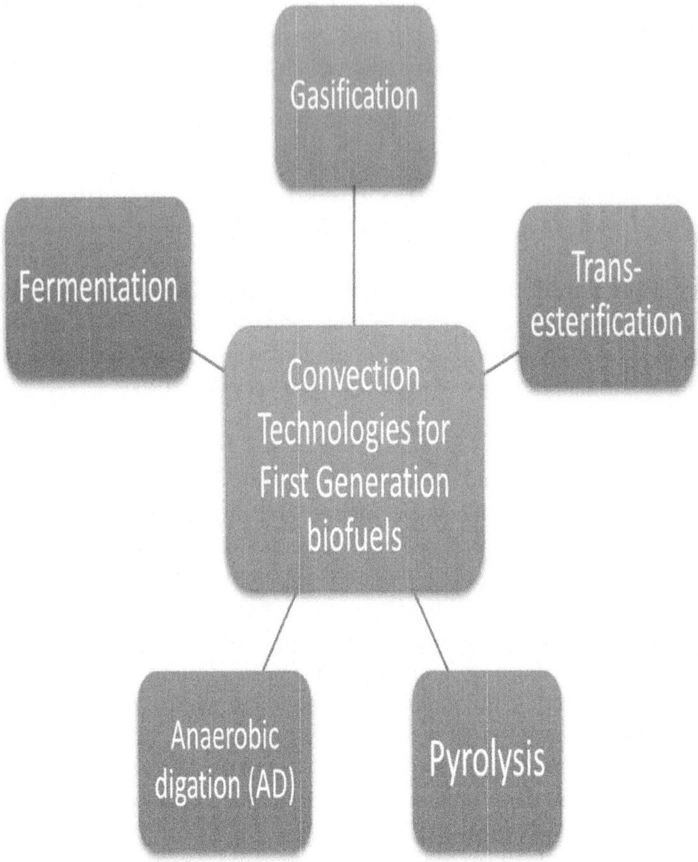

FIGURE 12.2 Flowchart representing the conversion technologies for first generation

fossil fuels, can be produced from animal biomass. Anaerobic digestion also lessens the amount of methane, a potent greenhouse gas, that is released from animal manure, enhancing environmental sustainability. Organic waste includes a variety of biodegradable substances, including leftover food, yard debris, and agricultural waste. Innovative methods can be used to harness these materials, which are frequently thrown away as rubbish and used to produce bioenergy. Composting, pyrolysis, and gasification are three technologies that can turn organic waste into biofuels, heat, and energy. Each technology has its own advantages and disadvantages.

12.2.3 Conversion Technologies for First Generation

Biofuels crops and organic materials are the sources of first-generation biofuels. These feedstocks are converted into biofuels using a variety of processes that turn biomass into usable energy sources.

12.2.3.1 Gasification

Gasification, a thermochemical process, transforms biomass into syngas, a combination of carbon monoxide, hydrogen, and methane gases. Through additional processing, this gas can be utilised to create liquid fuels, biogas, and synthetic natural gas (SNG). In this process, incomplete combustion of biomass takes place, which produces combustible gases called producer gases. Carbon monoxide (CO), methane (CH_4), and hydrogen (H_2) make up the producer gas. This gas has a calorific value that ranges from 4 MJ/Nm^3 to 6 MJ/Nm^3. It can only be utilised in small-scale operations, such as a fuel for petrol engines or in homes.

12.2.3.2 Fermentation

In this metabolic reaction, sucrose is hydrolysed and sugars are fermented. First, the enzyme breaks down the sucrose into glucose and fructose. Additionally, fructose and glucose undergo an enzymatic process to produce ethanol. After the enzymatic hydrolysis, fermentation, distillation, and dehydration take place to produce anhydrous bioethanol. Additionally, fermentation is widely employed economically in many nations to make ethanol from sugar and starch crops. Since there are bacteria present, the fermentation of carbohydrates has resulted in the production of hydrogen. Algae can also create hydrogen under certain circumstances.

12.2.3.3 Anaerobic Digestion (AD)

In the absence of oxygen, organic materials are broken down into biogas during anaerobic digestion. Methane, carbon dioxide, hydrogen sulphide, and traces of other "contaminant" gases are all present in biogas. In combined heat and power gas engines, this gas combination can be utilised directly for lighting and cooking, or it can be converted to biomethane, a type of natural gas. Anaerobic digestion can produce methane and power that can be utilised to replace energy sources like oil and gas. As a by-product, nutrient-rich digestant is also created and used as a fertiliser. $C_6H_{12}O_6$-$3CO_2 3CH_4$ is a condensed chemical formula for overall processes.

12.2.3.4 Pyrolysis

Pyrolysis is the term for the heat degradation of organic materials that takes place anaerobically, or when there is less oxygen present, to produce gases, liquids, and a variety of solids. This process is commonly utilised in the business world to produce chemicals, solvents, various fuels and other goods from organic materials. Additionally, the petroleum industry uses it. Microalgae were pyrolysed to produce oil known as "bio-oil," which was obtained similarly to fossil fuel.

12.2.3.5 Transesterification

When the enzyme is present, this procedure yields highly pure biodiesel. A by-product called glycerol is simple to separate from the main product. In this procedure, the enzyme does not move. This promotes the enzyme's reusability. Additionally, this procedure has the potential to significantly lower the temperature. The process's primary drawbacks are the high cost of the enzyme and the suppression of the reaction.

12.3 PROSPECTS OF FIRST-GENERATION BIOFUELS

12.3.1 Energy Security and Independence

First-generation biofuels have a lot to offer in terms of improving energy security and reducing reliance on limited fossil fuel supplies. Countries can lessen their sensitivity to shifting global oil prices and geopolitical unrest by relying on locally sourced biomass. This strategy also fosters energy independence, a sense of control over energy supplies and lowers the risks related to volatile oil markets. By 2050, it is anticipated that rising population density and shifts in eating patterns towards high-quality food in developing nations would result in a 60% rise in the need for food production. Unprecedented growth is also occurring, particularly in regions with historically low per capita demand for fossil fuels. Today, most people agree that both rich and developing nations need to boost agricultural productivity and efficiency. Crops that make greater use of scarce resources, such as land, water and fertiliser, are necessary for producing more food sustainably. Due to the fact that agroecology is no longer at the centre of the discussion, drawing a parallel between it and high-input agriculture, which cannot be sustained, is of little assistance. The key question is whether current understanding regarding agroecological practices can produce this pace of yield growth and, if not, if investment in research and innovation centred on this stream of practices rather than intensification will help produce them. Food stability is vital in addition to food security, and predictability is the main concern. It has become clear that the continuous emissions of greenhouse gases and the loss of carbon sinks are having an impact on the global climate as a result of rising energy demands (Popp et al. 2014).

12.3.2 Environmental Benefits and Carbon Mitigation

Comparing first-generation biofuels to traditional fossil fuels, significant reductions in greenhouse gas emissions are possible. The carbon dioxide absorbed by feedstock crops during their growing phase balances the carbon dioxide emitted during the combustion of biofuels. First-generation biofuels are positioned as a useful instrument in efforts to combat climate change and lessen the carbon footprint of energy production, thanks to their potential to be carbon-neutral or even carbon-negative. Compared to ethanol, biodiesel also produces less air pollution per unit of energy gained. These benefits of biodiesel over ethanol result from less intensive agricultural inputs and more effective feedstock-to-fuel conversion. Neither biofuel can substantially replace petroleum without affecting the food supply. Even if all maize and soybean output in the United States were devoted to biofuels, just 12% of the demand for petrol and 6% for diesel would be satisfied. Prior to recent increases in the price of petroleum, high production costs rendered biofuels unprofitable without government assistance. There are enough environmental benefits from biodiesel to justify subsidies. If made from low-input biomass cultivated on agriculturally marginal land or from waste biomass, transportation biofuels like synfuel hydrocarbons or cellulosic ethanol could offer significantly higher supply and environmental advantages than biofuels derived from food (Hill et al. 2006).

12.3.3 Economic Growth and Job Creation

By generating employment possibilities along the whole supply chain, from agricultural activities and crop harvesting through processing, distribution and research, the biofuel business has the potential to drive economic growth. This industry can boost possibilities for sustained economic development by bringing capital into rural economies, encouraging innovation and advancing technology. The majority of developing economies, and most of their populations, are directly dependent on natural resources. One or two primary commodities make up the majority of exports for many of these economies, and primary product exports account for the great majority of their export earnings (Barbier 2010). Nearly 80% of the labour force is employed in agricultural or resource-based industries, and agricultural value added makes up an average of 40% of GDP. By 2025, the rural population of the developing world will have expanded to roughly 3.2 billion, adding to these inequities and putting more strain on a depleting resource base.

12.3.4 Technological Advancements and Feedstock Optimisation

The future of first-generation biofuels is being shaped by ongoing technological breakthroughs. Crop yields and, consequently, the efficiency of biofuel production are being improved via enhanced agricultural practices, genetic alterations and optimised feedstock selection. These developments are essential to achieving biofuels' full potential as a mainstream energy source. Algae as a biofuel feedstock has advantages over other feedstocks that have been widely researched. Any algal biofuel production facility's total economic viability and environmental sustainability can be dramatically changed by choosing the best species or strain. It is difficult to select one algae species over others. The optimum species is defined differently depending on the desired biofuel type, the growth mode to be used, the culture circumstances, etc. Algal biofuels' water footprint and fertiliser requirements can be greatly decreased by using saline water and wastewater. For the purpose of producing biofuel, all three algal growth modes—photoautotrophic, heterotrophic, and mixotrophic—have been considered. The benefits of photoautotrophic growth using atmospheric CO_2 include the free availability of sunlight and CO_2, but they are outweighed by the fact that sunlight intensity and duration are subject to inherent spatio-temporal variability and that the atmospheric CO_2 concentration (around 400 ppm) is too low to support the best possible growth of algae. Flue gas from coal-fired thermal power stations and the cement industry can be used to supply CO_2 in concentrated form. One of the most frequently discussed advantages of using algae as a feedstock for biofuel is their capacity to sequester significant amounts of CO_2 with slightly better photosynthetic efficiency in comparatively very short periods of time. According to several studies, there are no net carbon dioxide emissions when algae are grown for biofuel. However, a number of variables affect the overall GHG balance of any algae biofuel production system.

The GHG balance, which takes into account all greenhouse gas inputs and outputs, shows how much CO_2 is trapped or released throughout the course of a product or process life cycle. All emissions of various GHGs are presented in kilogrammes of CO_2 equivalent per functional unit selected for an life cycle assessment (LCA) study (Kumar et al. 2017).

12.4 CHALLENGES AND LIMITATIONS OF FIRST-GENERATION BIOFUELS

First-generation biofuels made from organic materials and agricultural products have the potential to cut carbon emissions while also promoting rural development and energy security. However, this chapter digs further into the complexities of the difficulties and constraints that come with the creation and use of these biofuels. This chapter provides a thorough overview of the complicated context around first-generation biofuels by exploring ethical, environmental and economic considerations. Recent studies of alternative biofuels' energy and greenhouse gas performance have sparked a debate that might be best settled by following two straightforward concepts. In a world searching for answers to its energy, environmental and food problems, humanity cannot afford to miss out on the local and global environmental benefits that result from properly produced biofuels. However, society also cannot accept the negative effects of poorly implemented biofuels.

12.4.1 Competition with Food Production

First-generation biofuels' rivalry with food production is their biggest obstacle. Fuel production from crops like maize, sugarcane and oilseeds diverts resources away from the food supply chain, raising questions about food security, price volatility and the possibility of a rise in famine. The biofuels sector is poised for tremendous expansion. The accompanying strategy should plan for and support a biofuels sector that effectively and positively solves urgent sustainability and security issues. Only when they have a beneficial effect on four key goals—energy security, greenhouse gas emissions, biodiversity and the sustainability of the food supply—should biofuels be supported by policy as alternatives to fossil fuels. There is a need for performance-based policies that offer rewards in proportion to the benefits received. The massive projected increases in global energy and food consumption will significantly increase atmospheric greenhouse gas levels from fossil fuel combustion, land clearing and livestock production, and will result in immense biodiversity loss from habitat destruction and climate change unless new technologies and lifestyles are adopted globally over the coming decades. Human life will be compromised in quality. How the environmental effects and possible benefits of supplying the world's energy and food needs may be integrated into our economic systems, then, is a key concern for the ensuing decades. Simple answers and sound bites are insufficient to handle the complexity of this issue. Environmentalists, economists, technicians, the agriculture sector, involved individuals and governments from all around the world need to work together more.

12.4.2 Land Use Change and Deforestation

The increased interest in biofuels as a renewable energy source to minimise greenhouse gas emissions and lessen dependence on fossil fuels is the main driving force. One should be aware of the complex connection between land-use choices and their effects on the environment, though. The researchers recognise that even though using less fossil fuels could cut carbon emissions from biofuels, their expansion could unintentionally contribute to deforestation, thus negating the intended environmental benefits. The study's main premise is based on a phenomenon called indirect land-use change (ILUC), which describes the secondary effects of land-use changes brought on by the development of biofuels. The indirect effects of converting land for biofuel crops, such as deforestation brought on by rising agricultural demand for non-biofuel crops, can result in significant carbon emissions, even though the direct effects of doing so are frequently taken into consideration. Brazil is a major producer of biofuels and agricultural products worldwide. In some cases, the carbon benefits from using biofuels are outweighed by the carbon emissions brought on by deforestation. This unexpected result underlines the complexity of land-use choices connected to biofuels and the necessity for a comprehensive analysis of their environmental impact. Integrated strategies are required to ensure that the spread of biofuels does not unintentionally result in environmental deterioration, deforestation, or increased carbon emissions by taking into account both the direct and indirect effects of land-use changes. The results highlight the significance of implementing a thorough strategy that considers both direct and indirect impacts when evaluating the environmental advantages of biofuels. It is reminded that informed decision-making and coordinated regulations are necessary to prevent unanticipated negative outcomes linked with the spread of biofuels as the world searches for sustainable energy options (Lapola et al. 2010)

12.4.3 Environmental Impact of Cultivation

Growing crops for biofuel feedstock can have unforeseen negative effects on the environment, such as soil degradation, excessive water use and agrochemical pollution. These effects may negate the early environmental advantages of biofuels. Due to their ability to lower carbon emissions during burning, biofuels are frequently touted as a greener alternative to fossil fuels. However, this study tries to clarify the complex relationship between land-use changes, carbon emissions and the actual carbon savings obtained by biofuels. The authors use a thorough modelling strategy that incorporates information on land-use changes, carbon emissions and the ability of various ecosystems to sequester carbon to solve this issue. The study examines the carbon effects of converting grasslands, forests and wetlands into biofuel croplands with a focus on the United States, a significant player in the production of biofuels. The study's conclusions serve as both educational and advisory documents. The authors demonstrate that the carbon debt caused by changing carbon-rich ecosystems like forests and grasslands for biofuel cultivation can last for decades. In some circumstances, it might take centuries for the carbon emissions that arise from using biofuels to outweigh those that result from clearing land. This phenomenon casts doubt on the idea that biofuels will immediately reduce carbon emissions. A more comprehensive methodology should be used by policymakers and decision-makers when assessing the environmental effects of biofuels. It urges taking into account the total carbon budget linked to the production of biofuels, which includes both direct emissions from burning and indirect emissions from changes in land use. The study casts doubt on widely held beliefs regarding the immediate environmental benefits of biofuels by introducing the idea of the carbon debt. It emphasises the requirement for a thorough analysis of the sustainability of biofuels, taking into account the carbon effects of land-use changes and crop farming. This study serves as a helpful reminder that real carbon savings require a sophisticated understanding of the complex carbon dynamics involved in the production of biofuels, which is important as society works to find sustainable energy alternatives (Fargione et al. 2008).

12.4.4 Limited Carbon Savings

Even though switching to biofuels from fossil fuels may reduce carbon emissions overall, there may still be some emissions generated during production, transport and processing. First-generation biofuels' carbon footprints might not always be much smaller than those of conventional fuels. Producing biofuel may unintentionally increase greenhouse gas emissions rather than reduce them as anticipated. This phenomenon results from the altered land use needed for the production of the feedstock for biofuels, which can increase carbon emissions to levels greater than those reduced by the use of biofuels. The carbon savings attributable to the use of biofuels may be offset by the carbon emissions brought on by land-use changes, particularly when converting carbon-rich ecosystems like forests. In other circumstances, it may take decades or even centuries for carbon savings to equal the emissions caused by biofuel use. This fact casts doubt on the widely held belief that biofuels immediately reduce carbon emissions. The type of land converted, the feedstock grown and the amount of carbon emissions from changing land use are only a few of the variables that affect how carbon-efficient biofuels are. We require a thorough and comprehensive analysis of the sustainability of biofuels, taking into account both direct and indirect carbon emissions. It serves as a reminder that the complexity of the carbon cycle must be taken into account in order to achieve significant environmental advantages as society searches for feasible methods to address climate change (Searchinger et al. 2008).

12.4.5 Food Price Inflation

The complex relationship between rising food prices and their far-reaching socio-economic consequences has gained significant attention in recent years. Concerns about the multiple effects of the global rise in food costs on society, the economy and people's well-being are prevalent. Mitchell (2008) offers a perceptive examination of the elements causing the increase in food costs. The study emphasises the importance of both supply-side and demand-side factors, such as unfavourable weather, rising demand from emerging nations, greater energy prices and changes in food habits. Understanding these factors is necessary to develop efficient policy responses. Their budgets are heavily weighed down by food expenses, making them susceptible to inflationary pressures. According to Mitchell (2008), declining purchasing power has the potential to further impoverish already marginalised communities and widen economic disparities. Inflation in food prices has effects that go beyond simple financial ones. Access to sufficient and nutritious food is hampered by affordability, which causes malnutrition, especially in vulnerable communities. The effects of price increases are amplified by the global food market's interconnection. It describes how rising food prices can have a significant impact on international trade, food relief initiatives, and international cooperation. Additionally, it raises moral concerns about the fair distribution of resources. A multifaceted strategy that incorporates methods to control food prices and supplies, targeted social safety nets, improvements to agricultural productivity and funding for rural infrastructure is necessary.

12.4.6 Ethical Concerns

The moral consequences of using food crops as fuel are at the core of the ethical discussion around first-generation biofuels. When it affects food availability and worsens world famine, some claim that this practice is morally questionable. The distribution of land resources for the development of biofuels is a key ethical issue. Svensson (2011) emphasises the moral conundrum of prioritising land for the development of biofuel feedstock over food crops. This allocation poses a moral dilemma because it directly affects the availability and cost of food globally, raising concerns about the equal distribution of resources. The ethical tension between energy security and food security is a central theme in biofuel discussions. Morality is brought up for discussion by the effects biofuel production has on the environment. Although the production of biofuels has the potential to reduce emissions, doing so may result in habitat loss, deforestation and increased water use. Concerns are raised regarding how the expansion of biofuels may affect an equitable society. The displacement of indigenous tribes and smallholder farmers due to land conversion for biofuel crops raises moral concerns about resource rights and socio-economic inequalities. Human welfare and inclusivity must be given top priority in ethical biofuel manufacturing. The development of biofuels requires strong ethical governance systems. The need for measures that take into account both energy and food security is emphasised by Svensson (2011). To guarantee that the development of biofuels is beneficial to society as a whole, ethical governance should include open decision-making, adherence to human rights and fair involvement. Biofuels-related ethical issues transcend national borders. The necessity of international collaboration to fully address these difficulties should not be ignored. Collaboration across borders encourages the sharing of knowledge, skills and assets, harmonising with the moral imperatives of solidarity and shared responsibility. Energy security, food security, environmental sustainability and social equality must all be balanced, which calls for moral leadership, international cooperation and technological innovation. Biofuels can help create a more sustainable and morally responsible energy future by embracing ethical concerns.

12.4.7 Water Scarcity

Due to variables including population expansion, climate change and competing water demands, water shortage has become an urgent global issue. The imbalance between water supply and demand within a region is referred to as water shortage. Agriculture, industry and the generation of energy are just a few of the areas that causes serious difficulties for. As water resources are depleted, it is essential to distribute water wisely while taking socio-economic and environmental considerations into account. Bioenergy, which is promoted as a sustainable energy source, is essential for reducing global warming and improving energy security. Water resources, however, are impacted by the development of bioenergy, especially biofuels like ethanol. The amount of water needed for the production, processing and growing of feedstocks is included in the water footprint of bioenergy. The development of biofuel feedstocks can put a burden on local water supplies in areas with limited water supplies, like some parts of Brazil, with negative ecological and socio-economic effects. The study's conclusions show that although producing bioenergy can help ensure energy security, its effects on water resources should not be disregarded. The fact that producing maize for ethanol requires a lot of water highlights the necessity for thorough analyses that take both energy and water security into account. The water, energy and food nexus must be addressed, and sustainability issues are essential. Water-efficient biofuel feedstocks, better irrigation techniques and water resource management should be given priority in policies and initiatives. To make sure that bioenergy production does not increase water constraints or adversely affect other sectors, integrated measures are required. The work by Mao et al. (2013) provides insights into the intricacies of bioenergy-water interactions as water scarcity becomes a significant global concern. It emphasises how crucial it is to take into account how producing bioenergy will affect water within a larger sustainability framework. Collaboration and creative solutions are needed to strike a balance between socio-environmental issues, water availability and energy needs.

12.4.8 Technological Advances and Second-Generation Biofuels

Technology developments are crucial in determining the course of biofuel production, especially the transition to second-generation biofuel. Second-generation biofuels make use of non-food feedstocks like woody biomass, algae and crop wastes that do not directly compete with food production. Technological advances make it possible to efficiently convert these feedstocks into biofuels while minimising disruptions to the land's use and potential effects on food security. The creation of cellulosic ethanol is one of the technological highlights of second-generation biofuels. Complex cellulose and lignin structures can now be broken down into fermentable sugars and ethanol by the use of enzymes and biochemical processes. This innovation overcomes the drawbacks of conventional food-based feedstocks and boosts the effectiveness of biofuel production. A comprehensive examination of waste-to-energy conversio, a crucial component of second-generation biofuels, is provided in this work. Technological advancements enable the production of useful biofuels from organic waste products, such as agricultural and forestry leftovers. Hill et al. (2006) emphasise that these developments not only lessen the difficulties associated with trash disposal but also make use of resources that were previously underutilised for the production of sustainable energy. Algae have a high oil content and may grow in a variety of conditions, including effluent and non-arable land. The potential for algae to produce large amounts of renewable biodiesel has been unlocked by technological advances in growing, harvesting and oil extraction, relieving pressure on conventional crop-based feedstocks. Technology advancements in feedstock utilisation and conversion procedures help to minimise land-use changes and greenhouse gas emissions. But problems still exist, such as the demand for effective enzyme technology, environmentally friendly feedstock procurement and scalable production methods. Energy efficiency and energy balance have improved for second-generation biofuels. According to Hill et al. (2006), the production of biofuels from non-food feedstocks, waste products and algae can result in higher energy outputs than energy inputs. Technological advancements improve the net energy gain even more, adding to the sustainability of biofuel production as a whole.

12.5 FUTURE PERSPECTIVES

12.5.1 Technological Advancements

The manufacture of bioethanol and biodiesel, which are both considered first-generation biofuels, has undergone a radical change due to technological breakthroughs. The Smith et al. (2022) research explores the complex world of first-generation bioethanol production, showcasing a wide range of technological innovations that have completely transformed the entire procedure. The importance of genetic engineering and strain optimisation, which enable increased ethanol yield from feedstocks including corn, sugarcane and wheat, is emphasised by the authors. In order to convert starch and cellulose more effectively into fermentable sugars, enzyme hydrolysis techniques have been improved. This has sped up the fermentation process. Furthermore, increased productivity is a result of fermentation advancements, such as brand new yeast strains with higher ethanol tolerance. Smith et al. (2022) also highlight improvements in separation, purification and distillation methods that lead to higher-quality ethanol products and increased energy efficiency. Choudhury et al. (2021) present a thorough summary of technological advancements in the field of biodiesel production, which complements the advancements in bioethanol production. The analysis summarises how feedstock utilisation has changed over time, pointing out the shift from edible oils to non-edible feedstocks such as used cooking oil, animal fats and algae.

To improve the effectiveness and quality of oil extraction, new pre-treatment techniques such as supercritical fluid extraction and ultrasonication have been introduced. With an emphasis on heterogeneous catalysis utilising solid catalysts, which has benefits including faster reaction times and simpler catalyst recovery, the authors emphasise improvements in transesterification procedures. In addition, Choudhury et al. (2021) provided information on the use of enzymatic catalysts, providing additional environmentally friendly options for the synthesis of biodiesel. From feedstock source and conversion procedures through product purification and co-product utilisation, these developments span the whole value chain. The interdisciplinary nature of these innovations is highlighted by the use of genetic engineering, enzymatic hydrolysis, fermentation optimisation and innovative catalytic processes. These technical advancements greatly support the profitability and sustainability of the production of first-generation biofuels by improving efficiency, lowering prices and minimising environmental effects. These developments have the potential to progress bioethanol and biodiesel into a more significant position in the world's energy portfolio as the biofuels sector develops, enhancing both energy security and environmental well-being.

12.5.2 Feedstock Diversification

The production of first-generation biofuels is now more sustainable and resilient because of the cornerstone strategy of feedstock diversification. The importance of non-food biomass in resolving the frequently discussed food vs fuel conundrum is clarified by Gribben et al. (2025). The essay carefully examines a variety of non-food feedstock possibilities, such as forestry waste, agricultural wastes and special energy crops. These feedstocks, which are frequently disregarded because they do not compete with food crops, have enormous potential for reducing land-use conflicts and ensuring food security. According to Gribben et al. (2025), agricultural leftovers like corn stover, wheat straw and sugarcane bagasse can be used to harvest food production by-products without affecting the supply chain. The authors also explore the potential of energy crops like switchgrass and miscanthus to thrive on marginal soils and produce large amounts of biomass. This comprehensive analysis highlights the strategic importance of feedstock diversity for improving the sustainability and long-term viability of first-generation biofuel production. A compelling narrative about the exceptional potential of algae as a sustainable feedstock for biodiesel production is provided by Singh et al. (2023), which complements the discussion. The article explores the distinctive qualities of microalgae and macroalgae, emphasising their ability to flourish in a variety of conditions, such as ponds filled with sewage and non-arable terrain. It explores the complex biochemical mechanisms behind algal lipid accumulation, a step in the production of biodiesel.

Aiming to increase algal biomass output, new growing methods, such as photobioreactors and open ponds, are investigated. The authors specifically highlight how growing algae is regenerative because it uses less freshwater and absorbs carbon dioxide from the atmosphere. This ecological benefit makes algae a strong competitor in the search for sustainable and low-impact feedstocks for biodiesel, along with the possibility for co-product development. The studies by Gribben et al. (2025) and Gupta et al. (2020), taken together, emphasise the revolutionary potential of feedstock diversity in the manufacture of first-generation biofuels. Algae and non-food biomass indicate a development past the limitations of conventional feedstocks. These other sources provide a practical response to resource limitations as well as a considerate method of land usage. The biofuel business can navigate difficulties brought on by climate change, resource scarcity and ethical considerations by expanding the range of feedstock possibilities. The future of first-generation biofuel production is entwined with the sustainable future of energy through the use of non-food biomass and algae, illustrating the peaceful cohabitation of human advancement with environmental stewardship.

12.5.3 Sustainability Metrics and Certification

According to Brown et al. (2021), sustainability analyses of first-generation biofuels bring a variety of potentials and problems. The study emphasises the value of adopting a comprehensive strategy that takes into account a variety of factors, including socio-economic effects, conservation of biodiversity and land use. The authors provide insight into the use of life cycle assessment (LCA) frameworks that reveal the environmental impact of producing biofuels by quantifying greenhouse gas emissions, water use and other important variables. It probes the ethical ramifications of biofuel production by raising questions about matters such as food security, land rights and indigenous knowledge. The study highlights the potential for sustainability indicators to not only spot problem areas but also spur innovation and process improvement, allowing the market to move in a more responsible direction. By exploring the subject of certification and labelling for sustainable biofuel production, Miller et al. (2022) further this discussion. The authors explore the complex network of certification programmes designed to formalise ethical business practices in the biofuel sector. These models offer a reliable framework for evaluating and benchmarking the sustainability of biofuels and are frequently jointly created by industry players, non-governmental organisations and governmental entities. He uses the Sustainable Biodiesel Certification and the Roundtable on Sustainable Biomaterials

(RSB) certification as examples of how certification initiatives balance financial incentives with environmental stewardship. The research papers by Miller et al. (2022) and Brown et al. (2021) shed light on the critical role that sustainability measures and certification play in directing the development of first-generation biofuels in a morally and responsible direction. These frameworks capture the essence of sustainable development by recognising the interdependence of ecological integrity, social equality and economic viability. The biofuel business may actively participate in global efforts to mitigate climate change, protect biodiversity and promote inclusive growth if it is equipped with scientifically rigorous assessments and open certification processes. Incorporating sustainability factors, as promoted in this research, is a crucial step towards realising a future in which energy and environmental well-being coexist peacefully.

12.5.4 Benefits of First-Generation Biofuels

They can reduce greenhouse gas emissions. Biofuels produce fewer greenhouse gas emissions than fossil fuels when they are burned. This is because the carbon dioxide that is released when biofuels are burned is the same carbon dioxide that was absorbed by the plants from which the biofuels were made.

- They can improve air quality. Biofuels can help to reduce air pollution by emitting fewer pollutants than fossil fuels. This is because biofuels contain less sulphur and nitrogen than fossil fuels.
- They can create jobs. The production and use of biofuels can create jobs in the agricultural, transportation and energy sectors.
- They can help to reduce our reliance on foreign oil. Biofuels can be produced domestically, which can help to reduce our reliance on foreign oil.
- Here are some of the potential drawbacks of first-generation biofuels:
- They can compete with food crops for land and water resources. The production of first-generation biofuels can compete with the production of food crops for land and water resources. This can lead to higher food prices and food insecurity.
- They can contribute to deforestation. The production of first-generation biofuels can contribute to deforestation. This is because forests are often cleared to make way for the cultivation of biofuel crops.
- The production of first-generation biofuels can be energy-intensive. This means that they may not actually reduce greenhouse gas emissions when all of the energy inputs are considered.
- They can release pollutants into the environment. The production of first-generation biofuels can release pollutants into the environment, such as ammonia and nitrous oxide. These pollutants can contribute to smog and acid rain.

Furthermore, the first-generation biofuels have the potential to reduce greenhouse gas emissions and improve air quality. However, they also have some drawbacks, such as the potential to compete with food crops for land and water resources, contribute to deforestation and release pollutants into the environment. It is important to weigh the pros and cons of first-generation biofuels carefully before deciding about their use. In the future, it is likely that second- and third-generation biofuels will become more widely used. Second- and third-generation biofuels are made from non-edible biomass, such as grasses, trees and algae. They have the potential to be more sustainable than first-generation biofuels, but they are still in the early stages of development. The utilisation of biomass for biofuel production is a complex issue with both potential benefits and drawbacks. It is important to continue to research and develop new biofuel technologies in order to find the most sustainable and environmentally friendly solutions. One of the biggest challenges facing first-generation biofuels is the competition for land and water resources. The production of first-generation biofuels can compete with the production of food crops for land and water resources. This can lead to higher food prices and food insecurity. Another challenge facing first-generation biofuels is the potential for deforestation. The production of first-generation biofuels can contribute to deforestation. This is because forests are often cleared to make way for the cultivation of biofuel crops. The production of first-generation biofuels can also be energy-intensive. This means that they may not actually reduce greenhouse gas emissions when all of the energy inputs are considered. Finally, the production of first-generation biofuels can release pollutants into the environment. These pollutants can contribute to smog and acid rain. Despite these challenges, first-generation biofuels can play a role in reducing our reliance on fossil fuels and mitigating climate change. They can be used to blend with gasoline and diesel fuel, and they can be used to power vehicles and other machines.

The first-generation biofuels could be used for the generation of electricity and heat. The utilisation of biomass for first-generation biofuel production is a complex issue with both potential benefits and drawbacks. It is important to weigh the pros and cons of first-generation biofuels carefully before making a decision about their use. In the future, it is likely that second- and third-generation biofuels will become more widely used. Second- and third-generation biofuels are made from non-edible biomass, such as grasses, trees and algae. They have the potential to be more sustainable than first-generation biofuels, but they are still in the early stages of development. The utilisation of biomass for biofuel production is a rapidly evolving field. New technologies are being developed all the time, and it is likely that the future of biofuels will look very different from the present. It is important to stay up-to-date on the latest developments in this field in order to make informed decisions about the use of biofuels.

12.6 CHALLENGES

There is a need to develop more efficient production technologies and methods. Although, the production of first-generation biofuels can be energy-intensive. This means that they may not actually reduce greenhouse gas emissions when all of the energy inputs are considered. More efficient production methods are needed in order to reduce the environmental impact of first-generation biofuels. There is a need to develop crops that are more productive and less water-intensive. The production of first-generation biofuels can compete with the production of food crops for land and water resources. More productive and less water-intensive crops are needed in order to reduce the competition between food and fuel crops. There is a need to develop methods to mitigate the environmental impacts of biofuel production. The production of first-generation biofuels can release pollutants into the environment. Methods need to be developed to mitigate these pollutants in order to protect the environment. If these challenges can be addressed, first-generation biofuels have the potential to play a significant role in reducing our reliance on fossil fuels and mitigating climate change. However, it is important to remember that first-generation biofuels are not a perfect solution. They have some drawbacks that need to be considered before they can be widely adopted.

12.7 CONCLUSION

Biomass is a renewable resource that can be used to produce a variety of fuels, including first-generation biofuels. First-generation biofuels are made from edible biomass, such as corn, sugarcane and soybeans. They are the most common type of biofuel in use today, and they have the potential to reduce greenhouse gas emissions and improve air quality. However, first-generation biofuels also have some drawbacks. They can compete with food crops for land and water resources, and they can contribute to deforestation. Additionally, the production of first-generation biofuels can be energy-intensive, and they can release pollutants into the environment. Despite these drawbacks, first-generation biofuels can play a role in reducing our reliance on fossil fuels and mitigating climate change. They can be used to blend with gasoline and diesel fuel, and they can be used to power vehicles and other machines. First-generation biofuels can also be used to generate electricity and heat. The utilisation of biomass for first-generation biofuel production is a complex issue with both potential benefits and drawbacks. It is important to weigh the pros and cons of first-generation biofuels carefully before making a decision about their use.

REFERENCES

Ashokkumar, V., Venkatkarthick, R., Jayashree, S., Chuetor, S., Dharmaraj, S., Kumar, G., ... Ngamcharussrivichai, C. (2022). Recent advances in lignocellulosic biomass for biofuels and value-added bioproducts-A critical review. *Bioresource Technology*, *344*, 126195.

Barbier, E. B. (2010). Poverty, development, and environment. *Environment and Development Economics*, *15*(6), 635–660.

Bowyer, J. L., & Stockmann, V. E. (2001). Agricultural residues. *Forest Products Journal*, *51*(1), 10–21.

Bracmort, K., & Gorte, R. W. (2009, April). *Biomass: Comparison of definitions in legislation*. Congressional Research Service, Library of Congress.

Brown, B., Schoney, R., & Nolan, J. (2021). Assessing the food vs. fuel issue: An agent-based simulation. Energy Policy, 159, 112553.

Choudhury, N., Meghwal, M., & Das, K. (2021). Microencapsulation: An overview on concepts, methods, properties and applications in foods. Food Frontiers, 2(4), 426-442.

Dahman, Y., Syed, K., Begum, S., Roy, P., & Mohtasebi, B. (2019). Biofuels: Their characteristics and analysis. In *Biomass, biopolymer-based materials, and bioenergy* (pp. 277–325). Woodhead Publishing.

Fargione, J., Hill, J., Tilman, D., Polasky, S., & Hawthorne, P. (2008). Land clearing and the biofuel carbon debt. *Science*, *319*(5867), 1235–1238.

Gabrielle, B. (2008). Significance and limitations of first generation biofuels. *Journal de la Société de biologie*, *202*(3), 161–165.

Gribben, K. C., Johnson, K., Greenberg, P., Mencia, R., Taiba, J., Kintziger, K. W., ... & Bell, J. E. (2025). Environmental contamination associated with biofuel production involving pesticide-coated seed corn as feedstock: a survey of community environmental and health impacts. Environmental Health, 24(1), 17.

Hill, J., Nelson, E., Tilman, D., Polasky, S., & Tiffany, D. (2006). Environmental, economic, and energetic costs and benefits of biodiesel and ethanol biofuels. *Proceedings of the National Academy of Sciences*, *103*(30), 11206–11210.

Koçar, G., & Civaş, N. (2013). An overview of biofuels from energy crops: Current status and future prospects. *Renewable and Sustainable Energy Reviews*, *28*, 900–916.

Kumar, D., Singh, B., & Sharma, Y. C. (2017). Challenges and opportunities in commercialization of algal biofuels. *Algal Biofuels: Recent Advances and Future Prospects*, 421–450.

Lapola, D. M., Schaldach, R., Alcamo, J., Bondeau, A., Koch, J., Koelking, C., & Priess, J. A. (2010). Indirect land-use changes can overcome carbon savings from biofuels in Brazil. *Proceedings of the National Academy of Sciences*, *107*(8), 3388–3393.

Mao, Y., Zhang, H., Xu, N., Zhang, B., Gou, F., & Zhu, J. K. (2013). Application of the CRISPR–Cas system for efficient genome engineering in plants. *Molecular plant*, 6(6), 2008-2011.

Miller, J.H., Tifft, S.M., Wiatrowski, M.R., Benavides, P.T., Huq, N.A., Christensen, E.D., Alleman, T., Hays, C., Luecke, J., Kneucker, C.M. and Haugen, S.J. (2022). Screening and evaluation of biomass upgrading strategies for sustainable transportation fuel production with biomass-derived volatile fatty acids. IScience, 25(11).

Mitchell, H. (Ed.). (2008). Sweeteners and sugar alternatives in food technology. John Wiley & Sons.

Popp, J., Lakner, Z., Harangi-Rákos, M., & Fari, M. (2014). The effect of bioenergy expansion: Food, energy, and environment. *Renewable and Sustainable Energy Reviews*, *32*, 559–578.

Searchinger, T., Heimlich, R., Houghton, R. A., Dong, F., Elobeid, A., Fabiosa, J., ... Yu, T. H. (2008). Use of US croplands for biofuels increases greenhouse gases through emissions from land-use change. *Science*, *319*(5867), 1238–1240.

Smith, N. W., Fletcher, A. J., Millard, P., Hill, J. P., & McNabb, W. C. (2022). Estimating cropland requirements for global food system scenario modeling. Frontiers in sustainable food systems, 6, 1063419.

Singh, R. P., Yadav, P., Kumar, I., Solanki, M. K., Roychowdhury, R., Kumar, A., & Gupta, R. K. (2023). Advancement of abiotic stresses for microalgal lipid production and its bioprospecting into sustainable biofuels. Sustainability, 15(18), 13678.

Songstad, D. D., Lakshmanan, P., Chen, J., Gibbons, W., Hughes, S., & Nelson, R. (2009). Historical perspective of biofuels: Learning from the past to rediscover the future. *In Vitro Cellular & Developmental Biology-Plant*, *45*, 189–192.

Svensson, K. G., & Trulsson, M. (2011). Impaired force control during food holding and biting in subjects with tooth-or implant-supported fixed prostheses. *Journal of clinical periodontology*, 38(12), 1137–1146.

Yuan, J. S., Wang, X., & Stewart, C. N. (2011). Biomass feedstock: Diversity as a solution. *Biofuels*, 2(5), 491–493.

13 Current Status and Strategies for Second-Generation Biofuel Production Using Lignocellulosic Biomass

Rishi Richa, Radha Srivalli, Deepika Kohli and Mudasir Ali

13.1 INTRODUCTION

The fuel derived through a chemical process from biomass, as opposed to the gradual geological processes, is known as biofuel (Ganesan et al., 2020). The term "biofuel" is used to describe fuel derived from renewable, living materials, such as plants and animals (Rai et al., 2022). Earth's annual biomass production consists of around 100 billion tons of organic dry matter from land biomass and approximately 50 billion tons from aquatic biomass. A fraction of this biomass serves various purposes, such as food, feed, energy, and industrial raw materials. Notably, only 1.25% of the entire land biomass is utilized for food purposes (Naik et al., 2010).

As energy demand continues to rise, it's important for governments, industries, and individuals to focus on energy efficiency, renewable energy sources, and sustainable practices to mitigate the potential negative effects of increased energy consumption. Globally, energy consumption rose by 418 exajoules in 2019, and it is projected to experience an additional increase of 516 exajoules by the year 2040, representing a 23% growth (Wang et al., 2021; Srivastava et al., 2023). Furthermore, reports indicate that 33 gigatonnes (Gt) of CO_2 emissions were recorded in 2019. As per the International Energy Agency (IEA), the carbon footprint is projected to persist at 33 Gt in 2040. Achieving the target of net zero CO_2 emissions by 2050 necessitates the production and advancement of energy resources (Monir et al., 2021; Srivastava et al., 2023).

DOI: 10.1201/9781032650616-13

The Indian economy has been experiencing a growth rate of around 7% since the year 2000. According to the latest World Energy Outlook (WEO) report from the International Energy Agency (IEA), India's primary energy demand is expected to rise from 750 million tonnes of oil equivalent (Mtoe) to a range of 1258–1647 Mtoe between the years 2011 and 2035 (IEA, 2013; Purohit et al., 2015), i.e., it will most likely more than double over these 25 years. As per the International Energy Agency (IEA) in 2013, the demand for oil in India is projected to surpass 8 million barrels per day by 2035. However, the present domestic crude oil production has remained relatively constant over the years, satisfying only 18% of the country's overall demand (MPNG, 2012; Purohit et al., 2015). To offset this shortfall, the country relies on imports of approximately 172 million tons of crude petroleum products, incurring a cost of nearly US$140 billion in the fiscal year 2011–2012 (MPNG, 2012). Currently, India relies on imports for over 80% of its crude oil to fulfill the annual demand, exposing the growing economy to significant risks in terms of both price and quantity. This is particularly notable given that 90–95% of transportation in India is dependent on petroleum-based oil (Sumitra Ganguli et al., 2018). Hence, there is a clear need to thoroughly investigate and encourage the production of biofuels using lignocellulosic biomass. As part of India's Intended Nationally Determined Contribution (INDC), there is a decarbonization objective aiming for a 33–35% reduction in the CO_2 intensity of its Gross Domestic Product (GDP) from 2005 to 2030 (UNFCCC, 2017). Currently, around 80% of liquid biofuel production is derived from bioethanol, with the remaining portion being generated through biodiesel (Mahapatra et al., 2021; Figueiredo et al., 2009). Biofuels distinguish themselves from other petroleum feedstock based on their oxygen content.

13.2 BIOFUEL USE AND ITS APPLICATIONS

Certainly, biofuels have emerged as a notable alternative energy source, gaining recognition for their capacity to diminish greenhouse gas emissions and decrease dependence on fossil fuels. Biofuels encompass all fuels that are sourced and manufactured from organic matter, including plants and their remnants, agricultural crops, and secondary products, which can effectively serve as alternatives to fuels derived from petroleum. Biofuels are available in various forms:

1. Liquid biofuel: Liquid biofuels have numerous benefits compared to solid and gaseous biofuels due to their elevated energy density, which renders them well-suited for purposes such as transportation, storage, and retrofitting. Liquid biofuel includes bioethanol and biodiesel.
2. Solid biofuel: These biofuels are comprised of solid, organic biomass derived from biological sources rather than fossils. These biomass materials find important uses in generating energy and creating electricity. These are produced from charcoal, wood pellets, and fuel wood.

3. As gases, exemplified by biogas and biohydrogen. Biofuels derive from renewable sources through the photosynthetic conversion of solar energy into chemical energy. In contrast, petroleum products originate as derivatives of crude fossil fuel, obtained through processing in oil refineries (Rai et al., 2022). While the categorizations of biofuel crops differ substantially, the source materials can generally be grouped into three main categories: First-generation, Second-generation, Third-generation, and Fourth-generation feedstocks.

13.3 FIRST-GENERATION BIOFUEL

First-generation biofuels are predominantly produced from crops such as barley, wheat, maize, potatoes, sugarcane, and oilseeds. Biodiesel, on the other hand, is typically manufactured from sunflower and soybean (Malode et al., 2021). First-generation biofuels are recognized for their ability to reduce CO_2 emissions, but they face substantial criticism for issues related to land use and potential food shortages (Ganesan et al., 2020).

13.4 SECOND-GENERATION BIOFUEL

In contrast to first-generation biofuel production, the second generation represents an enhanced approach that emphasizes both heightened fuel recovery and the generation of secondary raw materials. Unlike the first-generation process, the second generation places emphasis on producing valuable fuels while simultaneously reducing overall energy costs and minimizing waste production, rendering it an economically viable method (Mahapatra et al., 2021). It is widely recognized that non-edible lignocellulosic biomass, abundant in the natural ecosystem, including vegetable grasses, forest residues, agricultural waste, etc., can be utilized as a feedstock for biofuel production (Rai et al., 2022).

Second-generation biomass comprises byproducts from the food processing industry and wood factories, encompassing inedible components such as dry wood, corn, and wheat stalks. Resources like used oil products from restaurants, animal waste, and oil crops such as jojoba, jatropha, and sea mango are also recognized as available materials for this next generation (Ganesan et al., 2020)

13.5 THIRD-GENERATION BIOFUELS

Biofuels of the third generation, produced from algae, have garnered significant attention due to their potential for large-scale production, CO_2 absorption capabilities, and comparatively straightforward refining processes. The release of sludge, wastewater, and substantial quantities of CO_2 into the atmosphere poses a severe threat to the ecosystem. Given this scenario, the cultivation of algal biomass using wastewater is regarded as an environmentally friendly solution for generating renewable energy (Ambaye et al., 2021). Third-generation (algal) biofuels present a potential solution to the challenges of food competition and land use, as microalgae can thrive on non-arable land, in wastewater, saline or brackish water, and exhibit rapid growth (Jeswani et al., 2020).

13.6 FOURTH-GENERATION BIOFUEL

The production of fourth-generation biofuels relies on genetically modified organisms with modified metabolic pathways, enhanced CO_2 fixation capabilities, and the post-genome technology of microalgae (Carere et al., 2008; Dragone et al., 2011; Dutta et al., 2014; Lü et al., 2011; Ambaye et al., 2021).

13.7 NEED OF SECOND GENERATION BIOFUEL

The utilization of fossil fuels, coupled with their combustion in automobiles, results in the emission of greenhouse gases (GHGs), significantly contributing to climate change and air pollution. Hence, it is imperative to produce alternative energy sources, particularly oxygenated biofuels, from lignocellulosic biomass waste and various residual materials to address land use concerns (Ambaye et al., 2021). The notable benefits of employing biomass include its renewable nature, making it well-suited for sustainable use, posing a less damaging threat to the environment with a low net release of carbon dioxide (CO_2) and sulfur. Furthermore, biomass proves to be economically more feasible compared to fossil fuels, which may see increased costs with rising demand (Mahapatra et al., 2021; Sharma et al., 2014). Meanwhile, the utilization of first-generation feedstocks like corn has generated significant controversy, primarily due to the competition they create with food cultivation and the apprehensions about shifting agricultural land toward fuel manufacturing. The increasing demand for agricultural products raises the risk of heightened deforestation and the utilization of land with significant biodiversity value to meet this demand. This, along with the associated use of freshwater, fertilizers, and pesticides, has adverse consequences on the environment. Second-generation biofuels also contribute to addressing the issue related to the disposal of crop waste (Gupta et al., 2020). The production of second-generation biofuels presents even greater challenges compared to first-generation biofuels, primarily due to the intricate nature of the biomass and the various issues associated with producing, harvesting, and transporting low-density biomass to central biorefineries. According to Carriquiry et al. (2011), second-generation biofuels outperform fossil fuels in energy output, reducing competition on land. They can be derived from a broader range of feedstock options and exhibit significantly lower environmental impacts. Researchers have considered various types of plant biomass for utilization in biofuel production (Robak et al., 2018) (Figure 13.1).

13.8 LIGNOCELLULOSIC BIOMASS AS FEED STOCK

Lignocellulose consists of three types of biopolymers: cellulose, hemicellulose, and lignin that are firmly connected by non-covalent forces and covalent cross-linkages.

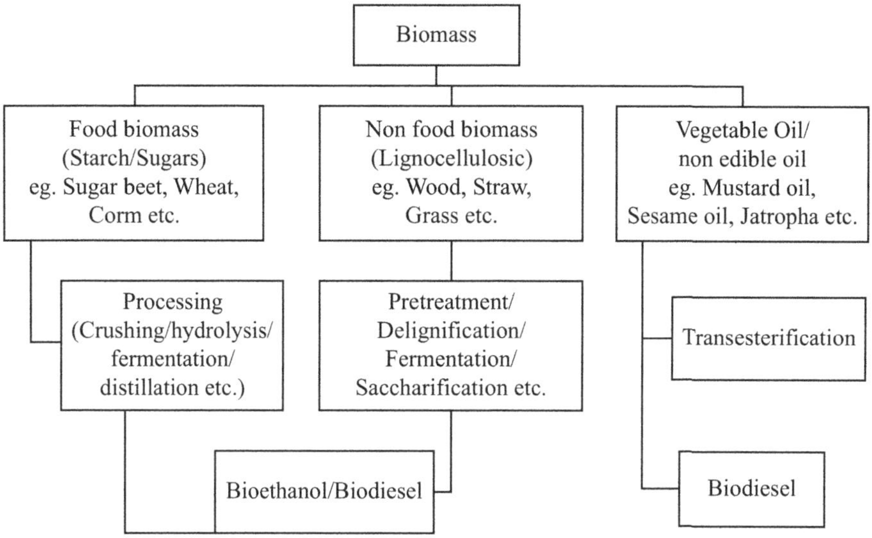

FIGURE 13.1 Second-generation biofuel

It has a relatively low content of monosaccharides, starch, protein, or oils shown in Figure 13.2 (Abdeshahian et al., 2010). Broadly bioethanol derived from agricultural and forest residues containing lignocellulose generally exhibits a lower global warming potential compared to bioethanol obtained from energy crops. This is largely due to N_2O emitted during the cultivation of energy crops, related to the use of fertilizers (Jeswani et al., 2020). Additionally, woody lignocellulose biomass (LCB) stands out as a crucial source for bio refining applications due to its advantages such as higher lignin content, season independence, lower ash content, and greater bulk density (Srivastava et al., 2023; Rodionova et al., 2022; Yan et al., 2017). Ethanol derived from lignocellulosic biomass, such as agricultural residues (e.g., rice straw and wheat straw) and grasses like switchgrass, is referred to as biomass ethanol. Lignocellulosic materials serve as a diverse range of inputs for advanced biofuels and can be acquired through two primary processes: hydrolysis and fermentation, leading to bioethanol production, or gasification, giving rise to fuels like Fischer-Tropsch biodiesel, bio-DME, and bio-SNG. Biodiesel production plants utilizing lignocellulosic biomass are limited to only a few countries. The primary producers are the United States (soybean oil), Europe and Canada (rapeseed oil), India and Southeast Asia (Jatropha oil), and Indonesia and Malaysia (palm oil) currently relying on vegetable oils for biofuel production (Duren et al., 2015 ; Mukherjee et al., 2014).

Edible oils can be converted into biodiesel; this shows that food resources are converted into automotive fuels. Large-scale production of biodiesel from edible oils may bring global imbalance to the food supply and demand market and may increase deforestation. Therefore, there is also a need to produce biodiesel from non-edible oil as well. According to Bhatia et al., 2017 to actualize the production of

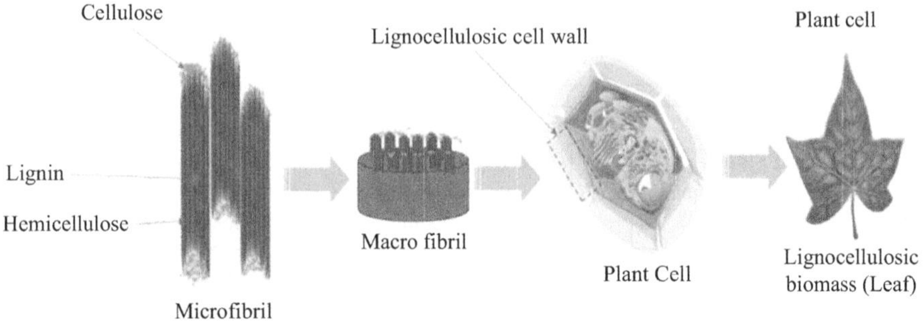

FIGURE 13.2 Structure of lignocellulosic biomass

second-generation biofuels, overcoming specific technological limitations is essential. Feedstock cost, accounting for more than 45% of the fermentation process, plays a crucial role in determining the economic viability of the overall process. The technologies employed in biomass pre-treatment are both inefficient and expensive. Further research is necessary to address this issue, aiming to enhance sugar yield and decrease the production of inhibitory byproducts (Le et al., 2017; Bhatia et al., 2017).

13.9 PRODUCTION OF BIOETHANOL

Production of bioethanol is a four-step process i.e., pre-treatment, hydrolysis, delignification, fermentation, and separation or distillation. Anhydrous ethanol as a final product is formed after this process (Figure 13.3).

The main objective to give pre-treatment and begin the biofuel production is (i) size reduction of raw material, (ii) exposure of carbohydrates (hemicelluloses, cellulose, and starch) before the hydrolysis to obtain improved reducing sugars, (iii) to provide proper assessment of enzyme for the hydrolysis of carbohydrates into fermentable sugar, and (iv) reduction of crystalline cellulose matrix. Pre-treatment is an important process as fermentable sugars are obtained as a product, and it prevents the premature degradation of obtained sugars, formation of inhibitors before the hydrolysis and fermentation; reduces operation cost and reduces the consumption of conventional energy (Graf et al., 2000; Mosier et al., 2005; Aditiya et al., 2016)

Pre-treatment in lignocellulosic material is done to deform the rigid structure of lignin, cellulose, and hemicelluloses, which reduces the degree of crystalline structure and is a required condition before hydrolysis. Reducing the size of feed mechanically is a sufficient pre-treatment for lignocellulosic material. But, chemically pre-treating the feed will ease the hydrothermal process in later stages. Combining two pre-treatment methods i.e., chemical and mechanical, will improve the yield of reducing sugar at the end of hydrolysis.

The purpose of selecting pre-treatments from the available methods is to break the crystalline structure of lignocellulosic material into an amorphous form of cellulose, so that it can be acted upon by the enzymes in the hydrolysis stage and converted into

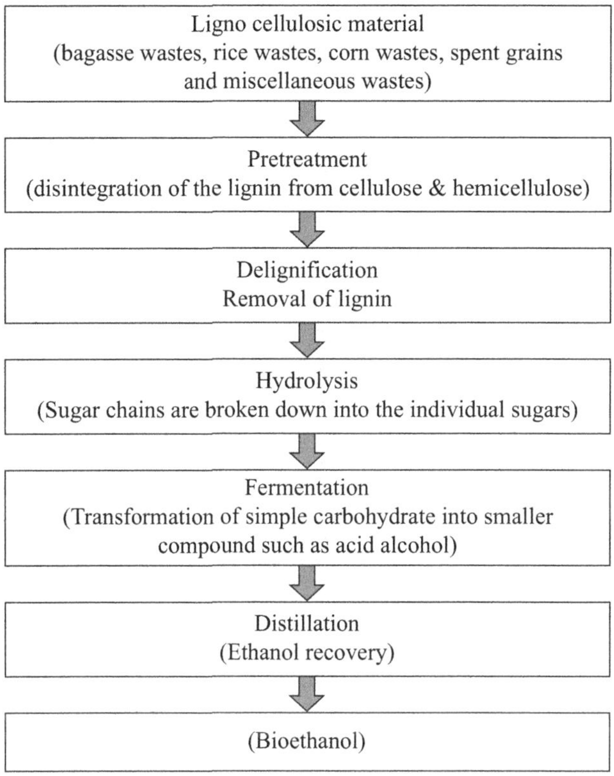

FIGURE 13.3 Process flow chart for bioethanol

sugar monomers (Aditiya et al., 2016). Conversion of complex sugars (carbohydrate) into sugar monomers in lignocellulosic materials is an energy-consuming process. The available pre-treatment methods are classified into biological, chemical, physical, and physicochemical.

13.9.1 Biological Pre-treatment

In this process, microorganisms are used for the pre-treatment work. The selection of microorganisms mainly depends on their capability to work on lignocellulosic feed to convert it into an amorphous form (Prasad et al., 2007). Brown rot, soft rot, and white rot fungi are some microorganisms that are used for this purpose. Brown rot can only act on cellulose, while soft and white rot can work on both cellulose and lignin (Prasad et al., 2007). Among these available fungi, white rot fungi are mostly used for biological pre-treatment. The ability of fungi to act on lignocellulosic structure depends on their hydrolytic system and ligninolytic system. These two systems are responsible for the production of hydrolases to liberate the polysaccharide and for breaking the complex lignin structure. These pre-treatments are environmentally friendly and sustainable in nature and consume less energy as they require mild conditioning. The fungi require no additional chemical substances as they have a built-in ability to disintegrate lignocellulosic walls, which makes it safe from harmful effects. However, microorganisms require time to act on biological materials, making this method economically not feasible for large-scale or industrial levels due to slow production rates. The productivity of this pre-treatment can be increased if we use a combination of other methods available (Vitousek et al., 2009).

13.9.2 Chemical Pre-treatment

Specific chemicals are added to disintegrate lignocellulosic or starchy material of the feed for better hydrolysis. Among all the available pre-treatment methods, treating feedstock with dilute acid is the most common chemical method that is being used at a commercial scale. Chemicals used for pre-treatment are not affected by technological factors like enzymes and are easy to obtain, durable, and less hassle in storage. Whereas engineered enzymes are not cost-friendly. The chemicals degrade the walls of lignocelluloses and complex carbohydrate chains of starch, which requires less energy (in the form of heat). This pre-treatment process is time-consuming and yields fewer carbohydrate monomers (for alkali pre-treatment). Several inorganic acids like hydrochloric acid, phosphoric acid, nitric acid, and sulfuric acid and organic acids like peracetic acid, maleic acid, lactic acid, and acetic acid are used for the disintegration of lignocelluloses (Balat & Balat, 2009). Acids are used in both forms, i.e., concentrated and diluted, for the chemical pre-treatment. Shorter time and mild temperature are required for concentrated acids to obtain sugar monomers, whereas diluted acids produce the same result but the nature of the reaction is different than concentrated. There are two methods by which diluted acid pre-treatment can be done, i.e., batch and continuous. In batch type, low temperature (usually less than 433K or 160°C) with high load of substrate (10–40 w/w %) is used, whereas for continuous type, high temperature (more than 433K or 160°C) and low substrate (5–10 w/w %) (Iranmahboob, 2002; Silverstein, 2005). Dilute acid pre-treatment does not require acid recovery and has negligible acid loss. This pre-treatment is the most common for the production of bioethanol from lignocellulosic biomass.

13.9.3 Mechanical Pre-treatment

In this method, biomass is physically reduced by cutting, chopping, or any other material-breaking method and aims to break the crystalline structure of lignocellulose to enhance the bioethanol production process. To reduce the size of biomass mechanically, milling methods like wet disk milling, ball milling, vibratory ball milling, compression milling, hammer milling, and roll milling are used (Aditiya et al., 2016).

13.9.4 Microwave Irradiation

An electromagnetic field is applied to pre-treated biomass to heat the sample internally and to reorganize the internal structure. The working principle of a microwave is that it uses the dielectric properties of liquid and the heat generated due to friction caused by aligning dipoles to the applied electrical field. This causes the fracture of lignocellulosic structure and provides a larger surface for enzymes to work on. Treatment with microwave improves the effectiveness of pre-treatment methods. Reducing the size of the feed to the powder may form lump (Shukla et al., 2023).

13.9.5 Extrusion

It is an advanced mechanical comminution method to reduce the size of biomass. It creates multiple openings for enzymes to work more efficiently and effectively on biomass. It is a HTST process that causes physical as well as chemical deformation in feedstock due to the presence of high temperature. Material passing through the extruder undergoes both shearing and mixing. This pre-treatment is controlled by the speed of the screw, temperature, and compression ratio (Shukla et al., 2023).

13.9.6 Physicochemical Pre-treatment

Steam explosion is a type of physicochemical pre-treatment in which the combined effect of hydrothermal treatment and sudden pressure switch is used to treat biomass. Initially, biomass undergoes high temperature and pressure for a short duration, usually up to a few seconds to minutes, which ends after pressure is removed from biomass when the fibrous structure disintegrates. Breaking of long fibers using steam is the mechanical part of this process. Whereas, the acetyl group present in hemicelluloses undergoes self-hydrolysis and is converted to acetic acid. An acidic environment is initiated when water is put at a high temperature. Treating hemicellulose-rich biomass with steam explosion is beneficial and efficient. High xylose recovery (about 45–65%), less energy requirement, and environmentally friendly are some other advantages of steam explosion in bioethanol production. The main disadvantage of this process is the formation of inhibitors that may create hindrance and act as a hurdle to microbial activity in the following stages, i.e., fermentation (Neves et al., 2007; Shukla et al., 2023).

13.9.7 Hydrolysis

It is the separation of long chains of carbohydrates by adding water and is catalyzed with the help of an enzyme or acid. The quality of hydrolysate is affected by the consequent fermentation process; it is a vital stage in bioethanol production. It is connected to the ethanol's quality. Hydrolysis is needed as microorganisms can only digest simple sugars derived from complex carbohydrates. Enzymatic hydrolysis has economic challenges and is considered impractical for commercial usage. However, when compared with acid hydrolysis, less equipment maintenance cost is required as the enzymes are working in a mild environment. The disposal system for acid hydrolysis is important to consider, and it requires extra costs while taking into consideration. One of the major problems is the ability of acid to degrade the sugar monomer (formed by an acidic environment at high temperature) (Ferreira et al., 2009).

Enzymatic hydrolysis of lignocellulosic biomass, which is majorly composed of cellulose, hemicellulose, and lignin, is degradable enzymatically with respect to the major constituents. Cellulose is a glucose chain that is connected by β-1,4 linkages. Cellulose has a certain degree of crystallinity, which is formed by hydrogen bonds. Cellulose is formed by hydrogen bonds along the chain and has a specific amount of degree of crystallinity. The presence of cellulose crystallinity slows the hydrolysis rate as well as lowers enzyme adsorption when crystallinity is high (Yang et al., 2011). Amorphous zones are easier to degrade than rigid crystalline areas, but a mechanical pre-treatment method, which suits the purpose, is used to reduce the crystallinity of the biomass. Cellulose is degraded by a specific enzyme, namely cellulose. The classification of cellulases' enzymatic activity is divided by their specific tasks, such as endoglucanase, exoglucanase, and β-glucosidase, which produce glucose as the sugar monomer (Jayasekara & Ratnayake, 2019).

Hemicellulose structure enzyme is complex but specific in its task; in contrast to cellulose, hemicelluloses are easier to hydrolyze due to their amorphous nature. Hemicellulose has xylan, which is a polysaccharide with xylose as the singular unit. Xylan accounts for 10–15% of the hemicelluloses in softwood and 10–35% in hardwoods (Asif, 2009).

13.10 DELIGNIFICATION/LIGNIN REMOVAL PROCESS

Lignin creates strong bonds of the cellulosic fibers, but it lessens the accessibility of cellulose to the cellulases. Lignin counteracts cellulase enzymatic activities, which lowers hydrolysis quality and make the hydrolysis process unfavorable. Hence, the removal of lignin becomes a must to maximize the effectiveness of hydrolysis by cellulase. In the lignin removal process, fibrils of microcellulose are openly exposed and neatly separated. This could help to provide better access for the cellulases. Hence, this delignification process is required for obtaining better fermentable sugars. It can be carried out by a chemical delignification method, using either aqueous or organic solvent (Wyman et al., 2005; Wu et al., 2020)

The ability of hemicellulases enzymes to degrade hemicellulose structure into sugar monomers, apparently, is not so favorable for cellulose hydrolysis. Xylooligosaccharides, xylose, mannose, and galactose are products of hemicellulose hydrolysis. These act as strong inhibitors to cellulase and β-glucosidase (Ximenes et al., 2011). Cellulose-degrading enzymes, known as cellulases, are highly effective for materials rich in cellulose. Similarly, it is recommended to use hemicellulolytic enzymes specifically for hemicellulose-rich materials. This is because sugars derived from hemicellulose can potentially slow down the activity of cellulase enzymes. Therefore, it is economically prudent to focus solely on cellulose degradation. By adhering to this principle, the overall production cost can be estimated based on the nature of the materials. Furthermore, understanding the nature of the material aids in selecting the appropriate microorganism for the subsequent fermentation process, as specific microorganisms are required to digest specific sugars. In hardwood, hemicellulose primarily consists of pentose subunits, specifically xylose.

On the other hand, softwoods contain hexose subunits like glucose, mannose, and galactose in their hemicellulose composition.

13.11 FERMENTATION

Ethanol is produced from the metabolic activity of the fermentation agent. To digest the respective sugar compounds, hydrolysate is introduced to a specific fermentation agent (yeast or bacteria). Glucose-fermenting bacteria need glucose-rich hydrolysate as food, and as a product, they leave ethanol compounds. Sugar monomers could be fractions of different monomers and several other oligosaccharides with probable inhibitors or indigestible substances. One kg of glucose and xylose has the capacity to produce 0.49 kg CO_2 and 0.51 kg of ethanol (Chang et al., 2018). The temperature and pH range should be optimum for fermentation so that the microbes can sustain. 303–311°K temperature is favorable for mesophilic organisms, most fermentation agents. A pH of 6.5–7.5 is required for bacteria-type fermentation microbes, and a pH of 3.5–5.0 is required for fungi-type fermentation. Microbial growth rate, tolerance of inhibitors, genetic stability, osmosis and alcohol, productivity, and the yield of ethanol are other factors that affect the fermentation process (Hettenhaus, 1998; Demirbaş, 2004; Balat & Balat, 2009). There are many other bacteria also that are suitable as fermentation microorganisms. Zymomonas mobilis fermentation bacteria can convert sucrose, glucose, and fructose into ethanol. Similarly, Saccharomyces cerevisiae yeast can naturally consume hexose sugars such as glucose and fructose (Parapouli et al., 2020).

13.12 DISTILLATION

Fuel-grade bioethanol is obtained through the distillation process. A portion of ethanol gets separated from a miscible liquid mixture. This liquid mixture is a fermentation broth. Consecutive selective evaporation and condensation processes are done for distillation. Distillation is based on a difference in volatility. Ninty-six percent concentrated ethanol requires a huge amount of energy and costs, as the post-fermentation mixture has a high water content. A rectification column is used to concentrate the product from about 37% bioethanol to a concentration of about 95%. Later, it is dehydrated to contain a minimum of 99.5% ethanol by volume (Robak & Balcerek, 2018). Adsorption distillation, azeotropic distillation, membrane processes, extractive diffusion distillation, vacuum distillation, and chemical dehydration are a few methods that are used for distillation (*Broda* et al., 2022).

13.13 BIODIESEL

Biodiesel, a renewable and biodegradable liquid fuel, serves as a substitute for petroleum-derived diesel in the transportation industry (Uthandi et al., 2022). The limitations imposed on the production of first-generation biodiesel have led to the rise of second and third-generation biodiesel, which is derived from inedible oils (such as used cooking oil), oils from plants that thrive in non-arable land, or microbial oils

(López-Fernández et al., 2021). Lignocellulosic biomass is utilized as a raw material to produce biodiesel from agricultural waste like rice bran, inedible oilseeds, coffee grounds, and oil-free cakes due to their residual oil content (Koul et al., 2022). The second method is to use lignocellulosic biomass as a substrate for oleaginous microorganisms (OMs) to generate lipids (Koul et al., 2022; Uthandi et al., 2022). Under specific environmental conditions, these microorganisms can accumulate oils within their cells, ranging from 20% to 70% (Uthandi et al., 2022). Moreover, OMs have the potential to enhance the circular economy and facilitate cost-effective process development (Chintagunta et al., 2021). Similar to other second-generation liquid biofuels, the pre-treatment of LCB is essential to release lignin and improve the accessibility of cellulose for enzymatic hydrolysis. This is followed by fermentation with OMs to produce lipids, which are then transesterified into biodiesel (Mujtaba et al., 2023).

Modern diesel engines are designed to run on petroleum (fossil fuel) diesel but can run on plant oils with low viscosity. The oil is concentrated in seeds, nuts, or the germ of the plants. Oil can be extracted from these plants and converted into biodiesel. Biodiesel is a mixture of long-chain monoalkyl esters of fatty acids derived from vegetable oil or animal fat. One of the most common methods for biodiesel production is a base-catalyzed batch reaction of oil with methanol called transesterification.

13.14 CONCLUSION

Lignocellulosic biomass is the most abundant bioresource on earth. First-generation biofuels are those that are produced from edible energy crops. Second-generation biofuels are needed because of the restricted amount of feedstock. This second-generation biomass includes non-food crops, including the waste from food crops, agricultural residue, wood chips, and waste cooking oil. This biomass contains lignocellulose. Hence, there is a need for improvements in the production of lignocellulosic biofuel for improving the cost-effectiveness and for commercialization from the laboratory to the industrial/commercial scale. Though production of these biofuels is increasing, there is still a need to increase the efficiency of the fermentation process to the point where all sugars (pentoses and hexoses) released during the pre-treatment and hydrolysis steps get fermented into bioethanol. Variable composition of biomass, generation of inhibitors during pre-saccharification treatment, end-product inhibition, osmotic and oxidative stress, and ethanol accumulation are some constraints that need solutions for optimization of biofuel production. Biofuel is helpful for maintaining a sustainable environment. Bioethanol is mostly used in petrol to improve the gas emissions. Low-level blends of biodiesel can also be used.

REFERENCES

Abdeshahian, P., Dashti, M. G., Kalil, M. S., & Yusoff, W. M. W. (2010). Production of biofuel using biomass as a sustainable biological resource. Biotechnology, 9(3), 274–282.

Aditiya, H. B., Mahlia, T. M. I., Chong, W. T., Nur, H., & Sebayang, A. H. (2016). Second-generation bioethanol production: A critical review. Renewable and Sustainable Energy Reviews, 66, 631–653.

Ambaye, T. G., Vaccari, M., Bonilla-Petriciolet, A., Prasad, S., van Hullebusch, E. D., & Rtimi, S. (2021). Emerging technologies for biofuel production: A critical review on recent progress, challenges and perspectives. *Journal of Environmental Management, 290*, 112627.

Asif, M. (2009). Sustainability of timber, wood and bamboo in construction. In F. Pacheco-Torgal, S. Jalali, & A. Fucic (Eds.), *Sustainability of construction materials* (pp. 31–54). Woodhead Publishing.

Balat, M., & Balat, H. (2009). Recent trends in global production and utilization of bio-ethanol fuel. *Applied Energy, 86*(11), 2273–2282.

Bhatia, S. K., Kim, S. H., Yoon, J. J., & Yang, Y. H. (2017). Current status and strategies for second generation biofuel production using microbial systems. *Energy Conversion and Management, 148*, 1142–1156.

Broda, M., Yelle, D. J., & Serwańska, K. (2022). Bioethanol production from lignocellulosic biomass—challenges and solutions. *Molecules, 27*(24), 8717.

Carere, C. R., Sparling, R., Cicek, N., & Levin, D. B. (2008). Third generation biofuels via direct cellulose fermentation. *International Journal of Molecular Sciences, 9*(7), 1342–1360.

Carriquiry, M. A., Du, X., & Timilsina, G. R. (2011). Second generation biofuels: Economics and policies. *Energy Policy, 39*(7), 4222–4234.

Chang, Y. H., Chang, K. S., Chen, C. Y., Hsu, C. L., Chang, T. C., & Jang, H. D. (2018). Enhancement of the efficiency of bioethanol production by Saccharomyces cerevisiae via gradually batch-wise and fed-batch increasing the glucose concentration. *Fermentation, 4*(2), 45.

Chintagunta, A. D., Zuccaro, G., Kumar, M., Kumar, S. J., Garlapati, V. K., Postemsky, P. D., & Simal-Gandara, J. (2021). Biodiesel production from lignocellulosic biomass using oleaginous microbes: Prospects for integrated biofuel production. *Frontiers in Microbiology, 12*, 658284.

Das Neves, M. A., Kimura, T., Shimizu, N., & Nakajima, M. (2007). State of the art and future trends of bioethanol production. *Dynamic Biochemistry, Process Biotechnology and Molecular Biology, 1*(1), 1–14.

Demirbaş, A. (2004). Ethanol from cellulosic biomass resources. *International Journal of Green Energy, 1*(1), 79–87.

Dragone, G., Fernandes, B. D., Abreu, A. P., Vicente, A. A., & Teixeira, J. A. (2011). Nutrient limitation as a strategy for increasing starch accumulation in microalgae. *Applied Energy, 88*(10), 3331–3335.

Dutta, K., Daverey, A., & Lin, J. G. (2014). Evolution retrospective for alternative fuels: First to fourth generation. *Renewable Energy, 69*, 114–122.

Ferreira, S., Duarte, A. P., Ribeiro, M. H., Queiroz, J. A., & Domingues, F. C. (2009). Response surface optimization of enzymatic hydrolysis of Cistus ladanifer and Cytisus striatus for bioethanol production. *Biochemical Engineering Journal, 45*(3), 192–200.

Figueiredo, M. K., Romeiro, G. A., Damasceno, R. N., & Franco, A. P. (2009). The isolation of pyrolysis oil from castor seeds via a Low Temperature Conversion (LTC) process and its use in a pyrolysis oil–diesel blend. *Fuel, 88*(11), 2193–2198.

Ganesan, R., Manigandan, S., Samuel, M. S., Shanmuganathan, R., Brindhadevi, K., Chi, N. T. L., ... Pugazhendhi, A. (2020). A review on prospective production of biofuel from microalgae. *Biotechnology Reports, 27*, e00509.

Ganguli, S., Somani, A., Motkuri, R. K., & Bloyd, C. N. (2018). *India alternative fuel infrastructure: The potential for second-generation biofuel technology* (No. PNNL-28283). Pacific Northwest National Lab(PNNL).

Graf, A., & Koehler, T. (2000). *Oregon cellulose-ethanol study*. Oregon Office of Energy.

Gupta, V., Singh, N. V. V., Shukla, A., Singh, N., & Mishra, S. K. (2020). Sustainable biofuel production using lignocellulosic biomass as a raw material. International Journal of Current Microbiology and Applied Sciences, 9(7), 3339–3350.

Hermwille, L., Obergassel, W., Ott, H. E., & Beuermann, C. (2017). UNFCCC before and after Paris–what's necessary for an effective climate regime?. *Climate policy*, 17(2), 150-170.

Hettenhaus, J. (1998). Ethanol fermentation strains: Present and future requirements for biomass to ethanol commercialization. United States Department of Energy and National Renewable Energy Laboratory.

IEA. (2013). *World energy Outlook 2013*. International Energy Agency (IEA).

Iranmahboob, J., Nadim, F., & Monemi, S. (2002). Optimizing acid-hydrolysis: A critical step for production of ethanol from mixed wood chips. *Biomass and Bioenergy*, *22*(5), 401–404.

Jayasekara, S., & Ratnayake, R. (2019). Microbial cellulases: An overview and applications. *Cellulose*, *22*, 92.

Jeswani, H. K., Chilvers, A., & Azapagic, A. (2020). Environmental sustainability of biofuels: A review. *Proceedings of the Royal Society A*, *476*(2243), 20200351.

Koul, B., Yakoob, M., & Shah, M. P. (2022). Agricultural waste management strategies for environmental sustainability. *Environmental Research*, *206*, 112285.

Le, R. K., Wells Jr, T., Das, P., Meng, X., Stoklosa, R. J., Bhalla, A., & Ragauskas, A. J. (2017). Conversion of corn stover alkaline pre-treatment waste streams into biodiesel via Rhodococci. *RSC Advances*, *7*(7), 4108–4115.

López-Fernández, J., Benaiges, M. D., & Valero, F. (2021). Second-and third-generation biodiesel production with immobilised recombinant Rhizopus oryzae lipase: Influence of the support, substrate acidity and bioprocess scale-up. *Bioresource Technology*, *334*, 125233.

Lü, J., Sheahan, C., & Fu, P. (2011). Metabolic engineering of algae for fourth generation biofuels production. *Energy & Environmental Science*, *4*(7), 2451–2466.

Mahapatra, S., Kumar, D., Singh, B., & Sachan, P. K. (2021). Biofuels and their sources of production: A review on cleaner sustainable alternative against conventional fuel, in the framework of the food and energy nexus. *Energy Nexus*, *4*, 100036.

Malode, S. J., Prabhu, K. K., Mascarenhas, R. J., Shetti, N. P., & Aminabhavi, T. M. (2021). Recent advances and viability in biofuel production. *Energy Conversion and Management: X*, *10*, 100070.

Monir, M. U., Hasan, M. Y., Ahmed, M. T., Aziz, A. A., Hossain, M. A., Woobaidullah, A. S. M., Biswas, P. K., & Haque, M. N. (2021). Optimization of fuel properties in two different peat reserve areas using surface response methodology and square regression analysis. Biomass Conversion and Biorefinery.

Mosier, N., Wyman, C., Dale, B., Elander, R., Lee, Y. Y., Holtzapple, M., & Ladisch, M. (2005). Features of promising technologies for pretreatment of lignocellulosic biomass. *Bioresource Technology*, *96*(6), 673–686.

MPNG. (2012). *Basic statistics on Indian petroleum & natural gas 2011–12*. Ministry of Petroleum and Natural Gas (MPNG), Government of India.

Mujtaba, M., Fraceto, L., Fazeli, M., Mukherjee, S., Savassa, S. M., de Medeiros, G. A., & Vilaplana, F. (2023). Lignocellulosic biomass from agricultural waste to the circular economy: A review with focus on biofuels, biocomposites and bioplastics. *Journal of Cleaner Production*, 136815.

Mukherjee, I., & Sovacool, B. K. (2014). Palm oil-based biofuels and sustainability in Southeast Asia: A review of Indonesia, Malaysia, and Thailand. Renewable and Sustainable Energy Reviews, 37, 1–12.

Naik, S. N., Goud, V. V., Rout, P. K., & Dalai, A. K. (2010). Production of first and second generation biofuels: A comprehensive review. *Renewable and Sustainable Energy Reviews*, *14*(2), 578–597.

Parapouli, M., Vasileiadis, A., Afendra, A. S., & Hatziloukas, E. (2020). Saccharomyces cerevisiae and its industrial applications. *AIMS Microbiology*, *6*(1), 1.

Prasad, S., Singh, A., & Joshi, H. C. (2007). Ethanol as an alternative fuel from agricultural, industrial and urban residues. *Resources, Conservation and Recycling*, *50*(1), 1–39.

Purohit, P., & Dhar, S. (2015). Promoting low carbon transport in India: Biofuel roadmap for India.

Rai, A. K., Al Makishah, N. H., Wen, Z., Gupta, G., Pandit, S., & Prasad, R. (2022). Recent developments in lignocellulosic biofuels, a renewable source of bioenergy. *Fermentation*, *8*(4), 161.

Robak, K., & Balcerek, M. (2018). Review of second generation bioethanol production from residual biomass. *Food Technology and Biotechnology*, *56*(2), 174.

Rodionova, M. V., Bozieva, A. M., Zharmukhamedov, S. K., Leong, Y. K., Lan, J. C. W., Veziroglu, A., Veziroglu, T. N., Tomo, T., Chang, J. S., & Allakhverdiev, S. I. (2022). A comprehensive review on lignocellulosic biomass biorefinery for sustainable biofuel production. International Journal of Hydrogen Energy, 47(3), 1481–1498.

Sharma, S., Meena, R., Sharma, A., & Goyal, P. (2014). Biomass conversion technologies for renewable energy and fuels: A review note. Journal of Mechanical and Civil Engineering, 11(2), 28–35.

Shukla, A., Kumar, D., Girdhar, M., Kumar, A., Goyal, A., Malik, T., & Mohan, A. (2023). Strategies of pretreatment of feedstocks for optimized bioethanol production: Distinct and integrated approaches. *Biotechnology for Biofuels and Bioproducts*, *16*(1), 44.

Srivastava, N., Singh, R., Singh, P., Ahmad, I., Singh, R. P., Rai, A. K., Asiri, M., & Gupta, V. K. (2023). Recent advances on lignocellulosic bioresources and their valorization in biofuels production: Challenges and viability assessment. *Environmental Technology & Innovation*, 31, 103037.

Uthandi, S., Kaliyaperumal, A., Srinivasan, N., Thangavelu, K., Muniraj, I. K., Zhan, X., & Gupta, V. K. (2022). Microbial biodiesel production from lignocellulosic biomass: New insights and future challenges. *Critical Reviews in Environmental Science and Technology*, *52*(12), 2197–2225.

van Duren, I., Voinov, A., Arodudu, O., & Firrisa, M. T. (2015). Where to produce rapeseed biodiesel and why? Mapping European rapeseed energy efficiency. Renewable Energy, 74, 49–59.

Vitousek, P. M., Naylor, R., Crews, T., David, M. B., Drinkwater, L. E., Holland, E., ... & Zhang, F. S. (2009). Nutrient imbalances in agricultural development. *Science*, *324*(5934), 1519-1520.

Wang, F., Ouyang, D., Zhou, Z., Page, S. J., Liu, D., & Zhao, X., 2021. Lignocellulosic biomass as sustainable feedstock and materials for power generation and energy storage. *Journal of Energy Chemistry*, *57*, 247–280.

Wu, J., Chandra, R. P., Takada, M., Liu, L. Y., Renneckar, S., Kim, K. H., & Saddler, J. N. (2020). Enhancing enzyme-mediated cellulose hydrolysis by incorporating acid groups onto the lignin during biomass pretreatment. *Frontiers in Bioengineering and Biotechnology*, *8*, 608835.

Wyman, C. E., Dale, B. E., Elander, R. T., Holtzapple, M., Ladisch, M. R., & Lee, Y. Y. (2005). Coordinated development of leading biomass pretreatment technologies. *Bioresource Technology*, *96*(18), 1959–1966.

Ximenes, E., Kim, Y., Mosier, N., Dien, B., & Ladisch, M. (2011). Deactivation of cellulases by phenols. *Enzyme and Microbial Technology*, *48*(1), 54–60.

Yan, X., Wang, Z., Zhang, K., Si, M., Liu, M., Chai, L., & Shi, Y. (2017). Bacteria-enhanced dilute acid pretreatment of lignocellulosic biomass. *Bioresource Technology*, *245*, 419–425.

Yang, B., Dai, Z., Ding, S. Y., & Wyman, C. E. (2011). Enzymatic hydrolysis of cellulosic biomass. *Biofuels*, *2*(4), 421–449.

14 Third-Generation Biofuel
Microalgae and Its Utilization

Sheetal Mane, Anupama Singh, Rachna Gupta, Shivani Desai, and Purnima Kumari

14.1 INTRODUCTION

Biofuel is a fuel containing hydrocarbons produced from organic matter (living or non-living matter) in a short duration (days, weeks, or conceivably months). Bhore (2019) predicts a 40% surge in fossil fuel requirements from 2010 to 2040, necessitating the exploration of alternative energy sources to meet our energy demands. Renewable sources of energy include wind, solar, and biomass-made biofuels. Biofuels and bioproducts have been generated over the last few decades to fulfill the growing need for fossil fuels. Biofuels are available in a variety of forms, including biodiesel, bioethanol, biogas, etc. Biofuels are primarily categorized by generations: first-generation, second-generation, and third-generation. Agrofuel or first-generation biofuel has been produced by fermenting plant sugars (for example, sugar beet, sugarcane, palm, maize, soybean, and sweet sorghum) or starches using yeast to produce bioethanol and subsequently to produce biodiesel. But first-generation biofuels raise substantial environmental, financial, and social issues since the massive feedstock production needs more arable agricultural land, which reduces the available acreage for growing food for humans and livestock. In addition, the food and water industries are also seriously impacted by these production techniques. Hence, interest in 1G biomass has decreased. And the researchers have concentrated on second-generation (2G) biofuels, obtained from different non-edible sources such as cellulosic biomass, jatropha, inedible plant parts, manure, straw, leftover cooking oil, sawdust, and wood. However, because of the requirement of expensive and modern techniques for biofuel production, second-generation biofuels are not economically feasible (Brennan and Owende 2010; Dragone et al. 2010). Third-generation biofuels encompass algal biofuels and synthetic biology. In Figure 14.1, the distinguishing characteristics of first-, second-, third-, and fourth-generation biofuels are shown. Algal biofuels offer two noteworthy benefits: high productivity and do not require agricultural land or potable water; in fact, the best location for their cultivation is nearby CO_2 sources. Synthetic biology is in its early stages, aiming to employ genetic modifications for the creation of plants or microorganisms capable of synthesizing drop-in fuels within their cells.

DOI: 10.1201/9781032650616-14

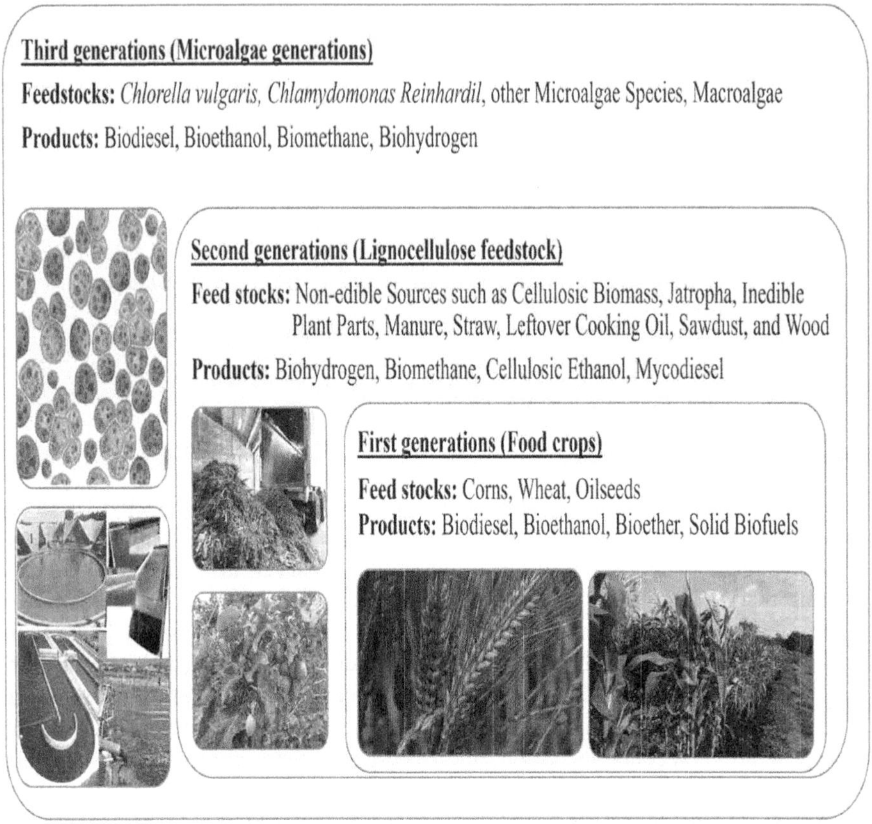

FIGURE 14.1 Distinctive characteristics of first, second, and third generations of biofuels

Henceforth, scientists have shifted their attention to the evolution of third-generation biomass using microalgae, which is now considered an acceptable choice for the renewable energy source of biofuel production, by tackling the shortcomings of first and second-generation biofuels.

14.2 MICROALGAE SOURCE OF BIOFUEL: CHARACTERISTICS AND VARIETIES

In contrast to conventional agriculture, photosynthetic microalgae species may develop quickly and produce significant amounts of biomass on a given area of land (Yap et al. 2021). Spolaore et al. (2006) reported that Nihon Chlorella, a Japanese individual, initiated the initial large-scale cultivation of *Chlorella* species during the 1960s. At the moment, several species of microalgae are employed as a source of biofuel that may be applied in various ways. Microalgae, a type of biofuel feedstock belonging to the third generation, are reportedly non-competitive with food and land

TABLE 14.1
Common microalgae species used in biofuel production

Microalgal Species	Targeted Biofuel	Cultivation Condition
C. protothecoides	Biodiesel	Heterotrophy; nitrogen limitation
S. obliques	Biohydrogen	Indirect process; light biophotolysis; (Fe–Fe) enzymes
Nannochloropsis salina	Biogas	Photobioreactor, large scale, 35°C
Chlorococum sp.	Bioethanol	Fermentation
Spirulina sp.	Biomethanol	Gasification/anaerobic fermentation
Microalgal consortium	Biochar	Hydrothermal liquefaction

crops. A few examples of microalgae that have high concentrations of lipids and hydrocarbons are *Botryococcus braunii, Dunaliella primolecta, Nannochloropsis* sp., *Crypthecodinium cohnii,* and *Chlorella* sp. Green colonial microalgae called *Botryococcus braunii* have a high biomass-to-hydrocarbons ratio (Borowitzka et al. 2010). They can also produce additional substances of commercial value, such as polysaccharides and carotenoids (Fernandez et al. 2021). Oil content in microalgae species can reach 80%, with ranges between 20% and 50% being relatively normal (Halder et al. 2021). The microalga *Chlorella* contains up to 50% lipids, with *Botryococcus braunii* having the greatest oil concentration at about 80%. Specific biofuels can be found in significant quantities in several common microalgal species, according to Xu et al. (2006) (Table 14.1).

14.3 MICROALGAE CULTIVATION

Microalgae are a common type of plant-like, oxygen-producing photosynthetic organisms that may thrive in a variety of aquatic environments, such as marine and freshwater, as well as diverse wastewaters, including industrial, agricultural, municipal, and more. In addition to their biological significance, microalgae have a wide range of industrial uses, including the storage of carbon, the production of oxygen through photosynthesis, and nutrient utilization, like nitrogen and phosphorus from wastewater (Chew et al. 2018). As a result, the combination of microalgae culture with wastewater treatment might be considered a potential strategy for cultivating algae while assisting the wastewater treatment process (Cruz et al. 2018). Importantly, a higher nutrient concentration (nitrogen, phosphorus) in the wastewater is necessary for microalgal culture since high carbon concentrations would accelerate development (Harun et al. 2010). Algal growth is also aided by light quality, light intensity, and photo-bioreactors (whether open or closed) (Cuaresma et al. 2011). Carbon fixation, which accelerates algae development, is the primary purpose of light in algal culture (Hosseini et al. 2006). According to (Qiang et al. 1994), microalgae can have a variety of metabolic processes, including phototrophic, heterotrophic, photoheterotrophic, and mixotrophic ones (Table 14.2). As a result, choosing an effective

TABLE 14.2
Attributes of various microalgae growth conditions

Cultivation Nature	Heterotrophic	Phototrophic	Photoheterotrophic	Mixotrophic
Source of energy	Organic material	Light	Light	Light and organic material
Carbon source	Organic material	Inorganic	Organic material	Organic and inorganic materials
Light availability requirement	Not mandatory	Mandatory	Mandatory	Not mandatory
Cell density	High	Low	Medium	Medium
Constraints and challenges	Contamination risk; high substrate cost	Low cell concentration; high condensation rate	Contamination risk; high substrate cost; costly equipment's	High substrate cost; costly equipment; contamination risk

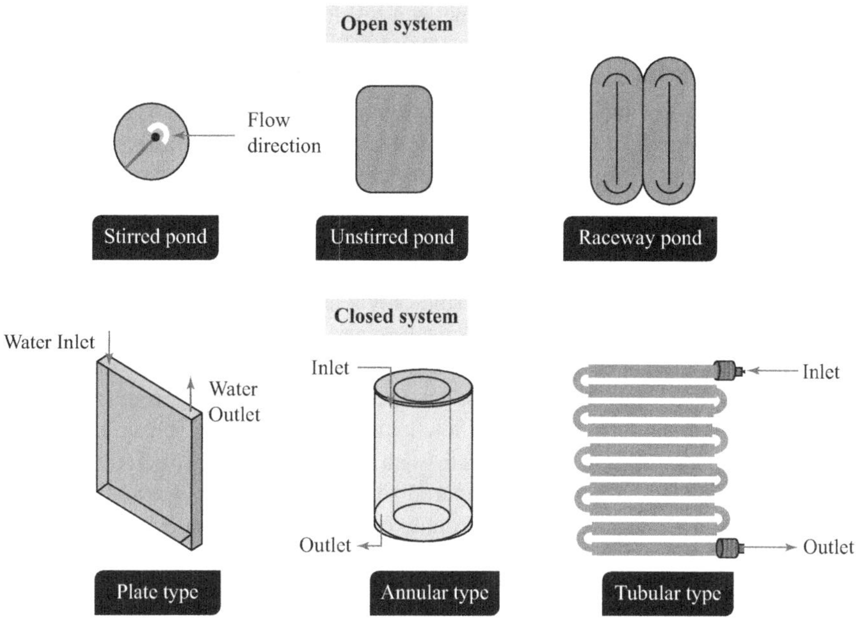

FIGURE 14.2 Distinctive characteristics of first, second, and third generations of biofuels

cultivation system for microalgae is a key step in determining their growth rate and the effectiveness of the chosen process. Microalgae have been grown on a significant scale for extracting valuable chemicals. Different cultivation technologies, including open pond, closed photobioreactor, and a hybrid system with two-stage growth, have been utilized for this purpose (Figure 14.2).

14.3.1 Cultivation Systems

14.3.1.1 Open Ponds

Open ponds or raceway ponds, such as shallow circular tanks, have been extensively studied as the most convenient microalgae cultivation methods since the 1950s, mainly because of their lower initial investment needs (Cruz et al. 2018). Tan et al. (2018) defined the Open (raceway) Pond as a shallow water pool with a depth ranging from 0.2 to 0.5 m, designed to provide adequate light and promote the rapid growth of microalgae. For the industrial-scale culture of microalgae and cyanobacteria like *Arthrospira platensis, Anabaena* sp., *Chlorella* sp., *Dunaliella salina, Phaeodactylum tricornotum, Nannochloropsis,* and *Pleurochrysis carterae* (Jorquera et al. 2010). Paddle wheels are further utilized to guarantee uniform distribution of the microalgae medium along the circulation channels. For example, Moazami et al. (2012) noticed an augmentation in both the lipid content and biomass of the microalgae medium by raising the paddle wheel speed for an indoor open pond system utilizing Nannochloropsis from 1.4 to 2.1 rad s^{-1}. Microalgae cells at the pond's bottom are in the dark zone as a result of the raceway ponds' intrinsic light reduction regime caused by the paddle wheel's decreased velocity (Rogers et al. 2014). Li et al. (2014) discovered that this design ultimately influences the biomass and lipid synthesis of microalgae. Therefore, continuous agitation in raceway ponds is crucial for reducing the duration during which microalgae are deprived of light. Nevertheless, as an increase in mixing speed causes a cubical increase in the paddle wheel power consumption, additional energy expenditures can be incurred (Rogers et al. 2014). Microalgae cultivation in raceway ponds presents several drawbacks, including excessive water evaporation, increased risk of culture contamination, and substantial land requirements. As a result, only a few species of microalgae may be effectively cultivated in this system. *Spirulina* and *Dunaliella* are two examples of organisms that can survive in environments with high alkalinity and high salinity. To successfully execute this system, it is imperative to choose the right site, preferably one with abundant sunlight and few natural catastrophes (Harun et al. 2010). In general, contamination is a problem with outdoor scale-up microalgal culture for raceway ponds. As a result, compared to other cultivation methods, they can be operated most effectively in autotrophic cultivation circumstances. In addition, raceway ponds are constrained by a few crucial development factors, including temperature, pH, and dissolved oxygen content (Harun et al. 2010). Despite its simpler setup and operation compared to closed photobioreactor systems, addressing issues related to low biomass concentration caused by uneven light distribution and lack of uniformity is crucial for the successful commercialization of microalgae-derived biodiesel production techniques. Therefore, complete life cycle analyses are essential to enhance this farming system, specifically in terms of biomass yield and system design.

14.3.1.2 Closed Photobioreactor

The best physiological and biological properties of microalgae cultures may be achieved by using photobioreactors to more effectively manage growth parameters, mainly temperature, pH, light intensity, CO_2, and nutrient content (Salama et al. 2017). A closed system called a photobioreactor enables the monoculture of microalgae over extended periods of time with high biomass output, rapid harvesting, and little chance of contamination (Tan et al. 2018). This method is typically employed to grow microalgae on a large scale to synthesize fatty acids, proteins, pigments, antioxidants, and animal feed. Several microalgal species, including *Chlorella sarkiniana*, *Chlorella vulgaris*, *Arthrospira platensis*, *Porphyridium cruentum*, *Nannochloropsis* sp., *Selenastrum capricornutum*, *Phaeodactylum tricornotum*, and *Haematococcus pluvialis*, have been investigated for their growth potential in photobioreactors (Pugliese et al. 2020). Specifically, based on the microalgal strains and scales utilized, photobioreactors provide increased volumetric productivity for the improved collection of radiant energy. Therefore, it is anticipated that the photobioreactor would be able to produce more oil per hectare than a raceway pond (136.9 m^{-3} ha^{-1} based on 70 wt.% oil in biomass) (Koru et al. 2014). Photobioreactors, which can operate in either batch or continuous mode, can be located indoors with artificial light or outside under natural sunlight (Duan et al. 2014).

Additional options for photobioreactor architecture include flat-plate and tubular constructions. For outdoor microalgae culture, tubular photobioreactors are favored because they may be built in a variety of ways, including vertically, horizontally, or inclinedly, depending on the system's requirements. The horizontal design is scalable but needs a large land area, whereas the vertical form offers greater gas-liquid mass transfer and uses less energy (Gupta et al. 2015). A flat-plate photobioreactor, on the other hand, has a small light pathway and is cuboidal in shape. These photobioreactors have a high surface-to-volume ratio and may be manufactured from transparent materials, including polycarbonate, glass, and plexiglass (Duan et al. 2014). Despite the limited research on flat-plate photobioreactors, it has been suggested that their wide lighting area and minimal accumulation of dissolved oxygen (O_2) enable efficient mass production of microalgae. The utilization of flat-plate photobioreactors has led to a lower biomass yield per unit area versus tubular photobioreactors, primarily because of inadequate access to light and absence of light dispersion, resulting in microalgal growth inhibition (Gupta et al. 2015).

14.3.1.3 Two-Stage Hybrid Cultivation System

A two-phase hybrid cultivation approach has been suggested to address limitations in closed-culture and open-culture methods. This technique aims to enhance microalgae biomass productivity while reducing production and operation expenses (Tan et al. 2018). In addition, the hybrid system, which consists of two-stage cultures, can boost lipid accumulation in microalgae (Narala et al. 2016). The initial stage of culture takes place in a photobioreactor, where growth conditions are managed to promote uninterrupted cell division and reduce contamination from other microorganisms (Zakariah et al. 2015). The photobioreactor's developing microalgae cells are transported to open ponds in the second phase, where they are exposed to nutrient reduction and various environmental challenges. The goal is to improve lipid storage in microalgae cells and obtain a specific fatty acid makeup for biodiesel production (Kadir et al. 2018). To ensure microalgae dominance in the culture medium and prevent the presence of other bacteria, a substantial inoculum is necessary during the second phase. Astaxanthin (a component utilized in salmon feed) and *Haematococcus pluvialis* oil were produced using a hybrid system (Huntley and Redalje 2007). As a result, depending on the local meteorological circumstances, using *Haematococcus pluvialis*, this system produced lipids on an annual average of 10 to 24 tonnes ha^{-1}. The research suggests that using microalgae species like *Chlorella vulgaris*, which have high lipid content and efficient photosynthesis, could potentially achieve a yield of 76 tonnes per hectare per year under identical conditions. In addition, Rodolfi et al. (2009) carried out a conceptual analysis of this two-stage cultivation method, finding that 90 kg ha^{-1} of lipid may be produced every day (10 kg ha^{-1} in the first phase and 8 kg ha^{-1} in the second phase). Rodolfi et al. (2009) achieved this by offering

22% of the production facility to cultivate biomass under nitrogen-rich environments while allocating the rest of the facility to lipid synthesis under nitrogen-limited conditions. In Mediterranean regions, the lipid production of this system reached 20 tonnes per hectare, while in hot tropical climates, it reached up to 30 tonnes per hectare. The microalgae culture system is crucial for achieving a high biomass output.

14.3.2 Growth Factors Affecting Microalgae Culture System

Microalgal growth is affected by several culture factors, including nutrient availability, light intensity, salinity, pH, mixing, CO_2, temperature, and dissolved oxygen levels. Hence, optimizing these parameters is crucial for maximizing biomass production.

14.3.2.1 Light Intensity

A significant limiting element in the production of microalgae has been thought to be light intensity. Algae can be cultivated using either natural (sun) or artificial light sources, like LED bulbs or tube lights, with specific intensities. Light span and intensity have a direct impact on microalgal photosynthesis, which also affects the biochemical makeup and microalgal biomass yield (Gonçalves et al. 2016). As demonstrated in research on Scenedesmus abundans microalgae (Mandotra et al. 2016), higher light intensities help to accelerate the photosynthesis rate until a state of equilibrium is reached between photoinhibition and photorespiration. Besides light intensity, photoperiod and the wavelength that cells are exposed to are also crucial factors for achieving optimal algae growth (Iasimone et al. 2018). A study conducted by Chew et al. (2017) experimentally exhibited variations in biomass production and growth rate of a particular strain of algae when exposed to diverse light intensities and durations. Depending on the variety of algae being employed, different light intensities are reportedly needed to promote optimal development. The bulk of research has found that the optimal environment for microalgae development is 16 hours of light followed by 8 hours of darkness (Chew et al. 2017).

14.3.2.2 Temperature

The control of temperature is essential for biological and microalgal development processes (Wang et al. 2019). The ideal temperature for each kind of microalgae species varies, and any deviation from temperature, whether lower or higher, may impede or halt the growth and functioning of microalgae, thereby reducing biomass production (Huang et al. 2019). Temperature has an impact on several factors linked to microalgal development, including pH value and the solubility of CO_2 in the medium (Xu et al. 2019). According to Suthar et al. (2018), most microalgae species thrive at temperatures between 20°C and 30°C. On the other hand, thermophile algae can endure greater temperatures (up to 80°C) (Suparmaniam et al. 2019). Temperature can be utilized as a stress stimulus to induce the production of beneficial metabolites (Khan et al. 2018). *Chlorella vulgaris* exhibited higher levels of lipid and carbohydrate production at 25°C compared to 30°C (Converti et al. 2009).

14.3.2.3 pH and Salinity

Changes in the medium pH affect photosynthesis and the microalgae growth rate. According to Qiu et al. (2017), microalgal organisms have a specific optimal pH and salinity. Due to the ease with which CO_2 may be absorbed from the atmosphere when microalgae are cultivated in alkaline settings, biomass growth is increased (Ren et al. 2014). Although researchers observed a reduction in chlorophyll levels in microalgae as the pH was raised to 9.5 (Maizatul et al. 2017). Most microalgae were shown to grow well in pH ranges of 6 to 9 (Khan et al. 2018). *Chlorella vulgaris* is an exception to the rule since it can thrive in a pH range of 4 to 10, with pH 9 to 10 producing the maximum levels of biomass production (Tham et al. 2019). To prevent the growing system from becoming contaminated with deadly fungus, it is also advisable to cultivate microalgae in acidic circumstances (pH range of 3 to 4) (Hwang et al. 2019). As the pH rises, the salinity of the culture medium also rises, which is extremely detrimental to algal cells (Hwang et al. 2019).

14.3.2.4 Mixing (Agitation)

Mixing is crucial for maintaining the appropriate balance of gases and pH, CO_2, and temperature in a system. Sufficient turbulence in microalgae cultures prevents the formation of gradients that can hinder the cells' performance. Thus, mixing in the culture broth diminishes nutrient gradients, prevents cell settling, and compels cell migration between light and dark regions, thereby boosting photosynthesis (Eriksen et al. 2008; Brindley et al. 2004). It promotes efficient heat transfer and prevents uneven distribution of heat by ensuring equal light and nutrient exposure for all cells in the population. According to research by Abu et al. (2016), the flashing light impact enhances both the amount and quality of algal biomass while also increasing microalgal photosynthesis. According to Iluz et al. (2016), increasing the L/D cycle frequency to 60 Hz significantly enhanced biomass output. On the other hand, excessive mixing can lyse and destroy cells in microalgae due to its shear power (Acien et al. 2017). For this reason, microalgae cultures require consistent agitation to ensure the buoyancy and light exposure of all cells.

14.3.2.5 Nutrients

Although various microalgae species have diverse dietary needs, all of them have the same fundamental ones (Wang et al. 2019). Nitrogen, phosphorus, and carbon are essential macronutrients required for development, making up the backbone of microalgae (Khan et al. 2018). Nitrogen is the nutrient that is most crucial for assisting in the formation of biomass. Based on the kind of microalgae and the nitrogen supply, the nitrogen concentration of biomass varies but typically ranges from 1% to more than 10% (Grobbelaar et al. 2013). Lack of nitrogen in the culture might hinder the development of microalgae and reduce biomass output (Khan et al. 2018). Nitrate (NO_3^-), urea, and ammonia (NH_4^+) are the three sources of nitrogen that are most often ingested (Xin et al. 2010). The little amounts of the micronutrients Co,

Mo, K, Mn, Fe, Zn, Mg, and B that are needed for microalgal development significantly impact various enzyme-related processes in algal cells (Gardner et al. 2017). Reduced biomass is the result of nutrient deprivation, which significantly impacts the growth rate of microalgae (Ito et al. 2013). The availability of nutrients has a significant impact on how lipid and carbohydrate production and accumulation occur in microalgae (Ito et al. 2013). In order to hasten algae development, the proper nutrients must be supplied. Because they may supply vital nutrients, certain bacteria can speed up the growth of microalgae. Bacteria may transform nutrients into ammonia and nitrate forms, which microalgae can take (Zhu et al. 2011).

14.4 HARVESTING

Microalgae harvesting is crucial concerning microalgal biofuel's economics and energy balance. However, no universal method has been documented for microalgae harvesting; depending on the specific requirements and characteristics of the algae species, it can be harvested using a single method or in combination, as discussed below.

14.4.1 Sedimentation

Sedimentation is a relatively simple and cost-effective method of harvesting. The gravitational forces set apart the algal biomass from the liquid using the principle of density difference. The growth medium is left undisturbed, allowing the algae cells to sediment at the bottom due to gravity's effect. The clear supernatant is then decanted or siphoned off, and the accumulated algae slurry at the base is collected. It can be defined by Stokes Law, which states that settling velocity is proportional to the square of cell radius and density difference between algal species cell and medium used as described in Equation 14.1 (Milledge and Heaven 2013).

$$\textit{Settling velocity} = \frac{2}{9} g \frac{r^2}{\eta} \left(\rho_s - \rho_l \right) \tag{14.1}$$

where r is the cell's radius, η is the dynamic viscosity of the fluid, and ρ_s and ρ_l are the solid and liquid densities, respectively.

The density difference is the most critical factor in sedimentation. Harvesting is prolonged if the density difference is low. The different species of microalgae possess different densities; some showed a density close to that of water and salt water, i.e., 1,025 kg m^{-3} (Millero and Lepple 1973), as described in Table 14.3.

The microalgae settlement depends on factors like the characteristics of algal species being cultivated, initial algal concentration, settling tank design, settling time, light intensity, nutrient deficiency, growth stage, environmental conditions, flocculants, or coagulants used, etc. (Milledge and Heaven 2013). The gravity-driven straightforward settling makes this method an economical and accessible option for small as well as large-scale cultivation and can be adapted without significant modifications. Despite these advantages, certain limitations are associated

TABLE 14.3
Densities of different microalgal species

Species	Density (kg m^{-3})
Marine micro-algae	1,030–1,100
Cyanobacteria	1,082–1,104
Marine diatom and dinoflagellates	1,030–1,230
Freshwater green microalgae	1,040–1,140

Source: Milledge and Heaven, 2013.

with sedimentation. It may not be suitable for low-density algal species and requires additional steps to improve harvesting efficiency.

14.4.2 Filtration

Filtration involves passing the algal culture through a filtration system to separate the cells from the growth medium. It is typically used after flocculation or coagulation to increase harvesting efficiency. Various filters, such as screen or membrane, vacuum, and belt filters, can retain the algae while allowing the medium liquid to pass through using prolonged pressure. By considering the pore size, membrane filters can be categorized as macrofiltration (>10 μm), microfiltration (0.1–10 μm), ultrafiltration (0.02–0.2 μm), and reverse osmosis (<0.001 μm). Filter aids and flocculants would help filtration and minimize equipment operational energy needs. Still, supplementary material increases cost and may require removal from both microalgal biomass and growth medium (Milledge and Heaven 2013).

14.4.3 Centrifugation

Centrifugation is the prevailing technique for harvesting, in which greater centrifugal forces replace gravitational forces to drive the separation. It involves the spinning of algal culture at high speed to allow the denser algal cells to sediment at the bottom of the tube, followed by the separation of supernatant and the collection of concentrated algae. It offers a rapid separation of algal biomass from culture. The rotational speed, centrifugation temperature, and time vary with the species characteristics, cell size, desired concentration of biomass, etc. The disc stack centrifuge is commonly used in commercial plants and algal biofuel pilot plants at a force of 4,000 to 14,000 times greater than the gravitational force (Milledge and Heaven 2013). But its high capital cost limits its application in high-value commodities like highly unsaturated FAs, medicinal products, etc.

14.4.4 Flocculation

Flocculation aggregates individual algae cells into larger clumps or flocs, which can be easily separated from the liquid medium. Various chemical or biological agents, such as polymers or coagulants, are added to the culture to promote aggregation and clumping of algal cells. Once the algae have formed larger flocs, they can be settled or filtered more efficiently (Mata et al. 2010). Flocculants are generally high molecular weight compounds that facilitate the binding of cells, whereas coagulants neutralize the surface charge of the cell and allow them to clump together. Several mechanisms are associated with flocculation, such as neutralization, adsorption, bridging, etc. The selection of flocculants is a critical factor that requires consideration of various factors, such as cost, compatibility, effectiveness, environmental factors, target application of biomass, etc. The dosage of flocculants or coagulants, mixing intensity, mixing duration, and pH adjustment can influence the effectiveness of flocculation and separation efficiency.

14.4.5 Flotation

Flotation methods utilize the differential buoyancy properties of microalgae and gas bubbles. It involves the introduction of air bubbles into the algal culture; the bubbles attach to the algae cells, forming a bubble-algae aggregates, causing them to rise to the surface of the culture medium. The concentrated algae layer or foam at the surface can then be skimmed off. Bubbles can be introduced using injecting air or gas spraying systems. The formation of bubble-algae aggregates depends on adherence, electrostatic interactions, and chemical interactions between the bubble and cell surface. Compared to medium buoyancy, the reduction in aggregate density causes it to rise to the surface (Milledge and Heaven 2013). Concentrated foam at the surface is collected using skimming, scraping, or suction, followed by further processing for the separation of gas bubbles and liquid residues. The stability of the foam layer also influences the separation efficiency.

14.4.6 Electrocoagulation-Flotation

Electrocoagulation-flotation (ECF) method involves an electrochemical reaction to induce coagulation. The method requires an electrode setup of two opposite charges in the culture medium. The ECF involves the integration of two principal mechanisms, i.e., electro-flotation and electrocoagulation. First, the formation of microbubbles through electrolysis of water causes flotation of flocs (electro-flotation), whereas the metal cations or hydroxide ions generated during the electrochemical reactions act as coagulants and lead to the formation of larger aggregates or flocs (electrocoagulation) (Visigalli et al. 2021). When an electric field is applied to the algal culture, it causes destabilization and coagulation of the algae cells, which are then floated by microbubbles (Ghernaout 2019). The formed flocs can be separated from the liquid medium through subsequent sedimentation or filtration. Then biomass can be dewatered, and residual electrolytes or coagulants are separated.

The different factors are associated with this method, viz., electrode material, electrolyte composition, current density, process time, algal species, growth stages, cell morphology, surface charge, pH, temperature, salinity, etc. which influence overall separation efficiency (Visigalli et al. 2021).

14.4.7 Harvesting Using Bio-filters

The culture medium containing microalgae can be passed through specially designed filters or membranes, allowing the algae cells to capture and retain on the surface and the liquid medium to pass through. Once the desired growth is achieved, the filters or membranes with attached algae can be harvested by scraping or rinsing off the cells.

It is worth noting that each harvesting technique has advantages and disadvantages regarding efficiency, cost, and scalability (Table 14.4). The technique selection depends on factors such as the algae species, application of algal biomass, scale of production, and available resources. Using a combination of two-stage separations is customary to increase harvesting efficiency.

14.5 BIOFUELS FROM MICROALGAE

The first-generation biofuels produced from food crops create competition for land, water, and other resources, potentially leading to food insecurity and a decline in biodiversity. In comparison, the second-generation biofuels produced from non-food sources such as feedstock or agricultural residues impose significant technical and economic challenges, such as collecting, transporting, and storing agricultural residues and low conservation efficiency. Neither is a sustainable choice as there is a trade-off between food and fuel; the biofuel industries have shifted to developing third-generation biofuels through algae or biofuels through synthetic biology approaches to overcome the constraints associated with earlier generations of biofuels and reliance on these finite resources. It is an upcoming feedstock for next-generation biofuels and has attracted attention due to its ability to capture, convert, and utilize carbon. Extensive research is going on to enhance biomass production and biofuel yield to make it a better alternative and cost-competitive fuel and green energy synthesis. Different biofuels produced from microalgae are discussed in Figure 14.3.

14.5.1 Biodiesel

Biodiesel is a sustainable alternative to conventional diesel fuel produced from the lipid content present in the cells of microalgae. It can be produced from lipids through a transesterification process in which lipids react with alcohol to produce fatty acid methyl esters (FAMEs) with crude glycerol, which shows chemical similarity to diesel. The process of biofuel production is straightforward; however, growing microalgae with a high lipid content is a considerable challenge. The composition of FAME, density, viscosity, cetane and iodine number, heating value, and oxidative stability are the essential parameters that determine the quality of energy fuel (Lanjekar and Deshmukh 2016; Arora et al. 2021). Conventionally, lipid (oil) is extracted through solvent extraction using a solvent like hexane (ex-situ method), followed by transesterification to produce biofuel. Recently, biomass can be directly trans-esterified (in situ) to reduce the long, expensive, cumbersome

TABLE 14.4
Summary of different harvesting methods

Harvesting Method	Dry Solid Output (%)	Specific Conditions	Advantages	Limitations
Sedimentation	0.5–3	• Settlement velocity: 0.1 m day^{-1} for spherical shape green microalgae • Min. cluster diameter for settlement: 60 μm	• Low operational and maintenance costs • Minimal energy input • Easy to scale up • Effective for both motile and non-motile species • High recovery for dense and fast-settling algal species • Minimal chemical usage • Flexibility in design • Easy to operate and maintain, therefore accessible to a wide range of users	• Prolonged process limits the productivity and throughput of the system. • Inefficient for low-density and slow-settling algal species • Algal species specific • Microalgae rich in lipids are less likely to settle due to lower density • Influenced by environmental variables like temperature and pH • Prone to clogging in narrow and small channels
Filtration	2–27		• Wide range of filters and membranes available	• Throughput depends on an algal variety • Low output • Primarily suitable for large algal cells • Issues of clogging and fouling
Microfiltration		Membrane size: 0.1–10 μm	• Suitable pore size for prevalent species • Used for the retrieval of microalgal cells from aquaculture	• Less economic at the commercial scale • Inadequate recovery of microalgal species with diameters of less than 10 μm
Ultrafiltration	1.5–4	Membrane size: 0.02–0.2 μm	• Best suited for delicate and fragile cells	• High capital and operational costs • Chances of rapid clogging due to extracellular organic matter • High energy input • Low output microalgal suspension concentration

(Continued)

Table 14.4 (Continued)

Harvesting Method	Dry Solid Output (%)	Specific Conditions	Advantages	Limitations
Centrifugation	10–22		• Handle most of the algal varieties • Efficient and rapid harvesting • High separation efficiency • Beneficial for commercial scale, where time is of the essence. • Reduce the water usage	• High energy requirement • Need for regular maintenance • High rotational speed causes damage to algal cells • High capital, operational, and maintenance costs
Flocculation	3–8	Species-specific flocculant Dosage of non-organic flocculants: 1 g L^{-1}	• Reliable and cost-effective method • Availability of a wide range of flocculants • Can handle large quantities of algal suspension • Increases rate of settling and separation efficiency • Higher biomass concentration • Naturally occurred in certain microalgae in response to environmental stress • Allows reuse or recycling of water	• Used in combination with other harvesting methods, so incur additional cost • Flocculation conditions may vary with algae species • Chances of chemical contamination • Recovery of flocculants is tedious
Flotation	7	Promoted by the addition of bubbles	• Faster than sedimentation • Higher separation efficiency for buoyant algal species • Can be operated in a continuous manner	• Species specific • High capital and operational costs • Energy-intensive to create tiny bubbles
Electrocoagulation	–		• Can achieve effective coagulation • Potentially reduces the reliance on chemical coagulants • Potential for a continuous algae cultivation system	• High operational and maintenance costs • High energy consumption • Electrolyte management is tedious

Source: Milledge and Heaven, 2013.

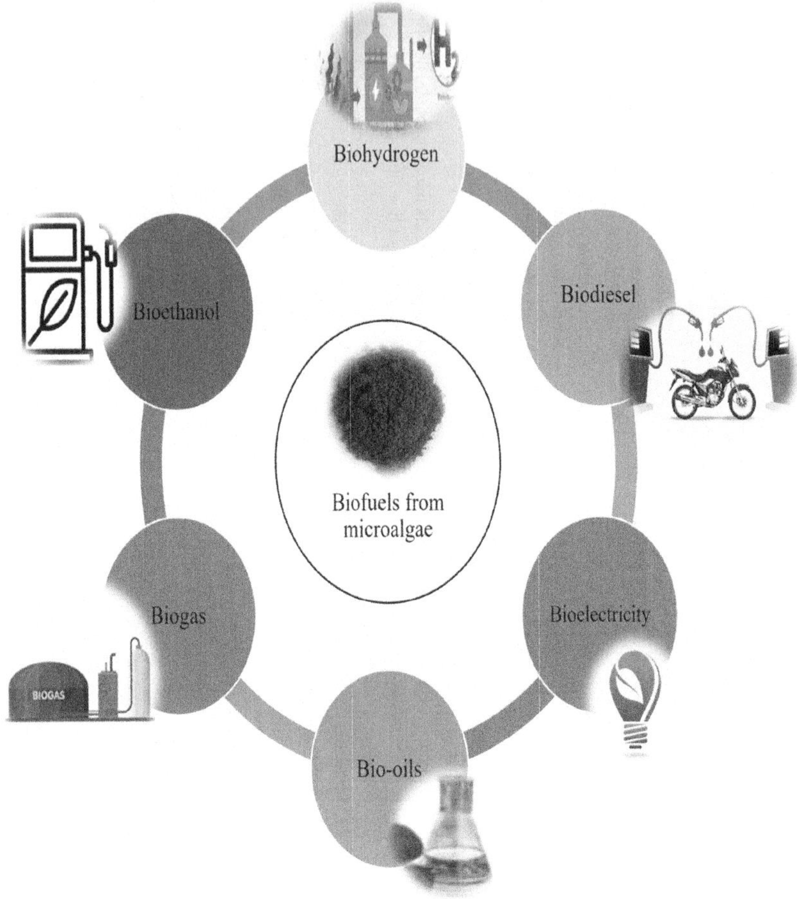

FIGURE 14.3 Biofuels from microalgae

protocol. Biodiesel from microalgae is an eco-friendly and more effective renewable choice to produce fuel. The limitation arises due to its high processing and production costs. To overcome these constraints, research is focused on modifications of algal strains with high biomass and enhanced lipid synthesis. The advancement in nanotechnology showed that the incorporation of nanoparticles used in culture medium amplifies the yield (Hossain et al. 2019).

14.5.2 Bio-oils

Bio or pyrolysis oils are produced from the two techniques of pyrolysis and hydrothermal liquefaction (HTL). The process of pyrolysis involves the thermochemical conversion of organic material at higher temperatures (350–530°C) in the absence of oxygen and can replace petroleum and offer a sustainable alternative. Whereas in

HTL, wet biomass is subjected to higher pressure (10 MPa) at a low temperature of 300°C (Arora et al. 2021). As the HTL conversion process involves water use, pre-dried algal biomass is not required. Also, the dissolution of inorganic substances in water offers cleaner fuel production (Jena and Das 2011; Panahi et al. 2019).

The researchers revealed the potential of different algal species employed in bio-oil production with varying yield ranges viz., *Spirulina* – 35-41% (Jena and Das 2011), *Scenedesmus* – 24-45% (Bordoloi et al. 2016), *Dunaliella* – 64.68% (Chen et al. 2012), *Desmodesmus* – 72.62%, etc. In contrast, macroalgae (*Laminaria saccharina*) can produce a 79% yield with HTL (Bach et al. 2014). The pyrolysis process produces high-density oil, making it a suitable energy source. Certain advancements in the pyrolysis process, such as catalytic, co-pyrolysis, hydrolysis, and microwave-assisted pyrolysis, reduce nitrogen and oxygenates and processing time with enhanced yield and favorable syngas production. Crude bio-oils can be further processed (esterification, emulsification, and hydrotreating) to enhance fuel properties and stability and to produce various biofuels like gasoline, diesel, aviation fuels, and other value-added products of biochemicals (Panahi et al. 2019).

14.5.3 Bioethanol

Bioethanol or ethyl alcohol is the output of microbial fermentation of feedstock with high starch or carbohydrate (Arora et al. 2021). The rapid growth, high oil content, and polysaccharides in algal cells favor its utilization as a feedstock for cellulosic production. Among different algal species, carbohydrate-rich brown algae are the most potential feedstock for bioethanol production. Microalgae undergo various pretreatments before ethanol production to break down the cell wall and enhance the accessibility of carbohydrates in the subsequent fermentation process. The pretreatment includes mechanical disruption using ultrasonication, chemical treatment (acid or alkali), enzymatic hydrolysis, or a combination of two treatments (Hossain et al. 2019). Pretreated microalgae are then added with the *Saccharomyces cerevisiae* to initiate fermentation through the conversion of carbohydrates or sugars into ethanol, followed by distillation to separate and concentrate the produced ethanol. Bioethanol can be directly used as a fuel or blended with gasoline in proportion. The sugar concentration, pH, and inoculum size significantly affect ethanol yield (EL-Sayed et al. 2016). Still, limitations are associated with bioethanol produced from microalgae, such as low vapor pressure, flame luminosity, the energy density of ethanol, harvesting efficiency, scaling up, etc.

14.5.4 Biogas

Biogas is generated when wet biomass or organic matter (carbohydrates, lipids, and proteins) undergoes anaerobic digestion in the absence of oxygen. The anaerobic breakdown of microalgae can produce 50–70% methane with carbon, H_2, and H_2S of 30–45%, $<$2%, and $<$3.5%, respectively, which is much higher than the plant residues (Vanegas et al. 2013). Therefore, it is an exciting area of research in the field of renewable energy. Further, upgrading must be performed after the production to remove impurities (carbon dioxide and hydrogen sulfide). The produced biogas can be utilized for biogas generators, electricity generation, or heat. Researchers reported that the different pretreatments, including thermal, microwave, enzymatic, and incorporation of nanoparticles in pretreated microalgae for augmentation of biomethane, resulted in higher methane content. *Desertifilum tharense*, *Phormidium animale*, *Chlorella* sp., *Anabeana variabilis*, and *Chlorophyta* showed 308, 293, 242, 229, and 230 mL g^{-1} VS biochemical methane potential (BMP), respectively (Perendeci et al. 2019). The factors like Kjeldahl nitrogen, glucose, and chlorophyll content of algal species have a positive correlation with BMP production (Perendeci et al. 2019). Harvesting efficiency, algal concentration, pretreatments, and operational cost are the significant factors that bottleneck the energy return on energy invested (EROI) (Collet et al. 2015). Efforts are ongoing to surmount these obstacles and explore the potential of microalgae as a sustainable source material for biogas production.

14.5.5 Biohydrogen

The hydrogen produced through biomass has enormous potential to serve as a clean or pollution-free fuel. The highest heating value (142 MJ/kg) of hydrogen makes it compatible with combustion (Chen et al. 2011; Nagarajan et al. 2017) and can be utilized directly to create electricity from fuel cells (Kim and Kim 2011). Biohydrogen is produced through different methods, directly and indirectly, such as thermochemical conversion, water splitting, biophotolysis (direct and indirect), fermentation (dark or photo fermentation), etc. (Li et al. 2022). The thermochemical conversion includes steam reforming of methanol, natural gas, coal, and ethanol, using heat and chemical reactions. But the process generates large amounts of greenhouse gases. The photosynthetic capability of microalgae is utilized in the photolysis method. In this method, microalgae are exposed to light and allowed to grow in anaerobic conditions to convert CO_2 into biomass and hydrogen.

14.5.5.1 Bio-photolytic

Direct biophotolysis converts solar energy into chemical energy through direct biophotolysis from water (Equation 14.2).

$$H_2O + Light\ Energy \rightarrow 2H_2 + O_2 \tag{14.2}$$

The transitional reaction in microalgae fixes light energy by PSI and PSII, which reduces water through oxygenic photosynthesis to produce H_2 (Li et al. 2022).

14.5.5.2 Indirect Bio-photolytic

In contrast, O_2 inhibition is avoided in indirect biophotolysis, in which CO_2 is fixed in carbohydrates prior to H_2 generation (Equation 14.3).

$$6H_2O + 6CO_2 + Light\ Energy \rightarrow C_6H_{12}O_6 + 6O_2 \tag{14.3}$$

The two-stage process used in indirect biophotolysis for the segregation of H_2 and photosynthesis. The photosynthesis and production of biomass are achieved in the first stage, whereas H_2 and O_2 segregation is achieved in the subsequent stage (Li et al. 2022). The light source, temperature, pH, nutrient availability, carbon source, and type of bioreactor are the significant factors responsible for the development of biohydrogen in the two-stage process of indirect biophotolysis (Rashid et al. 2013).

14.5.5.3 Photo Fermentation

The limitation of low conversion efficiency is conquered in this method. Biohydrogen is produced through a light-driven fermentation process in which algal and residual algal biomass undergo microbial fermentation (Equation 14.4) (Bhatia et al. 2019).

$$CH_3COOH + 2\,H_2O + light \rightarrow 4\,H_2 + 2\,CO_2 \tag{14.4}$$

The process only takes place under light, as light facilitates the process of photosynthesis to produce energy, avoiding pyruvate degradation and respiratory ATP generation.

14.5.5.4 Dark Fermentation

The above methods use sunlight as a source of energy. In contrast, dark fermentation produces H_2 anaerobically, where anaerobic bacteria break down the algal biomass in the absence of light (Show et al. 2019). The metabolic activity of hydrogen-producing bacteria converts biomass into biohydrogen through various stages of hydrolysis, acidogenesis, and acetogenesis (Show et al. 2019). The algal biomass undergoes different pretreatments to break down the complex carbohydrates and lipids into simple compounds and increase the accessibility of stored energy to enhance H_2 production. The different pretreatments include physical (bead milling), thermal (microwave, autoclave, steam explosion, hydrothermal), non-thermal (ultrasonication, pulsed electric field), chemical, or enzymatic (Li et al. 2022).

The advantages and limitations associated with the different methods of H_2 production are depicted in Table 14.5.

The complex operation of biohydrogen production depends on multiple parameters, including hydrogen-producing algal strains, micro and macronutrients, pH, substrate concentration, temperature, agitation, aeration, a configuration of the bioreactor, etc.

Recently, intensive studies have been going on into advanced approaches to increase enzyme stability and hydrolytic efficiency to commercialize H_2 generation from microalgae. These advancements include (1) genetic engineering (manipulation and modifications of genes) to alter the amino acid residues to inhibit O_2 activity; (2) co-cultivation of microalgae and bacteria to utilize the metabolic activity of bacteria to reduce O_2, enhance growth, break down the biomass, and improve hydrogenase activity; (3) hybrid biological and electrochemical processes involving the conversion of products of biological oxidation of organic feedstock into energy resources in microbial electrolysis cells (MECs); and (4) incorporation of nanomaterials to enhance enzyme stability and hydrolytic efficiency through intracellular electron

TABLE 14.5
Advantages and limitations of different methods of H_2 production

Method of Production	Advantages	Limitations
Direct biophotolysis	• Environmentally accessible as there is no accumulation of CO_2 • High availability of energy source	• Demand for customized photobioreactors • High cost of bioreactor • Low H_2 production • Complex, uncontrolled pathway • Long-term process • O_2 sensitivity of hydrogenase reactions
Indirect biophotolysis	• Avoid the difficulty of O_2 inhibition • Abundant availability of energy source	• Exterior light source required • Light transformation/conversion efficiency is very less • Less H_2 yield
Photo fermentation	• High efficiency of photoconversion to H_2	• Requirement for an external light source • Process limitation if sunlight is utilized as a light source • Extremely low light conversion/ transformation efficiency
Dark fermentation	• High energy efficiency • High H_2 yield • Environmental sustainability • Better carbon balance	• Low substrate transformation efficiency • Poor H_2 production • Limitations of kinetics and thermodynamics reaction • Separation of H_2-CO_2 is tedious

Source: Buitrón et al., 2017.

transfer, enhance hydrogenase activity, and improve the bioactivity of hydrogen-producing microalgae (Li et al. 2022). Besides the advantages of carbon neutrality and energy sustainability, technical challenges arise with scalability, conversion efficiency, operational cost, and mass storage.

14.5.6 Biobutanol

Biobutanol is an ameliorated version of bioethanol, produced from a similar process to bioethanol. *Clostridium* sp. has greater potential to produce biobutanol through anaerobic ABE (acetone, butanol, and ethanol) fermentation (Arora et al. 2021). The light intensity, acid hydrolysis are the factors responsible for higher yields of biobutanol.

14.5.7 Bioelectricity

Bioelectricity generation through Microbial Fuel Cells (MFC) from microalgae is an attractive technology recently gained much attention. The bioremediation approaches can be used for the bioelectricity generation from microalgae. In MFC, microbes act as catalysts to convert chemical energy to electrical energy (Arora et al. 2021). The microbial degradation of biomass produces CO_2, water, and energy through different metabolic pathways (Bhatia et al. 2021). The photosynthetic MFC is composed of two chambers, anodic and cathodic, and microalgae can be used either in the anodic or cathodic chamber as a biocatalyst. The oxidative breakdown of a substrate in an anodic chamber produces electrons and protons. The produced electrons travel through the external circuit to harness the energy, where O_2 produced from photosynthesis acts as an acceptor. In contrast, protons travel through the proton exchange membrane to the cathodic chamber to react with oxygen (reducing agent) and produce water to complete the cathodic cycle. Higher dissolved oxygen favors higher bioelectricity generation in algal-based (biocathode) MFC than in air-based cathode. Still, the growth of algae imposes a limitation on the performance of MFC (Bhatia et al. 2021).

14.6 EXTRACTION AND CONVERSION OF BIOFUELS FROM BIOMASS

14.6.1 Extraction or Fractionation of Biochemical Components of Microalgae

Extraction is the initial step in isolating the target natural component from the source materials. According to the extraction principle, solvent extraction, distillation, and sublimation are some of the extraction procedures that can be applied. The most often employed technique is solvent extraction. Several steps are involved in the extraction process: (1) solvent penetration into the solid matrix; (2) dissolution of solute into the solvent; (3) diffusion of solute from the solid matrix to the solvent; and (4) collection of solutes. Any element that increases the diffusivity and solubility of the preceding processes will make the extraction easier. Some current or environmentally friendly extraction procedures include pressurized liquid extraction (PLE), supercritical fluid extraction (SCFE), microwave-assisted extraction (MAE), and ultrasound-assisted extraction used in natural product extraction, and it has numerous benefits, like less demand for organic solvents, faster extraction, and more selectivity (Zhang et al. 2018).

Microalgae can serve as a valuable biological resource for generating significant biomass in third-generation biofuel production. Depending on the species, algal biomass contains 9.5–42% lipid, 20–50% protein, and 17–57% carbohydrate (dry weight basis), with some species containing up to 79.5% oil (Raheem et al. 2015). Triacylglycerols (TAGs) are a type of storage lipid synthesized by photosynthetic microalgae by fixing CO_2 in the presence of photon energy. Microalgae produce oil at a rate (L/ha) that is significantly greater (91%) than other feedstocks, including oil palm (3%), coconut (1.5%), avocado (1.4%), jatropha (1.2%), and rapeseed/canola (1%) (Maity et al. 2014). Various microalgal species accumulate considerable amounts of lipids. *Dunaliella tertiolecta, Chlorella emersonii, Botryococcus braunii*, and *Neochloris oleoabundans*, all had very high lipid concentration (>60% of dry weight (dw) biomass). The wide variety of microalgae such as *Chlorella vulgaris, Chlorella protothecoides, Dunaliella salina, Nannochloropsis oculata NCTU-3, Scenedesmus dimorphus*, and *Phaeodactylum tricornutum* had a moderate lipid content (40–60% of dw). In contrast, it was discovered that the lipid content in *Ankistrodesmus* sp., *Chaetoceros muelleri, Dunaliella primolecta, Cylindrotheca* sp., *N. oculata, Spirulina platensis*, and *Tetraselmis suecica* was less than 40% (dw) of the biomass (Maity et al. 2014). Therefore, choosing an algal strain is a crucial first step in the manufacture of algae-based biofuels. When comparing different microalgal varieties, green microalgae produce lipid at a rate (mg/L/day) that is noticeably higher than that of red microalgae. Algal oil can be extracted using standard processes (hexane extraction, oil press and expeller, and SCF extraction) and processed into enhanced hydrocarbon fuels like biogasoline and biodiesel. To make the extraction cost-efficient, the wet oil extraction method is used, where wet microalgae cake can be directly treated with solvent. Triglycerides, the main component of oils obtained from the feedstock used to make biodiesel, can be transformed into biofuels using three primary processing methods: pyrolysis (thermal cracking), transesterification, and micro-emulsification. The algal protein is also a good substrate for the production of biofuels. The alkaline process is generally used to extract the protein from microalgae biomass. Some protein may be lost during lipid extraction; thus, protein extraction should be the initial phase in a biorefinery workflow (Muñoz et al. 2015). Microalgae are likewise high in carbohydrates, and there are many biochemical conversion processes that utilize biocatalysts, including microbes or enzymes, in regulated environments to convert biomass carbohydrates into biofuels like biobutanol, bioethanol, and biohydrogen (Chia et al. 2018).

14.6.2 Conversion Methods

Biofuel production involves utilizing biological conversion processes to produce energy fuels. The three forms of energy conversion reactions for biomass are biochemical, thermochemical, and combustion process (Amin 2009). Biochemical conversion includes fermentation, anaerobic digestion, and other techniques that produce fuel from the metabolisms of organisms. Biogas or natural gas, as well as biohydrogen, are gaseous fuels that can be generated by dark fermentation, anaerobic digestion, and photobiological processes of microalgal biomass. Thermochemical processes such as gasification, pyrolysis, and hydrothermal liquefaction could produce biogas and bio-oils. Selecting the right conversion technology is essential for achieving economic viability and environmental sustainability in biofuel production (Raheem et al. 2015). A variety of biofuels can be derived from microalgae reliant on the conversion process, for instance, biomethane from anaerobic digestion, biohydrogen from photobiological processes, bioethanol from fermentation, liquid oil from thermal liquefaction, and biodiesel from transesterification (Akubude et al. 2019) (Figure 14.4).

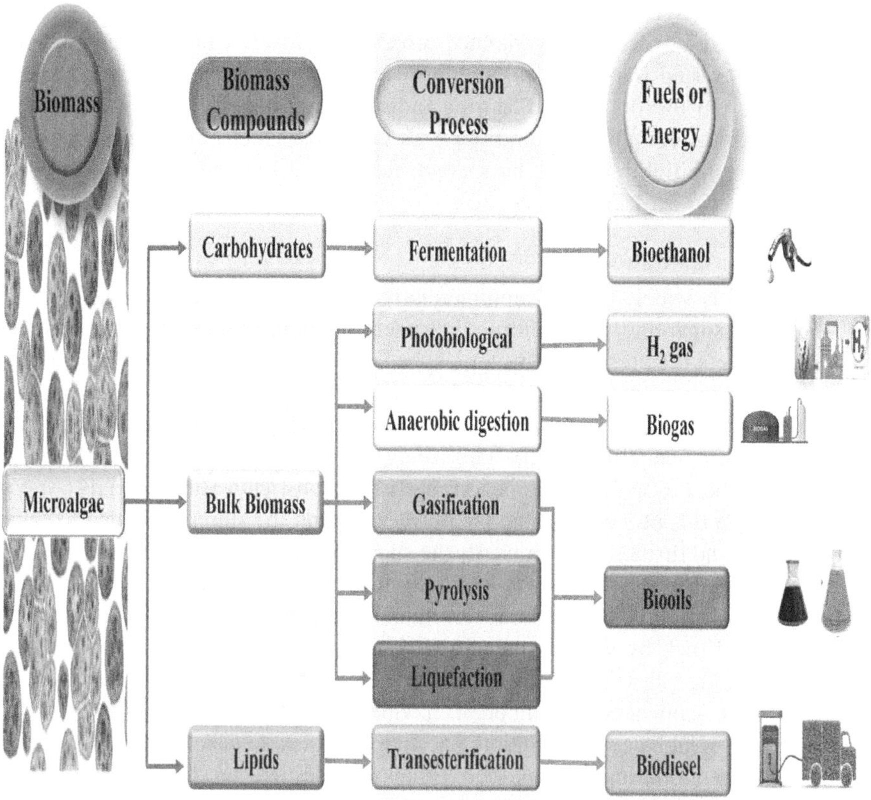

FIGURE 14.4 Extraction and conversion of biofuels from microalgal biomass

14.6.2.1 Photobiological

The photobiological synthesis of chemicals and fuels from microalgae is likely one of the most vital aspects toward developing a significant renewable energy supply. The idea behind microalgal bioconversion is the exploitation of the photosynthetic process to produce biomass that can be used as a source of energy and chemicals (Amin 2009). An excellent first step toward creating sustainable fuels using renewable solar energy should be harnessing sunlight to make hydrogen (Chatzitakis et al. 2013). Thus, producing hydrogen from biomass and hydrogen energy has numerous benefits, for instance, waste management, the substitution of fossil fuels with sustainable biofuel, and the reduction of CO_2 emissions (Schipper et al. 2019; Zebda et al. 2018). Light is required to control biological and physiological processes in biohydrogen synthesis (Musazade et al. 2018). Biohydrogen is a naturally occurring secondary metabolite produced by photosynthetic microalgae. There are several biophotolysis techniques to produce biological hydrogen. Direct biophotolysis, which occurs in the cells of higher plants and algae, is the same process as photosynthesis (Bharathiraja et al. 2016). Utilizing the photosynthetic process of microalgae, biohydrogen is generated directly from water. In this method, solar energy is converted into chemical energy to produce molecular hydrogen (Azwar et al. 2014). On the other hand, during indirect biophotolysis, microalgae create hydrogen from stored glycogen and starch. It entails two steps: first, producing carbohydrates using light energy; and second, fermenting the carbohydrates to make hydrogen (Anwar et al. 2019).

14.6.2.2 Fermentation

Fermentation is widely utilized commercially in several countries to manufacture ethanol from sugar and starch crops. Hydrogen is produced during fermentation in the dark without the presence of water, oxygen, or sunlight. Complex organic matter is broken down into a mixture of alcohol, organic acid, and hydrogen by microorganisms that have the ability to ferment. The production of hydrogen and other significant byproducts, including volatile fatty acids, acetic acid, and butyric acid, can be achieved by the dark fermentation of a range of carbon-rich waste materials (Anwar et al. 2019). In this process, biomass is pulverized, and the starch is turned to sugar by enzymes, and further, yeast converts the sugar to ethanol. The distillation process for purifying ethanol is an energy-intensive phase (McKendry 2002a). Microalgae carbohydrates can be converted into bioethanol, an eco-friendly and commercially feasible liquid fuel, by yeast fermentation (Ha et al. 2020). Using microalgae as a raw material, the following procedure can be utilized to manufacture ethanol. In the first phase, mechanical equipment or an enzyme is used to liberate microalgae starch from the cells. Upon cell degradation, Saccharomyces cerevisiae yeast is introduced to the biomass, initiating fermentation. Ethanol is the final product of this process. Subsequently, the ethanol is transferred to a holding tank before being fed to a distillation facility (Amin 2009).

14.6.2.3 Anaerobic Digestion

Microalgae can be employed for biofuel production, specifically biomethane via anaerobic digestion. Microalgae contain enormous amounts of nitrogen and phosphate, which may have unsustainable environmental and economic effects. Anaerobic digestion breaks down organic nitrogen and phosphorus in algal waste, converting them into phosphate and ammonium. These nutrients can serve as a microalgae growth medium (Olgun 2000; Phang et al. 2000). This process efficiently recycles phosphate and nitrogen while also generating methane when employed to handle algal waste. Anaerobic disintegration or phytoplanktonic cell digestion is a natural process that occurs in aquatic environments. Algae cells die and disintegrate as they enter the anoxic and aphotic zones. Nutrient remineralization is an important process that recycles nutrients in these anoxic layers of aquatic ecosystems. Ammonium and phosphate are released as a result, and these nutrients can eventually support the development of phytoplanktonic communities. This decomposition process is a time-consuming and unreliable procedure. The first inquiry on the anaerobic digestion of algal biomass was reported in 1957 by Golueke et al. (1957). They introduced an integrated method in 1960 that linked the microalgae production in an open pond for sewage water treatment with the anaerobic digestion of algal biomass to recover energy (Oswald and Golueke 1960). Microalgae possess mineral content that fulfills the nutrient needs of anaerobic microflora. Apart from essential elements like carbon, nitrogen, and phosphorus, microalgae also contain oligo nutrients like zinc, cobalt, and iron, which play a crucial role in promoting methanogenesis (Richmond 2004; Speece 1996). The algal biomass's high protein content causes a large ammonium release, which inhibits the anaerobic bacteria. Whereas, high salt concentrations, on the other hand, may affect the anaerobic process in marine species (Sialve et al. 2009).

14.6.2.4 Gasification

Gasification is a chemical transformation of carbonaceous substances (hydrocarbons) into synthesis gas (syngas) by partially oxidizing them with oxygen, air, and/or steam at high temperatures, often in the 800–1000°C range (Brennan and Owende 2010; Sanchez-Silva et al. 2013). Different processes, including biomass drying, pyrolysis, combustion, and reduction, are involved in the chemical reactions that lead to the gasification of carbonaceous materials. Biomass is broken down thermochemically during the first stage of gasification. Following the gasification process, syngas is created, which principally contains H_2 (65.5%), CO (8.3%), and CH_4 (2.26%) as a co-product (Raheem et al. 2015). Particle size, temperature, char porosity, and gasifying agents' partial pressure are some of the operational factors that have the biggest effects on biomass gasification, along with the type of biomass. The biomass is appropriate for gasification when the moisture content is less than 15% (McKendry 2002b). Algal biomass, though, is suitable at moisture levels as high as 40%. Elevated moisture levels within the biomass diminish both the energy content of the syngas and the gasifier's efficiency. A novel energy-generating system based on microalgae has been proposed, which combines nitrogen cycling with microalgae gasification at low temperatures. High-moisture biomass is gasified directly to methane-rich fuel gas without drying. Furthermore, nitrogen in the biomass is transformed to ammonia during the reaction (Raheem et al. 2015).

14.6.2.5 Pyrolysis

Pyrolysis is the thermal conversion of biomass into charcoal, biofuel, and gaseous components to temperatures between 200 and 800°C without air or with a catalyst, at a high heating rate of 103 to 104 K/s and with a brief gas residence time. The biomass is then quickly cooled to liquid, where it is then used as a fuel or for other purposes (Amin 2009; Raheem et al. 2015; Azizi et al. 2018). Biochar, bio-oil, and a gaseous combination of carbon dioxide and methane are the end products of microalgae pyrolysis (W. H. Chen et al. 2015). Pyrolysis can be classified into five subgroups based on the heating rate, heating method, carrier gas, and presence of catalysts: (1) slow pyrolysis, (2) fast pyrolysis, (3) catalytic pyrolysis, (4) microwave pyrolysis, and (5) hydropyrolysis. Slow pyrolysis produces char by allowing vapor to stay in the reactor for 10–60 minutes. It takes place at a low heating rate (between 0.1 and 1°C/s) for a long residence time. Char is the main byproduct of slow pyrolysis, and at low temperatures of around 300°C, the production of biochar can reach a level of 66%, compared to the average yield of bio-oil, which is 40% (Azizi et al. 2018). Fast pyrolysis techniques for biomass have gotten a lot of interest in recent years for maximizing liquid yields. It involves rapidly heating biomass to the pyrolysis temperature before it is completely degraded. Fast pyrolysis stands out due to its elevated heating rate, brief residence time, high reaction temperature, efficient heat transfer, and rapid pyrolysis vapor cooling. It has the advantage of directly producing liquid fuel. This process produces 19–57% bio-oil (end product) and char from biomass (Miao et al. 2004). A catalyst can improve the pyrolysis efficiency. Bio-oils derived from algal biomass, for example, typically have a high oxygen concentration. Adequate oxygen levels are essential. Reduced to improve oil stability (avoid polymerization and condensation processes), energy density, and acidity. In addition, by utilizing a catalyst, the high nitrogen content of bio-oil might be reduced. Catalytic pyrolysis has been found to be an alternative for improving the properties of microalgae bio-oil (Anand et al. 2016; Thangalazhy-Gopakumar et al. 2012; Zainan et al. 2015). Microwave irradiation is a recently investigated alternative heating method in microalgae pyrolysis. Biogas yield is more in MAP than in conventional pyrolysis (Azizi et al. 2018). Hydropyrolysis (HyPy) is a biomass degradation process that occurs with high-pressure hydrogen. It yields a larger amount of hydrocarbons with improved structural integrity than ordinary pyrolysis (Marcilla et al. 2013).

14.6.2.6 Liquefaction

Microalgal cell precipitates with a high moisture content generated from centrifugation are thus suitable raw materials for liquefaction. Wet biomass material can be transformed into liquid fuel by using a process known as direct hydrothermal liquefaction (Patil et al. 2008). The major purpose is to manufacture low molecular weight liquid fuels from high molecular weight organic compounds at high temperatures and pressures. The liquefaction takes place at a temperature of about 300°C and a pressure of 10 MPa without the presence of a reducing gas like carbon monoxide or hydrogen (Minowa et al. 1995). The benefit of this process is that feedstock drying is not necessary. Owing to the high latent heat of water vaporization and the high moisture content of microalgae biomass, a substantial amount of heating energy is needed for the drying process. Hydrothermal liquefaction uses a lot less energy than other processes, making the process substantial in terms of energy consumption. To achieve liquefaction, a stainless-steel autoclave with mechanical mixing can be employed. Algal cells are added to the autoclave, and then the air is purged using nitrogen. The autoclave is heated to a specified temperature, and more nitrogen pressure is applied to begin the reaction. The temperature was maintained for 5 to 60 minutes without changing before being lowered with an electric fan. The reaction is extracted with dichloromethane or chloroform for oil fraction separation. Filtration was used to remove the dichloromethane/chloroform from the reaction mixture. The remaining dichloromethane/chloroform was filtered and subjected to low-pressure evaporation at 35°C, resulting in the formation of a dark brown, thick substance referred to as "the oil." The aqueous phase (insoluble fraction) is purified by filtering and washing with water after extraction (Amin 2009; Shuping et al. 2010). During hydrothermal liquefaction, fat as well as other organic substances, including protein, fiber, and carbohydrates, along with the algal cells themselves, are transformed into bio-oil (Shuping et al. 2010).

14.6.2.7 Transesterification

Transesterification is the leading and commercially used technology for biodiesel production, proving successful for a diverse feedstock. During the transesterification reaction, the fats or oils react with alcohols in the presence of a catalyst, converting the fatty acids to the equivalent fatty acid alkyl esters, which are vital in the production of biodiesel. In this process, the hydrocarbon chains are broken down to produce biodiesel and glycerin. Two techniques were used to produce biodiesel from wet microalgae cake: first, wet oil extraction was performed, and the produced microalgal oil was transesterified using a homogeneous or heterogeneous catalyst. Process 2: the wet microalgae cake was employed immediately for transesterification, without the need for oil extraction. The direct transesterification process is simple and requires only the addition of alcohols to biomass along with catalysts and sometimes co-solvents. In this process, a high solvent-to-microalgae biomass ratio, high catalyst loading, or high reaction temperature (e.g., >90°C) are necessary to generate a high biodiesel production (C. L. Chen et al. 2015). The catalyst type and concentration significantly impact the transesterification process. The most widely utilized catalytic concentration ranges from 1 to 12% wt. Both homogeneous and heterogeneous catalysts are applicable in this reaction for biodiesel generation (Kumar Pathaka et al. 2017). The oil and alcohol transesterification with a homogeneous catalyst is an eminent technique for biodiesel production. Homogeneous alkali catalysts efficiently transform triglycerides into fatty acid methyl esters (FAMEs) with high yield, rapid reaction rates, and cost-effectiveness. On the contrary, isolating the catalyst from the product mixture for reuse is technically challenging. Following the reaction, the catalyst should be neutralized or eliminated with a significant volume of hot water, resulting in a substantial volume of industrial wastewater. Since the homogeneous catalyst has numerous disadvantages, including difficulty in product isolation, a high amount of water requirement, and environmental pollution from liquid wastes, solid heterogeneous catalysts are employed to resolve these issues, due to their noncorrosive, harmless, and ease of separation for recycling (Ali et al. 2014). However, heterogeneous catalytic techniques are challenging for mass transfer, time-intensive, and inefficient. Recently, a new trend in biodiesel synthesis has been the use of "nanocatalysts," which combine the advantages of both homogeneous and heterogeneous catalytic systems. A nanocatalytic system ensures faster, selective chemical transformations with high product yield, as well as easy catalyst separation and recovery (Feyzi et al. 2013). Some of the common catalysts used in transesterification are: sodium hydroxide (NaOH), potassium hydroxide (KOH), an alkaline catalyst, sulfuric acid (H_2SO_4) and hydrochloric acid (HCl) as acid catalysts, CaO and MgO supported on alumina, CaO on Al_2O_3, metallic oxides composed of SrO, ZnO, TiO_2-ZnO, ZrO, TiO, and Al_2O_3 (Kumar Pathaka et al. 2017; Park et al. 2014) (Table 14.6).

TABLE 14.6
Microalgae biochemical composition and their conversion to biofuels

Conversion Process	Microalgae Species	Biochemical Composition (%)	Biofuel Type	% Yield	References
Photobiological	*Chlamydomonas reinhardtii*	–	H_2 gas	54	Hemschemeier et al. (2009)
	Chlorella pyrenoidosa	–	H_2 gas	0.37 µmol	Chen et al. (2022)
Fermentation	*Chlamydomonas mexicana*	Lipids – 21 Protein – 20 Carbohydrates – 50	Bioethanol	42	Ha et al. (2020)
	Chlamydomonas pitschmannii	Lipids – 50 Protein – 20 Carbohydrates – 23	Bioethanol	36	
	Pseudochlorella sp.	Lipids – 30 Protein – 24 Carbohydrates – 36	Bioethanol	45	
Anaerobic digestion	*Nannocholoropsis gaditana*	Lipids – 17.6 Fatty acid – 12.6 Proteins – 24.1	Biogas (CO_2, CO, CH_4, and H_2)		Sanchez-Silva et al. (2013)
Gasification	*Botryococcus braunii*	Lipids – 17.7 Protein – 30.4 Carbohydrates – 15.5	Bio-oil	38.3	Muñoz et al. (2015)
Pyrolysis	*Nannochloropsis gaditana*	Lipids – 12.7 Protein – 30.7 Carbohydrates – 21.6	Bio-oil	39.7	Babich et al. (2011)

(*Continued*)

Table 14.6 (Continued)

Conversion Process	Microalgae Species	Biochemical Composition (%)	Biofuel Type	% Yield	References
	Chlorella sp.	Lipids – 7 Protein –34 Carbohydrates – 15.5	Bio-oil	40	Rizzo et al. (2013)
	Chlorella sp.	Lipids – 9–13 Protein – 29.6 Carbohydrates – 15–16.5	Bio-oil	34	
	Dunaliella tertiolecta	Lipids – 2.87 Protein – 61.32 Carbohydrates – 21.69	Bio-oil	25.8	Shuping et al. (2010)
Liquefaction	*Spirulina platensis*	Lipids – 13.30 Protein – 48.4 Carbohydrates – 30.2	Bio-oil	41	Jena and Das (2011)
Transesterification	*Tetraselmis suecica*	Lipids – 23	Biodiesel (fatty acid methyl esters)	18	Wahlen et al. (2011)
	Chlorella sorokiniana	Lipids – 23.5	Biodiesel (fatty acid methyl esters)	18	

14.7 POTENTIAL USES (APPLICATIONS IN FOOD AND NON-FOOD SECTORS)

The algae industry has made significant progress in utilizing algae for various applications such as food and the non-food sector. Microalgal biomass contains three primary components: proteins, lipids (oils), and carbohydrates, which have the potential to be transformed into diverse bioproducts. Lipids and carbohydrates have the potential to produce various biofuels, while proteins and carbohydrates can form biofertilizers, animal feed, and food supplements. Hence, choosing the right strain is crucial, considering the economic prospects and the type of final products. Many green algae species have been consumed as food since the beginning of time (Jensen et al. 2001). Microalgae are an excellent food source and can provide vital bioproducts like natural antibiotics (Becker W. 2004). Therefore, the identification of sustainable and renewable sources for the generation of biofuels has led to the exploration of microalgae as a novel research area for fuels along with other useful products.

Algae have massive usage in addition to the energy sector, yet most of these are unexplored because of the latent potential of this renewable resource. Many algal products are available in the market, including cosmetics, pharmaceuticals, and food. Further extensive study is still needed in this sector.

14.7.1 Food Products

Algae have various fascinating characteristics, including higher nutritive value, sustainability, and health benefits. Algae are an abundant source of protein, carbs, enzymes, and fiber in addition to a wide range of vitamins (like Vit. C, A, B1, B2, B6, and niacin) and minerals, including iodine, iron, potassium, magnesium, and calcium. Therefore, at present, it serves as a primary food source worldwide and exclusively in Asian nations such as Japan, Korea, and China. Asian nations have employed green microalgae as food products or nutritional supplements for hundreds of years (Borowitzka). There are roughly 500 species that humans eat. The green algae species Ulva and Monostroma resemble lettuce leaves and are consumed in soups or as salads, relishes, and fish or meat dishes. Brown algae (*Laminaria* species) are also consumed with fish or meat and in soups as well. Apart from this, algae-based compounds are applied as food additives, namely, alginates, agar, and carrageenan in the food industry. Recently, because of their gelling and thickening abilities, the utilization of such compounds has substantially increased (Bhattacharjee 2016) (Figure 14.5).

14.7.1.1 Food Additives

Alginates, also referred to as alginic acids, are isolated from brown seaweeds, particularly *Macrocystis*, *Laminaria*, and *Ascophyllum*. They have applications in both the food and pharmaceutical sectors. In the food industry, they are mostly used for obtaining a smooth texture and to inhibit ice formation in ice cream (due to their chelating properties that form a highly viscous solution), fillers and emulsifiers in

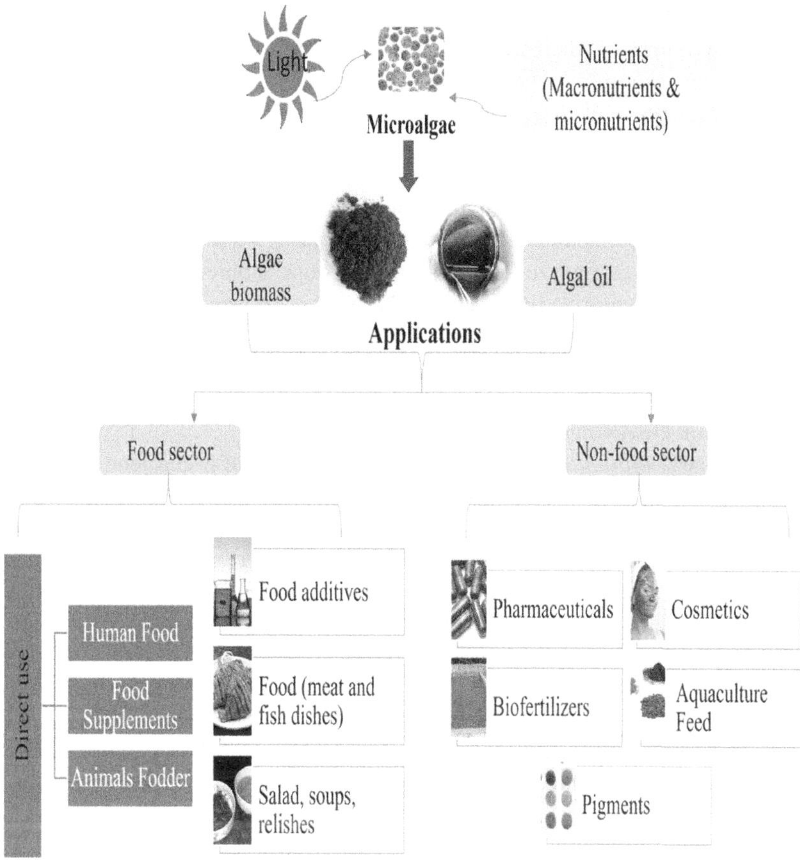

FIGURE 14.5 Potential use of microalgae

salad dressing, and as thickeners in candies and syrups. Moreover, they are often used for cotton yarn sizing in the textile sector (Arora et al. 2021; Goswami et al. 2015)

Agar is obtained from various red algae species, especially *Gracilaria* and *Gelidium,* which are released after boiling. It has wider applications in the food industry, like candies, frozen food, bakery icings, fruit juices, dessert gels, meringues, etc. Likewise, it can also be employed in textile printing, paper sizing, adhesives, casting, and impressions, and especially for the preparation of biological culture media (Goswami et al. 2015).

Carrageenan is a water-soluble polysaccharide used more frequently than agar as an emulsifier and stabilizer for several food items. It is produced from a variety of red algae that can be found across the world, including *Chondrus crispus* in the United States, *Eucheuma* in the Philippines, and *Iridaea* in Chile and Canada's Maritime Provinces. Ice cream, soy milk, yogurt, cottage cheese, salad dressings, jams, jellies, meat products, pet foods, and different processed foods specifically use carrageenan to improve their texture. Apart from this, due to its antiviral, antitumor, anti-coagulant, and immunomodulation activities, it can also be employed for pharmaceuticals (Arora et al. 2021; Goswami et al. 2015).

14.7.1.2 Pigments

Algae are an abundant source of natural pigments that exhibit numerous remarkable qualities and high economic value as well. Depending on their chemical structures, these pigments are categorized into two main groups: isoprenoids and tetrapyrroles. Carotenoids are categorized as isoprenoids, whereas phycobiliproteins and chlorophyll are classified as tetrapyrroles. In addition to this primary photosynthetic pigment chlorophyll, it also harnesses other valuable pigments, including phycocyanin, lutein, β-carotene, phycoerythrin, astaxanthin, zeaxanthin, phycobiliproteins, and canthaxanthin, etc. (Von Lintig 2018; Zhang et al. 2016).

Microalgae *Dunaliella salina* has extensively grown for β-carotene extraction, as it is the precursor of vitamin A and aids in improving the eyesight of retinitis pigmentosa (RP) patients and the immune system as well (Von Lintig 2018). According to Jayappriyan et al. (2013), research on prostate cancer has indicated that natural beta-carotene has the potential to induce cell death (apoptosis) and stimulate the activation of immune cells like natural killer cells, monocytes, and lymphocytes. This, in turn, enhances the immune response against tumor growth. Additionally, beta-carotene shields membrane lipids from peroxidation, which has been associated with several diseases, including Parkinson's disease, cardiovascular disease, and atherosclerosis (Lobo et al. 2010). On the other side, astaxanthin pigments extracted from microalgae *Haematococcus pluvialis* have excellent antioxidant characteristics and can work as a scavenger of toxic free radicals that are generated by excessive consumption of junk food or environmental pollution. This pigment is commonly utilized as a feed additive in the poultry industry and food colorant. Astaxanthin is presently available as gelatin-encased nutraceuticals, primarily for its antioxidant properties, anti-aging benefits, and its potential therapeutic applications in treating eye diseases, metabolic disorders, and neurodegenerative conditions (Ambati et al. 2014; Khalid et al. 2014).

14.7.2 Non-food Sectors

14.7.2.1 Cosmetics

In the cosmeceutical field, both macro and microalgae are extensively used due to their moisturizing, anti-aging, antioxidant, and anti-tanning qualities. Mainly, in industries, sunscreens, anti-wrinkle creams, skin sensitizers, thickening agents, hair dressing products, and moisturizers are formulated with algal extracts to avoid contamination chances (Joshi et al. 2018). *Chlorella* and *Arthrospira* microalgae strains are widely used for skin care products, by companies such as Daniel Jouvance, LVMH (Moet Hennessy Louis Vuitton), and Carnac (Cardozo et al. 2007). Additionally, other species, including *Ascophyllum nodosum, Chondrus crispus,*

Alaria esculenta, Nannochloropsis oculata, Spirulina platensis, Dunaliella salina, and *Chlorella vulgaris,* have found applications in the cosmetics industry. For example, Ferenes Cosmetics' spirulina whitening facial mask has been available in the market for improving skin tone and reducing aging signs (Schuurmans et al. 2015).

14.7.2.2 Pharmaceuticals

The extracts and bioactive compounds derived from living plants and animals are the paramount sources of the pharmaceutical industry (Atanasov et al. 2015). Recently, a notable transformation has been marked from such expensive-to-create pharmaceutical vaccines, proteins, and nutrients from these sources to a fast-growing algae population (Chu and Phang 2019). The majority of health-promoting compounds found in pharmaceuticals are in the form of tablets or liquids. However, a recent market trend offers extracts from many microalgae species as a supplement to other active ingredients (Vijayavel et al. 2007). Various algae species have various nutritional benefits; for instance, certain species serve as sources of vitamins such as thiamine, riboflavin, tocopherol, and ascorbic acids. Chlamydomonas pyrenoidosa and Chlorella vulgaris are two microalgae species that have antibacterial characteristics that are effective against gram-positive as well as gram-negative bacteria. Ochromonas sp. and Prymnesium parvum, on the other hand, can produce certain beneficial toxins for pharmaceutics (Jayshree and Jayshree 2016). Additionally, a variety of algal strains exhibit antiviral, antifungal, and cancer-fighting abilities.

14.7.2.3 Biofertilizer

A biofertilizer is a nutrient-rich compound that enhances soil fertility and stimulates plant and algae growth. Algae, as mesmerizing organisms, can be employed in modern farming because of their capacity to replenish soil nutrients and improve the utilization of micro and macronutrients (Swarnalatha 2021). Algae biofertilizer is made from the nitrogen-rich biomass that remains after utilizing carbohydrates for bioethanol production, oil for biodiesel production, and other bioactive compounds for pharmaceutics or nutraceuticals. Blue-green algae (BGA), specifically *Anabaena, Nostoc*, and *Tolypothrix genera*, aid in a gradual accumulation of soil carbon and nitrogen content by fixing atmospheric nitrogen (Singh et al. 2016). The research demonstrates that using de-oiled microalgal biomass waste from *Scenedesmus* sp. along with chemical fertilizer (MA50 + CF50) significantly improves and sustains the growth of rice crops (cv. IR36), revealing the superior performance with regard to biomass, tiller number, plant height, and grain yield. Also, in addition to this, harvest results showed that this combination produced the maximum plant dry weight, 1,000-grain weight, and panicle weight in contrast to other treatments (Nayak et al. 2019). Similarly, another study using Azolla biofertilizer substituting synthetic N fertilizer showed better NUE and rice crop yield (Yao et al. 2018).

14.7.2.4 Aquaculture Feed

Aquaculture typically uses algae as a feed to raise a variety of fish, including larvae, young fish, finfish, bivalves, and shrimps. For aquaculture feed, blends of algae involving Pavlova, *Tetraselmis*, *Phaeodactylum*, *Skeletonema*, *Nannochloropsis*,

Thalassiosira, and primarily *Spirulina* and *Chlorella* are used, whereas protein-rich algae like *Cryptonemia crenulata* and *Hypnea cervicornis* are considered suitable as a food source for shrimp (Sirakov et al. 2015). Research conducted on the cultivation of *Isochrysis zhangjiangensis* in an outdoor floating photobioreactor with a 1,000 L capacity showed the highest biomass productivity of 0.115 g L^{-1} d^{-1} and maximum cell density of 17.8 × 10^6 cells per L, which could be used as seed inoculums and feed for large-scale aquaculture water bodies at the least possible cost (Zhu et al. 2019).

The choice of certain algae for the feed used in aquaculture relies on the nutritional profile, algal cell size, growth speed, digestibility, and pigmentation. Studies performed by Ruffell et al. (2017) report that microalgae species chrysophyte *Boekelovia hooglandii* showed comparatively higher biomass yield with nutrients (lipid (34 FAs), protein, pigment (7), and carbohydrate) that suited as aquaculture feed for juvenile oysters and bivalve larvae.

14.7.2.5 Bioplastics

Recently, scientific investigations have focused on bioplastics as a viable strategy for developing sustainable products and mitigating environmental impacts (Talan et al. 2022). Algal derivatives, particularly starch from microalgae, polysaccharides from marine and freshwater macroalgae, and polyhydroxyalkanoates (PHAs) from cyanobacteria, have been suggested as a viable renewable biomass source for bioplastics production. At present, there are numerous polymeric bio-based plastic products on the market. The most often used ones are made of starch, protein, cellulose, polyhydroxyalkanoates (PHA), and polylactic acid (PLA). Algae biomass-derived biopolymers play a vital role in industrial fields like textiles, pharmaceuticals (drug delivery), cosmetics, food packaging, medicine, automobiles, consumer goods, etc. (Kalia et al. 2021; Kartik et al. 2021; Mehta et al. 2018). Cereplast and similar companies are leveraging seaweeds to produce agar, which serves as a biopolymer. Additionally, the residual crude glycerol, a byproduct of biofuel production, can be effectively employed as a plasticizer for manufacturing plastics (Jariyasakoolroj et al. 2020). Considering the higher cost associated with agar, starch is typically the favored choice as a biopolymer.

Microalgae produce polyhydroxyalkanoates (PHA) as internal granules when exposed to conditions of limited nutrients and abundant carbon sources. Algal metabolism utilizes biomolecules to produce acetyl-CoA, which is subsequently utilized in the synthesis of polyhydroxyalkanoates (PHA) through three distinct pathways: acetoacetyl-CoA, TCA cycle, and fatty acid metabolism pathways. In addition, a study on *Microcystis* sp. demonstrated that it has the maximum amount of polyhydroxybutyrate (PHB) compared to other algae found in high-rate algal ponds (Abdo and Ali 2019). Algae-based bioplastics are increasingly prevalent in modern times, although their large-scale production is limited. Numerous countries are embracing this environmentally friendly approach to advance sustainable development and combat the issue of plastic pollution.

14.7.2.6 Bio-Ink

Ink is a petroleum-based liquid containing dyes or pigments used for coloring surfaces, creating images, and writing. It typically consists of 80% petroleum products obtained from underground wells, involving the removal of vegetation and causing significant environmental damage. The remaining 20% consists of pigments, often carbon black, obtained from deep mines, which are highly toxic. These artificial inks possess both carcinogenic properties and are resistant to natural decomposition (https://energy central. com/c/ec/ ink- waste- environmental- impact- printer-cartridges). Algae are recognized for their ability to generate oils and contain living pigments. Hence, various algae species with diverse vibrant pigments are ideal for producing natural ink, as demonstrated by Scott Fulbright, CEO and cofounder of Living Inks (https://livin gink. co). These inks possess attributes of being natural, non-hazardous, cost-effective, and biodegradable, making them the optimal eco-friendly alternatives to conventional inks.

14.7.2.7 Papermaking

Algae, due to their high availability of cellulose and hemicellulose, coupled with the absence of lignin in their cell wall, render them an ideal raw material for paper production (Mukherjee and Keshri 2018). *Gelidium corneum* and *Gelidium amansi*, two commonly used red algae in paper manufacturing, are valued for their reddish cortical cells, rhizoidal filaments, and medullary cells that are rich in mucilaginous carbohydrates. These carbohydrates are extracted and processed to create bleached pulps, which are expected to yield highly bright paper (Seo et al. 2010).

Dr Phang from the University of Malaya suggests that producing paper from seaweeds is a significantly greener alternative to using wood pulp. Unlike wood, seaweeds do not contain lignin, which requires additional time and temperature during the papermaking process. By adopting this approach globally, we can decrease reliance on wood pulp, address environmental problems associated with deforestation and global warming, and promote sustainable practices. (https:// www. algae industrymagazine. com/ making- paper- seaweed)

14.8 ECONOMIC PROSPECT OF MICRO ALGAL BIOFUELS

The properties of biofuel produced from algae could be conceivable as an alternative and potential renewable energy source. However, the economic feasibility still does not compete with the first- and second-generation biofuels. The economic viability of producing algal biofuels relies heavily on the cost of TAG feedstock. Factors such as price, yield, and various other considerations will determine the competitiveness of algal-derived lipids in the market. Techno-Economic Analysis (TEA) has emerged as the key method for assessing algal biofuels' cost and economic viability. TEA primarily comprises (1) capital cost or total capital investment (TCI), encompassing equipment, engineering, contingency, land, and working capital expenses. (2) Operating costs encompass expenses related to various elements such as raw materials, workforce, overheads, services, waste management, maintenance, taxes, and insurance. Several TEAs have explored the feasibility of different approaches for generating algal biofuels. Open raceway pond cultivation has received greater focus than photobioreactors (PBRs) due to the significant cost difference, with PBRs being 2 to 2.5 times pricier than open ponds. Several studies have indicated that the present price per liter of biofuel derived from microalgae remains higher compared to traditional fossil fuels. Nevertheless, the expense can be reduced through enhanced and systematic production and conversion technology in the near future.

14.9 CONCLUSION AND FUTURE TRENDS

Based on the above review of literature, it can be concluded that third-generation biofuels generated using microalgae shows the potential to be more environmentally friendly and suitable in comparison to first- and second-generation biofuels. Specific species and characteristics of microalgae can be used for producing third-generation biofuels as these microalgae shows the potential to reduce the carbon footprints by mitigating greenhouse gas emissions. They can also be used as a sustainable resource of biomass for third-generation biofuel production under suitable growth conditions. Various technologies can be implemented to extract lipids, which can later be converted into biofuel. Thus, the chapter highlights the techniques, microalgae characteristics, cultivation, harvesting, conversion process, and utilization. Despite all the potential, microalgae have certain challenges regarding third-generation biofuel production; i.e., scaling up the process, efficient cultivation systems, techniques for maximum extraction of lipids from biomass, conversion into biofuels, and better utilization of co-products need to be addressed for commercialization and enhancement of the overall economics.

REFERENCES

Abdo SM, Ali GH. Analysis of polyhydroxybutyrate and bioplastic production from microalgae. Bull Natl Res Cent. 2019;43:1–4.

Abu-Ghosh S, Fixler D, Dubinsky Z, Iluz D. Flashing light in microalgae biotechnology Bioresour Technol. 2016;203:357–363.

Acien FG, Molina E, Reis A, Torzillo G, Zittelli GC, Sepúlveda C. In: Gonzalez-Fernandez C, Munoz R, editors. Microalgae-Based Biofuels and Bioproducts. Woodhead Publishing; 2017. p. 1–44.

Akubude VC, Nwaigwe KN, Dintwa E. Production of biodiesel from microalgae via nanocatalyzed transesterification process: A review. Mater Sci Energy Technol. 2019;2(2):216–225.

Ali A, Khullar P, Kumar D. Sodium impregnated zinc oxide as a solid catalyst for biodiesel preparation from a variety of triglycerides. Energy Sources Part A Recovery Util. Environ. Eff. 2014;36(18):1999–2008.

Ambati RR, Siew MP, Ravi S, Aswathanarayana RG. Astaxanthin: sources, extraction, stability, biological activities and its commercial applications-a review. Mar Drugs. 2014;12:128–152

Amin, S. Review on biofuel oil and gas production processes from microalgae. Energy Convers Manag. 2009;50(7):1834–1840.

Anand V, Sunjeev V, Vinu R. Catalytic fast pyrolysis of Arthrospira platensis (spirulina) algae using zeolites. J Anal Appl Pyrolysis. 2016;118:298–307.

Atanasov AG, Waltenberger B, Pferschy-Wenzig EM, Linder T, Wawrosch C, Uhrin P, Temml V, Wang L, Schwaiger S, Heiss EH. Discovery and resupply of pharmacologically active plant-derived natural products: a review. Biotechnol Adv. 2015;33:1582–1614.

Anwar M, Lou S, Chen L, Li H, Hu Z. Recent advancement and strategy on bio-hydrogen production from photosynthetic microalgae. Bioresour Technol. 2019;292: 121972.

Azizi K, Moraveji MK, Najafabadi HA. A review on bio-fuel production from microalgal biomass by using pyrolysis method. Renew Sust Energy Rev. 2018;82:3046–3059.

Azwar MY, Hussain MA, Abdul-Wahab AK. Development of biohydrogen production by photobiological, fermentation and electrochemical processes: a review. Renew Sust Energy Rev. 2014;31:158–173.

Arora K, Kumar P, Bose D, Li X, Kulshrestha S. Potential applications of algae in biochemical and bioenergy sector. 3 Biotech. 2021;11(6): 296.

Babich IV, Van der Hulst M, Lefferts L, Moulijn JA, O'Connor P, Seshan K. Catalytic pyrolysis of microalgae to high-quality liquid bio-fuels. Biomass Bioenergy. 2011;35(7):3199–3207.

Becker W. Microalgae in human and animal nutrition. In: Richmond A, editor. *Handbook of microalgal culture*. Blackwell; 2004. p. 312–51.

Bharathiraja B, Sudharsanaa T, Bharghavi A, Jayamuthunagai J, Praveenkumar R. Biohydrogen and Biogas–An overview on feedstocks and enhancement process. Fuel. 2016;185:810–828.

Bach QV, Sillero MV, Tran KQ, Skjermo J. Fast hydrothermal liquefaction of a Norwegian macro-alga: screening tests. Algal Res. 2014;6:271–276.

Bhatia SK, Bhatia RK, Jeon JM, Kumar G, Yang YH. Carbon dioxide capture and bioenergy production using biological system–A review. Renew Sust Energy Rev. 2019;110:143–158.

Bhatia SK, Mehariya S, Bhatia RK, Kumar M, Pugazhendhi A, Awasthi MK, Atabani AE, Kumar G, Kim W, Seo OS, Yang YH. Wastewater based microalgal biorefinery for bioenergy production: Progress and challenges. Sci Total Environ. 2021;751: 141599.

Bhattacharjee M. Pharmaceutically valuable bioactive compounds of algae. *Asian J Pharm Clin Re*. 2016;9:43–47.

Bhore N. Energy outlook: A view to 2040. In Robert Gardner and Theodore W. Pirog (eds.): *Detroit .automotive petroleum forum*. Exxon Mobil Corporation; 2019..

Borowitzka MA. Algae Oils for Biofuels: Chemistry, Physiology, and Production. In: Cohen Z, Ratledge C, editors. Single Cell Oils (2nd ed.). AOCS Press; 2010. p. 271–289.

Bordoloi N, Narzari R, Sut D, Saikia R, Chutia RS, Kataki R. Characterization of bio oil and its sub-fractions from pyrolysis of Scenedesmus dimorphus. Renew Energy. 2016;98:245–253.

Buitrón G, Carrillo-Reyes J, Morales M, Faraloni C, Torzillo G. Biohydrogen production from microalgae. In: Microalgae-based biofuels and bioproducts. Woodhead Publishing; 2017. p. 209–234.

Brennan L, Owende P. Biofuels from microalgae—a review of technologies for production, processing, and extractions of biofuels and co-products. Renew Sust Energy Rev. 2010;14(2):557–577.

Brindley C, Garcia-Malea MC, Acien FG, Fernandez JM, Garcia JL, Molina E. Influence of power supply in the feasibility of Phaeodactylum tricornutum cultures. Biotechnol Bioeng. 2004;87:723–733. https://doi.org/10.1002/bit.20179.

Chen CL, Huang CC, Ho KC, Hsiao PX, Wu MS, Chang JS. Biodiesel production from wet microalgae feedstock using sequential wet extraction/transesterification and direct transesterification processes. Bioresour Technol. 2015;194:179–186.

Chen CY, Liu CH, Lo YC, Chang JS. Perspectives on cultivation strategies and photobioreactor designs for photo-fermentative hydrogen production. Bioresour Technol. 2011;102(18):8484–8492.

Chen J, Li Y, Li M, Shi J, Wang L, Luo S, Liu H. Chemical flocculation-based green algae materials for photobiological hydrogen production. ACS Appl Bio Mater. 2022;5(2):897–903.

Chen Y, Wu Y, Zhang P, Hua D, Yang M, Li C, Liu J. Direct liquefaction of Dunaliella tertiolecta for bio-oil in sub/supercritical ethanol–water. Bioresour Technol. 2012;124:190–198.

Chew KW, Yap JY, Show PL, Suan NH, Juan JC, Ling TC, et al. Microalgae biorefinery: high value products perspectives. *Bioresour Technol.* 2017;229:53–62.

Chew KW, Chia SR, Show PL, Yap YJ, Ling TC, Chang JS. Effects of water culture medium, cultivation systems and growth modes for microalgae cultivation: a review. J Taiwan Inst Chem Eng. 2018;91:332–344.

Cardozo KHM, Guaratini T, Barros MP, Falcao VR, Tonon AP, Lopes NP, Campos S, Torres MA, Souza AO, Colepicolo P et al. Metabolites from algae with economic impact. *Comp Biochem Physiol C Toxicol Pharmacol.* 2007;146:60–78.

Chatzitakis A, Nikolakaki E, Sotiropoulos S, Poulios I. Hydrogen production using an algae photoelectrochemical cell. Appl Catal B Environ. 2013;142:161–168.

Chen WH, Lin BJ, Huang MY, Chang JS. Thermochemical conversion of microalgal biomass into biofuels: A review. Bioresour Technol. 2015;184:314–327.

Chia SR, Chew KW, Show PL, Yap YJ, Ong HC, Ling TC, Chang JS. Analysis of economic and environmental aspects of microalgae biorefinery for biofuels production: A review. Biotechnol J. 2018;13(6): 1700618.

Chu WL, Phang SM. Bioactive compounds from microalgae and their potential applications as pharmaceuticals and nutraceuticals. In: Hallmann A, Rampelotto PH, editors. Grand challenges in algae biotechnology. 2019. p. 429–469.

Collet P, Hélias A, Lardon L, Steyer JP, Bernard O. Recommendations for life cycle assessment of algal fuels. Appl Energy. 2015;154:1089–1102.

Converti A, Casazza AA, Ortiz EY, Perego P, Del Borghi M. Effect of temperature and nitrogen concentration on the growth and lipid content of Nannochloropsis oculata and Chlorella vulgaris for biodiesel production. Chem Eng Process Process Intensif. 2009;48:1146–1151.

Cruz YR, Aranda DAG, Seidl PR, Diaz GC, Carliz RG, Fortes MM, da Ponte DAMP, de Paula RCV. Cultivation systems of microalgae for the production of biofuels. In: Biofuels–State of development. Krzysztof Biernatin; 2018. p. 199–219.

Cuaresma M, Janssen M, Vílchez C, Wijels RH. Horizontal or vertical photobioreactors? How to improve microalgae photosynthetic efficiency. Bioresour Technol. 2011;102:5129–5137.

Dragone G, Fernandes BD, Vicente AA, Teixeira JA. Third generation biofuels from microalgae. Curr Res Technol Educ Top Appl Microbiol Microb Biotechnol. 2010;2:1355–1366.

Duan Y, Shi F. Bioreactor design for algal growth as a sustainable energy source. In Fan Shi (eds.) Reactor and process design in sustainable energy technology. Elsevier; 2014. p. 27–60.

Eriksen N. The technology of microalgal culturing. *Biotechnol Lett.* 2008;30:1525–1536.

El-Sayed WMM, Ibrahim HAH, Abdul-Raouf UM, El-Nagar MM. Evaluation of bioethanol production from Ulva lactuca by Saccharomyces cerevisiae. J Biotechnol Biomater. 2016;6(226): 2.

Fernandez FGA, Reis A, Wijffels RH, Barbosa M, Verdelho V, Llamas B. The role of microalgae in the bioeconomy. New Biotechnol. 2021;61:99–107.

Feyzi M, Hassankhani A, Rafiee HR. Preparation and characterization of Cs/Al/Fe3O4 nanocatalysts for biodiesel production. Energy Convers Manag. 2013;71:62–68.

Gardner-Dale DA, Bradley IM, Guest JS. Influence of solids residence time and carbon storage on nitrogen and phosphorus recovery by microalgae across diel cycles. Water Res. 2017;121:231–239.

Ghernaout D. Electrocoagulation process for microalgal biotechnology—a review. Appl Eng. 2019;3(2):85–94.

Golueke CG, Oswald WJ, Gotaas HB. Anaerobic digestion of algae. Appl Microbiol. 1957;5(1):47–55.

Gonçalves AL, Pires JCM, Simoes M. The effects of light and temperature on microalgal growth and nutrient removal: an experimental and mathematical approach. *RSC Adv.* 2016; 6:22896–22907.

Grobbelaar J. Inorganic algal nutrition. In: *Handbook of microalgal culture: applied phycology and biotechnology*; 2013. p. 123–133.

Goswami G, Bang V, Agarwal S. Diverse applications of algae. Int J Adv Res Sci Eng. 2015;4(1):1102–1109.

Gupta PL, Lee SM, Choi HJ. A mini review: Photobioreactors for large scale algal cultivation. World J Microbiol Biotechnol. 2015;31:1409–1417.

Ha GS, El-Dalatony MM, Kim DH, Salama ES, Kurade MB, Roh HS, El-Fatah Abomohra A, Jeon BH. Biocomponent-based microalgal transformations into biofuels during the pretreatment and fermentation process. Bioresour Technol. 2020;302: 122809.

Halder P, Azad AK. Chapter 7 – Recent trends and challenges of algal biofuel conversion technologies. In: Azad AK, Rasul M, editors. Advanced Biofuels. Woodhead Publishing; 2021. p. 167–179.

Harun R, Singh M, Forde GM, Danquah MK. Bioprocess engineering of microalgae to produce a variety of consumer products. Renew Sustain Energy Rev. 2010;14:1037–1047.

Hossain N, Mahlia TMI, Saidur R. Latest development in microalgae-biofuel production with nano-additives. Biotechnol Biofuels. 2019;12:1–16.

Hosseini A, Shariati M. Pilot culture of three strains of Dunaliella salina for β-carotene production in open ponds in the center region of Iran. World J Microbiol Biotechnol. 2006;22:1003–1006.

Huang J, Hankamer B, Yarnold J. Design scenarios of outdoor arrayed cylindrical photobioreactors for microalgae cultivation considering solar radiation and temperature. Algal Res. 2019;41:101515.

Huntley ME, Redalje DG. CO_2 mitigation and renewable oil from photosynthetic microbes: A new appraisal. Mitig Adapt Strateg Glob Chang. 2007;12: 573–608.

Hwang SW, Choi HI, Sim SJ. Acidic cultivation of Haematococcus pluvialis for improved astaxanthin production in the presence of a lethal fungus. Bioresour Technol. 2019;278:138–144

Hemschemeier A, Melis A, Happe T. Analytical approaches to photobiological hydrogen production in unicellular green algae. Photosynth Res. 2009;102: 523–540.

Iluz D, Abu-Ghosh S. A novel photobioreactor creating fluctuating light from solar energy for a higher light-to-biomass conversion efficiency. Energy Convers Manag. 2016;126:767–773.

Iasimone F, Panico A, de Felice V, Fantasma F, Iorizzi M, Pirozzi F. Effect of light intensity and nutrient supply on microalgae cultivated in urban wastewater: Biomass production, lipids accumulation and settleability characteristics. J Environ Manage. 2018;223:1078–1085. https://doi.org/10.1016/j.jenvman.2018.07.024

Ito T, Tanaka M, Shinkawa H, Nakada T, Ano Y, Kurano N, et al. Metabolic and morphological changes of an oil accumulating trebouxiophycean alga in nitrogen-deficient conditions. *Metabolomics* 2013; 9:178–187.

Jariyasakoolroj P, Leelaphiwat P, Harnkarnsujarit N. Advances in research and development of bioplastic for food packaging. *J Sci Food Agric.* 2020; 100(14):5032–5045

Jayappriyan KR, Rajkumar R, Venkatakrishnan V, Nagaraj S, Rengasamy R. In vitro anti-cancer activity of natural β-carotene from *Dunaliella salina* EU5891199 in PC-3 cells. *Biomed Prev Nutr* 2013;3:99–105.

Jayshree A, Jayashree S, Thangaraju N. *Chlorella vulgaris* and *Chlamydomonas reinhardtii*: effective antioxidant, antibacterial and anticancer mediators. *Indian J Pharm Sci.* 2016;78(5):575–581.

Jena U, Das KC. Comparative evaluation of thermochemical liquefaction and pyrolysis for bio-oil production from microalgae. Energy Fuels. 2011;25(11):5472–5482.

Jensen GS, Ginsberg DI, Drapeau MS. Bluegreen algae as an immune enhancer and bio-modulator. *J Am Nutraceutical Assoc.* 2001;3:24–30.

Jorquera O, Kiperstok A, Sales EA, Embirucu M, Ghirardi ML. Comparative energy life-cycle analyses of microalgal biomass production in open ponds and photobioreactors. Bioresour Technol. 2010;101:1406–1413.

Joshi S, Kumari R, Upasani VN. Applications of algae in cosmetics: an overview. *Int J Inn Res Sci Eng Technol.* 2018;7:11.

Kadir WNA, Lam MK, Uemura Y, Lim JW, Lee KT. Harvesting and pre-treatment of micro-algae cultivated in wastewater for biodiesel production: A review. Energy Convers Manag. 2018;171:1416–1429.

Kalia VC, Singh Patel SK, Shanmugam R, Lee JK. Polyhydroxyalkanoates: trends and advances toward biotechnological applications. *Bioresour Technol.* 2021; 326:124737

Kartik A, Akhil D, Lakshmi D, Gopinath KP, Arun J, Sivaramakrishnan R, Pugazhendhi A. A critical review on production of biopolymers from algae biomass and their applications. Bioresour Technol. 2021;329: 124868.

Khan MI, Shin JH, Kim JD. The promising future of microalgae: current status, challenges, and optimization of a sustainable and renewable industry for biofuels, feed, and other products. *Microb Cell Fact* 2018;17:36.

Khalid N, Barrow CJ. Critical review of encapsulation methods for stabilization and delivery of astaxanthin. *J Food Bioact.* 2014;1:104–115.

Kim DH, Kim MS. Hydrogenases for biological hydrogen production. Bioresour Technol. 2011;102(18):8423–8431.

Koru E, Delice GG. Microalgae and their cultivation system to produce biodiesel. IBU J Sci Technol. 2014;2:73–84.

Kumar Pathaka P, Rajb J, Saxenab G, Shankar Sharmac U. A review on production of biodiesel by transesterification using heterogeneous nanocatalyst. Int J Sci Res Dev. 2017 Jan;5.

Lanjekar RD, Deshmukh D. A review of the effect of the composition of biodiesel on NO_x emission, oxidative stability and cold flow properties. Renew Sustain Energy Rev. 2016;54:1401–1411.

Li S, Li F, Zhu X, Liao Q, Chang JS, Ho SH. Biohydrogen production from microalgae for environmental sustainability. Chemosphere. 2022;291: 132717.

Li Y, Zhang Q, Wang Z, Wu X, Cong W. Evaluation of power consumption of paddle wheel in an open raceway pond. Bioprocess Biosyst Eng. 2014;37: 1325–1336.

Lobo V, Patil A, Phatak A, Chandra N. Free radicals, antioxidants and functional foods: impact on human health. *Pharmacogn Rev.* 2010; 4(8):118–1126.

Maizatul AY, Radin Mohamed RMS, Al-Gheethi AA, Hashim MKA. An overview of the utilisation of microalgae biomass derived from nutrient recycling of wet market waste-water and slaughterhouse wastewater. *Int Aquat Res.* 2017;9:177–193.

Maity JP, Bundschuh J, Chen CY, Bhattacharya P. Microalgae for third generation biofuel production, mitigation of greenhouse gas emissions and wastewater treatment: Present and future perspectives – A mini review. Energy. 2014;78:104–113.

Mandotra SK, Kumar P, Suseela MR, Nayaka S, Ramteke PW. Evaluation of fatty acid profile and biodiesel properties of microalga Scenedesmus abundans under the influence of phosphorus, pH and light intensities. *Bioresour Technol.* 2016;201:222–229.

Marcilla A, Catalá L, García-Quesada JC, Valdés FJ, Hernández MR. A review of thermochemical conversion of microalgae. Renew Sustain Energy Rev. 2013;27:11–19.

Mata TM, Martins AA, Caetano NS. Microalgae for biodiesel production and other applications: a review. Renew Sustain Energy Rev. 2010;14(1):217–232.

McKendry P. Energy production from biomass (part 2): conversion technologies. Bioresour Technol. 2002a;83(1):47–54.

McKendry P. Energy production from biomass (part 1): overview of biomass. Bioresour Technol. 2002b;83(1):37–46.

Mehta P, Singh D, Saxena R, Rani R, Gupta RP, Puri SK, Mathur AS. High-value coproducts from algae—An innovational way to deal with advance algal industry. Waste to Wealth. 2018;343–363.

Miao X, Wu Q, Yang C. Fast pyrolysis of microalgae to produce renewable fuels. J Anal Appl Pyrolysis. 2004;71(2):855–863.

Milledge JJ, Heaven S. A review of the harvesting of micro-algae for biofuel production. Rev Environ Sci Biotechnol. 2013;12:165–178.

Millero FJ, Lepple FK. The density and expansibility of artificial seawater solutions from 0 to 40°C and 0 to 21‰ chlorinity. Mar Chem. 1973;1(2):89–104.

Minowa T, Yokoyama SYa, Kishimoto M, Okakura T. Oil production from algal cells of Dunaliella tertiolecta by direct thermochemical liquefaction. Fuel. 1995;74(12):1735–1738.

Moazami N, Ashori A, Ranjbar R, Tangestani M, Eghtesadi R, Nejad AS. Large-scale biodiesel production using microalgae biomass of Nannochloropsis. Biomass Bioenergy. 2012;39:449–453.

Mukherjee P, Keshri JP. Present status and development of algal pulp for hand-made paper making technology: a review. Adv Plants Agric Res. 2018;8(1):10–18.

Muñoz R, Navia R, Ciudad G, Tessini C, Jeison D, Mella R, Rabert C, Azócar L. Preliminary biorefinery process proposal for protein and biofuels recovery from microalgae. Fuel. 2015;150:425–433.

Musazade E, Voloshin R, Brady N, Mondal J, Atashova S, Zharmukhamedov SK, Huseynova I, Ramakrishna S, Najafpour MM, Shen JR, Bruce BD. Biohybrid solar cells: Fundamentals, progress, and challenges. J Photochem Photobiol C Photochem Rev. 2018;35:134–156.

Nagarajan D, Lee DJ, Kondo A, Chang JS. Recent insights into biohydrogen production by microalgae–From biophotolysis to dark fermentation. Bioresour Technol. 2017;227:373–387.

Narala RR, Garg S, Sharma KK, Thomas-Hall SR, Deme M, Li Y, Schenk PM. Comparison of microalgae cultivation in photobioreactor, open raceway pond, and a two-stage hybrid system. Front Energy Res. 2016;4: 29.

Nayak M, Swain DK, Sen R. Strategic valorization of de-oiled microalgal biomass waste as biofertilizer for sustainable and improved agriculture of rice (Oryza sativa L.) crop. Sci Total Environ. 2019;682:475–548.

Olguín EJ. The cleaner production strategy applied to animal production. In: Environmental biotechnology and cleaner bioprocesses. CRC Press; 2000. p. 227–243.

Oswald WJ, Golueke CG. Biological transformation of solar energy. In: Advances in applied microbiology. Academic Press; 1960. p. 223–262.

Panahi HKS, Tabatabaei M, Aghbashlo M, Dehhaghi M, Rehan M, Nizami AS. Recent updates on the production and upgrading of bio-crude oil from microalgae. Bioresour Technol Rep. 2019;7: 100216.

Park JY, Park MS, Lee YC, Yang JW. *Advances in direct transesterification of algal oils from wet biomass*. 2014.

Patil V, Tran KQ, Giselrød HR. Towards sustainable production of biofuels from microalgae. Int J Mol Sci. 2008;9(7):1188–1195.

Perendeci NA, Yılmaz V, Taştan BE, Gökgöl S, Fardinpoor M, Namlı A, Steyer JP. Correlations between biochemical composition and biogas production during anaerobic digestion of microalgae and cyanobacteria isolated from different sources of Turkey. Bioresour Technol. 2019;281:209–216.

Phang SM, Miah MS, Yeoh BG, Hashim MA. Spirulina cultivation in digested sago starch factory wastewater. J Appl Phycol. 2000;12:395–400.

Pugliese A, Biondi L, Bartocci P, Fantozzi F. Selenastrum capricornutum a new strain of algae for biodiesel production. Fermentation. 2020;6:46.

Qiang H, Richmond A. Optimizing the population density in Isochrysis galbana grown outdoors in a glass column photobioreactor. J Appl Phycol. 1994;6:391–396.

Qiu R, Gao S, Lopez P, Ogden K. Effects of pH on cell growth, lipid production and CO2 addition of microalgae Chlorella sorokiniana. *Algal Res*. 2017;28:192–199.

Raheem A, Wan Azlina WAKG, Taufiq Yap YH, Danquah MK, Harun R. Thermochemical conversion of microalgal biomass for biofuel production. Renew Sustain Energy Rev. 2015;49:990–999.

Rashid N, Rehman MSU, Memon S, Rahman ZU, Lee K, Han JI. Current status, barriers and developments in biohydrogen production by microalgae. Renew Sustain Energy Rev. 2013;22:571–579.

Ren T. *Primary Factors Affecting Growth of Microalgae Optimal Light Exposure Duration and Frequency*. 2014.

Richmond A, editor. Handbook of microalgal culture: biotechnology and applied phycology. Oxford: Blackwell Science; 2004. Vol. 577.

Rizzo AM, Prussi M, Bettucci L, Libelli IM, Chiaramonti D. Characterization of microalga Chlorella as a fuel and its thermogravimetric behavior. Appl Energy. 2013;102:24–31.

Rodolfi L, Chini Zittelli G, Bassi N, Padovani G, Biondi N, Bonini G, Tredici MR. Microalgae for oil: Strain selection, induction of lipid synthesis and outdoor mass cultivation in a low-cost photobioreactor. Biotechnol Bioeng. 2009;102:100–112.

Rogers JN, Rosenberg JN, Guzman BJ, Oh VH, Mimbela LE, Ghassemi A, Betenbaugh MJ, Oyler GA, Donohue MD. A critical analysis of paddlewheel-driven raceway ponds for algal biofuel production at commercial scales. Algal Res. 2014;4:76–88.

Ruffell SE, Packull-McCormick SR, McConkey BJ, Muller KM. Nutritional characteristics of the potential aquaculture feed species *Boeckelovia hooglandii*. *Aquaculture* 2017;474:113–120.

Salama ES, Kurade MB, Abou-Shanab RA, El-Dalatony MM, Yang IS, Min B, Jeon BH. Recent progress in microalgal biomass production coupled with wastewater treatment for biofuel generation. Renew Sustain Energy Rev. 2017;79:1189–1211.

Sanchez-Silva L, López-González D, Garcia-Minguillan AM, Valverde JL. Pyrolysis, combustion and gasification characteristics of Nannochloropsis gaditana microalgae. Bioresour Technol. 2013;130:321–331.

Schipper K, Al Muraikhi M, Alghasal GSH, Saadaoui I, Bounnit T, Rasheed R, Dalgamouni T, Al Jabri HMS, Wijffels RH, Barbosa MJ. Potential of novel desert microalgae and cyanobacteria for commercial applications and CO2 sequestration. J Appl Phycol. 2019;31:2231–2243.

Schuurmans RM, van Alphen P, Schuurmans JM, Matthijs HCP, Hellingwerf KJ Comparison of the photosynthetic yield of cyanobacteria and green algae: different methods give different answers. *PLOS One* 2015;10:e0139061.

Seo Y-B, Lee Y-W, Lee C-H, You H-C Red algae and their use in papermaking. *Bioresour Technol.* 2010;101:2549–2553.
Show KY, Yan Y, Zong C, Guo N, Chang JS, Lee DJ. State of the art and challenges of biohydrogen from microalgae. Bioresour Technol. 2019;289:12174
Shuping Z, Yulong W, Mingde Y, Kaleem I, Chun L, Tong J. Production and characterization of bio-oil from hydrothermal liquefaction of microalgae Dunaliella tertiolecta cake. Energy. 2010;35(12):5406–5411.
Sialve B, Bernet N, Bernard O. Anaerobic digestion of microalgae as a necessary step to make microalgal biodiesel sustainable. Biotechnol Adv. 2009;27(4):409–416.
Singh JS, Kumar A, Rai AN, Singh DP Cyanobacteria: a precious bio-resource in agriculture, ecosystem, and environmental sustainability. *Front Microbiol.* 2016;7:529.
Sirakov I, Velichkova K, Stoyanova S, Staykov Y, Katya VC. The importance of microalgae for the aquaculture industry. *Int J Fish Aquat Stud.* 2015;2(4):81–84.
Speece RE. *Anaerobic biotechnology for industrial wastewaters* 1996. p. 394–400.
Spolaore P, Joannis-Cassan C, Duran E, Isambert A. Commercial applications of microalgae. *J Biosci Bioeng.* 2006;101:87–96.
Suparmaniam U, Lam MK, Uemura Y, Lim JW, Lee KT, Shuit SH. Insights into the microalgae cultivation technology and harvesting process for biofuel production: a review. Renew Sustain *Energy Rev.* 2019;115:109361.
Suthar S, Verma R. Production of Chlorella vulgaris under varying nutrient and abiotic conditions: a potential microalga for bioenergy feedstock. *Process Saf Environ Prot.* 2018;113:141–148.
Swarnalatha T. *Biofertilizer from algae.* Project Report Department Of Biotechnology School Of Bio And Chemical Engineering; 2021.
Talan A, Pokhrel S, Tyagi RD, Drogui P. Biorefinery strategies for microbial bioplastics production: Sustainable pathway towards circular bioeconomy. Bioresour Technol Rep. 2022;17:100875.
Tan X, Lam MK, Uemura Y, Lim JW, Wong CY, Lee KT. Cultivation of microalgae for biodiesel production: A review on upstream and downstream processing. Chin J Chem Eng. 2018;26:17–30.
Tham PE, Ng YJ, Sankaran R, Khoo KS, Chew KW, Yap YJ, et al. Recovery of protein from dairy milk waste product using alcohol-salt liquid biphasic flotation. *Processes* 2019;7.
Thangalazhy-Gopakumar S, Adhikari S, Chattanathan SA, Gupta RB. Catalytic pyrolysis of green algae for hydrocarbon production using H+ZSM-5 catalyst. Bioresour Technol. 2012;118:150–157.
Vijayavel K, Anbuselvam C, Balasubramanian MP. Antioxidant effect of the marine *Chlorella vulgaris* against naphthalene-induced oxidative stress in the albino rats. *Mol Cell Biochem.* 2007;303:39–44.
Vanegas CH, Bartlett J. Green energy from marine algae: biogas production and composition from the anaerobic digestion of Irish seaweed species. Environ Technol. 2013;34(15):2277–2283.
Visigalli S, Barberis MG, Turolla A, Canziani R, Zrimec MB, Reinhardt R, Ficara E. Electrocoagulation–flotation (ECF) for microalgae harvesting–A review. Sep Purif Technol. 2021;271:118684.
von Lintig J Metabolism of carotenoids and retinoids related to vision. *J Biol Chem.* 2018;287:1627–1633.
Wahlen BD, Willis RM, Seefeldt LC. Biodiesel production by simultaneous extraction and conversion of total lipids from microalgae, cyanobacteria, and wild mixed-cultures. Bioresour Technol. 2011;102(3):2724–2730.

Wang L, Addy M, Lu Q, Cobb K, Chen P, Chen X, et al. Cultivation of Chlorella vulgaris in sludge extracts: Nutrient removal and algal utilization. *Bioresour Technol.* 2019;280:505–510.

Xin L, Hong-ying H, Ke G, Ying-xue S. Effects of different nitrogen and phosphorus concentrations on the growth, nutrient uptake, and lipid accumulation of a freshwater microalga Scenedesmus sp. *Bioresour Technol.* 2010;101:5494–5500.

Xu H, Miao X, Wu Q. High quality biodiesel production from a microalga Chlorella protothecoides by heterotrophic growth in fermenters. *J Biotechnol.* 2006;126(4):499–507.

Xu X, Gu X, Wang Z, Shatner W, Wang Z. Progress, challenges and solutions of research on photosynthetic carbon sequestration efficiency of microalgae. *Renew Sustain Energy Rev.* 2019;110:65–82.

Yap JK, Sankaran R, Chew KW, Halimatul Munawaroh HS, Ho S-H, Rajesh Banu J, et al. Advancement of green technologies: A comprehensive review on the potential application of microalgae biomass. *Chemosphere* 2021;281:130886.

Yao Y, Zhang M, Tian Y, Zhao M, Zeng K, Zhang B, Yin B. Azolla biofertilizer for improving low nitrogen use efficiency in an intensive rice cropping system. Field Crops Res. 2018;216:158–164.

Zainan NH, Srivatsa SC, Bhattacharya S. Catalytic pyrolysis of microalgae Tetraselmis suecica and characterization study using in situ Synchrotron-based Infrared Microscopy. Fuel. 2015;161:345–354.

Zakariah NA, Rahman NA, Hamzah F, Jahi TM. Cultivation of system of green microalgae Botryococcus braunii: A review. J Technol. 2015;76:1–5.

Zebda A, Alcaraz JP, Vadgama P, Shleev S, Minteer SD, Boucher F, Cinquin P, Martin DK. Challenges for successful implantation of biofuel cells. Bioelectrochemistry. 2018;124:57–72.

Zhu L, Li Z, Ketola T. Biomass accumulations and nutrient uptake of plants cultivated on artificial floating beds in China's rural area. *Ecol Eng.* 2011;37:1460–1406.

Zhang D, Wan M, del Rio-Chanona EA, Huang J, Wang W, Li Y, Vassiliadis VS. Dynamic modelling of *Haematococcus pluvialis* photoinduction for astaxanthin production in both attached and suspended photobioreactors. *Algal Res.* 2016;13:69–78.

Zhang QW, Lin LG, Ye WC. Techniques for extraction and isolation of natural products: A comprehensive review. Chin Med. 2018;13:1–26.

Zhu C, Han D, Li Y, Zhai X, Chi Z, Zhao Y, Cai H. Cultivation of aquaculture feed *Isochrysis zhangjiangensis* in low-cost wave driven floating photobioreactor without aeration device. *Bioresour Technol.* 2019; 293:122018.

https://sist.sathyabama.ac.in/sist_naac/documents/1.3.4/bsc-microbiology-batchno-31.pdf

http://files.rushim.ru/books/biochemie/bioactive-marine-natural-products.pdf

15 Crude Microbial Enzyme Production Using Food Industry Wastes as Feedstock

Gintu Sara George and Wasiya Farzana

15.1 INTRODUCTION

In scientific literature, materials intended for human consumption that are subsequently lost, deteriorated, or polluted are frequently referred to as food loss and food waste. Food loss at the retail and consumption stages is referred to as food waste, and its production is influenced by customer and seller behavior (UNEP, 2021).

Several biological catalysts are called enzymes in every living system. Enzymes, which are proteinaceous in origin, catalyze a variety of processes. For many years, enzymes have been utilized accidentally in the production of goods in the manner of microorganisms or extracts from plants. But the widespread use of refined enzymes in manufacturing processes only began a few decades ago (Adrio & Demain, 2014).

Enzymes are highly specialized biomolecular catalysts that quicken the pace at which substrates are converted into products. These biomolecular catalysts can either be catalytic RNA molecules, often known as ribozymes, or proteinaceous macromolecules. The molecular weights of enzymes range from 10 kDa to 2000 kDa, and they are composed of monomeric amino acids that are joined by peptide bonds to form very massive macromolecules. The expense involved in utilizing enzymes in industrial processes is a significant problem. The synthesis of enzymes on a large scale requires a lot of capital, and the use of enzymes during various stages of manufacturing has an indirect impact on product pricing. Plant installation and equipment account for a large portion of an enzyme production facility's annual operating expenses. However, raw materials contribute to 28% of the operational costs (Klein-Marcuschamer et al., 2011)

The demand for energy is growing more quickly, putting pressure on the petroleum industry and creating a fuel crisis. On Earth, people use energy mostly for transportation, heating, and producing electricity. According to a recent study, fossil fuels like coal and petroleum accounted for 79.5% of all energy usage (Bhushan et al., 2023).

DOI: 10.1201/9781032650616-15

Humans have been using fermentation for thousands of years to give food flavor, preservation, enrichment, and better textural quality. With the advancement of bioprocess technologies and biotechnology, fermentation has expanded beyond its original use as a technique of producing fermented foods to become a crucial process for the food business. Consumers also worry about the safety of chemical food additives, so they try to eat as many natural foods as they can (Martí-Quijal et al., 2021).

The food ingredients produced through fermentation, such as enzymes, sweeteners, vitamins, organic acids, stabilizers, thickening agents, and amino acids, are of interest to many companies and researchers. Additionally, lowering costs and enhancing sustainability are two additional primary justifications for producing these chemicals through fermentation. Enzymes, for instance, are frequently employed in a variety of food industry activities, including cheesemaking, baking, brewing, etc. Additionally, the market for enzymes now exceeds USD 6.1 billion yearly (Zhang et al., 2018).

Lignocellulose has been used for many years as the starting point for processing to produce a variety of commodities of great value. It is regarded as a cheap carbohydrate (Ravindran & Jaiswal, 2016). The vast majority of the food industry's processing waste is inedible and lignocellulosic in composition (Maitan-Alfenas et al., 2015). Because of its intricate structure, lignocellulose is extremely resistant and impractical for use in microbial processes. However, pretreatments and enzymatic hydrolysis of lignocellulose result in the release of fermentable sugars that enzyme-producing bacteria can use for growth and maintenance (Meng & Ragauskas, 2014).

15.1.1 Possibility in Market Conditions

The enzyme market is expanding quickly on a global scale. It was worth about $4.8 billion in 2013, and by 2018, it's anticipated to be worth an astounding $7.1 billion. Enzymes are primarily utilized in the creation of a number of goods that we consume on a daily basis.

Enzyme usage in different industrial processes can be regulated using recombinant DNA technology and protein modification. A significant portion of the industrial production of biomolecules involves biocatalytic processes, which find extensive use in the manufacturing of medicines, chemicals, and food additives (Castiglioni et al., 2020). In contrast to traditional chemical processes, enzyme catalysis enables the synthesis of high-value goods in an eco-friendly and waste-free manner (Bornscheuer, 2017).

15.1.2 Different Food Waste

The equilibrium, safety, and profitability of manufacturing/process development constitute the central concept of loss utilization for the development of products. Effects of wheat bran's starch, protein, and soluble oligosaccharide contents on the extracellular biomass-hydrolyzing enzyme activities produced by Penicillium decumbens mycelia cultured in batch fermentations were investigated, and the further result has shown the increase in the amount of starch content, which is directly related to an increasing amount of activity of amylase but inversely related to the levels of secreted cellulase and xylanase. Food industry wastes are good sources of media for microbial growth due to their high amount of carbohydrates as well as supplementary nutrients.

Based on where they come from, losses from the processing of food can be divided into six kinds: fruit peels, vegetable trimmings, starch-based waste materials, leftover wheat, leftover seafood, animal trash, and dairy production waste. The scope of this essay does not include meat, fish, or dairy waste. The additional groups are distinct from other categories in terms of variables like arrangement, molecular makeup, quantity of moisture, etc.; yet, they are all high in carbohydrates. The majority of the composition is made up of lignin and non-starchy carbohydrates; however, some grain waste varieties also have considerable amounts of lipids, amino acids, carbohydrate starch, and glucans (Sun et al., 2007). Depending on the harvest season, different grain kinds have a different exact percentage makeup (Saunders, 1985). Food-related losses like pomace from apples are a significant source of dietary fiber, polyphenols, and bioactive molecules, in addition to carbs, proteins, and lipids (Egüés et al., 2021).

15.1.3 The Value of Food Waste

During the preharvest and immediately following harvest phases of the food development process, the waste is produced. Both edible and inedible food is included in this waste. Food fiber loss also comes in a wide variety of forms and originates from various governments (Skaf & others, 2021). Contrary to watery diets made from plant waste, such as cereals and fruits, food waste derived from animals is frequently energy-rich. When developing beneficial end uses for multiple purposes, the selected method of the food waste valuation technique depends on the waste's chemical makeup and amount of moisture (Moult et al., 2018).

15.1.4 Waste in the Food Industry (Global Status)

The result of the non-product movement of unprocessed ingredients is food waste. These items have been discarded as trash because it would be more expensive to collect and treat them for reuse than it would be worth. Approximately 89 million tonnes of waste from food production are created yearly in the European Union (EU)-27, particularly industries such as manufacturing, accounting for more than 38% of this total (Lin et al., 2013). It is simpler to collect, distribute, and value-add waste from the manufacturing and agricultural sectors since it is produced in concentrated amounts. Innovative methods for the valorization of food waste have been developed as a result of industry efforts to increase sustainability, processing efficiency, and green credentials (Pfaltzgraff et al., 2013). The European Union is making significant efforts by enacting tight legislation and charging a high price for disposing of food waste in landfills, which promotes the use of this waste in the creation of products with additional value (Lin et al., 2013).

Food waste often happens at every stage of the food supply chain, including in the retail, domestic, and food services sectors, as well as in the post-harvest, processing, and distribution industries. Both edible and inedible food waste (FW) are produced by it. Food waste (FW) is defined as the uneaten portions of food and the food that is thrown away. Additionally, food is wasted because of a lack of markets, bad infrastructure, outdated technology, inadequate logistics, and managerial capabilities. The amount of waste from food production generated by the retail and food processing sectors was 244 and 118 million cubic meters, respectively.

The most common source of FW in the post-consumer stages is food services. For instance, 50% of the food served at tables is wasted. Just 10 to 15% of the food that remains can be provided without risk due to regulations regarding food safety. Therefore, by diverting FW, which is generally disposed of through waste disposal facilities, efficient and creative use of the FW might fulfill two different goals: creating value-added goods that agricultural manufacturing or other industries might utilize, which would aid in reducing the management of waste expenses as well as thereby decreasing greenhouse gas emissions (Cozzio et al., 2021).

A large portion of wastes generated by the agricultural processing sector is composed of lignocellulosic materials. The significant polysaccharide known as cellulose, which is composed of repeating glucose units, is present in lignocellulose. Depending on the kind of waste product made from plants, food that remains is also high in pectin and inulin, in addition to cellulose. Because of its high polysaccharide content, residual lignocellulosic biomass is a natural option as a raw material for the production of lignocellulose-degrading enzymes (Ragauskas et al., 2014). Every component of carbohydrates has an enzyme produced by microorganisms for the purpose of providing energy to facilitate cell growth and maintenance. The effectiveness of employing lignocellulose as a naturally unprocessed product for industrial microbial processes has been extensively studied because it is a cheap source of carbohydrates (Yasin et al., 2013). Lignocellulosic food wastes have been investigated as a possible substrate element for the manufacture of a number of significant enzymes used in industry, especially those that are used to break down cellulose, due to their high concentration of cellulose. The production of lignocellulose-degrading enzymes is one of the most significant uses of waste from lignocellulosic plants. The bioremediation process, paper and pulp manufacturing, the textile industry, and the petroleum industry all use the lignin-degrading laccase enzyme (Rodríguez Couto & Toca Herrera, 2006). On the other hand, cellulase and hemicellulase are used in the manufacturing of second-generation biofuels, particularly bioethanol. The generation of cellulase on the spot, together with the breakdown of feedstock at the same location, is a crucial technique for reducing expenses in the growth of sustainable ethanol manufacturing (Liao et al., 2015). Prior to being employed as a substrate for manufacturing operations, lignocellulosic substrates must go through a number of upstream procedures. The lignocellulose's structural limitations and component fractions prevent bacteria from proliferating in sufficient numbers to ferment the material of interest (Figure 15.1).

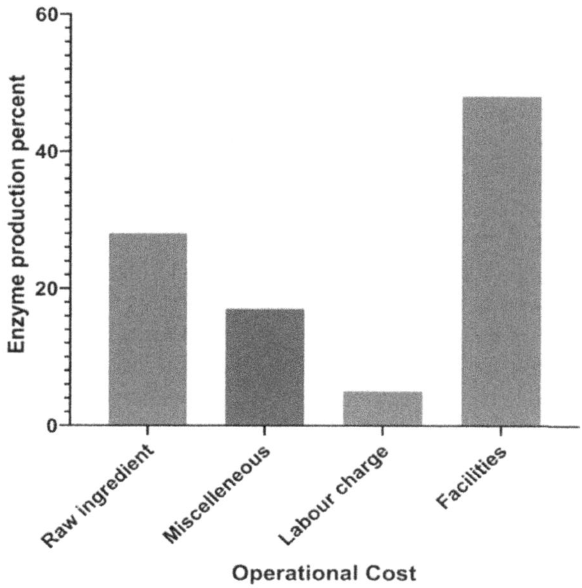

FIGURE 15.1 Enzyme production with respect to the operational cost (Ravindran & Jaiswal, 2016a)

15.2 CHALLENGES OF PRODUCING ENZYMES FROM FOOD WASTE

15.2.1 Agriculture Industrial Food Waste

The industries of agriculture and food processing, as a consequence of the global population's constant rise, rapid urbanization, and lifestyle changes, result in substantial quantities of animal, plant, and agricultural leftover wastes. Due to a lack of effective waste management methods, a significant concern for humanity is the vast accumulation of such natural food waste. Food loss is the term used to describe the abandonment, wastage, discarding, and burning of food from the harvest or slaughter source. The majority of food waste is created through its growing, collection, shipping, industrial processing, and consumption (Singh et al., 2022). The cellulose backbone makes up the majority of agricultural waste's lignocellulosic biomass (LCB), with hemicellulose and lignin serving as additional essential carbonaceous fractions. It has been investigated whether these LCB polymers could serve as substrates for lignocellulosic biorefineries that produce second-generation biofuels and other useful commodities (Sharma et al., 2020).

15.2.1.1 Waste from Food Processing

Many food manufacturing and packaging companies produce millions of food products to satisfy the food market's rising demand. However, it is inevitable that unprocessed food manufacturing, unused food, degradation, and impurities of unprocessed food components will occur in the food industries. The main categories of food waste are waste made of starch, waste from dairy products, waste from meat and fish, waste from vegetables and fruits, spent grains, and waste from pulp. Traditionally, the waste of food from the processing of food has been landfilled, consumed, put into compost, or anaerobically degraded to improve the soil. Due to their high nutritional content, these food wastes do, nevertheless, offer precursors for microbial growth and the manufacture of a variety of enzymes. Additionally, the above waste may be used as low-cost raw materials to produce both secondary and primary metabolites employing microorganisms (Yaashikaa et al., 2022).

Fruit, as well as vegetable waste (FVW), is created in the food processing industry by slicing, thermal manufacturing, handling, preparation, packaging, transportation, and naturally ripening (Sharma et al., 2020). Additionally, wastes are created by a number of biochemical activities, including oxidation, phenolic compounds, enzymes, antioxidants, microbial assault, and other chemical reactions like discoloration (Arya et al., 2022). An alternative strategy that can help to indirectly reduce environmental pollution and further improve soil nutrition is the anaerobic digestion of FVW into biofertilizers. In order to address the current shortage of animal feed, FVW, which includes pineapples, papaya fruits, leaves of bananas, peels of oranges, the spinach plant, leaves of sugarcane plants, cabbage, etc., is turned into feed for livestock in many underdeveloped nations (Sahoo et al., 2021).

During the many stages of the manufacture of edible oils, such as neutralization, degumming, deodorization, the bleaching process, and hydrolytic/oxidative rancidity, cooking or oil waste is produced. It is brought about by the hydrolysis oxidation of old frying oil by the oxidation of lipids, water, aging, the process, and the incorporation of effluents from businesses that make plenty of protein molecules, fatty acids, and carbohydrates (Okino-Delgado et al., 2017). Waste cooking oil is a liquid that remains after many deep-frying operations and is dark in color. It contains a variety of compounds that render it inappropriate for human consumption, including unbound short-chain fatty acids, ketone compounds, aldehydes, mono/di-glycerides, and volatile biochemical polymer compounds. Food waste and commercial consumable oil waste can co-ferment to create a cheap medium for microbial lipid production. Waste cooking oil is a liquid that remains after many deep-frying processes and is dark in color. Industrial edible oil waste can be co-fermented with food waste to generate a low-cost substrate for microbial lipid synthesis (Gao et al., 2022) (Table 15.1) (Figure 15.2).

15.3 PROCESSING TO PRODUCE AN ENZYME

15.3.1 Raw Material

Lignocellulose, often known as lignin, is a complex polymer primarily composed of polysaccharides. Lignin is the most refractory component of lignocellulose, which makes up all plant matter. Lignin prevents lignocellulose from being successfully broken down through enzymes or utilized by microorganisms to create fermentable sugars. Due to this, lignin must be effectively removed using a number

TABLE 15.1
Conditions of bioprocessing with respect to the production of enzyme (Ravindran & Jaiswal, 2016b)

Enzymes Produced	Conditions of Bioprocessing
Amylase	Hydrolysate is added as a supplement to the amylase generation medium during the submerged fermentation process, which lasts for 30 h.
Cellulase	Solid state fermenting method, 48–72 h of fermentation time, initial temperature of 30°C, and initial moisture content of 75%.
Inulinase	Temperature: 30°C, pH: 5, fermentation period: 7 days, immersed batch fermentation mode,
	Mode of solid-state fermentation, 65% moisture content, 35°C, and 72 h of fermentation K_2HPO_4, $ZnSO_47H_2O$, 60% moisture content, in solid-state fermentation mode, 30°C, 48 h of fermentation.
Invertase	A solid phase development mode, 30°C, 72 h of fermentation
	A solid phase fermentation mode, 30°C, 32 h of fermentation, initial pH of 5.5.
Lactase	Mode of fermentation in submerged, media supplemented with 0.75% lactose and 1% ragi, fermentation period of 12 h, pH 5.5.

of pretreatment processes (Chang & Holtzapple, 2000). The existence of crystallized cellulose, which is resistant to enzyme hydrolysis unless it is changed into an unstructured state, gives the biomass material its crystallinity (Bansal et al., 2010). The plant cell wall contains two substances, cellulose and hemicellulose, which, when degraded, yield fermentable carbohydrates like glucose, xylose, and galactose. The primary carbon source for microorganisms that make enzymes is glucose. Therefore, a possible lignocellulosic substrate should have a high cellulose concentration. In these circumstances, hemicellulose can inhibit the synthesis of glucose, the breakdown of cellulose, and the absorption of nutrients (DeMartini et al., 2013). However, the presence of a specific precursor in the media's composition may have an impact on the generation of various enzymes. For instance, it has been observed that xylans and arabinoxylans stimulate the production of xylanase in substrates (Low & Yee, 2021). Inulin can also be produced by a wide range of microbial species more efficiently than uncommon sugars such as fructooligosaccharides and fructans (Cavalitto et al., 2013).

15.3.2 Pretreatment

In order to use cellulose in bioenergy as a source of carbon, enzymes must be adsorbed on the surface of the substrate, other protein components must be hydrolyzed cooperatively, and the hydrolyzed product must be released into the liquid at large. Pectin and acetyl groups are a couple of the compositional elements that

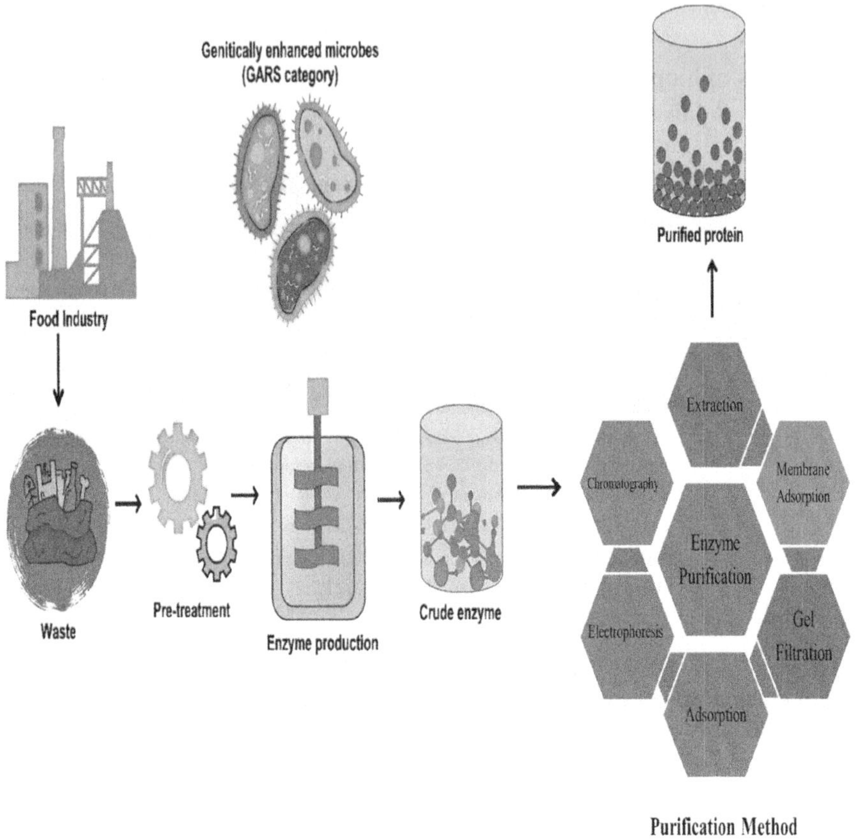

FIGURE 15.2 Graphical Representation of enzyme production (Ravindran & Jaiswal, 2016b)

affect lignocellulose breakdown, in addition to lignin and hemicellulose (Xiao & Anderson, 2013). The elimination of pectin coming from lignocellulose employing pectinase enzymes together with cellulolytic enzymes improves the utilization of biomass (Fakkaew & Polprasert, 2021). It takes initial treatments to alter the lignocellulose's chemistry and structure in order to address these problems. To remove resistant lignocellulose components, pretreatment techniques use high temperatures, high pressures, or a combination of the two (Singh et al., 2015). Ingenious disruption methods, including ultrasonic treatment, microwaves being exposed, treatments with harmful substances like chemicals and alkaline compounds, and enzymatic hydrolysis, are additional effective pretreatment techniques. However, pretreatment methods that involve heat and acid hydrolysis produce enzyme and microbial growth inhibitors that could impede the production process (Ravindran & Jaiswal, 2016a).

15.3.3 Selection of Microbes

Different research has been carried out to reveal various types of microorganisms that belong to the group of fungi, yeast, bacteria, and actinomycetes. Microorganism screening was one of the most effective methods used in the business to find novel enzymes. This is indeed applicable to the microorganisms which belong to the thermophilic type of category, which are isolated from the external locations and thus the extraction of enzymes is carried out. One benefit of choosing microorganisms that are thermophilic during the enzyme activation is that the possibility of contamination is minimized as a result of bioprocessing procedures being performed at higher temperatures. Further, the higher temperatures lead to low viscosity and increase the solubility of substrates. Simultaneously, increasing the yield of the product is considerable due to the shifts in the equilibrium during reactions that are endothermic (Ravindran & Jaiswal, 2016a).

Commercial enzyme preparations of microbial origin used in the industry are offered. These preparations include the necessary enzyme in addition to other production strain metabolites, as well as extra food-grade preservatives and stabilizers that also meet regulatory requirements. The safety of the production strain is still the most crucial factor to consider when assessing an enzyme's efficiency. For the manufacture of industrial enzymes, strains that have undergone thorough characterization to prove they are neither contagious nor toxic, especially those with a history of protection, are appropriate possibilities (Pariza & Johnson, 2001). A relatively small number of bacterial and fungal strains, such as the bacteria subtilis variant *Bacillus licheniformis*, a particular type of fungus called *Aspergillus niger*, as well as Fungi *ryzae*, have produced a significant number of industrially significant enzymes. It has been established that these bacteria are effective for industrial-scale production because they have traditionally been employed for the commercial manufacture of a variety of metabolites. This has allowed for a complete study of their features and metabolic processes. They are renowned for having the ability to overexpress specific proteins in the media for fermentation and for being simple to genetically manipulate. These features make microorganisms extremely alluring hosts for a large range of heterologous enzymes. In addition, genetic engineering has been successful in using a variety of microorganisms—including *Escherichia coli* (*E. coli*), *Fusarium venenatum*, as well as *Pseudomonas fluorescens* that were not previously used in the commercial production of native enzymes as hosts for the development of essential enzymes for the sector (Olempska-Beer et al., 2006). The wide range of extracellular enzymes produced by wild-type strains may spontaneously create enzymes with economic significance. Discovering such microbes is typically done using the technique of bioprospecting. It gathers microorganisms from certain niches in the environment and assesses how well they can hydrolyze various substrates. After checking the existence of a crucial enzyme, the most effective one is then selected. Utilizing metagenomic technologies to analyze genetic composition is another approach. To identify novel enzymes, probes, and group-specific primers are used. The main qualitative flaw with this approach is that the metabolic potential cannot be assessed because the microbe is not isolated or cultured. The quick screening of potential microorganisms is made possible by comparative genomic analysis of microbial strains. This makes it easier to assess the microorganism's proteome (Youssef et al., 2013).

Using conventional genetic modifications, selection, as well as recombinant methods, a system of metabolic pathways linked to the creation of a desirable chemical molecule during the fermentation process in a cell is used to overexpress a gene. Through genetic modification, the lab "evolved" cells to become resistant to high product concentrations by changing the genes and enzymes involved in the system and removing the usual genetics as well as metabolic regulation of such genes and enzymes. Lastly, a reliable fermentation procedure is created that enables the mass manufacture of the desired chemical (Keasling, 2012). It has been successful to create engineered microbes using traditional metabolic pathway engineering techniques, but new strategies are needed to produce microbes and manufacturing methods that are cost-competitive with the current high-volume, inexpensive chemical manufacturing processes. Researchers have significantly contributed to improving metabolic engineering's capacity to create novel, highly designed organisms with outstanding productivity efficiency and employ substrates generated from natural resources to further cut production costs over the past several decades. This work has been tremendously aided by methods like sequencing genomes and new academic disciplines like biological informatics, systems biology, and metabolomics (Rajesh & Gummadi, 2022).

15.3.4 Enzyme Immobilization by Agric-Food Waste

Utilizing food and agricultural waste for innovative immobilization or aids for enzymatic offers cost-effective options for industrial bioprocesses while also addressing the issues of disposing of waste and environmental contamination. Agro-industrial food and agricultural wastes are definitely characterized by their huge area of coverage, significant porosity, and their abundance of various chemical groups. The potential uses for these traits are exciting (Girelli et al., 2020).

To help in the biological degradation of anthracene, the ligninolytic enzyme laccase was immobilized on the outermost layer of rice straw charcoal. The laccase demonstrated a 60% effective yield of immobilization, excellent stability for up to six cycles, and an approximate forty percent retention of initial activity.

15.3.5 Biopolymers for Immobilizing Enzymes

Due to their "green" properties, including biological compatibility, breakdown, biofunctionality, and bio-stability, biopolymers are incredibly helpful as a foundation for enzymatic immobilization. In addition, biopolymers can be readily obtained, possess a variety of chemical characteristics and structural shapes, and can be produced at reasonable temperatures. The two outstanding immobilization carriers being researched as materials of support for immobilizing enzymes are chitin and chitosan. Dextran polyaldehyde was used as the macromolecular cross-linking substance in experiments to immobilize pectinase into chitin nanoparticles with

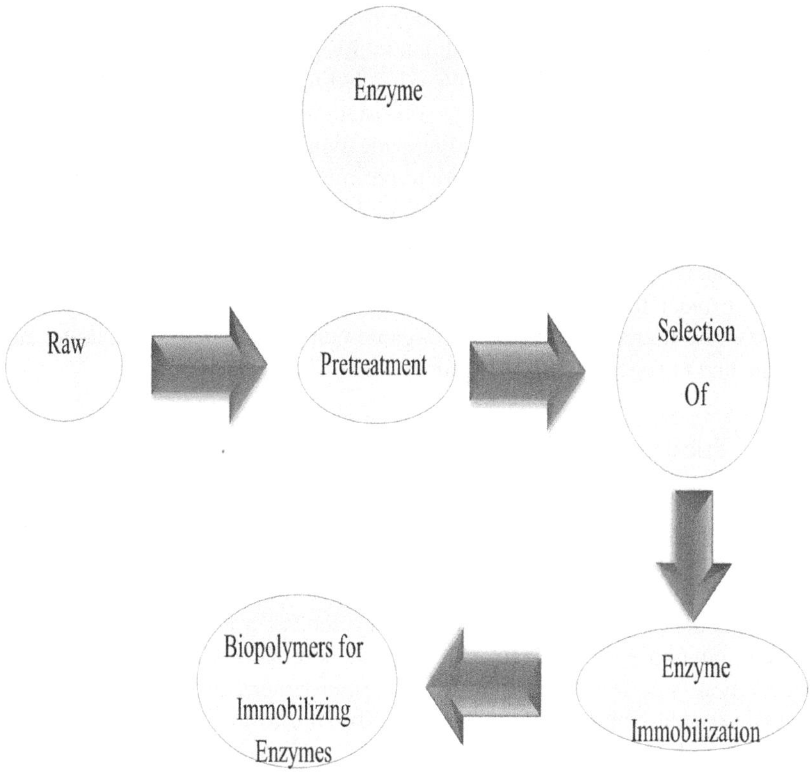

FIGURE 15.3 Production of enzyme

magnetic properties (Sojitra et al., 2017). After a total of seven tries, along with 15 days of storage, respectively, the immobilized pectinase maintained 85% and 89% of its peak activity, demonstrating remarkable longevity and reusability (Benucci et al., 2020) (Figure 15.3).

15.4 STRATEGIES FOR FERMENTATION

15.4.1 Fermentation in the Solid State

Several industrial enzymes have been produced by solid-state fermentation. On solid surfaces, this kind of fermentation encourages the growth of fungi as well as the production of conidiospores (Armenta-López et al., 2021). The majority of industrial enzymes come from fungi. Fungi release their genes differently during submerged fermentation compared to solid-state fermentation (Barrios-González, 2012). However, due to the difficulty in standardizing processes and the low repeatability of findings, the technique of solid-substrate fermentation has not been implemented for large-scale manufacture. Without adequate controls, temperatures can soar to previously unheard-of heights, deactivating the reactor's produced enzymes. Aeration has been suggested as a solution to this problem.

Progress in fermentation technology has led to some success in the creation of methods to make fermentation using solid-state technology a manageable process. Even culture development and rapid enzyme synthesis with great repeatability are promoted by the membrane's capacity to permit water vapor to escape while maintaining the substrate dry (Ito et al., 2013). There is minimal required subsequent processing thanks to solid-state fermentation's great volumetric productivity, high product concentrations, and low effluent discharge (Singhania et al., 2010). But a research project that employed solid-state fermentation to create enzymes from organic waste noticed the release of organic compounds that are volatile, such as methane and nitrogen (Maulini-Duran et al., 2015).

15.4.2 Fermenting while Floating

Submerged fermentation is the method used most frequently to produce enzymes on a large scale. Another of the subsurface fermentation techniques used to make enzymes using lignocellulose as a substrate is a stirred tank reactor. It allows for oxygenation and movement inside the container while using a water-based medium to regulate the pH and temperature. The reactor has a low risk of contamination and can maintain a sterile environment (Roukas & Kotzekidou, 1997). Submerged fermentation is carried out in a consistent environment, and the reactor enhances process variable control (Gibbs et al., 2000). Due to the need for big, expensive equipment, submerged fermentation is less cost-effective than solid-state fermentation. According to comparison research involving the manufacture of biodiesel, submerged fermentation required a 78% higher capital expenditure than the solid-state approach (Zhang et al., 2011).

15.4.3 Different Fermentation Techniques

In order to create an enzyme from lignocellulosic substrates, biomass must first be pretreated before using microorganisms designed to produce the desired enzyme. In some instances, a crucial stage in this procedure is the enzymatic hydrolysis of pretreatment biomass, which yields a hydrolysate with fermentable sugars. The manufacturing of enzymes can then use this hydrolysate as an addition in combination with other media elements. Cellulase and hemicellulase are a pair of enzymes that are used in the hydrolysis of lignocellulose, which releases glucose and xylose as the two primary reducing sugars. Mannose, galactose, and arabinose are among the other sugars that are released. The hydrolysate of sugar cane bagasse creates -amylase. However, the type of microbial digestion of the biomass will determine whether the enzymatic hydrolysis phase is included. Enzymes are created by specific bacteria to degrade the polymeric structure of lignocellulose. In these circumstances, predigesting lignocellulose with enzymes is a step that is unnecessary and renders the production process unfeasible from an economic standpoint. Researchers have been using successive solid-substrate and immersed fermentation processes more frequently lately. This technique uses lignocellulosic material as the substrate for solid-state pre-culture fermentation. To reduce the biomass's moisture content to 70%, it is treated with a nourishing broth. The solid mass content is then supplemented with spores in exact ratios. The liquid state fermenting takes place for 24 h after the medium used for fermentation is provided with a larger amount of nutritional fluid containing higher amounts of glucose, and the next steps are carried out under submerged conditions (Cunha et al., 2012).

15.5 SEPARATION, PURIFICATION, AND RESTORATION OF ENZYME

Enzymes are created in nature as crude extracts once the production process is complete. Despite the fact that the crude extract demonstrates the required activity of the enzyme, it cannot be used as a commercial product due to its slow reaction rates and the presence of pollutants such as metabolites and poisons produced by the microorganisms and components of the media used for microbial cultivation. Isolation and purification are therefore the last stages of any procedure for manufacturing enzymes. During purification, the specific activity of the enzyme is boosted, and any unwanted elements are removed from the final product. Techniques for purifying proteins have been the subject of extensive research because of how important the product is from a commercial aspect. The final product must be obtained through arduous processes that include cell killing (based on the origin), the extraction process, fractionation, and the last step, the product.

15.5.1 Production of Enzyme

All biological systems contain enzymes, but there is a good chance that their quantities, activity, stabilities, accessibility, and the presence of inhibitory chemicals will all vary greatly. With the development of recombinant DNA technology, traditional plants, animals, and microorganisms have made way for genetically engineered creatures. Instances of bacteria that have eukaryotic proteins that have been copied and expressed in them include *E. coli* and *Bacillus subtilis*. These proteins can exist either extracellularly or in the cytoplasm or periplasm of cells. The bacterium *Escherichia coli* and *Bacillus subtilis* are two examples of bacteria that have proteins from eukaryotic cells replicated and expressed in them (Fakruddin et al., 2013). However, using recombination procedures, it is possible to add "affinity labels" to the target gene. However, it is feasible to use recombination techniques incorporating an "affinity" tag to the genetic material of interest, assisting with the separation an enzyme. Later, by applying highly specific proteolytic enzymes, this tag can be eliminated (Young et al., 2012).

15.5.2 Immobilization of Biocatalyst

Enabling enzymatic segregation from the process media and their favorable reuse in the following steps, immobilization, a method in which enzymatic processes are connected to or enclosed in a substrate that is solid, improves catalyst reuse (Lindeque & Woodley, 2019). Increasing the parameters of kinetics and enzyme stability, which are important prerequisites for the long-term viability of industrial biocatalysis, is another advantage of this technique (Nwagu et al., 2021). Enzyme immobilization is a key improvement enabling enzyme recovery and reuse because it enables the binding of catalysts for change onto insoluble solid substrates that are easy to seperate from the medium (Fryszkowska & Devine, 2020).

Immobilization is a very efficient technique for maintaining the catalytic function of immobilized derivatives, and because it improves the biocatalyst's operational stability, it can also significantly reduce operating costs. Due to these benefits, enzyme immobilization is now often used in numerous large-scale operations in the food, medicine, chemical, and textile industries (Taheri-Kafrani et al., 2020). The properties of the screen-printed material for support have an impact on the choice of enzyme immobilization technique. When selecting an immobilization technique, factors including surface area, chemical composition, and permeability are crucial. For an intended procedure, which could differ depending on the kind and kind of interactions between the support and the enzyme, a suitable biocatalyst, support, and immobilization method must be chosen (Romero-Fernández & Paradisi, 2020). Although techniques like covalent bonding and cross-linking produce highly stable immobilized enzymes, they also carry a risk of causing enzyme inactivation due to the chemical reaction on the enzyme surface (Shen et al., 2019). The weak connections between enzymatic processes and porous materials, which could lead to enzyme leakage during reactions, make adsorption approaches generally limited (Xie & Zang, 2018); entrapment methods usually have limitations due to the weakened mechanical characteristics of immobilized enzymes produced by those techniques (Liu et al., 2020). Immobilization efficiency also depends on the kind of enzymatic activity required to be immobilized and its particular features. For example, the act of adsorption of lipases deviates from the usual hydrophobic process because of the pseudo-affinity the enzyme has for the hydrophobic part of the support due to its distinct functional properties (Remonatto et al., 2022). Thus, the technique of lipase immobilization most commonly used is physical adsorption. Covalent bonding is frequently employed to immobilize proteases; however, by encouraging the development of robust connections between the structure of the substrate as well as the biocatalyst, this method lowers the likelihood of enzymatic desorption. Immobilization techniques like cross-linkage, capture, or encapsulation are typically avoided because they impede mass transfer or have diffusion constraints as a drawback (Tacias-Pascacio et al., 2021).

15.5.3 Electrophoresis

According to the total electrical charge in macropore gels or 1–2% agarose, proteins are separated by electrophoresis. When the amount of the target protein fraction is too low, it is a fantastic method for protein separation. Proteins are separated based on their molecular size using gels like polyacrylamide gel. Using the correct blotting procedures, proteins that are isolated using this method are eluted. Various forms of electrophoresis have developed depending on their intended use; for instance, the addition of a mild cleaning agent (for example, sodium dodecyl sulfate, or SDS) to the electrophoresis gel enables us to study distinct protein molecule subunits. Without using SDS, the protein's native form will be retrieved by electrophoresis. According to their isoelectric point, proteins are sorted by isoelectric focusing. There are certain restrictions to this approach; yet it can resolve the target protein. Proteins precipitate at an isoelectric point whenever the Sephadex™ bed serves to act as an anti-convection fluid in the experiment, which may harm other bands. This problem can be resolved by creating distinct compartments within the separating chamber (čeřovská, 1997).

15.5.4 Adsorption in Expansion Bed

Large volumes of cellular homogenates can be purified of cells and cell debris in one step using expansion bed adsorption. In essence, it is a protein separation fluidized-bed reactor modified. Adsorbent granules remain stationary and segregated by an upward movement of liquid in a fluidized bed. The space between the particles of the adsorbent is used by the cells and cell debris to pass through. Only thorough mixing results in a significant amount of target molecule adsorption onto particles of adsorbent in a fluidized-bed reactor. This method was utilized to eliminate galactosidase in the form of expanded bed adsorption using streamline-diethylamino ethanol (DEAE). The specific activity increases by 12.6 times, where 65% of the total quantity of enzymes was eluted (Boeris et al., 2012).

15.5.5 Immobilization and Purification in a Single Step

Methods that simplify the process have increased in popularity. The enzymes can be simultaneously immobilized and purified in a single step. Usually, all that is required to stabilize an enzyme is one reaction between it and the support substance. But occasionally, a large number of contacts may be necessary for the protein molecule to keep its link with the support intact. However, a multipoint or multi-subunit attachment should be the goal of enzyme immobilization in order to increase enzyme stability (García-Galán et al., 2011). One of three techniques can be used to accomplish enzyme immobilization and purification in one step:

- Stasis at a single location.
- The emergence of new realms.
- Use of supports with a range of functionalities to immobilize.

In order to immobilize a specific enzyme via multipoint attachment, personalized supports that are specific to the target protein are designed based on specific structural characteristics (Singh et al., 2019). Then, domains with a preference for the heterobifunctional supports are inserted within the target protein molecule via site-directed mutagenesis. The interfacial activation-induced immobilization of lipases, which are on hydrophobic supports, is one unusual circumstance. The antigen-antibody interaction is a type of single-point immobilization. Since just the desired protein is immobilized, this method is extremely sensitive and selective and further uses monoclonal or polyclonal antibody types. The target protein must be immobilized on the surface of the antibody in order for the technique to be successful, and any unwanted adsorption must be avoided. How this enzyme works can be chosen by the user.

When using monoclonal antibodies, the position of the enzyme in relation to the support surface can be chosen by the user. This feature is important since it allows the user to select a pathway where the enzyme's active site is fully accessible (Hernandez & Fernandez-Lafuente, 2011). If the place of connection is a weak spot that is prone to inhibitor deactivation, this may also help to safeguard the enzyme. Following this procedure, the immobilized enzyme purity is almost 100% and the immobilization yield is 100%. In some instances, an antibody will selectively bind to a specific portion of a target protein that has gone through site-directed mutagenesis rather than the entire target protein itself. Use this strategy to ensure that the intended peptide is oriented properly. On this support, any protein with the necessary sequence can likewise be immobilized. Since full enzyme purity has been attained, the immobilization is also almost 100% in this instance; however, stabilization is not very significant as there is no relationship between the enzyme and the support substance (Wang & others, 2001).

The second approach entails the introduction of numerous domains that result in the formation of a bond between the molecule of protein and the support. These domains are composed of short peptides like the cellulose-binding site as well as synthetic peptides having protein capabilities like the his-tag. In the instance of lipases, some catalytic processes are advantageous. Interfacial activation is the process through which lipases bond to the outermost layer of oil droplets. The lipase's extremely hydrophobic active center is covered by a polypeptide chain but becomes visible whenever its enzyme exists in a solution of water. The lipase is effectively closed by this confirmation. The open form is another form in which the hydrophobic aspect of the catalytic core is exposed; the two major forms alternate and maintain equilibrium. In order to disclose the hydrophobic nature of the catalysis base and the state of equilibrium, which is now located in an open confirmation, the lid is shifted aside whenever a tiny drop of oil is present. According to van Tilbeurgh et al. (1993), the ability of lipases to attach to hydrophobic surfaces is known as interfacial activation (Van Tilbeurgh et al., 1993).

Any hydrophobic item, such as molecules that are hydrophobic, hydrophobic support structures, hydrophobic protein molecules, and oil drops, can be used to produce interfacial activation. The purification of lipases has made considerable use of all these materials. Immobilizing the lipases with the hydrophobic substrate at low ionic strength is a frequent phase in the purification procedure that is used to stabilize the open structure of the enzyme in one step (Fernandez-Lorente et al., 2008). Only extremely weak ions can be used in this technique. A strong ion concentration promotes the lipase enzyme to be in the packed form, as well as the adsorption of more proteins onto the matrix while immobilizing or purifying. Compared to the traditional method of lipase adsorption, interfacial activity is more efficient.

The different lipase enzyme behavior in terms of adsorption on hydrophobic substrates is determined by this criterion. The different lipases' behavior in terms of adsorption on hydrophobic substrates is determined by this criterion. Other lipases require extremely hydrophobic supports, whereas likely lipases are able to be adsorbed on intermediate supports. The kind of unit on the support could be crucial for lipase purification. Supports that are frequently used to boost enzyme durability by an amount of 1,000 include octyl agarose as well as sepabeadsdecaoctyl (Palomo et al., 2002). This protocol is appropriate for use in industrial settings and is comparatively simple to use. The enzyme may covalently connect with supporting groups after adsorption to increase stability. When the pH is elevated, the nucleophilic compounds on the outermost layer of the enzyme's structure can grow more reactive toward the support (Pedroche et al., 2007).

15.5.6 Enzyme Isolation

The quality and quantity of the protein that has been taken out of a solid source are compromised. The optimum isolation method should allow the target enzyme to be released while preventing problematic contaminants such as bacteria, viruses, and nucleic acids. Additionally, care should be taken to avoid degrading or denaturing the protein during the extraction process. The most widely used technique for removing proteins from the cellular milieu is homogenization. Ultrasonication is another technique for disrupting cell structures. Cells are more easily broken up by ultrasonication, and intracellular proteins are made visible to the growth media. High-frequency vibrations are used in ultrasonication procedures to create cavitation on the microbial cell wall and subsequently disintegrate it.

The protein liberated during cell lysis can be altered by extended ultrasonic exposure. Consequently, sonication cycles need to be optimized in a certain way. Prior to column chromatography, the extract is often clarified by centrifugation or filtration after the cell disruption procedure (Liu et al., 2014). The parameters necessary for the stability of the released protein determine the characteristics of the separation medium. The pH level, chemical compounds, reducing agents like chelators or metal ions, proteolytic inhibitory substances, and bacterial contamination are the key variables impacting whether the isolation medium is created. Usually, the pH is changed to correspond to the range of values where the enzyme functions best. However, this may not consistently be the case. According to Pečová et al. (2013), trypsin is most active between pH 8 and pH 9 and remains most stable within pH 3 to 8 (Pečová et al., 2013). In order to free the enzyme from the membrane's hydrophobic bonds, detergents are used. Most of the detergents used for isolation do not denature the globular proteins or alter their catalytic function. The isolation medium is often the only place where detergent is used. Detergents produce "micelles" as they approach the critical concentration of micelle (CMC) because of their amphipathic nature. This may hinder the purifying process in column chromatography. As a result, the concentration of detergent used during separation needs to be smaller than CMC (Malinsky et al., 2013). Many enzymes expose thiol groups when a protein is released from the cytosol and deposited in the development medium during separation, which may lead to an oxidation process.

The internal disulfide bonds can typically be maintained with concentrations of these reducing agents as low as 10–25 mM (Mathieson & Thomas, 2013). Three issues that arise during enzyme separation include metal ions, proteases, and bacterial contamination. Metal ions can be problematic because they can complicate compounds with certain groups and enhance the oxidation of thiol groups. Heavy metals can be eliminated from a solution using chelating substances like EDTA, or ethylenediaminetetraacetic acid, and the solvent ethylene glycol tetraacetic acid (EGTA). A buffer's pH level can be altered by the addition of EDTA. As a result, after injecting EDTA, caution should be taken to correct the pH (Campden et al., 2015). The proteins of interest are seriously endangered by proteases that are present in the cellular lysate naturally. Adding protease inhibitors is the simplest way to stop the proteolytic action. As inhibitors of protease are so expensive, this could present an issue in large-scale procedures. Adsorbing proteases, which are hydrophobic absorbents, and altering pH levels such that proteases lose their efficiency are two further practical answers to this issue. Following sterility measures is essential for preventing the growth of microorganisms during the manufacture of enzymes. Certain buffers as well, like phosphate and acetate, among others, are more likely to promote bacterial growth at neutral pH values. Always incorporating antimicrobial compounds into buffers is an additional strategy to prevent contamination

15.5.7 Separation Process (Fractionation)

The adsorption and precipitation processes were the first fractionation techniques. Electrophoresis and chromatographic techniques have since replaced these. Because every protein has different properties, specific purification procedures must be developed. Every aspect of the protein of interest, including its properties and the types of contaminants it contains, must be gathered and examined. However, the various protein purification procedures must be carried out sequentially, one after the other. The most recent methods and advancements in protein purification are covered in this section. The cell lysate is clarified by high-speed centrifugation, which also removes all cell pieces and other particles. This is possible in the lab, where fundamental cell preparations may be produced at high speeds of 40,000 to 5,000 g and kept chilled. Microfiltration and ultrafiltration are advancements in the removal of contaminated insoluble materials used in conjunction with centrifugation. Protein fractions can be separated using the ultrafiltration process based on the molecular weight of the protein, which can range from >1 to 300 kDa. This method is rapid, dependable, and cost-effective because it isolates sodium as well as other small molecules from fragments of protein with a greater molecular weight (He et al., 2013). By including polymers, salts, and organic solvents, altering the pH level, as well as raising or lowering the solution's temperature, a chosen protein can be precipitated. The most popular salting-out chemicals are antichaotropic salts. As a result of their interactions with water molecules, the hydrophobic effect is amplified, and protein molecules clump together. Ammonium sulfate is the antichaotropic salt that is used most frequently to precipitate proteins (Mirica et al., 2012). Protein precipitation can also be done with the use of naturally occurring chemicals like acetone and ethanol. Organic solvents and organic polymers have similar characteristics, and the most widely used organic polymer for use in this field is polyethylene glycol (PEG). The solution can be diluted using the buffering agent because it gets viscous at high concentrations. Because PEG is uncharged, it can be employed directly in ion exchange chromatography, which is used to separate proteins.

Liquid-liquid extraction, also known as aqueous two-phase extraction, is a distinct technique for separating proteins. In a liquid-liquid extraction experiment, a polymer like dextran or a certain salt is frequently used in a single stage and PEG in the other. In a perfect world, the contaminating substances, such as extra proteins and other particles, would accumulate in the bottom stage while the proteins of interest would migrate to the highest stage (PEG). Then, the same can be removed using centrifugation. Pineapple enzymes have been extracted using aqueous two-phase techniques. Bromelain and polyphenol oxidase were separated using a mixture of polyethylene glycol and potassium phosphate (18% PEG with 1,500 molecular mass, or 14% phosphate). While bromelain revealed a 228% efficiency increase and a 4.0-fold increase in purity, isolated polyphenol oxidase exhibited just 90% activation and a 2.7-fold increase in purity (Babu et al., 2008).

A phase that is stationary and a mobile phase, two separate components, work together to separate substances in chromatography. The stationary phase in protein purification is frequently a solid substance shaped into a column. Typically, the phase that moves is a liquid, like solutions of salt or buffer with certain pH values. Size exclusion chromatography, chromatography with ion exchange, hydrophobic interaction, immobilized metal ion affinity, and affinity chromatography are some of the most widely used chromatography techniques for protein purification (Schmidt, 1983).

Protein molecules are separated using size exclusion chromatography based on their hydrodynamic volume. The porous beads used in size exclusion chromatography typically have varying pore diameters that are specific to each column. The phase of mobile particles that move through the column will have various hydrodynamic volumes. The larger particles will pass through the column more slowly, though. As a result, particle separation takes place within the mobile phase. A size exclusion column used to be built with materials like starch and dextran. Substances like S-200 Sephacryl beads have since replaced this (Kunji et al., 2008).

The technique of ion exchange (IE) chromatography is frequently used to eliminate proteins from mixtures. In IE chromatography, biological molecules in the mobile phase are electrostatically interacting with chargeable compounds in the stationary phase. The degree of contact among the protein molecules relies on the reversed binding of proteins and their suitable ligands, just like the antigen-antibody interaction. Due to the great specificity of the interaction, the chosen protein can be removed after separation by using chemicals that cause denaturation or by adjusting the pH or ionic strength. Affinity chromatography can use pre-activated materials like Affi-Gel 10 and 15 as well as Profinity™ Epoxide substance, UltraLink

Iodoacetyl resin, and CarboLink Coupling resin. The use of recombinant DNA techniques has increased the significance of affinity chromatography over time. It will be considerably simpler to recognize, isolate, and purify the target protein if DNA sequences that encode for protein tags are included in the target gene. The histidine, which is a tag, composed of a configuration of six histidine residues at the end of the desired protein's sequence, is one of the most often used affinity tags (Kapust & Waugh, 2000). A large number of tags appear to prevent proteins from folding properly or serving their biological purpose. Fusion tags were created as a solution to this issue, improving expression yields and promoting stability and native fold.

15.6 CHALLENGES FACED BY THE INDUSTRY

Low enzyme productivity, low enzymatic titer, and low enzyme yield are mostly caused by technical problems that obstruct microorganism development on lignocellulosic hydrolysates. Technical challenges include low sugar content, the existence of 5-C and 6-C sugars, which can lead to the development of diauxics, and the challenge of controlling the ratio of C to N (Roukas T, 2015). Due to the presence of microbial inhibitors, including furfural, hydroxymethyl furfural, acetic acid, various phenolic substances, and numerous other chemical species, using lignocellulose for fermentation is primarily problematic (Jönsson et al., 2013). As a result of lignin degradation during abrasive pretreatments, several byproducts are produced. We are yet unsure of the exact way that phenolic chemicals prevent development (Jönsson et al., 2013). The main techniques for getting rid of inhibitors' negative effects include (1) preventing their creation during pretreatment and hydrolysis, (2) detoxifying the hydrolysate, (3) creating bacteria that can tolerate them, and (4) neutralizing the poisonous substances. Careful lignocellulosic material selection and the use of gentle pretreatments can stop the development of hazardous chemicals. It is possible to detoxify the hydrolysate using certain chemicals. One of the most affordable methods for detoxification is over-liming, which uses $Ca(OH)_2$ to pretreat lignocellulose (Ranatunga et al., 2000).

It has been known to use enzymes to remove harmful substances. The phenolic compounds in hemicellulose hydrolysate have been successfully removed by Trametes versicolor laccase and peroxidase enzymes. Enzyme inhibitors can be metabolized by certain bacteria. A fungus called *Coniochaeta ligniaria* NRRL30616 has the ability to break down phenolic, furfural, and HMF compounds. The poisonous compounds could either be transformed into biomass or into less dangerous substances by *C. ligniaria* NRRL30616. A greater cellulose conversion rate was additionally noted (Cao et al., 2013)

One of the most popular physical techniques for removing inhibitors is vacuum evaporation, which has the advantage of completely removing volatile fractions like acetic acid, furfural, and vanillin but the disadvantage of concentrating non-volatile compounds like lignin derivatives. The lignocellulose hydrolysate's phenolics were what caused the enzyme activity to be reduced. Accordingly, full enzyme activity was recovered after the removal of the same. Either ethyl acetate, activated charcoal, or PEG surfactant can be used to remove the phenolic chemicals.

15.7 CONCLUSION

One of the most urgent problems facing modern society is the management of agricultural and food waste. Green technology must be used to properly repurpose agro-food wastes to minimize the harmful and destructive effects of waste disposal and help implement the circular economy by producing compounds with added value. Compared to typical chemical catalysts, microbial enzymes are crucial in enhancing the value of agro-industrial produce and food waste. It has a great deal of potential for productive waste processing and suitable biocatalytic systems with high conversion efficiencies, enabling the achievement of sustainable development objectives. Microorganisms exploit agro-food waste to produce economically important enzymes. New, inexpensive enzyme immobilization supports and altered enzymes can also demonstrate improved catalytic activity when applied in industrial environments.

REFERENCES

Adrio, J. L., & Demain, A. L. (2014). Microbial enzymes: tools for biotechnological processes. *Biomolecules*, 4(1), 117-139.

Armenta-López, S. E., Valenzuela-Solano, C., & Hernández-Martínez, R. (2021, April 29). Identification and molecular analysis of races of Fusarium oxysporum f. sp. lycopersici isolated from tomato in Baja California, Mexico. *Revista Mexicana De Fitopatología, Mexican Journal of Phytopathology*, 39(2).

Arya, P. S., Yagnik, S. M., Rajput, K. N., Panchal, R. R., & Raval, V. H. (2022, October). Valorization of agro-food wastes: Ease of concomitant-enzymes production with application in food and biofuel industries. *Bioresource Technology*, 361, 127738.

Babu, B. R., Rastogi, N., & Raghavarao, K. (2008, January). Liquid–liquid extraction of bromelain and polyphenol oxidase using aqueous two-phase system. *Chemical Engineering and Processing: Process Intensification*, 47(1), 83–89.

Bansal, P., Hall, M., Realff, M. J., Lee, J. H., & Bommarius, A. S. (2010, June). Multivariate statistical analysis of X-ray data from cellulose: A new method to determine degree of crystallinity and predict hydrolysis rates. *Bioresource Technology*, 101(12), 4461–4471.

Barrios-González, J. (2012, February). Solid-state fermentation: Physiology of solid medium, its molecular basis and applications. *Process Biochemistry*, 47(2), 175–185.

Benucci, I., Caso, M. C., Bavaro, T., Masci, S., Keršienė, M., & Esti, M. (2020, April). Prolyl endopeptidase from Aspergillus niger immobilized on a food-grade carrier for the production of gluten-reduced beer. *Food Control*, 110, 106987.

Bhushan, S., Jayakrishnan, U., Shree, B., Bhatt, P., Eshkabilov, S., & Simsek, H. (2023, June). Biological pretreatment for algal biomass feedstock for biofuel production. *Journal of Environmental Chemical Engineering*, 11(3), 109870.

Boeris, V., Balce, I., Vennapusa, R. R., Rodríguez, M. A., Picó, G., & Lahore, M. F. (2012, July). Production, recovery and purification of a recombinant β-galactosidase by expanded bed anion exchange adsorption. *Journal of Chromatography B*, 900, 32–37. https://doi.org/10.1016/j.jchromb.2012.05.024

Bornscheuer, U. T. (2017, November 27). The fourth wave of biocatalysis is approaching. *Philosophical Transactions of the Royal Society A: Mathematical, Physical and Engineering Sciences*, 376(2110), 20170063.

Campden, R., Audet, N., & Hébert, T. E. (2015, February). Nuclear G protein signaling. *Journal of Cardiovascular Pharmacology*, 65(2), 110–122.

Cao, G., Ximenes, E., Nichols, N. N., Zhang, L., & Ladisch, M. (2013, October). Biological abatement of cellulase inhibitors. *Bioresource Technology*, 146, 604–610.

Castiglioni, G. Z., Bettio, G., Matte, C. R., Jacques, R. A., Dos Santos Polidoro, A., Rosa, C. A., & Ayub, M. A. Z. (2020, March 21). Production of volatile compounds by yeasts using hydrolysed grape seed oil obtained by immobilized lipases in continuous packed-bed reactors. *Bioprocess and Biosystems Engineering*, 43(8), 1391–1402.

Cavalitto, S. F., Chesini, M., Neila, L. P., Fratebianchi de la Parra, D., Rojas, N. L., Contreras Esquivel, J. C., Ghiringhelli, P. D., & Hours, R. A. (2013, May 15). Aspergillus kawachii produces an inulinase in cultures with yacon (Smallanthus sonchifolius) as substrate. *Electronic Journal of Biotechnology*, 16(3).

Chang, V. S., & Holtzapple, M. T. (2000). Fundamental factors affecting biomass enzymatic reactivity. *Applied Biochemistry and Biotechnology*, 84–86(1–9), 5–38.

Cozzio, C., Tokarchuk, O., & Maurer, O. (2021, September 14). Minimising plate waste at hotel breakfast buffets: An experimental approach through persuasive messages. *British Food Journal*, 123(9), 3208–3227.

Cunha, F., Esperança, M., Zangirolami, T., Badino, A., & Farinas, C. (2012, May). Sequential solid-state and submerged cultivation of Aspergillus niger on sugarcane bagasse for the production of cellulase. *Bioresource Technology*, 112, 270–274.

DeMartini, J. D., Pattathil, S., Miller, J. S., Li, H., Hahn, M. G., & Wyman, C. E. (2013). Investigating plant cell wall components that affect biomass recalcitrance in poplar and switchgrass. *Energy & Environmental Science*, 6(3), 898.

Egüés, I., Hernandez-Ramos, F., Rivilla, I., & Labidi, J. (2021, June 22). Optimization of ultrasound assisted extraction of bioactive compounds from apple pomace. *Molecules*, 26(13), 3783.

Fakkaew, K., & Polprasert, C. (2021, June). Air stripping pre-treatment process to enhance biogas production in anaerobic digestion of chicken manure wastewater. *Bioresource Technology Reports*, 14, 100647.

Fakruddin, M., Mohammad Mazumdar, R., Bin Mannan, K. S., Chowdhury, A., & Hossain, M. N. (2013, September 13). Critical factors affecting the success of cloning, expression, and mass production of enzymes by RecombinantE. coli. *ISRN Biotechnology*, 2013, 1–7.

Fernandez-Lorente, G., Cabrera, Z., Godoy, C., Fernandez-Lafuente, R., Palomo, J. M., & Guisan, J. M. (2008, October). Interfacially activated lipases against hydrophobic supports: Effect of the support nature on the biocatalytic properties. *Process Biochemistry*, 43(10), 1061–1067.

Fryszkowska, A., & Devine, P. N. (2020, April). Biocatalysis in drug discovery and development. *Current Opinion in Chemical Biology*, 55, 151–160.

Gao, Z., Ma, Y., Liu, Y., & Wang, Q. (2022, January). Waste cooking oil used as carbon source for microbial lipid production: Promoter or inhibitor. *Environmental Research*, 203, 111881.

Garcia-Galan, C., Berenguer-Murcia, N., Fernandez-Lafuente, R., & Rodrigues, R. C. (2011, November). Potential of different enzyme immobilization strategies to improve enzyme performance. *Advanced Synthesis & Catalysis*, 353(16), 2885–2904.

Gibbs, P. A., Seviour, R. J., & Schmid, F. (2000, January). Growth of filamentous fungi in submerged culture: Problems and possible solutions. *Critical Reviews in Biotechnology*, 20(1), 17–48.

Girelli, A. M., Astolfi, M. L., & Scuto, F. R. (2020, April). Agro-industrial wastes as potential carriers for enzyme immobilization: A review. *Chemosphere*, 244, 125368.

He, R., Girgih, A. T., Malomo, S. A., Ju, X., & Aluko, R. E. (2013, January). Antioxidant activities of enzymatic rapeseed protein hydrolysates and the membrane ultrafiltration fractions. *Journal of Functional Foods*, 5(1), 219–227.

Hernandez, K., & Fernandez-Lafuente, R. (2011, February). Control of protein immobilization: Coupling immobilization and site-directed mutagenesis to improve biocatalyst or biosensor performance. *Enzyme and Microbial Technology*, 48(2), 107–122.

Ito, K., Gomi, K., Kariyama, M., & Miyake, T. (2013, November). Rapid enzyme production and mycelial growth in solid-state fermentation using the non-airflow box. *Journal of Bioscience and Bioengineering*, 116(5), 585–590.

Jönsson, L. J., Alriksson, B., & Nilvebrant, N. O. (2013, January 28). Bioconversion of lignocellulose: inhibitors and detoxification. *Biotechnology for Biofuels*, 6(1).

Kapust, R. B., & Waugh, D. S. (2000, July). Controlled intracellular processing of fusion proteins by TEV protease. *Protein Expression and Purification*, 19(2), 312–318.

Keasling, J. D. (2012, May). Synthetic biology and the development of tools for metabolic engineering. *Metabolic Engineering*, 14(3), 189–195.

Klein-Marcuschamer, D., Oleskowicz-Popiel, P., Simmons, B. A., & Blanch, H. W. (2011, November 21). The challenge of enzyme cost in the production of lignocellulosic biofuels. *Biotechnology and Bioengineering*, 109(4), 1083–1087.

Kunji, E. R., Harding, M., Butler, P. J. G., & Akamine, P. (2008, October). Determination of the molecular mass and dimensions of membrane proteins by size exclusion chromatography. *Methods*, 46(2), 62–72.

Liao, H., Fan, X., Mei, X., Wei, Z., Raza, W., Shen, Q., & Xu, Y. (2015, March). Production and characterization of cellulolytic enzyme from Penicillium oxalicum GZ-2 and its application in lignocellulose saccharification. *Biomass and Bioenergy*, 74, 122–134.

Lin, C. S. K., Pfaltzgraff, L. A., Herrero-Davila, L., Mubofu, E. B., Abderrahim, S., Clark, J. H., Koutinas, A. A., Kopsahelis, N., Stamatelatou, K., Dickson, F., Thankappan, S., Mohamed, Z., Brocklesby, R., & Luque, R. (2013). Food waste as a valuable resource for the production of chemicals, materials and fuels. Current situation and global perspective. *Energy & Environmental Science*, 6(2), 426.

Lindeque, R., & Woodley, J. (2019, March 14). Reactor selection for effective continuous biocatalytic production of pharmaceuticals. *Catalysts*, 9(3), 262.

Liu, F., Ye, M., Pan, Y., Zhang, Y., Bian, Y., Sun, Z., Zhu, J., Cheng, K., & Zou, H. (2014, July 2). Integration of cell lysis, protein extraction, and digestion into one step for ultrafast sample preparation for phosphoproteome analysis. *Analytical Chemistry*, 86(14), 6786–6791.

Liu, J., Shen, X., Zheng, Z., Li, M., Zhu, X., Cao, H., & Cui, C. (2020, December). Immobilization of laccase by 3D bioprinting and its application in the biodegradation of phenolic compounds. *International Journal of Biological Macromolecules*, 164, 518–525.

Low, Y. W., & Yee, K. F. (2021, November). A review on lignocellulosic biomass waste into biochar-derived catalyst: Current conversion techniques, sustainable applications and challenges. *Biomass and Bioenergy*, 154, 106245.

Maitan-Alfenas, G. P., Visser, E. M., & Guimarães, V. M. (2015, February). Enzymatic hydrolysis of lignocellulosic biomass: Converting food waste in valuable products. *Current Opinion in Food Science*, 1, 44–49.

Malinsky, J., Opekarová, M., Grossmann, G., & Tanner, W. (2013, April 29). Membrane microdomains, rafts, and detergent-resistant membranes in plants and fungi. *Annual Review of Plant Biology*, 64(1), 501–529.

Martí-Quijal, F. J., Khubber, S., Remize, F., Tomasevic, I., Roselló-Soto, E., & Barba, F. J. (2021, July 7). Obtaining antioxidants and natural preservatives from food by-products through fermentation: A review. *Fermentation*, 7(3), 106.

Mathieson, W., & Thomas, G. A. (2013, February). Simultaneously extracting DNA, RNA, and protein using kits: Is sample quantity or quality prejudiced? *Analytical Biochemistry*, 433(1), 10–18.

Maulini-Duran, C., Abraham, J., Rodríguez-Pérez, S., Cerda, A., Jiménez-Peñalver, P., Gea, T., Barrena, R., Artola, A., Font, X., & Sánchez, A. (2015, March). Gaseous emissions during the solid state fermentation of different wastes for enzyme production at pilot scale. *Bioresource Technology*, 179, 211–218.

Meng, X., & Ragauskas, A. J. (2014, June). Recent advances in understanding the role of cellulose accessibility in enzymatic hydrolysis of lignocellulosic substrates. *Current Opinion in Biotechnology*, 27, 150–158.

Mirica, K. A., Lockett, M. R., Snyder, P. W., Shapiro, N. D., Mack, E. T., Nam, S., & Whitesides, G. M. (2012, January 26). Selective precipitation and purification of monovalent proteins using oligovalent ligands and ammonium sulfate. *Bioconjugate Chemistry*, 23(2), 293–299.

Moult, J., Allan, S., Hewitt, C., & Berners-Lee, M. (2018, May). Greenhouse gas emissions of food waste disposal options for UK retailers. *Food Policy*, 77, 50–58.

Nwagu, T. N., Okolo, B., & Aoyagi, H. (2021, February). Immobilization of raw starch saccharifying amylase on glutaraldehyde activated chitin flakes increases the enzyme operation range. *Bioresource Technology Reports*, 13, 100645.

Okino-Delgado, C. H., Prado, D. Z. D., Facanali, R., Marques, M. M. O., Nascimento, A. S., Fernandes, C. J. D. C., Zambuzzi, W. F., & Fleuri, L. F. (2017, October 26). Bioremediation of cooking oil waste using lipases from wastes. *PLoS One*, 12(10), e0186246.

Olempska-Beer, Z. S., Merker, R. I., Ditto, M. D., & DiNovi, M. J. (2006, July). Food-processing enzymes from recombinant microorganisms—a review. *Regulatory Toxicology and Pharmacology*, 45(2), 144–158.

Palomo, J. M., Muñoz, G., Fernández-Lorente, G., Mateo, C., Fernández-Lafuente, R., & Guisán, J. M. (2002, December). Interfacial adsorption of lipases on very hydrophobic support (octadecyl–Sepabeads): immobilization, hyperactivation and stabilization of the open form of lipases. *Journal of Molecular Catalysis B: Enzymatic*, 19–20, 279–286.

Pariza, M. W., & Johnson, E. A. (2001, April). Evaluating the safety of microbial enzyme preparations used in food processing: Update for a new century. *Regulatory Toxicology and Pharmacology*, 33(2), 173–186.

Pečová, M., Šebela, M., Marková, Z., Poláková, K., Čuda, J., Šafářová, K., & Zbořil, R. (2013, March 6). Thermostable trypsin conjugates immobilized to biogenic magnetite show a high operational stability and remarkable reusability for protein digestion. *Nanotechnology*, 24(12), 125102.

Pedroche, J., del Mar Yust, M., Mateo, C., Fernández-Lafuente, R., Girón-Calle, J., Alaiz, M., Vioque, J., Guisán, J. M., & Millán, F. (2007, April). Effect of the support and experimental conditions in the intensity of the multipoint covalent attachment of proteins on glyoxyl-agarose supports: Correlation between enzyme–support linkages and thermal stability. *Enzyme and Microbial Technology*, 40(5), 1160–1166.

Pfaltzgraff, L. A., De bruyn, M., Cooper, E. C., Budarin, V., & Clark, J. H. (2013). Food waste biomass: A resource for high-value chemicals. *Green Chemistry*, 15(2), 307.

Ragauskas, A. J., Beckham, G. T., Biddy, M. J., Chandra, R., Chen, F., Davis, M. F., Davison, B. H., Dixon, R. A., Gilna, P., Keller, M., Langan, P., Naskar, A. K., Saddler, J. N., Tschaplinski, T. J., Tuskan, G. A., & Wyman, C. E. (2014, May 16). Lignin valorization: Improving lignin processing in the biorefinery. *Science*, 344(6185).

Rajesh, R., & Gummadi, S. N. (2022, February 10). Production of multienzymes, bioethanol, and acetic acid by novel Bacillus sp. PM06 from various lignocellulosic biomass. *Biomass Conversion and Biorefinery.*

Ranatunga, T. D., Jervis, J., Helm, R. F., McMillan, J. D., & Wooley, R. J. (2000, August). The effect of overliming on the toxicity of dilute acid pretreated lignocellulosics: The role of inorganics, uronic acids and ether-soluble organics. *Enzyme and Microbial Technology*, 27(3–5), 240–247.

Ravindran, R., & Jaiswal, A. K. (2016a, January). A comprehensive review on pre-treatment strategy for lignocellulosic food industry waste: Challenges and opportunities. *Bioresource Technology*, 199, 92–102.

Ravindran, R., & Jaiswal, A. K. (2016b, January). Exploitation of food industry waste for high-value products. *Trends in Biotechnology*, 34(1), 58–69.

Remonatto, D., Miotti Jr., R. H., Monti, R., Bassan, J. C., & de Paula, A. V. (2022, March). Applications of immobilized lipases in enzymatic reactors: A review. *Process Biochemistry*, 114, 1–20.

Rodríguez Couto, S., & Toca Herrera, J. L. (2006, September). Industrial and biotechnological applications of laccases: A review. *Biotechnology Advances*, 24(5), 500–513.

Romero-Fernández, M., & Paradisi, F. (2020, April). Protein immobilization technology for flow biocatalysis. *Current Opinion in Chemical Biology*, 55, 1–8.

Roukas, T. (2015, January 20). The role of oxidative stress on carotene production by Blakeslea trisporain submerged fermentation. *Critical Reviews in Biotechnology*, 1–10.

Roukas, T., & Kotzekidou, P. (1997, September). Pretreatment of date syrup to increase citric acid production. *Enzyme and Microbial Technology*, 21(4), 273–276.

Sahoo, A., Sarkar, S., Lal, B., Kumawat, P., Sharma, S., & De, K. (2021, June). Utilization of fruit and vegetable waste as an alternative feed resource for sustainable and eco-friendly sheep farming. *Waste Management*, 128, 232–242.

Saunders, R. M. (1985, January). Rice bran: Composition and potential food uses. *Food Reviews International*, 1(3), 465–495.

Schmidt, B. (1983, November). Proteine protein purification - principles and practice. Von R. Scopes. Springer-Verlag, Berlin - Heidelberg - New York 1982. XIII, 282 S., 145 Abb., DM 79. *Nachrichten Aus Chemie, Technik Und Laboratorium*, 31(11), 914–914.

Sharma, P., Gaur, V. K., Kim, S. H., & Pandey, A. (2020, March). Microbial strategies for bio-transforming food waste into resources. *Bioresource Technology*, 299, 122580.

Sharma, S., Sharma, V., Nargotra, P., & Bajaj, B. K. (2020, March 21). Bioprocess development for production of a process-apt xylanase with multifaceted application potential for a range of industrial processes. *SN Applied Sciences*, 2(4).

Shen, X., Yang, M., Cui, C., & Cao, H. (2019, May). In situ immobilization of glucose oxidase and catalase in a hybrid interpenetrating polymer network by 3D bioprinting and its application. *Colloids and Surfaces A: Physicochemical and Engineering Aspects*, 568, 411–418.

Singh, A., Singhania, R. R., Soam, S., Chen, C. W., Haldar, D., Varjani, S., Chang, J. S., Dong, C. D., & Patel, A. K. (2022, September). Production of bioethanol from food waste: Status and perspectives. *Bioresource Technology*, 360, 127651.

Singh, J., Suhag, M., & Dhaka, A. (2015, March). Augmented digestion of lignocellulose by steam explosion, acid and alkaline pretreatment methods: A review. *Carbohydrate Polymers*, 117, 624–631.

Singh, K., Mishra, A., Sharma, D., & Singh, K. (2019, February 19). Nanotechnology in enzyme immobilization: An overview on enzyme immobilization with nanoparticle matrix. *Current Nanoscience*, 15(3), 234–241.

Singhania, R. R., Sukumaran, R. K., Patel, A. K., Larroche, C., & Pandey, A. (2010, June). Advancement and comparative profiles in the production technologies using solid-state and submerged fermentation for microbial cellulases. *Enzyme and Microbial Technology*, 46(7), 541–549.

Skaf, L., Franzese, P., Capone, R., & Buonocore, E. (2021, August). Unfolding hidden environmental impacts of food waste: An assessment for fifteen countries of the world. *Journal of Cleaner Production*, 310, 127523.

Sojitra, U. V., Nadar, S. S., & Rathod, V. K. (2017, February). Immobilization of pectinase onto chitosan magnetic nanoparticles by macromolecular cross-linker. *Carbohydrate Polymers*, 157, 677–685.

Sun, X., Liu, Z., Qu, Y., & Li, X. (2007, September 22). The effects of wheat bran composition on the production of biomass-hydrolyzing enzymes by Penicillium decumbens. *Applied Biochemistry and Biotechnology*, 146(1–3), 119–128.

Tacias-Pascacio, V. G., Morellon-Sterling, R., Castañeda-Valbuena, D., Berenguer-Murcia, N., Kamli, M. R., Tavano, O., & Fernandez-Lafuente, R. (2021, October). Immobilization of papain: A review. *International Journal of Biological Macromolecules*, 188, 94–113.

Taheri-Kafrani, A., Kharazmi, S., Nasrollahzadeh, M., Soozanipour, A., Ejeian, F., Etedali, P., Mansouri-Tehrani, H. A., Razmjou, A., Yek, S. M. G., & Varma, R. S. (2020, July 25). Recent developments in enzyme immobilization technology for high-throughput processing in food industries. *Critical Reviews in Food Science and Nutrition*, 61(19), 3160–3196.

UNEP. (2021). *Food waste index report 2021.* United Nations Environment Programme;

Van Tilbeurgh, H., Egloff, M. P., Martinez, C., Rugani, N., Verger, R., & Cambillau, C. (1993, April). Interfacial activation of the lipase–procolipase complex by mixed micelles revealed by X-ray crystallography. *Nature*, 362(6423), 814–820.

Wang, J., Bhattacharyya, D., & Bachas, L. G. (2001, August 23). Orientation specific immobilization of organophosphorus hydrolase on magnetic particles through gene fusion. *Biomacromolecules*, 2(3), 700–705.

Xiao, C., & Anderson, C. T. (2013). Roles of pectin in biomass yield and processing for biofuels. *Frontiers in Plant Science*, 4.

Xie, W., & Zang, X. (2018, August). Lipase immobilized on ionic liquid-functionalized magnetic silica composites as a magnetic biocatalyst for production of trans -free plastic fats. *Food Chemistry*, 257, 15–22.

Yaashikaa, P., Senthil Kumar, P., & Varjani, S. (2022, January). Valorization of agro-industrial wastes for biorefinery process and circular bioeconomy: A critical review. *Bioresource Technology*, 343, 126126.

Yasin, N. H. M., Mumtaz, T., Hassan, M. A., & Abd Rahman, N. (2013, November). Food waste and food processing waste for biohydrogen production: A review. *Journal of Environmental Management*, 130, 375–385.

Young, C. L., Britton, Z. T., & Robinson, A. S. (2012, January 10). Recombinant protein expression and purification: A comprehensive review of affinity tags and microbial applications. *Biotechnology Journal*, 7(5), 620–634.

Youssef, N. H., Couger, M. B., Struchtemeyer, C. G., Liggenstoffer, A. S., Prade, R. A., Najar, F. Z., Atiyeh, H. K., Wilkins, M. R., & Elshahed, M. S. (2013, August). The genome of the anaerobic fungus orpinomyces sp. Strain C1A reveals the unique evolutionary history of a remarkable plant biomass degrader. *Applied and Environmental Microbiology*, 79(15), 4620–4634.

Zhang, B., Weng, Y., Xu, H., & Mao, Z. (2011, November 15). Enzyme immobilization for biodiesel production. *Applied Microbiology and Biotechnology*, 93(1), 61–70.

Zhang, Y., He, S., & Simpson, B. K. (2018, February). Enzymes in food bioprocessing novel food enzymes, applications, and related techniques. *Current Opinion in Food Science*, 19, 30–35.

16 Biodegradable Packaging Materials from Agro-Food Industry Waste

Vivek Kumar, Moirang them Kalpana Devi

16.1 INTRODUCTION

The global consumption of plastics for packaging has escalated significantly in recent decades, leading to severe environmental consequences. Traditional plastic packaging materials, mainly derived from fossil fuels, are non-biodegradable, resulting in long-lasting pollution and detrimental impacts on ecosystems. As concerns about environmental sustainability and waste management intensify, there is a growing demand for alternative packaging materials that are biodegradable and environmentally friendly (Lee and Liew, 2020).

The agro-food industry, encompassing agriculture, food processing, and distribution sectors, generates substantial amounts of organic waste. This waste comprises agricultural residues, fruit and vegetable peels, husks, shells, stems, seeds, and byproducts from food processing. Improper disposal of agro-food waste poses environmental challenges, including greenhouse gas emissions and soil and water contamination (Kumar and Thakur, 2017).

In response to the pressing need for sustainable packaging solutions, researchers and industries have turned their attention to utilizing agro-food waste as a valuable resource to produce biodegradable packaging materials. The concept of transforming these organic residues into biodegradable materials has gained traction due to their potential to reduce waste, minimize the environmental burden, and promote circular economy practices.

Biodegradable packaging materials offer a promising solution, as they can break down naturally over time, avoiding the persistence of plastic waste in the environment. Various biodegradable materials, such as polylactic acid (PLA), polyhydroxyalkanoates (PHA), starch-based films, cellulose derivatives, and chitosan, have shown potential for replacing conventional plastics in packaging applications (Polman et al., 2021).

DOI: 10.1201/9781032650616-16

16.2 AGRO-FOOD INDUSTRY WASTE AS A RESOURCE

The agro-food industry, while essential for sustaining our growing global population, often generates significant amounts of waste throughout its various stages – from production and processing to distribution and consumption. Natural biodegradable polymers include polysaccharides like cellulose derivatives, chitosan, and starch, as well as protein-based polymers such as collagen, gelatin, and albumin. In contrast, synthetic biodegradable polymers consist of aliphatic polyesters, polyanhydrides, poly (alkylcyanoacrylates), polyamino acids, phosphorous-based polymers, and acrylic polymers. Biodegradable polymers derived from petroleum resources encompass polyglycolide (PGA), PLA, poly(lactide-co-glycolide) (PLGA), polycaprolactone (PCL), poly(butylene succinate) (PBS), poly (p-dioxanone) (PPDO), polycarbonate, polyamides, poly (esteramides), polyurethanes, polyanhydrides, and vinyl polymers. Additionally, PHA and poly (hydroxybutyrate-co-hydroxyvalerate) [P(3HB-co-3HV)] are examples of biodegradable polymers produced through microbial fermentation (Ranganathan et al., 2020).

Biopolymers such as cellulose, starch, pectin, chitin/chitosan, and their variations are typically obtained from agricultural waste sectors. Table 16.1 provides

TABLE 16.1
Application of biopolymers produced from different agro-industrial waste

S. No.	Food Industries	Source of Food Waste	Biopolymer	Application
1.	Grain processing industry	Broken grains, fine dust, husk, malt dust, grain separator waste, straw, and bran	PHA	Food packaging Drug delivery
			Cellulose	Food packaging
2.	Dairy processing and beverage industries	Whey hydrolysates and cheese residue, molasses, carbonation sludge, bagasses, coffee grounds, and spent grains	PHA	Medicine Food additives
			Xanthum gum	Drug delivery Entertainment industry
3.	Fruits and vegetable industries	Vegetable washing water, peel, pomace, molasses, beet pulp, and distillates	Pectin Lignin	Food packaging Biomedical application
			PLA	Food industries
			PHA	Skin care
4.	Meat, poultry, and sea food processing industries	Bones, tendons, skin, blood, gastrointestinal tract content, and internal organs, egg shells, and shells of seafoods.	Collagen Gelatin Chitin chitosan	Paper and textile industries Biomedical Waste water treatment Cosmetic preparation Biocatalyst Agricultural drugs

an overview of waste products specific to various food industries. Polysaccharides stand out as the most prevalent natural biopolymers present in plants, animals, and microorganisms.

Lignocellulosic biomass comprises cellulose (35–50%), hemicelluloses (15–35%), lignin (15–25%), and various inorganic substances. The lignocellulosic material is sourced from agro-industrial residues like sugarcane bagasse, corn cobs, wheat straw, banana, soybean hull, oil palm mesocarp, rice straw, soybean straw, and hulls (Tapia-Bl´ácido et al., 2017). Additionally, food wastes such as peanut husk, citrus peel, straw, and corn stover also exhibit high cellulose concentrations (Sanchez-Vazquez et al., 2013). The physical properties of cellulosic materials are significantly influenced by the formation of a hydrogen bond network among the hydroxyl groups of cellulose. These networks display notable resistance to oxidizing agents and strong alkali. Given their robustness, cellulose fibers are currently employed to enhance the stability of thermoplastics or thermosetting polymer matrices, gradually replacing glass fibers (Sanchez-Vazquez et al., 2013).

Xanthan gum, a significant biopolymer, is manufactured globally at a volume of approximately 30,000 tons per year (Stredansky and Conti, 1999). This heteropolysaccharide is generated by the *Xanthomonas* genus and is widely applied in the food, pharmaceutical, and petrochemical sectors. Typically, *Xanthomonas campestris* is employed in industries for xanthan gum production (Mesomo et al., 2009). Additionally, food byproducts such as cheese whey, spent malt grains, and citrus waste are recognized as viable sources of xanthan gum.

Hemicellulose ranks as the most abundant renewable plant material following cellulose. These compounds form hydrogen bonds with cellulose moieties and are covalently linked to lignin (Peng et al., 2009). Lignin, another polymer, comprises three primary monomers: p-hydroxyphenyl alcohol (H), coniferyl alcohol (G), and synapyl alcohol (S). Lignin behaves as a thermoplastic polymer, with softwood containing G-units, herbaceous plants having G-S-H units, and hardwood featuring G-S units. This biopolymer imparts mechanical rigidity to the cell wall and offers protection against microbial attacks (Huang et al., 2019).

Collagen, a plentiful protein-based biopolymer sourced from mammals, is primarily obtained from cattle and pig skins, as well as their bones, mainly sourced from abattoirs. Collagen production also utilizes fish waste like scales and bones, along with eggshells and chicken processing waste (Kaya et al., 2015). These collagen sources find applications in various industries, including food, pharmaceuticals, cosmetics, and leather. Type I collagen, a fibrillar type responsible for the tensile strength of the extracellular matrix in bones, constitutes approximately 90% of the total collagen. It is predominantly present in the animal's skin, tendon, ligament, as well as the organic portion of calcified tissues in bones and teeth (Arunmozhivarman et al., 2017). Its versatile applications span the pharmaceutical, biomedical, and food sectors.

Gelatin is a widely employed biopolymer in the fields of food, pharmaceuticals, and photography (Killekar et al., 2012). Approximately 95% of commercially derived gelatin comes from porcine and bovine sources. It is prominently present in the bones, tendons, skin, and cartilage of animals, obtained through the partial hydrolysis of fibrous collagen (Sockalingam and Abdullah, 2015). Another abundant biopolymer is chitin, composed of N-acetyl-glucosamine units linked through β (1 → 4) linkage (Cheba et al., 2011). This substance serves as a major structural component in the exoskeletons of marine zooplankton, such as corals, jellyfish, shrimp, crab, prawn, and lobsters (Kandile et al., 2018), as well as in the cell walls of fungi and the cuticles of insects. The alkaline deacetylation of chitin yields chitosan, widely employed in the food, medicinal, pharmaceutical, textile, agriculture, cosmetics, and water treatment industries (Al-Manhel et al., 2018).

PHA represents another category of biodegradable polyesters produced by various microorganisms as materials for intracellular carbon and energy storage. Some domains of archaea are also capable of synthesizing PHA (Ranganathan et al., 2020). The production involves utilizing agro-industrial byproducts that contain carbon and nitrogen sources. Microorganisms synthesize and accumulate PHA as inclusions when there is an excess of carbon but a limited supply of nitrogen source (Salgaonkar and Bragança, 2017). PHA accumulation is observed across a wide range of Gram-positive and Gram-negative bacterial species, as well as archaea (Koller et al., 2017). Bacterial fermentation is the primary method for PHA synthesis, involving microbes such as *Alcaligenes*, *Azotobacter*, *Agrobacterium*, *Azospirillum*, *Aeromonas*, recombinant *E. coli*, *Pseudomonas* sp., *Cupriavidus*, *Rhizobium*, *Rhodobacter*, *Bacillus*, and *Sphaerotilus* (Sanchez-Vazquez et al., 2013; Obruca et al., 2018).

16.3 PRODUCTION OF BIOPOLYMERS FROM DIFFERENT FOOD WASTES

16.3.1 Cellulose

To produce cellulose, a mild acid-alkali condition is employed to fractionate fruit and vegetable pomace. Fruit pomace, containing higher levels of cellulose, hemicelluloses, pectin, and starch, serves as a carbon source for fermentation. Cucumber pomace, for instance, exhibits a high cellulose content (16.13%), while tomato and apple pomace have lower levels (8.60% and 8.81%, respectively) (Tapia-Bl´ácido et al., 2017). This information guides the potential applications of constituents derived from these fruits and vegetables, based on their cellulose content. The initial step involves boiling the substrates for 10 minutes to eliminate phenolic compounds, sugars, and water-soluble polysaccharides. Subsequently, HCl treatment removes pectin polysaccharides, and alkali treatment follows to eliminate hemicelluloses. The residues undergo bleaching with a sodium hypochlorite solution, resulting in the recovery of cellulose as a precipitate (Szyman´ska-Chargot et al., 2017). A similar process is employed for cellulose isolation from rice husk, involving chemical treatment with sodium hydroxide and neutralization with HCl (Shukla, 2013). This approach for pomace component isolation is characterized by its efficiency, as it avoids prolonged processing times and the use of toxic chemicals, thereby reducing costs and environmental hazards.

16.3.2 Production of Xanthum Gum

Xanthan gum is predominantly manufactured in various formulations, viscosities, and quantities using different strains of *X. campestris*, often employing cheese whey and citrus waste. Apple pomace is considered a suitable substrate for xanthan production due to its high concentration of soluble sugars, as well as other substances like pectin and organic acids that are essential for sustaining *X. campestris*. While grape pomace has been utilized as a substrate, its low absorption capacity and sugar content result in a notably low yield (Stredansky and Conti, 1999). The same bacteria have also been employed for xanthan gum production using jackfruit seed powder in combination with peptone, citric acid, K_2HPO_4, and KH_2PO_4, leading to a product yield of 51.62 g L^{-1} (Katherine et al., 2017).

In a potato chip production facility, starch is regularly discarded as waste, primarily generated during slicing and washing processes. The production of biopolymers from potato processing involves enzymatic hydrolysis. The highest glucose concentration is achieved by utilizing a combination of potato starch waste and malt in a 90:10 ratio, converting approximately 96% of the starch in the waste (Ranganathan et al., 2020). Potato waste has also been employed as an alternative substrate for xanthan gum production using the *X. campestris* NRRL B-1003 bacterial strain through solid-state fermentation, submerged fermentation, and semisolid state fermentation. Experimental results indicate that xanthan yields in solid and semisolid fermentation surpass those in submerged state fermentation (Bilanovic et al., 2010).

Bacteria have utilized hydrolysate from kitchen waste to produce 11.73 g L^{-1} of xanthan through batch fermentation (Li et al., 2016). Additionally, to boost the product yield, chicken feather peptone has been introduced as an enhancing supplement in the production media, resulting in a 1.73-fold increase in xanthan production (Ozdal and Kurbanoglu, 2018). The potential for xanthan gum production has also been explored using coconut shells and cocoa husks by *X. campestris*, with minimal supplementation of urea and potassium (da Silva et al., 2018).

16.3.3 Lignin and Hemicelluloses

Sugarcane bagasse stands out as a primary source of lignocellulose, being an economical by-product of the sugarcane industry that comprises cellulose, hemicelluloses, lignin, and ash (Peng et al., 2009). Lignin, a derivative of lignocellulose, demonstrates greater resistance to various biological stresses compared to cellulose and hemicellulose. Reports indicate that lignin extraction is more successful with a 40% NaOH solution in distilled water (yielding 20.4%) than with a 40% NaOH solution in 50% ethanol (yielding 9%) (Jonglertjunya et al., 2014). The fibrous residues present in sugarcane bagasse impede microbial degradation, making pretreatment necessary to enhance substrate availability and the fermentation process. Alkali treatment is employed to extract hemicelluloses from cereal straw and bagasse, with higher concentrations of alkali facilitating the cleavage of the α-ether linkage between lignin and hemicelluloses in bagasse (Peng et al., 2009). Alkali treatment with 1% and 3% NaOH resulted in a 25.1% yield of hemicelluloses, equivalent to 74.90% of the original hemicellulose content in bagasse. Sequential extraction of the hemicellulosic fraction has been successfully achieved through graded precipitation with ethanol (Peng et al., 2009). Researchers have found alkali treatment to be effective in fractionating alkali-soluble lignin and hemicelluloses from bagasse and straw. Hydrogen peroxide (H_2O_2) is another agent utilized for both lignification and solubilization of hemicelluloses (Brienzo et al., 2009). According to studies, a 94.5% yield of hemicelluloses from sugarcane bagasse was attained under 4% alkaline H_2O_2 treatment at 40°C for a 10-hour reaction time. Conversely, minimal degradation of hemicelluloses was observed at optimal conditions of 6% H_2O_2 for 4 hours at 20°C, resulting in hemicelluloses with a lignin content ranging from 4.6% to 14.1% (Brienzo et al., 2009).

16.3.4 Collagen

Collagen extraction is performed using discarded parts of chicken, egg, and fish, primarily through acid and enzymatic hydrolysis. The hydrolysis of non-collagenous proteins at cross-linking sites is achieved using NaCl, acetic acid, and pepsin. The yield of collagen depends significantly on acetic acid concentration, pepsin content, and the duration of hydrolysis (Araújo et al., 2018). Nagai and Suzuki (2000) demonstrated that acetic acid (0.5 M) is effective for extracting collagen from fish skin and fins, while bone collagen can be decalcified with 0.5 M EDTA (ethylenediaminetetraacetic acid), resulting in a collagen yield ranging from 36% to 54% from these fish waste sources.

To investigate the impact of salts on collagen extraction from poultry wastes, Zhou and colleagues utilized 0.45 M NaCl for salt-soluble collagen, 0.5 M acetic acid for acid-soluble collagen, and 0.1% pepsin with a ratio of 1:80 (w/v) for pepsin-soluble collagen extraction from chicken feet skin. Notably, pepsin-soluble collagen exhibited the highest yield at 49.10%, followed by acid-soluble collagen (14.49%) and salt-soluble collagen (1.13%) (Zhou et al., 2016). In a separate study, the removal of non-collagenous proteins involved mixing the sample with 0.1 N NaOH, followed by defatting with butyl alcohol, and organic matter removal through soaking the defatted sample in 0.1 N HCl. These pretreatments reportedly increased collagen yield by 38.7% (Munasinghe et al., 2015).

Chicken skin collagen extraction has been achieved by treating it with 0.5 M acetic acid and 1% pepsin; fat and non-collagenous proteins were removed with 20% ethanol and 0.1 N NaOH. The yield of type I collagen in this process ranged from 10% to 12% (Arunmozhivarman et al., 2017). An optimization study for collagen extraction from chicken feet revealed that the highest yield (72.98%) was achieved at 0.3 mol/L acetic acid, 0.2% pepsin, and 12 hours of hydrolysis (Araújo et al., 2018). Further enhancements in process efficiency could be explored through considerations of unit operations, particle size, the effect of sample moisture, and similar parameters.

16.3.5 PHA and PHB

Microbial strains commonly employed for the production of PHB and PHA include *Bacillus* sp., *Comamonas p.*, *Cupriavidus* sp., *Pseudomonas* sp., *Hydrogenophaga* sp., *Azotobacter* sp., *Burkholderia* sp., *Acinetobacter* sp., *Haloferax* sp., and *Azohydromonas* sp. A pretreatment is necessary to convert food waste into PHA, accomplished through either chemical or biological processing for the digestion of the sample matrix. Following this, PHA-producing microorganisms utilize the available carbon source to synthesize PHA (Nielsen et al., 2017; Rodriguez-Perez et al., 2015). It's crucial to note that waste substrates used for PHA production contain varying concentrations of nitrogen and phosphorus. To achieve optimal PHA production, a suitable ratio of carbon to nitrogen to phosphorus is required, as an excess concentration of any of these nutrients may lead to inhibition (Rodriguez-Perez et al., 2015).

The properties of PHA, such as hardness and brittleness, depend on its structure, and the structure of monomers varies based on the bacteria and substrate used. Poly 3-hydroxybutyric acid (PHB) is the most widely recognized microbial PHA, characterized as a homopolymer of 3-hydroxybutyrate (Ranganathan et al., 2020). This homopolymer is both brittle and biodegradable, exhibiting thermoplastic properties; however, its industrial use has been limited due to the reported high production cost of PHB to date (Reis et al., 2008; Johnston et al., 2018).

However, a paradigm shift is occurring, transforming this waste into a valuable resource through innovative approaches and sustainable practices.

Agro-food industry waste, which includes byproducts, leftovers, and discarded materials, can be harnessed to create a circular economy that benefits both the environment and the economy.

16.4 IMPORTANCE OF BIODEGRADABLE PACKAGING

Biodegradable packaging has emerged as a vital solution in addressing the environmental challenges posed by traditional non-biodegradable packaging materials. As our planet faces increasing concerns about pollution, waste management, and the depletion of natural resources, the importance of adopting biodegradable packaging cannot be overstated (Poletto et al., 2013).

- **Reduced Environmental Impact**: Biodegradable packaging materials are designed to break down naturally over time, significantly reducing their environmental impact compared to non-biodegradable alternatives. This helps to mitigate the long-lasting presence of plastic and other harmful materials in landfills and ecosystems (Reddy et al., 2013).
- **Resource Conservation**: Biodegradable packaging often utilizes renewable resources like plant-based materials, agricultural waste, or even mushroom-based materials. This reduces our reliance on fossil fuels and promotes the responsible use of resources, contributing to a more sustainable and circular economy.

- **Less Pollution**: Traditional plastic packaging contributes to visual pollution, clogs waterways, and harms marine life. Biodegradable packaging, on the other hand, is less likely to accumulate in the environment, minimizing harm to wildlife and ecosystems (Ross et al., 1991).
- **Waste Management**: Biodegradable packaging materials can be composted, which not only reduces the burden on landfills but also produces nutrient-rich compost that can enrich soil and support agricultural activities (Sanchez-Garcia and Lagaron, 2010).
- **Consumer Preferences**: With growing awareness about environmental issues, consumers are increasingly seeking eco-friendly products. Biodegradable packaging can appeal to environmentally conscious consumers, enhancing brand reputation and marketability.
- **Regulatory Compliance**: Many regions are implementing stricter regulations on single-use plastics and non-biodegradable materials. By adopting biodegradable packaging, businesses can stay ahead of these regulations and ensure long-term compliance (Munasinghe et al., 2015).
- **Innovation and Research**: The shift toward biodegradable packaging encourages innovation and research in material science and packaging design. This can lead to the development of new, more sustainable packaging solutions that benefit both businesses and the environment.
- **Climate Change Mitigation**: The production of traditional packaging materials often involves high energy consumption and greenhouse gas emissions. Biodegradable packaging, especially if produced from renewable sources, has a lower carbon footprint, contributing to efforts to mitigate climate change.
- **Educational Opportunities**: Embracing biodegradable packaging provides opportunities for educating consumers, businesses, and communities about the importance of sustainable choices and responsible consumption.
- **Long-Term Viability**: As concerns about the environmental impact of non-biodegradable materials continue to grow, businesses that invest in biodegradable packaging demonstrate their commitment to a more sustainable future. This can enhance brand loyalty and contribute to long-term success (Peng et al., 2009).

16.5 APPLICATIONS OF BIODEGRADABLE MATERIALS

- **Packaging**: Biodegradable packaging solutions include bags, wraps, containers, and even edible packaging, reducing the environmental impact of disposable materials (Rodriguez-Perez et al., 2015).
- **Textiles and Apparel**: Biodegradable fibers are used in clothing, accessories, and home textiles to reduce the environmental impact of fast fashion.
- **Agriculture**: Biodegradable materials are used in mulching films, seedling pots, and crop covers, promoting sustainable farming practices.

- **Medical and Healthcare**: Biodegradable materials find applications in sutures, wound dressings, tissue engineering scaffolds, and drug delivery systems (Nielsen et al., 2017).
- **Construction**: Bio-based composites and biodegradable polymers are used in construction materials such as insulation, panels, and flooring.

16.6 ADVANTAGES AND DISADVANTAGES OF BIODEGRADABLE PACKAGING

Biodegradable packaging presents a range of advantages and disadvantages that need to be carefully considered in the context of its environmental and practical implications. Here's an overview of the pros and cons of biodegradable packaging.

16.6.1 Advantages of Biodegradable Packaging

I. **Environmental Friendliness**: Biodegradable packaging materials break down naturally into harmless components, reducing their impact on landfills and ecosystems. This helps mitigate pollution and long-term environmental damage (Araújo et al., 2018).
II. **Reduced Carbon Footprint**: Many biodegradable materials are derived from renewable resources and have a lower carbon footprint compared to traditional nonbiodegradable materials, contributing to climate change mitigation.
III. **Resource Conservation**: Biodegradable packaging often utilizes renewable resources, such as plant-based materials, agricultural waste, or fungi. This reduces the demand for fossil fuels and non-renewable resources.
IV. **Waste Reduction**: Biodegradable packaging can alleviate the strain on waste management systems by reducing the volume of non-recyclable waste that accumulates in landfills (Li et al., 2016).
V. **Consumer Appeal**: In an increasingly eco-conscious market, products packaged in biodegradable materials may attract environmentally aware consumers, enhancing brand reputation and loyalty.
VI. **Innovation and Research**: The development of biodegradable packaging materials encourages innovation in material science and design, leading to the creation of new, sustainable packaging solutions.

16.6.2 Disadvantages of Biodegradable Packaging

- **Limited Durability**: Biodegradable packaging may have a shorter shelf life compared to traditional packaging materials, which could affect the product's integrity and usability.
- **Dependence on Conditions**: Biodegradation is influenced by environmental conditions, such as temperature, humidity, and microbial activity. In

certain environments, such as marine or anaerobic conditions, biodegradation may be slower or limited (Munasinghe et al., 2015).

- **Production Challenges**: The production of biodegradable materials may involve energy-intensive processes, and some bioplastics require specific conditions for effective degradation (Sanchez-Garcia and Lagaron, 2010).
- **Contamination Concerns**: Improper disposal of biodegradable packaging, such as mixing it with non-biodegradable waste, can lead to contamination of recycling streams and hinder the recycling process.
- **Land Use and Competition**: Some biodegradable materials, such as those derived from crops, may raise concerns about land use competition with food production, potentially affecting food security.
- **Market Acceptance**: While the demand for biodegradable packaging is growing, it may still face challenges in terms of market acceptance, availability, and cost-effectiveness (Araújo et al., 2018).
- **Misleading Claims**: Not all products labeled as "biodegradable" are created equal. Some materials may only degrade under specific conditions, and false claims could contribute to confusion and greenwashing.

In conclusion, biodegradable packaging offers several benefits for the environment and sustainability efforts. However, it also comes with challenges and limitations that need to be carefully addressed through proper design, production, and disposal processes. As the field of biodegradable materials continues to evolve, striking a balance between the advantages and disadvantages is crucial to maximizing the positive impact of these materials on our planet (Abushammala et al., 2015).

16.7 PROCESSING TECHNIQUES FOR BIODEGRADABLE PACKAGING MATERIALS

Processing biodegradable packaging materials involves various techniques that transform raw materials into usable forms, enabling the creation of sustainable and eco-friendly packaging solutions (Li et al., 2016). These techniques ensure that the materials maintain their desired properties while also being suitable for specific applications. Here's an overview of some key processing techniques for biodegradable packaging materials:

I. **Extrusion**: Extrusion is a common technique used to process biodegradable polymers into various forms, such as films, sheets, and profiles. In this process, the biopolymer is melted and forced through a die to create a continuous shape that can be subsequently cooled and solidified (Kaya et al., 2015).

II. **Injection Molding**: Injection molding is employed to produce three-dimensional objects from biodegradable materials. The material is heated until it becomes molten, then injected into a mold cavity, where it cools and solidifies to form the desired product (Araújo et al., 2018).

III. ***Blow Molding***: Blow molding is used to create hollow objects, such as bottles and containers, from biodegradable materials. The process involves inflating a heated parison (a hollow tube) within a mold to shape it into the desired form.

IV. **Thermoforming**: Thermoforming is suitable for creating packaging trays, containers, and lids. A sheet of biodegradable material is heated and then formed into the desired shape using a mold. The formed part is then cooled and trimmed.

V. **Compression Molding**: Compression molding is used for producing larger and thicker biodegradable products, such as packaging trays and tableware. The material is placed in a heated mold cavity and compressed until it takes the desired shape (Kaya et al., 2015).

VI. **Film Casting**: Film casting involves creating thin biodegradable films by spreading a molten material onto a flat surface, then cooling and solidifying it. These films are often used for flexible packaging applications (Fahmy et al., 2020).

VII. **Coating and Laminating**: Coating and laminating techniques involve applying thin layers of biodegradable materials onto substrates, such as paper or other films, to enhance their properties, such as barrier properties or appearance.

VIII. **Extrusion Coating**: This technique involves extruding a thin layer of biodegradable polymer onto a substrate, forming a protective and functional coating. It's commonly used to enhance the barrier properties of paper-based packaging (Li et al., 2016).

IX. **Fiber Spinning**: For producing biodegradable textiles and nonwovens, fiber spinning techniques like melt spinning, wet spinning, or electrospinning can be employed to create continuous fibers or fabrics.

X. **Additive Manufacturing (3D Printing)**: Additive manufacturing allows for the creation of intricate and customized biodegradable packaging products layer by layer. It's particularly useful for prototypes and small-scale production.

XI. **Chemical and Mechanical Modification**: Processing techniques can involve chemical and mechanical modifications to improve the properties of biodegradable materials, such as enhancing their strength, flexibility, or thermal stability.

XII. **Extrusion Foaming**: This technique produces lightweight and insulating materials by introducing a gas or chemical blowing agent into the biodegradable polymer during extrusion, resulting in a foam structure (Araújo et al., 2018).

It's important to note that the choice of processing technique depends on the specific biodegradable material being used, the intended application, and the desired properties of the final product. Proper processing is essential to ensure that biodegradable packaging materials maintain their eco-friendly attributes while meeting the functional requirements of packaging solutions.

16.8 CHARACTERISTICS AND PERFORMANCE OF AGRO-FOOD WASTE-BASED BIODEGRADABLE PACKAGING

16.8.1 Mechanical Properties

Agro-food waste-based biodegradable packaging has gained attention as a sustainable alternative to traditional packaging materials. These innovative packaging solutions utilize agricultural and food processing byproducts to create environmentally friendly packaging options (Fahmy et al., 2020). One critical aspect of assessing the feasibility and effectiveness of such packaging is understanding the mechanical properties, which play a vital role in determining the packaging's performance and applicability. Here, we delve into the characteristics and performance of agro-food waste-based biodegradable packaging from a mechanical standpoint.

Mechanical properties of agro-food waste-based biodegradable packaging:

- **Strength and Stiffness***:* The tensile strength and stiffness of packaging materials are crucial for their ability to withstand external forces and maintain structural integrity. Agro-food waste-based biodegradable packaging should exhibit adequate strength and stiffness to protect the contents and endure handling and transportation (Fahmy et al., 2020).
- **Elasticity and Flexibility**: Packaging materials should possess elasticity and flexibility to accommodate deformation without permanent damage. These properties are important for cushioning and absorbing impacts during handling and distribution.
- **Toughness and Impact Resistance**: Toughness reflects a material's ability to absorb energy before fracturing. Good impact resistance is essential to prevent damage during drop or impact events, ensuring the protection of the packaged products.
- **Creep Resistance**: Creep refers to the gradual deformation of a material under constant load over time. Packaging materials should exhibit low creep to maintain their shape and structural integrity throughout their intended use (Fahmy et al., 2020).
- **Tear and Puncture Resistance**: Resistance to tear and puncture is crucial for preventing accidental damage and maintaining the barrier properties of the packaging. This is particularly important for packaging materials used to package sharp or irregularly shaped products.
- **Compression Strength***:* Compression strength is relevant for stacking and storing packaged goods. The packaging should be able to withstand the weight of other packages or items without collapsing or deforming.
- **Barrier Properties**: While primarily related to the material's structure, barrier properties like oxygen and moisture permeability affect the mechanical performance indirectly. Proper barrier properties ensure the protection and preservation of the packaged products (Katherine et al., 2017).

16.8.2 Performance Considerations

- **Product Compatibility**: The mechanical properties of agro-food waste-based biodegradable packaging should align with the specific requirements of the packaged product. Delicate items may require different mechanical characteristics compared to robust goods.
- **Processing Techniques**: The chosen processing techniques during manufacturing can influence the mechanical properties of the packaging. The processing method should be selected carefully to achieve the desired attributes.
- **Environmental Conditions**: The packaging's mechanical performance can be affected by temperature, humidity, and exposure to sunlight. Considerations must be made to ensure that the packaging retains its mechanical integrity under varying environmental conditions (Katherine et al., 2017).
- **End-of-Life Handling**: Biodegradable packaging's mechanical properties should also be compatible with its intended end-of-life scenario. For instance, compostable packaging should maintain its strength throughout its usage life and degrade effectively during composting.
- **Regulatory Compliance**: The mechanical properties of agro-food waste-based biodegradable packaging should meet industry standards and regulatory requirements to ensure its suitability for various applications (Katherine et al., 2017).

In conclusion, understanding the mechanical properties of agro-food waste-based biodegradable packaging is essential for assessing its feasibility and performance as a sustainable packaging solution. A comprehensive evaluation of these properties ensures that the packaging effectively safeguards products while aligning with environmental and functional considerations.

Agro-food waste-based biodegradable packaging is a promising avenue for sustainable packaging solutions, addressing both environmental concerns and functional performance requirements. Several key characteristics and performance aspects play a vital role in evaluating the suitability and effectiveness of these innovative packaging materials (Katherine et al., 2017).

16.8.3 Thermal Properties

Thermal properties of agro-food waste-based biodegradable packaging materials are crucial for maintaining the stability and integrity of packaged goods during storage, transportation, and usage. These properties include

- **Melting Point and Glass Transition Temperature**: Understanding these temperatures helps determine the packaging's behavior under different temperature conditions. A high glass transition temperature can contribute to improved stability and reduced deformation under heat.

- **Heat Resistance**: The packaging should be able to withstand moderate temperatures without softening or losing structural integrity, ensuring that it retains its shape and functionality during typical usage scenarios (Fahmy et al., 2020).
- Barrier Properties
- Barrier properties influence the packaging's ability to protect contents from external factors such as moisture, oxygen, light, and odors. Effective barrier properties can extend the shelf life of perishable goods and maintain the quality of packaged products. Key barrier characteristics include:
  - **Oxygen Permeability**: Packaging with low oxygen permeability prevents spoilage, oxidation, and color changes in oxygen-sensitive products like fresh produce or meats.
  - **Water Vapor Permeability**: Low water vapor permeability helps maintain the moisture content and freshness of packaged goods.
  - **Light Barrier**: Some products are sensitive to light exposure, so effective light barrier properties prevent UV degradation and photo-induced reactions (Li et al., 2016).
  - **Aroma and Odor Barrier**: Effective barrier properties against aroma and odor transfer are important for maintaining the sensory quality of food products.

16.8.4 Biodegradability and Compostability

Biodegradability and compostability are fundamental characteristics of agro-food waste-based biodegradable packaging, contributing to its overall sustainability and reduced environmental impact:

- **Biodegradability**: Biodegradable packaging materials can naturally break down over time through biological processes, reducing their presence in landfills and ecosystems. Understanding the rate and extent of biodegradation is essential (Arora et al., 2016).
- **Compostability**: Compostable materials can fully biodegrade within a specific timeframe under controlled composting conditions. Compostable packaging can contribute to nutrient-rich compost, supporting soil health and plant growth.
- **Certifications**: Compliance with recognized standards and certifications (e.g., EN 13432, ASTM D6400) ensures that agro-food waste-based biodegradable packaging meets specific biodegradability and compostability criteria (Katherine et al., 2017).

16.8.5 Performance Considerations

- **Product Compatibility**: The characteristics of agro-food waste-based biodegradable packaging should align with the needs of the packaged product.

Perishable goods may require higher barrier properties, while non-food items may emphasize durability (Kaya et al., 2015).

- **Processing Techniques**: The choice of processing techniques and formulation during manufacturing can impact thermal, barrier, and biodegradability properties. Optimal processing methods should be selected to achieve the desired performance.
- **Environmental Impact**: The choice of agro-food waste feedstock, processing techniques, and end-of-life scenarios collectively determines the packaging's overall environmental impact and sustainability.
- **Regulatory Compliance**: Packaging materials must adhere to industry and regulatory standards to ensure their safety, performance, and environmental claims.

In conclusion, understanding the thermal properties, barrier properties, biodegradability, and compostability of agro-food waste-based biodegradable packaging materials is essential for designing effective and sustainable packaging solutions. Balancing these characteristics with functional requirements contributes to packaging that not only protects products but also contributes positively to environmental stewardship (Kaya et al., 2015).

Comparing agro-food waste-based biodegradable packaging materials with conventional packaging materials reveals the distinct advantages and disadvantages of each option. This comparison highlights the potential benefits of adopting biodegradable materials in packaging while considering the limitations they might present in certain contexts.

16.9 INNOVATIONS AND ENHANCEMENTS OF AGRO-FOOD WASTE-BASED BIODEGRADABLE PACKAGING

Agro-food waste-based biodegradable packaging continues to undergo remarkable innovations and enhancements, driving the evolution of sustainable packaging solutions. These advancements leverage the inherent properties of agricultural and food processing byproducts, transforming them into effective, eco-friendly alternatives. Here are some notable innovations and enhancements in this field (Priya et al., 2014).

16.9.1 Nanocomposite Reinforcement

Incorporating nanomaterials, such as nanoparticles and nanofibers, into agro-food waste-based biodegradable packaging materials enhances their mechanical strength, thermal stability, and barrier properties. Nanocomposites improve material performance, making them suitable for a wider range of applications (Dixit and Yadav, 2020).

Bio-Based Compatibilizers: Bio-based compatibilizers act as bridge molecules, improving the compatibility between different biodegradable components. This enhancement strengthens the material's structural integrity, enhancing mechanical properties and stability.

16.9.2 Biopolymer Blends

Blending different biopolymers derived from agro-food waste creates packaging materials with enhanced properties. Biopolymer blends offer a balanced combination of characteristics from various components, allowing for tailored mechanical strength and barrier performance (Benbettaieb et al., 2016).

Active and Intelligent Packaging: Incorporating additives such as antimicrobial agents, oxygen scavengers, and indicators extends shelf life, enhances product safety, and provides real-time information about package and product conditions. This innovation improves food preservation and quality (Koller, 2015).

16.9.3 Edible and Water-Soluble Films

Innovations have led to the development of edible films made from agro-food waste-derived materials. These films can be used to package certain foods and are consumed along with the product, reducing waste and enhancing convenience.

Smart Packaging: Sensors and RFID technology integrated into agro-food waste-based biodegradable packaging enable real-time monitoring of product quality and integrity. This innovation facilitates better supply chain management, reduces food waste, and ensures product safety (Arora et al., 2016).

16.9.4 Nanotechnology and Coatings

Nanoscale coatings can enhance barrier properties, improve moisture resistance, and provide surface functionalities to agro-food waste-based biodegradable packaging materials. This technology enhances protection and preservation capabilities (Xu et al., 2019).

16.9.5 Hybrid Materials

Combining agro-food waste-derived biopolymers with other sustainable materials, such as recycled fibers or bio-based resins, creates hybrid packaging solutions that harness the strengths of multiple materials. This approach enhances material performance and sustainability.

16.9.6 Design Optimization

Advanced modeling and simulation techniques are used to optimize the design of agro-food waste-based biodegradable packaging, ensuring optimal material usage, structural integrity, and performance.

16.9.7 Regulatory Compliance

Ongoing research ensures that agro-food waste-based biodegradable packaging materials meet industry standards and regulatory requirements, providing assurance of their safety, performance, and environmental compatibility (Arora et al., 2016).

These innovations and enhancements demonstrate the dynamic nature of agro-food waste-based biodegradable packaging, driving its effectiveness, versatility, and sustainability. By leveraging these advancements, packaging designers and manufacturers contribute to reducing environmental impact while meeting the functional and market demands of modern packaging applications (Li et al., 2016).

16.10 FUTURE PERSPECTIVES AND OPPORTUNITIES

The future perspective of agro-food waste-based biodegradable packaging holds tremendous promise as society increasingly embraces sustainable practices and seeks alternatives to conventional packaging materials. This emerging field offers numerous opportunities that can significantly impact the packaging industry and foster a more environmentally friendly approach to packaging solutions (Arora et al., 2016). Here are some future perspectives and opportunities for agro-food waste-based biodegradable packaging.

The future of agro-food waste-based biodegradable packaging is driven by continuous research and emerging technologies. Advancements in material science, nanotechnology, and processing techniques will lead to the development of biodegradable materials with enhanced mechanical properties, barrier performance, and tailored functionalities (Sahu M and Sahoo PK, 2017). Research will explore new feedstock sources, biopolymer blends, and biodegradation mechanisms to optimize material design and performance.

16.10.1 Policy and Regulatory Support for Biodegradable Packaging

Policy and regulatory support play a pivotal role in the adoption of agro-food waste-based biodegradable packaging. Governments and regulatory bodies can incentivize research and development, create standards for biodegradability and compostability, and establish waste management guidelines for sustainable packaging. Clear policies can create a level playing field and encourage industries to invest in greener alternatives (Katherine et al., 2017).

16.10.2 Potential Market Growth and Industry Adoption

The market for agro-food waste-based biodegradable packaging is poised for significant growth and adoption. As consumers become more environmentally conscious, demand for sustainable packaging solutions will rise. Companies that embrace biodegradable packaging will gain a competitive edge, appealing to environmentally aware consumers and meeting changing market preferences.

16.10.3 Circular Economy and Sustainable Practices

The future of agro-food waste-based biodegradable packaging aligns with the principles of the circular economy. Designing packaging for reusability, recycling, and compostability will contribute to waste reduction and resource efficiency

(Yang et al., 2018). This shift toward sustainable practices will drive industries to adopt greener alternatives and integrate circular economy initiatives into their business models (Guo et al., 2019).

16.10.4 Socioeconomic and Environmental Impact

Agro-food waste-based biodegradable packaging has the potential to create positive socioeconomic and environmental impacts. It can stimulate rural economies by providing a market for agricultural byproducts and reducing waste disposal costs. At the same time, it addresses environmental concerns by reducing greenhouse gas emissions, plastic pollution, and pressure on natural resources (Guo et al., 2019).

16.10.5 Consumer Education and Behavior Change

Future perspectives also involve consumer education and behavior change. Raising awareness about the benefits of agro-food waste-based biodegradable packaging and promoting proper disposal practices will drive responsible consumer behavior (Guo et al., 2019). Educated consumers will actively choose sustainable packaging, leading to broader adoption and a more eco-friendly packaging landscape.

16.10.6 Innovation and Market Competitiveness

Continuous innovation in agro-food waste-based biodegradable packaging will drive market competitiveness. Companies investing in research and development will lead the way in providing innovative, high-performance, and cost-effective solutions. This competitiveness will stimulate further advancements and broaden the applications of biodegradable packaging.

In conclusion, the future of agro-food waste-based biodegradable packaging presents exciting prospects and opportunities. Through research, collaboration, policy support, market growth, and consumer education, this sustainable packaging solution can play a significant role in reducing environmental impact, fostering circular economy practices, and promoting a greener, more sustainable future for the packaging industry.

16.11 CONCLUSION

Implications and importance of agro-food waste-based biodegradable packaging:

Agro-food waste-based biodegradable packaging carries significant implications for sustainability, waste reduction, and environmental protection. Its importance lies in addressing the environmental challenges posed by conventional plastic packaging. Here are some key implications and importance:

- Agro-food waste-based biodegradable packaging utilizes renewable agricultural byproducts that might otherwise be discarded or underutilized.

By transforming waste into valuable resources, it promotes sustainable resource management and reduces the burden on landfills.

- One of the most pressing environmental issues is plastic pollution. Agro-food waste-based biodegradable packaging offers a viable solution by reducing the reliance on petroleum-based plastics and providing packaging that naturally biodegrades, minimizing plastic waste in the environment.
- Compared to conventional plastics, agro-food waste-based biodegradable packaging often requires lower energy consumption during production, leading to reduced greenhouse gas emissions and contributing to climate change mitigation.
- Embracing biodegradable packaging aligns with the principles of the circular economy, where materials are designed to be reused, recycled, or composted. It promotes a closed-loop system that reduces waste and maximizes resource efficiency.
- Companies adopting agro-food waste-based biodegradable packaging can enhance their brand value by demonstrating a commitment to sustainability. Consumers increasingly prefer eco-friendly products, and such packaging can attract environmentally conscious consumers.

REFERENCES

Abushammala, H., Krossing, I., Laborie, M.P., 2015. Ionic liquid-mediated technology to produce cellulose nanocrystals directly from wood. *Carbohydr. Polym.* 134, 609–616.

Al-Manhel, A.J., Al-Hilphy, A.R.S., Niamah, A.K., 2018. EXtraction of chitosan, characterisation and its use for water purification. *J. Saudi Soc. Agric. Sci.* 17, 186–190.

Araújo ´I.B.D.S., Bezerra, T.K.A., Nascimento, E.S.D., Gadelha C.A.D.A., Santi- Gadelha, T., Madruga, M.S., 2018. Optimal conditions for obtaining collagen from chicken feet and its characterization. *Food Sci. Technol.* 38, 167–173.

Arora, D., Sharma, N., Sharma, V., Abrol, V., Shankar, R., Jaglan, S., 2016. An update on polysaccharide-based nanomaterials for antimicrobial applications. *Appl. Microbiol. Biotechnol.* 100, 2603–2615.

Arunmozhivarman, K., Abraham, R.J.J., Appa, V., Parthiban, M., 2017. EXtraction and molecular characterization of collagen from poultry meat processing by-product (chicken skin). *Int. J. Pure App. Biosci.* 5, 1085–1091.

Benbettaieb, N., Gay, J.P., Karbowiak, T., Debeaufort, F., 2016. Tuning the functional properties of polysaccharide–protein bio-based edible films by chemical or physical cross-linking. *Compr. Rev. Food Sci. Food Saf.* 15, 739–752.

Bilanovic, D. D., Malloy, S. H., Remeta, P., 2010. *U.S. Patent No. 7,727,747.* U.S. Patent and Trademark Office.

Brienzo, M., Siqueira, A.F., Milagres, A.M.F., 2009. Search for optimum conditions of sugarcane bagasse hemicellulose extraction. *Biochem. Eng. J.* 46 (2), 199–204.

Cheba, B.A., 2011. Chitin and chitosan: Marine biopolymers with unique properties and versatile applications. *Global Journal of Biotechnology & Biochemistry,* 6(3), 149–153.

da Silva, J.A., Cardoso, L.G., de Jesus Assis, D., Gomes, G.V.P., Oliveira, M.B.P.P., de Souza, C.O., Druzian, J.I., 2018. Xanthan gum production by Xanthomonas campestris pv. campestris IBSBF 1866 and 1867 from lignocellulosic agroindustrial wastes. *Appl. Biochem. Biotechnol.* 186, 750–763.

Dixit, S., Yadav, V.L., 2020. Biodegradable polymer composite films for green packaging applications. In Kharissova, O., Martinez, L., Kharisov, B. (Eds.), *Handbook of Nanomaterials and Nanocomposites for Energy and Environmental Applications*. Springer.

Fahmy, H.M., Salah Eldin, R.E., Abu Serea, E.S., Gomaa, N.M., AboElmagd, G.M., Salem, S.A., Shalan, A.E., 2020. Advances in nanotechnology and antibacterial properties of biodegradable food packaging materials. *RSC Adv.* 10, 20467–20484.

Guo, Z., Yan, N., Lapkin, A.A., 2019. Towards circular economy: Integration of bio-waste into chemical supply chain. *Curr. Opin. Chem. Eng.* 26, 148–156.

Huang, J., Fu, S., Gan, L. (Eds.), 2019. *Lignin Chemistry and Applications*. Elsevier, pp. 25–50.

Johnston, B., Radecka, I., Hill, D., Chiellini, E., Ilieva, V.I., et al., 2018. The microbial production of polyhydroxyalkanoates from waste polystyrene fragments attained using oxidative degradation. *Polymers* 10, 957.

Jonglertjunya, W., Juntong, T., Pakkang, N., Srimarut, N., Sakdaronnarong, C., 2014. Properties of lignin extracted from sugarcane bagasse and its efficacy in maintaining postharvest quality of limes during storage. *LWT - Food Sci. Technol.* 57, 116.

Kandile, N.G., Zaky, H.T., Mohamed, M.I., Nasr, A.S., Ali, Y.G., 2018. Extraction and characterization of chitosan from shrimp shells. *Open J. Org. Polym. Mater.* 33.

Katherine, R.F., Muthukumaran, C., Sharmila, G., Kumar, N.M., Tamilarasan, K., Jaiganesh, R., 2017. Xanthan gum production using jackfruit-seed-powder-based medium: optimization and characterization. *3 Biotech* 7 (4), 248.

Kaya, M., Baran, T., Karaarslan, M., 2015. A new method for fast chitin extraction from shells of crab, crayfish and shrimp. *Nat. Prod. Res.* 29, 1477–1480.

Killekar, V.C., Koli, J.M., Sharangdhar, S.T., Metar, S.Y., 2012. Functional properties of gelatin extracted from skin of black kingfish (Ranchycentron Canadus). *Indian J. Fundam. Appl. Life Sci.* 2, 106–116.

Koller, M., Braunegg, G., 2015. Biomediated production of structurally diverse poly (hydroxyalkanoates) from surplus streams of the animal processing industry. *Polimery* 60.

Koller, M., Maršálek, L., de Sousa Dias, M. M., & Braunegg, G. 2017. Producing microbial polyhydroxyalkanoate (PHA) biopolyesters in a sustainable manner. *New biotechnology*, 37, 24-38.

Kumar, S., Thakur, K., 2017. Bioplastics-classification, production and their potential food applications. *J. Hill Agric.* 8, 118.

Lee, A., Liew, M.S., 2020. Ecologically derived waste management of conventional plastics. *J. Mater. Cycles Waste Manag.* 22, 1–10.

Li, P., Li, T., Zeng, Y., Li, X., Jiang, X., Wang, Y., Xie, T., Zhang, Y., 2016. Biosynthesis of Xanthan gum by Xanthomonas campestris LRELP-1 using kitchen waste as the sole substrate. *Carbohydr. Polym.* 151, 684–691.

Mesomo, M., Silva, M.F., Boni, G., Padilha, F.F., Mazutti, M., Mossi, A., de Oliveira, D., Cansian, R.L., di Luccio, M., Treichel, H., 2009. Xanthan gum produced by Xanthomonas campestris from cheese whey: Production optimisation and rheological characterisation. *J. Sci. Food Agric.* 89, 2440–2445.

Munasinghe, K.A., Schwarz, J.G., Whittiker, M., 2015. Utilization of chicken by-products to form collagen films. *J. Food Process* 2015, 1–6.

Nagai, T., & Suzuki, N. 2000. Isolation of collagen from fish waste material—skin, bone and fins. *Food chemistry*, 68(3), 277-281.

Nielsen, C., Rahman, A., Rehman, A.U., Walsh, M.K., Miller, C.D., 2017. Food waste conversion to microbial polyhydroxyalkanoates. *Microb. Biotechnol.* 10, 1338–1352.

Obruca, S., Sedlacek, P., Koller, M., Kucera, D., Pernicova, I., 2018. Involvement of polyhydroxyalkanoates in stress resistance of microbial cells: Biotechnological consequences and applications. *Biotechnol. Adv.* 36, 856–870.

Ozdal, M., Kurbanoglu, E.B., 2018. Valorisation of chicken feathers for xanthan gum production using Xanthomonas campestris MO-03. *J. Genet. Eng. Biotechnol.* 16, 259–263.

Packiavathy, I. A. S. V., Priya, S., Pandian, S. K., & Ravi, A. V. 2014. Inhibition of biofilm development of uropathogens by curcumin–an anti-quorum sensing agent from Curcuma longa. *Food chemistry*, 148, 453-460.

Peng, F., Ren, J.L., Xu, F., Bian, J., Peng, P., Sun, R.C., 2009. Comparative study of hemicelluloses obtained by graded ethanol precipitation from sugarcane bagasse. *J. Agric. Food Chem.* 57 (14), 6305–6317.

Poletto, M., Zattera, A.J., Pistor, V., 2013. Structural characteristics and thermal properties of native cellulose. In van de Van, T., Godbout, L. (Eds.), *Cellulose Fundamental Aspects.* InTech.

Polman, E.M.N., Gruter, G.J.M., Parsons, J.R., 2021. Tietema, A. Comparison of the aerobic biodegradation of biopolymers and the corresponding bioplastics: A review. *Sci. Total Environ.* 753, 141953.

Ranganathan, S., Dutta, S., Moses, J.A., Anandharamakrishnan, C. 2020, September 16. Utilization of food waste streams for the production of biopolymers. *Heliyon.* 6 (9), e04891. doi: 10.1016/j.heliyon.2020.e04891.

Reddy, M.M., Vivekanandhan, S., Misra, M., Bhatia, S.K., Mohanty, A.K., 2013. Biobased plastics and bionanocomposites: current status and future opportunities. *Prog. Polym. Sci.* 38, 1653–1689.

Reis, K.C., Pereira, J., Smith, A.C., Carvalho, C.W.P., Wellner, N., Yakimets, I., 2008. Characterization of polyhydroxybutyrate-hydroxyvalerate (PHB-HV)/maize starch blend films. *J. Food Eng.* 89, 361–369.

Rodriguez-Pérez, S., Serrano, A., Pantión, A. A., & Alonso-Fariñas, B., 2015. Challenges of scaling-up PHA production from waste streams. A review. *J. Environ. Manag.* 205, 215–230.

Ross, P., Mayer, R., Benziman, M., 1991. Cellulose biosynthesis and function in bacteria. *Microbiol. Rev.* 55, 35–58.

Sahu, M., Sahoo, P.K., 2017. Bio polymers: Sustainable alternative for food packaging. *Int. J. Eng. Manag. Res.* https://www.ijemr.net/DOC/ BioPolymersSustainableAlternativeFo rFoodPackaging. pdf.

Salgaonkar, B.B., Bragança, J.M., 2017. Utilization of sugarcane bagasse by Halogeometricum borinquense strain E3 for biosynthesis of poly (3-hydroXybutyrate-co- 3-hydroXyvalerate). *Bioengineering* 4, 50.

Sanchez-Garcia, M., Lagaron, J., 2010. On the use of plant cellulose nanowhiskers to enhance the barrier properties of polylactic acid. *Cellulose* 17, 987–1004.

Sanchez-Vazquez, S.A., Hailes, H.C., Evans, J.R.G., 2013. Hydrophobic polymers from food waste: Resources and synthesis. *Polym. Rev.* 53, 627–694.

Shukla, S.K., 2013. Preparation and characterization of cellulose derived from rice husk for drug delivery. *Adv. Mater. Lett.* 4 (9), 714–719.

Sockalingam, K., Abdullah, H.Z., 2015. Extraction and characterization of gelatin biopolymer from black Tilapia (Oreochromis mossambicus) scales 020053. *AIP Conf. Proc.* 1669 (1), 020053.

Stredansky, M., Conti, E., 1999. Xanthan production by solid state fermentation. *Process Biochem.* 34, 581–587.

Szyman´ska-Chargot, M., Chylin´ska, M., Gdula, K., Kozioł, A., Zdunek, A., 2017. Isolation and characterization of cellulose from different fruit and vegetable pomaces. *Polymers* 9 (10), 495.

Tapia-Bl´acido, D.R., Maniglia, B.C., Martelli-Tosi, M., 2017. Biopolymers from sugarcane and soybean lignocellulosic biomass. *Sustain. Polym. Biomass* 227–253.

Xu, C., Nasrollahzadeh, M., Selva, M., Issaabadi, Z., Luque, R., 2019. Waste-to-wealth: Biowaste valorization into valuable bio(nano)materials. *Chem. Soc. Rev.* 48, 4791–4822.

Yang, W., Fortunati, E., Bertoglio, F., Owczarek, J.S., Bruni, G., Kozanecki, M., Kenny, J.M., Torre, L., Visai, L., Puglia, D., 2018. Polyvinyl alcohol/chitosan hydrogels with enhanced antioxidant and antibacterial properties induced by lignin nanoparticles. *Carbohydr. Polym.* 181, 275–284.

Zhou, C., Li, Y., Yu, X., Yang, H., Ma, H., Yagoub, A.E.G.A., Cheng, Y., Hu, J., Otu, P.N.Y., 2016. Extraction and characterization of chicken feet soluble collagen. *LWT - Food Sci. Technol.* 74, 145–153.

Index

G

H

I

L

M

N

For Product Safety Concerns and Information please contact our EU representative GPSR@taylorandfrancis.com Taylor & Francis Verlag GmbH, Kaufingerstraße 24, 80331 München, Germany

Printed and bound by CPI Group (UK) Ltd, Croydon, CR0 4YY
14/09/2026
02216256-0015